Gender and Tourism Sustainability

This book examines the relationship between gender and sustainability in tourism. Whilst an extensive body of work exists in the areas of gender and sustainability, these two fields of knowledge are seldom combined to examine tourism phenomena.

When we look at the evolution of tourism, we see that sustainability has become an essential element in educational programmes, policy making and strategic considerations for organisations and destinations. Whilst the beginnings of tourism sustainability were challenging, presently, its relevance is seldom questioned. However, this situation is not the case with gender research. Although gender theorising and research have existed for over a century, and a rich legacy of knowledge exists on this topic, meaningful and respectful engagement with this line of scholarship is thus far peripheral in tourism studies. The aim of this book is to reflect on and rethink the intersection of gender and tourism sustainability through the lens of gender theory and feminist epistemology to stay with the trouble and devise pathways for sustainability gender knowledge.

This book will be of great interest to students, researchers, and academics in tourism, gender, and sustainability, as well as tourism management. The chapters in this book were originally published as a special issue of the *Journal of Sustainable Tourism*.

Claudia Eger is Associate Professor at Copenhagen Business School, Denmark. Her research focuses on business ethics, gender, and sustainability in tourism. Her research in different Middle Eastern and African countries addresses concerns of relevance to international organisations and policy makers, striving to inform gender and development strategies.

Ana María Munar is Associate Professor at Copenhagen Business School, Denmark, and Co-chair of the Critical Tourism Studies network. Her research applies philosophical approaches to gender, higher education, and tourism. Her gender research combines academic reports and publications, with advocacy and action research projects.

Cathy H. C. Hsu is Chair Professor at School of Hotel & Tourism Management, The Hong Kong Polytechnic University. Her research foci include tourist behaviour and stereotypes, as well as resident sentiment. She is Editor-in-Chief of *Tourism Management*.

First published 2023
by Routledge
4 Park Square, Milton Park, Abingdon, Oxon OX14 4RN

and by Routledge
605 Third Avenue, New York, NY 10158

Routledge is an imprint of the Taylor & Francis Group, an informa business

British Library Cataloguing in Publication Data
A catalogue record for this book is available from the British Library

ISBN13: 978-1-032-35961-8 (hbk)
ISBN13: 978-1-032-35962-5 (pbk)
ISBN13: 978-1-003-32954-1 (ebk)

DOI: 10.4324/9781003329541

Typeset in Myriad Pro
by Newgen Publishing UK

Publisher's Note
The publisher accepts responsibility for any inconsistencies that may have arisen during the conversion of this book from journal articles to book chapters, namely the inclusion of journal terminology.

Disclaimer
Every effort has been made to contact copyright holders for their permission to reprint material in this book. The publishers would be grateful to hear from any copyright holder who is not here acknowledged and will undertake to rectify any errors or omissions in future editions of this book.

Gender and Tourism Sustainability

Edited by
Claudia Eger, Ana María Munar and Cathy H. C. Hsu

Routledge
Taylor & Francis Group
LONDON AND NEW YORK

Notes on Contributors

Derya Demirdelen Alrawadieh, Department of Hotel, Restaurant and Catering Services, Istanbul Ayvansaray University, Istanbul, Turkey.

Zaid Alrawadieh, Department of Hotel, Restaurant and Catering Services, Istanbul University-Cerrahpasa, Istanbul, Turkey.

Carla Barbieri, Department of Parks, Recreation & Tourism Management, North Carolina State University, Raleigh, NC, USA.

Gul Erkol Bayram, School of Tourism and Hotel Management, Sinop University, Sinop, Turkey.

Ana B. Casado- Díaz, Department of Marketing, University of Alicante, Alicante, Spain.

Donna Chambers, Faculty of Business, Law & Tourism, University of Sunderland, Sunderland, UK.

Hengxuan Chi, School of Hospitality Business Management, Carson College of Business, Washington State University, Pullman, WA, USA.

Katherine Dashper, School of Events, Tourism and Hospitality Management, Leeds Beckett University, Leeds, UK.

Anna de Jong, School of Interdisciplinary Studies, University of Glasgow, Dumfries, Scotland.

Gregory Denton, School of Hospitality Business Management, Carson College of Business, Washington State University, Pullman, WA, USA.

Claudia Eger, Department of Management, Politics and Philosophy, Copenhagen Business School, Frederiksberg, Denmark.

Cristina Figueroa-Domecq, School of Hospitality and Tourism Management, University of Surrey, Guildford, UK.

Bryan S. R. Grimwood, Department of Recreation and Leisure Studies, University of Waterloo, Waterloo, Canada.

Dogan Gursoy, School of Hospitality Business Management, Washington State University, Pullman, WA, USA.

Colin Michael Hall, Department of Management, Marketing and Entrepreneurship, University of Canterbury, Christchurch, New Zealand; School of Business and Economics, Linnaeus University, Kalmar, Sweden; Department of Geography, University of Oulu, Oulu, Finland; School of Tourism & Hospitality, University of Johannesburg, Johannesburg, South Africa.

Chapter 14

The gendered effects of statecraft on women in tourism: Economic sanctions, women's disempowerment and sustainability?
Siamak Seyfi, Colin Michael Hall and Tan Vo-Thanh
Journal of Sustainable Tourism, volume 30, issue 7 (2022), pp. 1736–1753

Chapter 15

The social, cultural, economic and political strategies extending women's territory by encroaching on patriarchal embeddedness in tourism in Nepal
Wendy Hillman and Kylie Radel
Journal of Sustainable Tourism, volume 30, issue 7 (2022), pp. 1754–1775

Ana Beatriz Hernández-Lara, Department of Business Management, Rovira i Virgili University, Reus, Spain.

Wendy Hillman, Social Sciences, Central Queensland University, North Rockhampton, Australia.

Susan Jakes, Community and Rural Development, NC State Extension, North Carolina State University, Raleigh, NC, USA.

Jess Sanggyeong Je, Department of Tourism, Sport and Hotel Management, Griffith University, Brisbane, QLD, Australia.

Cathy H. C. Hsu, School of Hotel & Tourism Management, The Hong Kong Polytechnic University, Kowloon, Hong Kong.

Onur Cuneyt Kahraman, Tourism Faculty, Bolu Abant Izzet Baysal University, Bolu, Turkey.

Catheryn Khoo, Department of Tourism, Sport and Hotel Management, Griffith University, Brisbane, QLD, Australia.

Albert Kimbu, School of Hospitality and Tourism Management, University of Surrey, Guildford, UK.

Clifford Lewis, School of Management and Marketing, Charles Sturt University, Bathurst, Australia.

Ana María Munar, Department of Management, Politics and Philosophy, Copenhagen Business School, Frederiksberg, Denmark.

Adel Nikjoo, Geography Research Unit, University of Oulu, Oulu, Finland.

Hossein G. T. Olya, Sheffield University Management School, Sheffield, UK.

Faith Ong, Business School, Tourism Cluster, The University of Queensland, Brisbane, Australia.

Kylie Radel, School of Business and Law, Central Queensland University, North Rockhampton, Australia.

Carla Rodriguez-Sanchez, Department of Marketing, University of Alicante, Alicante, Spain.

Shima Salehi, Shahid Bahonar University of Kerman, Kerman, Iran.

Franco Sancho-Esper, Department of Marketing, University of Alicante, Alicante, Spain.

Ann E. Savage, Department of Parks, Recreation & Tourism Management, North Carolina State University, Raleigh, NC, USA.

Ricardo Sellers-Rubio, Department of Marketing, University of Alicante, Alicante, Spain.

Siamak Seyfi, Geography Research Unit, University of Oulu, Oulu, Finland.

Jennie Small, Management Discipline Group, UTS Business School, University of Technology Sydney, Ultimo, Broadway, Australia.

Michela J. Stinson, Department of Recreation and Leisure Studies, University of Waterloo, Waterloo, Canada.

Hazel Tucker, Department of Tourism, University of Otago, Dunedin, New Zealand.

Jane Turner, School of Events, Tourism and Hospitality Management, Leeds Beckett University, Leeds, UK.

Tan Vo-Thanh, Excelia Business School, CERIIM & CEREGE (EA 1722), La Rochelle, France.

Oscar Vorobjovas-Pinta, School of Business and Law, Edith Cowan University, Joondalup, Australia.

Yana Wengel, Arizona State University Joint International Tourism College, Hainan University, Haikou, China.

Allan M. Williams, School of Hospitality and Tourism Management, University of Surrey, Guildford, UK.

Elaine Chiao Ling Yang, Department of Tourism, Sport and Hotel Management, Griffith University, Brisbane, QLD, Australia.

Mustafeed Zaman, Metis Lab, EM Normandie Business School, Le Havre, France.

Introduction: Gender and tourism sustainability

Claudia Eger, Ana Maria Munar and Cathy H. C. Hsu

ABSTRACT

It is long overdue for tourism research to move beyond the basic question of whether gender matters, because there is no humanity (or human phenomenon) without gender dimensions. Instead this article asks how does it matter? It does so by challenging tourism sustainability knowledges from the perspective of feminist epistemologies. It presents a broad and necessary conceptualization of gender which includes the spectrums of sex, sexuality, gender expression, and gender identity. Drawing on the philosophical conception of ideology by Elisabeth Anderson, this article invites to reimagine dominant models of the world in gender, culture and nature ideologies. It introduces the contributions and learnings of the special issue on "Gender and Tourism Sustainability". Finally, it states that a future agenda for gender and tourism sustainability research must highlight that being and knowing includes the non-human and a multiplicity of ecologies and cosmologies, that knowledges are multitude, and that they can be found beyond the written word and/or sanctioned instutionalized knowledge.

Introduction

This special issue examines the relationship between gender and sustainability in tourism. Whilst an extensive body of work exists in the areas of gender and sustainability, these two fields of knowledge are seldom combined to examine tourism phenomena. The aim of this special issue is to reflect on and rethink the intersection of gender and tourism sustainability through the lens of gender theory and feminist epistemology to stay with the trouble (Haraway, 2016). What trouble? If we look at the evolution of tourism, we will see that sustainability has become an essential element in educational programmes, policy making and strategic considerations for organisations and destinations. Whilst the beginnings of tourism sustainability were challenging, presently, its relevance is seldom questioned. However, this situation is not the case with gender research. Although gender theorising and research have existed for over a century, and a rich legacy of knowledge exists on this topic (e.g. Butler, 1990; De Beauvoir, 1973; Haraway, 1988; hooks, 2000/2015; Irigaray, 1985; Lorde, 2017; Woolf, 1929/2010), meaningful and respectful engagement with such thinkers is thus far peripheral in tourism studies (Figueroa-Domecq et al., 2015). We observe an increased interest in women and travel (Khoo-Lattimore & Wilson, 2017; Yang et al., 2017), masculinity (Porter et al., 2021; Schänzel & Smith, 2011; Thurnell-Read & Casey, 2014), sexuality (Carr & Poria, 2010), gendered work (Baum et al., 2016; Costa et al., 2017; Mooney et al., 2017), gender and entrepreneurship (Kimbu et al., 2019; Moswete & Lacey, 2015;

Zhang et al., 2020), gender paradigms (Denizci Guillet et al., 2019; Munar & Jamal, 2016; Pritchard et al., 2007) and gender in tourism academia (Chambers et al., 2017; Pritchard & Morgan, 2017). However, with a few exceptions (Camargo et al., 2016; Kato, 2019; Swain, 2016), these studies only scratch the surface.

The marginalisation of gender studies in tourism is generally due to the misconception of this field as dealing 'only' with 'women's issues'. Such a prejudice is a reflection of patriarchal bias and ignorance (Munar et al. 2017). Moving beyond the basic question of whether gender matters is long overdue in tourism research, because humanity (or human phenomena) cannot exist without gender dimensions. Instead, we should ask the following questions: How does gender matter? What is the relationship between gendered cultures, behaviours and worldviews? What is or will become of sustainability in tourism? Questioning how we can realise sustainability demands a reflection on how gender is done in, with, and through sustainability. Moreover, if sustainable and unsustainable ways of doing gender exist, then it entails a broad understanding of the conditions and consequences of gendered (un)sustainable doings.

This special issue was born as an invitation to contribute to this enquiry with the aim of moving away from universal answers, grand narratives or the catalogue of 'one-size-fits-all' ideas of what emancipatory tourism sustainability could be. This issue aims to consider the gendered complexity of the world and engage in contextualising knowledge production. Bright examples of such work include the paradigmatic proposal for an embodied cosmopolitanism of Swain (2016), the study on ecocultural justice by Jamal and Camargo (2014) and the research of Chambers and Buzinde (2015) on the connection between geopolitics and body politics or the problematisation of nature as 'both contested and promiscuous in an ontological sense', as presented in the work of Grimwood et al. (2018, p. 2). Additionally, this special issue calls for the reconsideration of the traditional division between knowledge and emotions and invites scholars to reflect on the epistemic value of attending to one's emotions in the process of knowledge creation (Jaggar, 1989).

Conceptualising gender

Gender is a core part of being human, and similar to other key social science concepts, its meaning is contested, varied and evolved over time. This special issue maintains an openness towards the different theoretical or disciplinary traditions that engage with gender as a phenomenon. For example, in disciplines characterised by quantitative methodologies, we can typically find surveys in which gender tends to follow a binary pattern, with participants divided into two gender categories. However, beyond this idea, the key aim of this introduction is to expand and refine gender terminologies and conceptualisation in tourism sustainability to firstly improve the future development of the field and secondly and more importantly to foster an academic practice that is inclusive and respectful towards diverse and different ways of being.

In this article we adopt the conceptualization of gender as spectrum inspired in poststructuralism while also recognizing the importance of corporeality and embodiment, as enlightened by studies of transgender and transsexual's lived experiences (Gender Spectrum, 2021; Monro, 2005). We understand gender in its broadest conceptualisation as a complex and fluid landscape that includes the spectra of sex, sexuality, gender expression and gender identity. *Sex* is defined as the biological reality of the human body and its sexual organs and features. The identity that infants receive at birth by their parents, caretakers and/or medical personnel belongs to the sex dimension. Traditionally, based on one's sexual anatomy (including chromosomes, gonads, hormones or genitals), one can be identified by others as being either *female, male or intersex* (i.e. characterised by sexual anatomy that does not fit the typical definition of female or male, which accounts for 1.7 percent of infants worldwide; United Nations Free & Equal, 2020). Later in life, we can or may not be able to identify with the primary identity assigned to us as infants. For example, *cisgender* individuals identify with the sex they were assigned at birth, *transsexuals* seek to transition from one sex identity to another and *transgenders* do not identify with the gender identity or expression traditionally fitting the biological sex they were assigned at birth (e.g. identified at birth as biologically male but do not identify themselves as men/male or with masculinity). The previous example shows that we should not confuse sex with gender identity, as these two concepts do not necessarily coincide. Gender identity corresponds with the personal and intimate feelings of identifying oneself as a man/male and with masculinity, woman/female and with femininity or something else (i.e. individuals who feel that they do not belong to either of the binary structure). Owing to historical prejudice, repression or fear of violence and persecution, this identification (as is the case with sexuality) is not always declared or made public. Gender expression refers to how we present ourselves or see others, including aspects such as the appearance, style, body language, tone of voice and physical traits that specific cultures and epochseras identify with a gender type (e.g. the use of high-heeled shoes or lipstick as a sign of female gender expression).

Meanwhile, sexuality refers to people's romantic, sexual and erotic desires, feelings and behaviours, with a broad spectrum of possibilities and representations, including heterosexual and LGBTQIA+ individuals. Noticing how thinking of sexuality in terms of spectra or fluidity challenges heteronormativity (i.e. considering heterosexuality as the default norm) is important. This short review of gender terminologies is not exhaustive or closed, as different cultures (i.e. indigenous people) use various terms for the fluidity and plurality of gender spectra (e.g. the third gender in Mexico).

Traditionally, scholarship was characterised by the simplification of seemingly challenging but beautiful complexities and the interplay and interdependence of the four spectra with one another into the two categories (in the worst case, essentialised with fix traits) of (a) woman/women, corresponding to the female sex, woman gender identity, female gender expression and heterosexuality and (b) man/men, the male sex, man gender identity, male gender expression and heterosexuality. Whilst such gender patterns are undoubtedly applicable to numerous individuals in society, they are far from representing the whole of humanity. The spectra can be combined to make multiple forms, such as the female sex, woman gender identity, male gender expression and lesbian sexuality, among many others. This practice is not merely the banal exercise of naming or tagging. Identities have and historically had vital consequences for the flourishing and becoming of individuals. Recognition of this plurality is important to scholarship that takes its ethical duties seriously.

The evolution of scholarship is analysed especially in this contribution. The article by Ong et al. (2022) titled 'LGBTIQ+ identities in tourism and leisure research: a systematic qualitative literature review' provides a review of the current state of knowledge. The authors employed an original methodological approach combining a machine learning technique using Leximancer with a qualitative method to examine the scope and topics of existing LGBTIQ+ tourism research. The method allowed the authors to conduct a sophisticated content analysis of the field tracking the evolution of the literature, thereby enabling them to provide novel insights into underlying trends. The article highlights the important role that research plays in challenging heteronormativity. In discussing the implications and future avenues for research, the authors called for the broad consideration of intersectionality and the multitudes of identities in LGBTIQ+ research, emphasising the importance of gaining a practical understanding of the LGBTIQ+ segment to create safe travel experiences.

Troubling tourism sustainability

A basic understanding of sustainability involves considering 'development that meets the needs of the present without compromising the ability of future generations to meet their own needs'

(WCED, 1987, p. 43). This well-known approach does not exist without its challenges. The institutionalisation of sustainable development relied on the imposition of Western/imperial knowledge, in which sustainability became a key mantra rather than a novel paradigm (Escobar, 1995). Sustainable tourism emerged from the alternative paradigm and gained prominence between the 1980s and 1990s (Telfer, 2002). However, it evolved without the significant questioning of contemporary tourism development models (Ferguson & Alarcón, 2015). A central debate involves the tension between ecocentric and anthropocentric worldviews (Yudina & Grimwood, 2016). The latter is characterised by second natures, capital accumulation and certain forms of environmentalism (Katz, 1998), focusing on neoliberal conservation processes and the conservation of natures and cultures in tourism (Mowforth & Munt, 2009). While sustainability thinking has been incorporated into tourism planning and development policymaking (Buckley, 2012), it tends to engage in line-drawing and categorisation approaches, which may themselves require a critical examination.

An example of critical examination through the lens of postmodernism is beautifully presented in the article by Tucker (2022) titled 'Gendering sustainability's contradictions: between change and continuity'. Tucker explored tourism-related 'sustainability fluxes' manifested as lived tensions and contradictions. The author adopted an ethnographic approach rooted in poststructural feminist thought to examine the contradictions between change and continuity in women's everyday lives in the Cappadocia region of Central Turkey. The analysis drew attention to the inherent ambiguities that underlie sustainability knowledge and the elusiveness of change if gender is not considered. This thought-provoking article advances our understanding of the multiplicity of sustainability knowledge and affects in tourism, which are not static but seem to shift their relational configuration continuously, increasing the importance of engaging in the uncertainty, doubt and ambivalence of attending to sustainability fluxes.

The sustainable tourism paradigm is characterised by resistance to the integration of gender equality considerations as core principle (Alarcón & Cole, 2019; Ferguson & Alarcón, 2015). When gender considerations are made in tourism and development, they often rely on an 'add women and stir' (Harding, 1995, p. 295) approach (Alarcón & Cole, 2019; Moser, 1993). This practice demands critical engagement with the ways in which gender is instrumentalised in development policies and practices, which has not led to substantial transformations at the personal, institutional and political levels. Rather, it increased unequal power structures as well as the environmental risks that communities, especially women, are exposed to (Cole, 2017; Shiva, 2016). Gender is often absent, silenced or sidelined in mainstream discussions. For example, Eger et al. (2020) highlighted the neglect of gender-based violence in dominant tourism discussions. Baum et al. (2016) noted the limited attention paid to workforce considerations and related gender issues in the sustainable tourism literature. This situation is not a naïve choice, but, as explained by the epistemology of ignorance, 'often the result not of a benign gap in our knowledge, but in deliberate choices to pursue certain kinds of knowledge while ignoring others' (Grasswick, 2011: xviii).

By introducing gender theorising and feminist epistemology into the overall body of knowledge of sustainability and tourism, we aim to transform the nature of what we understand sustainability to be, including our overall comprehension of the interface between tourism and sustainable development goals (SDGs; Alarcón & Cole, 2019; Kato, 2019). In addition, the prevailing administrative and instrumentalist perspectives of sustainability question its ability to contest dominant forms of perceiving and relating to others, including the nonhuman world. Ferguson and Alarcón (2015) identified the integration of gender as an isolated component as a key problem, which does not allow the rethinking and challenging of key assumptions in sustainable tourism projects. SDG 5, which aims to achieve gender equality and empower women and girls, is foundational for achieving the other SDGs (Alarcón & Cole, 2019). However, Baum et al. (2016) showed that gender aspects are not sufficiently considered and met in customary tourism employment in relation to different SDGs (i.e. 1, 3, 4, 5, 8, 10 and 16). Hence, gender mainstreaming remains a valuable concern in determining how to achieve a gender-aware framework for tourism (Ferguson, 2011; Kinnaird & Hall, 1996).

To create broad systemic changes, sustainability analyses should account for the complex intersections between gender and tourism sustainability, which are hybrid, political, historically and spatially variable and do not adhere to one single discourse or worldview. This combination of various SDGs is thematised and addressed in the following three articles. Casado-Díaz et al. (2022) reimagined the connection between the SDGs of gender equality and water conservation. Their article 'Tourists' water conservation behaviour in hotels: the role of gender' examines whether gender differences matter for sustainable behaviour. The authors used a face-to-face survey to gather data in hotel premises in Spain. Confirmatory factor analysis was employed for the statistical data analysis, and the hypotheses were tested via multigroup structural equation modelling. Specifically, the authors examined whether gender differences existed in the hotel guests' reported water conservation behaviours. The research findings suggested that factors such as attitude towards water conservation and normative and hedonic motives may affect gender differences.

Adding to this discussion but in terms of climate change, Denton et al. (2022) examined gendered patterns of travellers' behaviours in relation to carbon offsetting. In their article 'An examination of critical determinants of carbon offsetting attitudes: The role of gender', the authors employed a quantitative methodology to test their conceptual cognitive appraisal model in a pilot study. Next, the authors used covariance-based structural equation modelling to analyse the answers of a panel of respondents recruited via Amazon MTurk to assess the role of a series of factors in influencing gender differences in carbon offsetting attitudes. The interesting findings of this study revealed the factors behind relevant differences in behaviours between genders.

Gender differences are also observed in the article by Figueroa-Domecq et al. (2022) titled 'Sustainability through the tourism entrepreneurship journey: a gender perspective'. The authors adopted a post-structuralist framework to investigate the role of gender and sustainability in different stages of entrepreneurship in Spain. The authors used a mixed-methods research design, starting with a questionnaire survey of 539 tourism students, representing the latent and nascent entrepreneurship stages, followed by 19 interviews with established entrepreneurs. The quantitative data were analysed via structural equation modelling to test the hypothesised relations, whereas the qualitative data offered nuanced insights into gender influences. Broad gender differences were observed in terms of timing in engagement in sustainability, conceptualisation of sustainability and emphasis on different sustainability dimensions. Whilst societal perceptions of risk aversion were gendered, gender differences in personal attitudes and behavioural control were impalpable.

Models of the world: gender, culture and nature ideologies

Ideology as a model of the world

'Paradigm' and 'ideology' are terms that aim to express how a subject interprets and makes sense of the world. Whilst 'paradigm' is a useful term for examining scholarship traditions (Munar & Jamal, 2016), for investigating gender and tourism sustainability, we turn to the concept of ideology, as defined and applied by feminist philosopher Elisabeth Anderson (2017). Why? Ideology is a broad term representing individuals' and collectives' worldviews and not restricted to scientific communities. When we examine the dominance of the binary view in gender, the historical belief in the superiority or universal representativeness of man versus women or others and the separation of the body–mind and human–nature divide, we move beyond exclusive scientific debates into the large realm of society. Our ideological lenses of the world impact the ways we interpret and give meaning to gender and sustainability. Anderson adopted a normative/ethical view on ideology as being paradoxical. Ideologies can be positive, because they provide us with an 'abstract model to represent and cope with the social world' (Anderson, 2017, p. xx). As humans, we can experience the realities of our world only partially. Ideologies can be useful mechanisms for helping our cognitive limitations.

> An ideology is good if it helps us navigate [the world] successfully. To help us, it must identify the normatively important features of the world, and the main causal connections between these features to which people can respond, enabling them to discover effective means to promoting their goals. Ideologies also help us orient our current evaluations of the world, highlighting what we think is already good or bad in it. Finally, they are vehicles for our hopes and dreams. (Anderson, 2017, p. xxi).

The same features that make ideologies useful (i.e. their capacity to reduce and represent) can make them negative. This phenomenon occurs when ideologies mask or distort what is problematic about our world; specifically, they 'misrepresent the space of possibilities so as to obscure better options, their means to realizing them, or their merits' (Anderson, 2017, p. xxi). In other words, if abstract models of the world systematically undermine and discriminate against the interests or potential of specific groups of people seriously or gratuitously, then they must be changed. Anderson warned that to stop or modify such discrimination can be extremely difficult when 'the interests of those who dominate public discourse are already served by the dominant ideology' (Anderson, 2017, p. xxii). Ideologies can become tools for sustaining injustice, blind us to the possibility of becoming and block flourishing (e.g. Eger, 2021a, b). In this way, ideological models embed prejudice and stereotyping, which are psychological and social mechanisms (conscious and unconscious) that we use to categorise and make sense of our social world (Hardin & Banaji, 2013).

The negative aspects of ideology are intensified when combined with belief in objectivism. Objectivism is the unfortunate but common assumption that our reduced maps of the world (i.e. ideologies) are objective ways of knowing reality; this is not my/our way of making sense of the world but rather *the* world itself. Belief in objectivism is dominant in mundane and habitual relationships and present in the way's academics tend to approach discussions about gender (Munar et al., 2017). What characterises ideological objectivism is the belief that what is objectively real exists independently of who we are as researchers (the subject/object dichotomy); researchers attain the right knowledge of reality independent of their relation to it owing to 'a view from nowhere', which transcends time and space (aperspectivity); research representations are dictated by how things actually are and not by the knower (external guidance); to become an objective researcher demands emotional detachment and the adoption of an evaluative neutral attitude towards what is studied (value neutrality); and finally, that 'really' getting to know something is equal to being able to control that something (control),(Anderson, 2015).

A deeply problematic relationship exists between objectivism and ideology. Objectivism fosters a form of hubris in ideology, that is, the fantasy of human understanding as a perfect reflection of the nature of things/people/the world. This concept is one of the key epistemological problems identified by feminist philosophy. Although we cannot provide a detailed critique of the features of objectivism (see Anderson, 2015; Harding, 1993; Haslanger, 1993), the dangers of belief in aperspectivity are worth explaining.

The problem of aperspectivity involves the lack of awareness of how our backgrounds, biases and values impact our gaze and interpretation of the world (i.e. confusing what is historically/socially contingent with something permanent/necessary). Moreover, if we trust that what we understand is a perfect reflection of the nature of things/people and therefore treat such things/people accordingly, then we may be contributing to the reproduction of the oppression and misinterpretation of groups in subordinate positions. We believe that prejudice and stereotyping are truths and contribute to making them truths. In short, aperspectivity founds the false belief that what is attributed to culture, language, society and so on is an essential 'natural' or biological trait that cannot be modified. Therefore, aperspectivity chains our possibility of becoming, fuels stereotype and bias reproduction and limits our freedom, disguising oppression as biological fate.

Abstract models of the world include ideological understandings of what gender is, what it does and how it relates to tourism sustainability. In such models, a series of beliefs on the human–nature divide, gender binary divide, and culture–nature divide traditionally play a dominant role. Such a dominance systematically blinds us to the possibility of flourishing that may be achieved by and through tourism sustainability. The article by Donna Chambers (2022) titled 'Are we all in this together? Gender intersectionality and sustainable tourism' courageously challenges preconceived ideological notions of gender, race and sustainability by introducing other ways of doing knowledge and writing that disrupt the dominant Western tradition. Chambers employed critical race theory and black feminism to examine intersectional marginalisation. Moreover, the author's analysis of the fictional film *Heading South* relied on the storytelling technique to query the (re)presentation of black women in the film, drawing wide parallels with the ways in which such representation can shape popular cultural narratives in tourism and beyond. The author's original approach provides a source of inspiration to intersectionality studies, and her findings highlighted the material and political implications of diminishing the agency of black women as subjects in tourism and sustainable development. In the subsequent section, we expound on such crucial discussions and the critical challenges to dominant beliefs, that is, what we consider as alternative models of the world. Therefore, the following sections present a critique and an invitation to explore new ideological interpretations and combinations of ideals and hopes for a sustainable future.

Challenging the dichotomy of the human–nature divide

Early discussions on gender focused on the critical analyses and deconstructions of conceptual dichotomies such as culture/nature, mind/body and reason/emotion. Such discussions remain crucial to contemporary debates on not only gender but also sustainability. The latter is generally defined by a human–nature dualism and revealed and managed by apolitical natural sciences, specifically, 'Science as heroic quest and as erotic technique applied to the body of nature' (Haraway, 1991, p. 205). This debate continues to this day as the human–nature tension or divide. Nancy Fraser (2014, 2017) proposed a threefold perspective of nature to disentangle multiple overlapping and competing ecological discourses that were advanced to circumscribe sustainability. Nature can be viewed as *an object of climate science*, which assumes that nature operates independently of what we know and how we perceive it. This view is characterised by works concerned with providing a universal, objective and rationale perspective, removing the subjectivity of the knowledge producer. Environmental changes are primarily considered as a problem when humans are affected, alluding to the implicit anthropocentrism that continues to guide conservation ethics in sustainable tourism (Yudina & Grimwood, 2016).

In a capitalist logic, nature is made *the other of value*, building on the argument of Neil Smith (1984) on second natures, resulting in the appropriation and subsequent accumulation of natural capital in foreign markets. Capitalism portrays a gender-neutral society based on economic relations, which through their patriarchal structure and history formed a society marked by gender inequality (Sinclair, 1997). Reproductive conditions for a capitalist society rely on separating and valuing production over social reproduction. Feminist economists argued for the inclusion of caring activities in the concept of labour to account for the mutually interdependent values of (re)productivity, which remain neglected by economic theory (Biesecker & Hofmeister, 2010). The predominant association of women with paid and unpaid care work with the devaluation of the work as feminised labour problematises an unquestioned notion of labour in which the devaluation of women and nature is conceptually linked (Gaard, 1997).

Sustainability emerged as a cautionary tale of modernity and late modernity but remains predominantly gender blind when analysing what we experience and theorise as the other of nature. Clearly, we cannot divide nature from humans, and besides our deep concern for the planet, at the core of sustainability lies our responsibility of the possibility of humanity's flourishing. 'Human development is about enlarging freedoms so that all human beings can pursue choices that they value' (UNDP, 2016, p. 1). This conceptualisation builds on the capability approach developed by Sen (1992), in which he advocated embracing the complexities and ambiguities that underline our understanding of human flourishing and inequalities. Sen (1992, p. 125) proposed an understanding of gender inequality that builds on the intrinsic worth of functionings and capabilities, arguing that the 'issue of gender inequality is ultimately one of disparate freedoms' rather than disparate resources. The capability approach is referred to only peripherally in tourism and gender research (e.g. Eger et al., 2018) but exhibits vast potential for feminist enquiries.

Nature can also be defined based on individual worldviews, that is, ideologies of nature, understanding nature *as constructed*. Critical tourism scholars proposed a view of '"sustainability" as Indigenous and Native sovereignty' (Devine & Ojeda, 2017, p. 614), thereby challenging the hegemonic imposition of knowledge marginalising local populations. Ecology-based studies, such as queer ecology and ecofeminism, challenged the deterministic view of nature, considering the environment as no longer a one-dimensional force but a cultural landscape questioning the customary division between the social and natural. Warren and Cheney (1991) described ecofeminist ethics as a *quilt in the making*, in which the different patches represent the diverse socioeconomic conditions, cultures and histories of the quilters. The contours of the quilt are its spatiotemporal dimensions, though the actual pattern of the quilt, that is, *its interior*, stems from the life experiences of the quilters and their ethical concerns as 'an articulation of knowledge, and [...] a way of knowing the world' (Kato, 2019, p. 952). Camargo et al. (2016) applied an ecofeminist framework to rethink sustainable tourism from the perspective of an embodied paradigm that considers emotions, feelings and the ethics of care. Margaret Swain, a pioneer of gender scholarship in tourism, and Melissa Swain (2004) employed ecofeminism to provide a critical examination of ecotourism.

These perspectives influence the discursive formation of human–environmental interactions, as poetically illustrated in the exploration of affect as a form of defacing of Stinson and Grimwood (2022). The article questions the dominant masculine terrain of rock climbing and opens up the more-than-human embodied possibilities of sustainability knowledge. 'Defacing: affect and situated knowledges within a rock climbing tourismscape' offers a deeply connected, humble and poetic narrative unfolding through and with the more-than-human. Drawing on actor-network theory, the authors engaged with the more-than-human through the narrative capacities of rock climbing, illuminating the seamless in-betweenness of bodies and spaces. The ability to affect and be affected subverts dominant dualisms and instead points to the relational potentialities between humans and more-than-human. In the process, the authors unsettled the gendered demarcations of knowing and being in the world.

Whilst different critiques of ecofeminism were advanced, Haraway (1991, p. 199) contended that '[e]cofeminists have perhaps been the most insistent on some version of the world (nature/body) as active subjects, not as resources to be mapped or appropriated in bourgeois, Marxist or masculinist projects'. The nuanced representation of knowledge and capabilities, that is, whose voices and choices count, is essential to understanding the driving forces of territorial politics in the global South, as reflected in the feminist political ecology of water of Stroma Cole (2017) and the ecohumanities perspective of gender and sustainability in tourism of Kumi Kato (2019). This creates a new spectrum for theorising sustainability, which goes beyond pre-established thought processes, to focus on sensory, hybrid, spiritual and reciprocal aspects in human and other-than-human relations (Kato, 2019; Yudina & Grimwood, 2016), thereby reweaving the texture of sustainability gender knowledge.

Gender and social organising

The upheaval of social movements is a response to a politicised environment wherein spatial organising represents a new perspective of the personal as political, moving away from a modernist marginalising notion to the concept of moral economy and everyday resistance. Struggles against colonialism simultaneously become struggles against anti-environmentalism, as many indigenous groups do not follow the clear separation of humans and nature or culture and nature (Ramirez, 2021). This idea is illustrated in the Chipko movement in India, which brought people from different castes, ages and genders together to unite and rise up against resource deprivation and political marginalisation (Shiva, 2016; Swain & Swain, 2004). However, by interpreting the Chipko movement as a purely environmental movement outside the framework of postcolonial impositions, the actual intention behind the social mobilisation of women is distorted. Such conflicting views on sustainability also have gendered aspects and relate to masculine and feminine cosmologies, in which, as suggested by ecofeminism (Warren, 1997), the domination of nature is linked to the domination of women and other historically marginalised ethnic groups in tourism (Camargo et al., 2016). Recognising women's environmental identity as linked to production and livelihood aspects reduces the possibility of romanticising heritage and reinforcing traditional gender roles (cf. Jiménez-Esquinas, 2017).

Movements provoke sociocultural change and can unsettle and thus transform gender norms and roles. Such a change is equally important to organisations. In their article 'Gendering knowledge in tourism: Gender (in)equality initiatives in the tourism academy', Dashper et al. (2022) explored change through a methodological innovation. The authors described the preparations of a tourism department to apply for a gender equity charter accreditation using Ketso, which is a toolkit for creative engagement. The use of Ketso as a participatory action research tool facilitated the active contribution of the participants and broke down traditional hierarchies to explore issues of power and voice within groups and organisations. Workshops were conducted to collect qualitative evidence of gender (in)equality. Although the university's management decided to not proceed with the application, the process enabled the discussion of generally overlooked topics and exposed engrained gender–power undercurrents and structural challenges to reform. Guided by the concept of gendered organisations, the work illustrated numerous gendered practices in tourism academia and their limitations in overcoming entrenched gender inequity.

Another core organisational challenge related to gender roles and norms is sexism and sexual harassment. The latter is examined in the study of Alrawadieh et al. (2022), which focused on the sexual harassment of female tour guides. In their article 'Sexual harassment, psychological well-being, and job satisfaction of female tour guides: the effects of social and organizational support', the authors developed an original construct measured by multiple-item scales adopted from previous research on sexual harassment. An online-based self-administered survey was used to collect the data, and the hypotheses were tested using statistical analysis. The findings showed the negative impact of sexual harassment on job satisfaction and psychological well-being. Additionally, the results indicated that missing organisational support plays a significant and negative role in triggering sexual harassment.

The extensive literature review conducted by Je et al. (2022) reveals the lack of focus on the cultural and individual dimensions of gendering processes in tourism organisations whilst explicitly focusing on the organising system and developed economies. Their article 'Gender issues in tourism organisations: insights from a two-phased pragmatic systematic literature review' relies on a pragmatic systematic review to compare tourism academic and grey studies from NGOs and large public tourism companies. This relevant method allowed for the identification of under-researched areas and gaps to address issues of gender equalities. This extensive analysis comprises 102 academic and 122 grey studies to examine how academic research is translated into actual practices by tourism organisations.

The general tendency is to blame the individual when gender inequality, sexism or harassment issues emerge in organisations. According to previous research, sexism is a prevalent structural and societal problem that must be addressed by the leadership of an organisation (Eger, 2021a). Women often become the problem by pointing to the problem. It is time to move away from victim blaming towards collective responsibility and accountability through organisational action at the structural level (Einersen, 2021; Finniear et al., 2020).

Challenging the dichotomy of the gender binary

Sex/gender distinction traditionally imposes a passive perspective on the body, as in preceding discourse. Moreover, the body's association with the female sex restricts the latter to its embodiment, whereas the male sex 'becomes, paradoxically, the incorporeal instrument of an ostensibly radical freedom' (Butler, 1990, p. 16). The existence of a subject as the master (i.e. in control of, superior to and separated from nature) is symbolic of the predominant hegemony of the masculine subject (Haraway, 1991). Rather than a subject, this master–subject ideal reflects an ideological often taken-for-granted and therefore invisible position rationalised through discourse, which does not reflect the multiplicity of femininities and masculinities and their dynamic interrelations. The white corporate wealthy heterosexual cultural model tops this hierarchy (Kiesling, 1998, p. 71). In her article 'The Sustainability of Gender Norms: Women over 30 and their Physical Appearance on Holiday', Small (2022) questioned the embodiment of gender. The author explored the taken-for-granted freedom and laissez-faire attitude often associated with being on holiday by providing intimate insights into women's self-perceived body image. Small employed memory work to examine women's experiences of their body image whilst on holiday. The author's original methodological approach provided a counter narrative to the dominant focus on the 'ideal' body, which is at odds with the norm(al), thereby disclosing the intricate unsustainability of the gendered social norms of appearance. The author's focus on bodies as a social practice illuminated the ways in which memories of our holidays can affect our wellbeing before (preparing the body), during (bodily exposure) and after (images 'capturing the body forever') going on a trip. Rather than negotiating societal ideals, the women in the study tended to adjust their behaviours to these ideals, thereby emphasising the prominent role of body politics in tourism. Challenging bodily norms and representations requires an understanding of how these concepts were institutionalised throughout history, shaping power relations that encode and evaluate physical appearance.

Strong gendered stereotyping and prejudice related to specific professions can also be observed. Savage et al. (2022) challenged the gendered stereotypes of farmers. In their paper 'Cultivating success: personal, family and societal attributes affecting women in agritourism', the authors used a survey to identify personal, household and social factors contributing to women farmers' success in functional and life aspects. Mail and online questionnaires were used to collect data based on the snowball sampling technique. The data were analysed using descriptive statistics, exploratory factor analysis and multiple linear regressions. In addition, feminist and systems approaches were used to identify the influencers of women achievements in their role as farmers and entrepreneurs and in other facets of their lives. Besides their perception of moderate success, the participants' disposition to civic and conservationist values supports their preference for engaging in sustainable practices. The inclusion of values in the study advances the post-productivist research agenda and helps us reimagine the normativity of professions.

Norms and normativity also dominate the area of sexuality. The belief of heterosexuality as superior and the norm requires the complete rethinking of gender narratives to understand experiences that lie beyond the heterosexual matrix expressed through the coherence of the female and male form (Butler, 1990). Wittig (1993, p. 103) argued that 'One is not born a woman'. Performativity theory developed by Butler (1990) postulates that gender is formed and constantly remade through repetition, 'In other words, acts, gestures, and desire produce the effect of an internal core or substance, but produce this on the surface of the body, through the play of signifying absences that suggest, but never reveal, the organizing principle of identity as a cause' (Butler, 1990, p. 185). Gender, in all its spectra, is not an either–or but a both–and.

Gender is the becoming of a subject through embodiment and the overall discursive and normative means through which gender relations, including the construction of not only sex but also violence, are (re)produced (Eger, 2016; 2021a).

The increasing work on intersectionality shows that identity categories are fluid (Crenshaw, 1989; Mooney, 2016; Villesèche et al., 2018). Gender norms and categories do not exist in isolation but are in a complex interplay with other aspects of differences, such as race, ethnicity, disability, age, culture, religion and class. Inequality is experienced at the intersection of multiple differences, and looking at gender alone may cause scholars to miss certain underlying dynamics (Eger, 2021b). However, concepts such as empowerment, equality and sustainability often assume a certain level of stability (e.g. a heterosexual matrix) and an implicit (e.g. nonracialised and Western) standpoint. This idea implies that the concepts we employ in research are not always compatible with local structures, practices and gender norms. Whilst the West does not refer to a monolithic concept, and the multiplicity of Western feminism exists, it is 'possible to trace a coherence of effects resulting from the implicit assumption of "the west" (in all its complexities and contradictions) as the primary referent in theory and praxis' (Mohanty, 1988, pp. 61–62).

We highlight three case studies focusing on gender beyond the West. 'The contribution of all-women tours to well-being in middle-aged Muslim women' by Nikjoo et al. (2022) examines the experiences of Iranian middle-aged women who participated in all-women tours. The authors adopted the constructivist version of grounded theory and relied on participant observation and interviews. This original article reveals how this form of tourism can have social and personal benefits and contribute to women's overall wellbeing. The article of Seyfi et al. (2022) titled 'The gendered effects of statecraft on women in tourism: Economic sanctions, women's disempowerment and sustainability?' provides novel insights into the intricate relationship between economic sanctions and gender empowerment in the Iranian tourism industry. The authors employed Scheyvens' empowerment framework to elucidate the economic, psychological, social and political implications of this particular form of economic statecraft. The research draws on 28 interviews with women working in the Iranian tourism industry and analyses the data via thematic analysis. The findings highlighted the ways in which sanctions further augment gendered vulnerabilities, circumscribing possibilities for empowerment, making the latter contingent on its economic dimension to allow for the emergence of other empowerment dimensions.

Finally, the paper of Hillman and Radel (2022) titled 'The social, cultural, economic and political strategies extending women's territory by encroaching on patriarchal embeddedness in tourism in Nepal' uses a critical theoretical framework to investigate the ways through which Nepali women challenge longstanding gender and power structures through entrepreneurship. The authors interviewed 16 Nepali women entrepreneurs over a six-year period to capture experiences in a wide range of urban and rural settings. As critical constructivist researchers, the authors developed a reciprocity-based methodology by offering free business-skills workshops. Participation in the workshops was not contingent on involvement in the interviews for the research purpose. The interviews were conducted in English with the assistance of a Nepali female translator, when necessary. The women took part in not only traditionally female territories, such as homestays, but also male domains, such as trekking and tour guiding. Using a combination of strategies to combat patriarchy in tourism and navigate the complex positioning of entrepreneurship, the Nepali women are instrumental in their empowerment.

Knowledge on sustainability as a construct and organising trope can be advanced through non-Western postcolonial and decolonial epistemologies to deconstruct the self/other binary, that is, 'the rule of colonial difference' (Chatterjee, 1993), and challenge the gendered legacies of colonialism (Chambers & Buzinde, 2015). Gender stereotypes have long been held accountable for differences between modernity and tradition (Harding, 2011). This idea is reinforced through the division of an inner and outer sphere, 'a kind of private and public, in which men could safely emulate the ways of the West and appropriate its technologies in order to gain power as long as the home, with women its clearest representatives, could be preserved as a space of spirituality and cultural authenticity' (Abu-Lughod, 1998, p. 17). However, Westernisation is not synonymous with modernity. Postcolonial scholars emphasised the role that hybridisation, translation and dislocation plays in colonial encounters (Abu-Lughod, 1998), thereby opening up new ways of understanding and examining sustainability. Chambers and Buzinde (2015) further called for the epistemological decolonisation of tourism knowledge, recognising that the 'master's tools will never dismantle the master's house' (Lorde, 1983, p. 98).

Challenging the dichotomy of the culture–nature divide

Is gender a product of culture or nature? Although old, this debate, which at times is characterised by its virulence and dogmatism, continues to attract news media, with typical headlines announcing that 'women are like this and men are like that' (Fine, 2017). The question in itself is a trap, as it takes for granted that human nature without culture or human culture without nature exists. This divide is a classical one. In Western thought, the ideological mapping of nature and culture as two separate entities is related to the mind–body divide (Haraway, 1991; Swain, 2016). In this classical structure, the mind corresponds to culture and what humans 'learn' and become through socialisation (e.g. language, behaviours and so on), whereas the body corresponds to nature, our biological evolutionary being and inheritance (basic biological needs of survival and reproduction and processes such as ageing and so on). Such a division is generally linked to the Cartesian philosophy (I think therefore I am), though its precedence can be found much earlier in the ideas of the 'soul' and body from classical Hellenistic philosophy (the soul representing the human capacity of consciousness andlanguage). Such a division has long been questioned and rejected by 20th-century theoretical and philosophical traditions (i.e. phenomenology, psychoanalysis, linguistic turn, critical theory, feminist philosophy, standpoint theory) (Butler, 1990; Copjec, 1994/2015; Haraway, 1991; Pernecky, 2016).

Unfortunately, current philosophical knowledge has yet to transform the ideological mapping of much of public opinion. The allure of the nature–culture divide seems to be persistent, as it is resurrected occasionally by the latest research pointing to the existence of gender differences, with studies that can be epistemically problematic (Rippon, 2019) (see our previous comment on the problems of aperspectivity). The problem of common studies explaining that 'women tend to do/think/feel X and men tend to do/think/feel Y' is not that empirical research points to differences in behaviours/understandings/feelings in specific points in history. The epistemological error emerges when such a result is considered as an essentialised truth about the being of 'all women/men/others' for the rest of history (Anderson, 2015). It would be as if we recorded the number of women who participated in marathons in the 70 s compared with men and cemented the information as an ahistorical truth about women's running abilities or desires (for example in the Boston Marathon in 1972 there were 8 women from a total of 1219 runners (0,6%), while in 2019, 13.684 women joined the run, 45,2% of the runners (Boston Marathon, 2021)).

This divide can also be visualised as a continuum. At the extremes of this debate, we find those who believe in the essentialist characteristics of (a) gender(s) (predetermined and fixed by 'nature') versus those who defend the pure construction of gender, which exists only in the sphere of the imaginary or linguistic realm (owing to culture/nurture) without any point of fixation in nature or biology. The latest debate on gender being essentially 'fixed' by nature or the idea of the female brain versus the male brain was strongly contested by psychologist Cordelia Fine (2017) in her book *Testosterone Rex* and neuroscientist Gina Rippon (2019) in *The Gendered Brain*. We know (as brilliantly shown in the work of philosopher Martha Nussbaum, [2016]) that emotions do not occur beyond our minds, such as an opposition between emotion and reason or thinking. Complex emotions such as anger, love, jealousy and compassion are grounded in beliefs, reasoning (false and true) and cognition and language. This position is further confirmed by theory of constructed emotion of Barrett (2017) in her excellent and ground-breaking book *How emotions are made: The secret life of the brain*, wherein she provided an extensive critique of the simplistic nature–culture debate based on the latest research on the science of emotions. The same simplistic generalisation criticism can be said about the human–natural 'exterior' world: we are oxygen and water, matter and energy; we are multitudes.

Pathways for sustainability gender knowledge

With a deep sense of gratitude to each and every one who answered the call of this special issue, in this section, we expound on what lies ahead on the road to gender and tourism sustainability. After this special issue, we will be able to move beyond the idea that we can have a sustainable future without gender. This idea implies the profound transformation of the understanding of tourism sustainability. The contributions of the authors demonstrate that sustainability is multitudinous and ambiguous, we should be wary of imaginary ideas of sustainability that are neither contextualised nor contested. Sustainability is not a template to impose on others but a space and culture that allow for openings, messiness and experimentation, infused with the ethics of care.

We can learn from feminist epistemology, as it sheds critical light on taken-for-granted universalising truths that underpin and allow sustainability narratives to acquire global currency. Often, these narratives are portrayed with no political implications, but their current and past translation, hybridisation and dislocation represent political acts that increasingly define the new world order, which can often be exclusionary of minorities, indigenous knowledge and diverse others. The amalgamation of the political, cultural and physiological in knowledge production provides a justification for ancient as well as modern forms of domination based on naturalised differences seen as inevitable and consequently, moral (Haraway, 1991). This special issue challenges the missing reflection on and lack of academic rigour of the meaning of gender in tourism sustainability. This concept is a key problem that cuts across 'geopolitical boundaries and cultural constraints on who is imagining whom, and for what purpose' (Butler, 2004, p. 10). Challenging traditional and binary epistemic gender knowledge in tourism sustainability starts with critical engagement in the subject of feminism and aperspectivity.

However, using the right theories may not be sufficient. We must also challenge our ideologies. We strongly believe that a profound change in the way scholars engage and categorise the world is necessary. The change that this special issue calls for is a turn towards humbleness and compassion, in which what and who is studied is approached with dignity, care and a sense of wonder about others and ourselves. Such a turn will have methodological implications and destabilise the taken-for-granted positionalities of the researcher and researched and their power relations. Moreover, it will transform our understanding of agency in knowledge production and move us away from 'speaking *for*' towards 'speaking *with*'. A future agenda for gender and tourism sustainability must emphasise that being and knowing include and transcend humans through and with nonhumans and multiple ecologies and cosmologies, and knowledge is multitudinous and can be found beyond the written word and/or sanctioned institutionalised knowledge. This special issue shows that there is still a lot of suffering and inequality in tourism but also many voices that want to do something about it. This is a humble invitation to continue this collective effort.

Acknowledgements

We thank the editor, Xavier Font, for his valuable guidance and inspiration for this special issue and the anonymous reviewers for their time and insightful comments. We are especially grateful to all contributors, who collectively made this special issue possible and the inspiration they have provided for future work.

Disclosure statement

No potential conflict of interest was reported by the authors.

References

Abu-Lughod, L. (1998). Introduction: Feminist longings and postcolonial conditions. In L. Abu-Lughod (Ed.), *Remaking women: Feminism and modernity in the Middle East* (pp. 3–31). Princeton University Press.

Alarcón, D. M., & Cole, S. (2019). No sustainability for tourism without gender equality. *Journal of Sustainable Tourism, 27*(7), 903–919. https://doi.org/10.1080/09669582.2019.1588283

Alrawadieh, Z., Alrawadieh, D. D., Olya, H., Bayram, G. E., & Kahraman, O. C. (2022). Sexual harassment, psychological well-being, and job satisfaction of female tour guides: The effects of social and organizational support. *Journal of Sustainable Tourism, 30*(7).

Anderson, E. (2015). Feminist epistemology and philosophy of science. In *The Stanford Encyclopedia of Philosophy*. https://plato.stanford.edu/entries/feminism-epistemology/.

Anderson, E. (2017). *Private government*. Princeton University Press.

Barrett, L. F. (2017). *How emotions are made: The secret life of the brain*. Houghton Mifflin Harcourt.

Baum, T., Cheung, C., Kong, H., Kralj, A., Mooney, S., Nguyễn Thị Thanh, H., Ramachandran, S., Dropulić Ružić, M., & Siow, M. (2016). Sustainability and the tourism and hospitality workforce: A thematic analysis. *Sustainability, 8*(8), 809–821. https://doi.org/10.3390/su8080809

Biesecker, A., & Hofmeister, S. (2010). (Re)productivity: Sustainable relations both between society and nature and between the genders. *Ecological Economics, 69*(8), 1703–1711. https://doi.org/10.1016/j.ecolecon.2010.03.025

Boston Marathon. (2021). *Participation*. https://www.baa.org/races/boston-marathon/results/participation

Buckley, R. (2012). Sustainable tourism: Research and reality. *Annals of Tourism Research, 39*(2), 528–546. https://doi.org/10.1016/j.annals.2012.02.003

Butler, J. (1990). *Gender trouble: Feminism and the subversion of identity*. Routledge.

Butler, J. (2004). *Precarious life: The power of mourning and violence*. Verso.

Camargo, B. A., Jamal, T., & Wilson, E. (2016). Toward a critical ecofeminist research paradigm for sustainable tourism. In A. M. Munar & T. Jamal (Eds.), *Tourism research paradigms: Critical and emergent knowledges* (pp. 73–85). Emerald.

Carr, N., & Poria, Y. (Eds.) (2010). *Sex and the sexual during people's leisure and tourism experiences*. Cambridge Scholars Publishing.

Casado-Díaz, A. B., Sancho-Esper, F., Rodriguez-Sanchez, C., & Sellers-Rubio, R. (2022). Tourists' water conservation behaviour in hotels: The role of gender. *Journal of Sustainable Tourism, 30*(7).

Chambers, D. (2022). Are we all in this together? Gender intersectionality and sustainable tourism. Journal of Sustainable Tourism, 30*(7)*.

Chambers, D., & Buzinde, C. (2015). Tourism and decolonisation: Locating research and self. *Annals of Tourism Research, 51*, 1–16. https://doi.org/10.1016/j.annals.2014.12.002

Chambers, D., Munar, A. M., Khoo-Lattimore, C., & Biran, A. (2017). Interrogating gender and the tourism academy through epistemological lens. *Anatolia, 28*(4), 501–513. https://doi.org/10.1080/13032917.2017.1370775

Chatterjee, P. (1993). *The Nation and Its Fragments: Colonial and Postcolonial Histories*. Princeton University Press.

Cole, S. (2017). Water worries: An intersectional feminist political ecology of tourism and water in Labuan Bajo. *Annals of Tourism Research, 67*, 14–24. https://doi.org/10.1016/j.annals.2017.07.018

Copjec, J. (1994/2015). *Read my desire. Lacan against the historicists*. Verso.

Costa, C., Bakas, F. E., Breda, Z., Durão, M., Carvalho, I., & Caçador, S. (2017). Gender, flexibility and the 'ideal tourism worker'. *Annals of Tourism Research, 64*, 64–75. https://doi.org/10.1016/j.annals.2017.03.002

Crenshaw, K. (1989). Demarginalizing the intersection of race and sex: A black feminist critique of antidiscrimination doctrine, feminist theory and antiracist politics. *University of Chicago Legal Forum, 1*, 139–167.

Dashper, K., Turner, J., & Wengel, Y. (2022). Gendering knowledge in tourism: gender (in)equality initiatives in the tourism academy. *Journal of Sustainable Tourism, 30*(7).

De Beauvoir, S. (1973). *The second sex*. Vintage Books.

Denizci Guillet, B., Pavesi, A., Hsu, C. H. C., & Weber, K. (2019). Is there such thing as feminine leadership? Being a leader and not a man in the hospitality industry. *International Journal of Contemporary Hospitality Management, 31*(7), 2970–2993.

Denton, G., Chi, O. H., & Gursoy, D. (2022). An examination of critical determinants of carbon offsetting attitudes: The role of gender. *Journal of Sustainable Tourism, 30*(7).

Devine, J., & Ojeda, D. (2017). Violence and dispossession in tourism development: A critical geographical approach. *Journal of Sustainable Tourism, 25*(5), 605–617. https://doi.org/10.1080/09669582.2017.1293401

Eger, C. (2016). *Empowerment through education: Tour operators promoting gender equality through capacity building in destination communities* [Doctoral dissertation]. University of Surrey.

Eger, C. (2021a). Gender matters: Rethinking violence in tourism. *Annals of Tourism Research, 88*, 103143. https://doi.org/10.1016/j.annals.2021.103143

Eger, C. (2021b). Equality and gender at work in Islam: The Case of the Berber Population of the High Atlas Mountains. *Business Ethics Quarterly, 31*(2), 210–241. https://doi.org/10.1017/beq.2020.21

Eger, C., Miller, G., & Scarles, C. (2018). Gender and capacity building: A multi-layered study of empowerment. *World Development, 106*, 207–219. https://doi.org/10.1016/j.worlddev.2018.01.024

Eger, C., Vizcaino, P., & Jeffrey, H. (2020). Introducing critical debates on gender-based violence in tourism. In P. Vizcaino, H. Jeffrey, & C. Eger (Eds.), *Tourism and gender-based violence: Challenging inequalities* (pp. 1–13). CABI.

Einersen, A. F. (2021). Sexism in Danish higher education and research: Understanding, exploring, acting. Draft version, Copenhagen, March 2021. https://sexismedu.dk/wp-content/uploads/2021/03/SEXISM-IN-DANISH-HIGHER-EDUCATION-AND-RESEARCH_first-draft_March-8-2021.pdf

Escobar, A. (1995). *Encountering development: The making and unmaking of the Third World*. Princeton University Press.

Figueroa-Domecq, C., Kimbu, A., de Jong, A., & Williams, A. M. (2022). Sustainability through the tourism entrepreneurship journey: A gender perspective. Journal of Sustainable Tourism, 30*(7)*.

Ferguson, L. (2011). Promoting gender equality and empowering women? Tourism and the third Millennium Development Goal. *Current Issues in Tourism, 14*(3), 235–249. https://doi.org/10.1080/13683500.2011.555522

Ferguson, L., & Alarcón, D. M. (2015). Gender and sustainable tourism: Reflections on theory and practice. *Journal of Sustainable Tourism, 23*(3), 401–416. https://doi.org/10.1080/09669582.2014.957208

Figueroa-Domecq, C., Pritchard, A., Segovia-Pérez, M., Morgan, N., & Villacé-Molinero, T. (2015). Tourism gender research: A critical accounting. *Annals of Tourism Research, 52*, 87–103. https://doi.org/10.1016/j.annals.2015.02.001

Fine, C. (2017). *Testosterone rex: unmaking the myths of our gendered minds*. Icon Books.

Finniear, J., Morgan, N., Chambers, D., & Munar, A. M. (2020). Gender-based harassment in tourism academia: Organizational collusion, coercion and compliance. In P. Vizcaino, H. Jeffrey, & C. Eger (Eds.), *Tourism and gender-based violence: Challenging inequalities* (pp. 30–44). CABI.

Fraser, N. (2014). Can society be commodities all the way down? Post-Polanyian reflections on capitalist crisis. *Economy and Society, 43*(4), 541–558. https://doi.org/10.1080/03085147.2014.898822

Fraser, N. (2017). A triple movement? Parsing the politics of crisis after Polanyi. In M. Burchardt & G. Kirn (Eds.), *Beyond neoliberalism? Social analysis after 1989* (pp. 29–42). Palgrave.

Gaard, G. (1997). Toward a queer ecofeminism. *Hypatia, 12*(1), 114–137. https://doi.org/10.1111/j.1527-2001.1997.tb00174.x

Gender Spectrum. (2021). *Principles of gender-inclusive puberty and health education*. https://gender-spectrum.cdn.prismic.io/gender-spectrum%2F9ab3b6f1-314f-4e09-89d8-d5d8adc6511a_genderspectrum_2019_report_web_final.pdf

Grasswick, H. E. (Ed.) (2011). *Feminist epistemology and philosophy of science: Power in knowledge*. Springer.

Grimwood, B. S., Caton, K., & Cooke, L. (2018). Introduction: Tourism, nature and morality. In B. S. Grimwood, K. Caton, & L. Cooke, (Eds.). *New moral natures in tourism* (pp. 1–12). Routledge.

Haraway, D. (1988). Situated Knowledges: The Science Question in Feminism and the Privilege of Partial Perspective. *Feminist Studies, 14*(3), 575https://doi.org/10.2307/3178066

Haraway, D. (1991). *Simians, cyborgs and women: the reinvention of nature*. Free Association Books.

Haraway, D. (2016). *Staying with the trouble: Making kin in the Chthulucene*. Duke University Press.

Hardin, C. D., & Banaji, M. R. (2013). The nature of implicit prejudice: Implications for personal and public policy. In E. Shafir (Ed.), *The behavioral foundations of public policy* (pp. 13–31). Princeton University Press.

Harding, S. (1993). Rethinking standpoint epistemology: What is "strong objectivity"? In L. Alcoff & E. Potter (Eds.), *Feminist epistemologies* (pp. 49–82). Routledge.

Harding, S. (1995). Just add women and stir?. In Gender Working Group, United Nations Commission on Science and Technology for Development (Eds.), *Missing links: Gender equity in science and technology for development* (pp. 295–308). International Development Research Centre.

Harding, S. (2011). Interrogating the modernity vs. tradition contrast: Whose science and technology for whose social progress? In H. E. Grasswick (Ed.), *Feminist epistemology and philosophy of science: Power in knowledge* (pp. 85–108). Springer.

Haslanger, S. (1993). On being objective and being objectified. In L. Antony & C. Witt (Eds.), *A mind of one's own: Feminist essays on reason and objectivity* (pp. 85–125). Westview Press.

Hillman, W., & Radel, K. (2022). The social, cultural, economic and political strategies extending women's territory by encroaching on patriarchal embeddedness in tourism in Nepal. *Journal of Sustainable Tourism, 30*(7).

hooks, b. (2000/2015). *Feminism is for everybody: Passionate politics*. Routledge.

Irigaray, L. (1985). *This sex which is not one*. Cornell University Press.

Jaggar, A. (1989). Love and knowledge: Emotion in feminist epistemology. *Inquiry, 32*(2), 151–176. https://doi.org/10.1080/00201748908602185

Jamal, T., & Camargo, B. A. (2014). Sustainable tourism, justice and an ethic of care: Toward the just destination. *Journal of Sustainable Tourism, 22*(1), 11–30. https://doi.org/10.1080/09669582.2013.786084
Je, J. S., Khoo, C., & Yang, E. C. L. (2022). Gender issues in tourism organisations: Insights from a two-phased pragmatic systematic literature review. *Journal of Sustainable Tourism, 30*(7).
Jiménez-Esquinas, G. (2017). "This is not only about culture": On tourism, gender stereotypes and other affective fluxes. *Journal of Sustainable Tourism, 25*(3), 311–326. https://doi.org/10.1080/09669582.2016.1206109
Kato, K. (2019). Gender and sustainability–exploring ways of knowing–an ecohumanities perspective. *Journal of Sustainable Tourism, 27*(7), 939–956. https://doi.org/10.1080/09669582.2019.1614189
Katz, C. (1998). Whose nature, whose culture? Private productions of space and the 'preservation' of nature. In B. Braun & N. Castree (Eds.), *Remaking reality: Nature at the millennium* (pp. 46–63). Routledge.
Kiesling, S. F. (1998). Men's identities and sociolinguistic variation: The case of fraternity men. *Journal of Sociolinguistics, 2*(1), 69–99. https://doi.org/10.1111/1467-9481.00031
Kimbu, A. N., Ngoasong, M. Z., Adeola, O., & Afenyo-Agbe, E. (2019). Collaborative Networks for sustainable human capital management in women's tourism entrepreneurship: The role of tourism policy. *Tourism Planning & Development, 16*(2), 161–178. https://doi.org/10.1080/21568316.2018.1556329
Kinnaird, V., & Hall, D. (1996). Understanding tourism processes: A gender-aware framework. *Tourism Management, 17*(2), 95–102. https://doi.org/10.1016/0261-5177(95)00112-3
Khoo-Lattimore, C., & Wilson, E. (Eds.). (2017). *Women and travel: Historical and contemporary perspectives*. Apple Academic Press.
Lorde, A. (1983). The Master's tools will never dismantle the Master 's house. In C. Moraga & G. Anzaldúa (Eds.), *This bridge called my back: Writings by radical women of color* (pp. 98–101). Kitchen Table.
Lorde, A. (2017). *Your silence will not protect you*. Silver Press.
Mohanty, C. T. (1988). Under Western eyes: Feminist scholarship and colonial discourses. *Feminist Review, 30*(1), 61–88. https://doi.org/10.1057/fr.1988.42
Monro, S. (2005). Beyond male and female: Poststructuralism and the spectrum of gender. *International Journal of Transgenderism, 8*(1), 3–22. https://doi.org/10.1300/J485v08n01_02
Mooney, S. (2016). Nimble' intersectionality in employment research: A way to resolve methodological dilemmas. *Work, Employment and Society, 30*(4), 708–718. https://doi.org/10.1177/0950017015620768
Mooney, S., Ryan, I., & Harris, C. (2017). The intersections of gender with age and ethnicity in hotel careers: still the same old privileges? *Gender. Work & Organization, 24*(4), 360–375. https://doi.org/10.1111/gwao.12169
Moser, C. (1993). *Gender planning and development: Theory, practice and training*. Routledge.
Moswete, N., & Lacey, G. (2015). "Women cannot lead": Empowering women through cultural tourism in Botswana. *Journal of Sustainable Tourism, 23*(4), 600–617. https://doi.org/10.1080/09669582.2014.986488
Mowforth, A., & Munt, I. (2009). *Tourism and sustainability: Development, globalisation and new tourism in the Third World* (3rd ed.). Routledge.
Munar, A. M., Khoo-Lattimore, C., Chambers, D., & Biran, A. (2017). The academia we have and the one we want: on the centrality of gender equality. *Anatolia, 28*(4), 582–591. https://doi.org/10.1080/13032917.2017.1370786
Munar, A. M., & Jamal, T. (Eds.). (2016). *Tourism research paradigms: Critical and emergent knowledges*. (Tourism Social Science Series, Vol. 22). Emerald Group Publishing Limited.
Nikjoo, A., Zaman, M., Salehi, S., & Hernández-Lara, A. B. (2022). The contribution of all-women tours to well-being in middle-aged Muslim women. *Journal of Sustainable Tourism, 30*(7).
Nussbaum, M. (2016). *Anger and forgiveness: Resentment, generosity, justice*. Oxford University Press.
Ong, F., Vorobjovas-Pinta, O., & Lewis, C. (2022). LGBTIQ+ identities in tourism and leisure research: A systematic qualitative literature review. *Journal of Sustainable Tourism, 30*(7).
Porter, B. A., Schänzel, H. A., & Cheer, J. M. (Eds.) (2021). *Masculinities in the field: Tourism and transdisciplinary research*. Channel View Publications.
Pernecky, T. (2016). *Epistemology and metaphysics for qualitative research*. Sage.
Pritchard, A., & Morgan, N. (2017). Tourism's lost leaders: Analysing gender and performance. *Annals of Tourism Research, 63*, 34–47. https://doi.org/10.1016/j.annals.2016.12.011
Pritchard, A., Morgan, N., Ateljevic, I., & Harris, C. (Eds.) (2007). *Tourism and gender: Embodiment, sensuality and experience*. CABI.
Ramirez, J. (2021). Contentious dynamics within the social turbulence of environmental (in)justice surrounding wind energy farms in Oaxaca, Mexico. *Journal of Business Ethics, 169*(3), 387–404. https://doi.org/10.1007/s10551-019-04297-3
Rippon, G. (2019). *The gendered brain: The new neuroscience that shatters the myth of the female brain*. Random House.
Savage, A. E., Barbieri, C., & Jakes, S. (2022). Cultivating success: Personal, family and societal attributes affecting women in agritourism. *Journal of Sustainable Tourism, 30*(7).
Schänzel, H. A., & Smith, K. A. (2011). The absence of fatherhood: Achieving true gender scholarship in family tourism research. *Annals of Leisure Research, 14*(2-3), 143–154. https://doi.org/10.1080/11745398.2011.615712
Sen, A. (1992). *Inequality reexamined*. Oxford University Press.

Seyfi, S., Hall, C. M., & Vo-Thanh, T. (2022). The gendered effects of statecraft on women in tourism: Economic sanctions, women's disempowerment and sustainability? *Journal of Sustainable Tourism, 30*(7).
Shiva, V. (2016). *Staying alive: Women, ecology, and development* (3rd ed.). North Atlantic Books.
Sinclair, T. (1997). Issues and theories of gender and work in tourism. In T. Sinclair (Ed.), *Gender, work and tourism* (pp. 1–15). Routledge.
Small, J. (2022). The sustainability of gender norms: Women over 30 and their physical appearance on Holiday. *Journal of Sustainable Tourism, 30*(7).
Smith, N. (1984). *Uneven development: Nature, capital and the production of space*. Blackwell.
Stinson, M. J., & Grimwood, B. S. R. (2022). Defacing: Affect and situated knowledges within a rock climbing tourismscape. *Journal of Sustainable Tourism, 30*(7).
Swain, M. B., & Swain, M. T. B. (2004). An ecofeminist approach to ecotourism development. *Tourism Recreation Research, 29*(3), 1–6. https://doi.org/10.1080/02508281.2004.11081451
Swain, M. B. (2016). Embodying cosmopolitan paradigms in tourism research. In A. M. Munar & T. Jamal (Eds.), *Tourism research paradigms: Critical and emergent knowledges* (pp. 87–111). Emerald Group Publishing Limited.
Telfer, D. J. (2002). The evolution of tourism and development theory. In R. Sharpley & D. J. Telfer (Eds.), *Tourism and development: Concept and issues* (pp. 35–78). Channel View Publications.
Thurnell-Read, T., & Casey, M. (Eds.) (2014). *Men, masculinities, travel and tourism*. Palgrave Macmillan.
Tucker, H. (2022). Gendering sustainability's contradictions: Between change and continuity. *Journal of Sustainable Tourism, 30*(7).
United Nations Free & Equal. (2020). *United Nations for intersex awareness*. Retrieved January 21, 2021, from https://www.unfe.org/intersex-awareness/
UNDP. (2016). *Human development report 2016: Human development for everyone*. United Nations Development Programme.
Villesèche, F., Muhr, S. L., & Sliwa, M., (2018). From radical black feminism to postfeminist hashtags: Re-claiming intersectionality. *Ephemera, 18*(1), 1–16.
Warren, K. J., & Cheney, J. I. M. (1991). Ecological feminism and ecosystem ecology. *Hypatia, 6*(1), 179–197. https://doi.org/10.1111/j.1527-2001.1991.tb00216.x
Warren, K. J. (1997). *Ecofeminism – Women, culture, nature*. Indiana University Press.
Wittig, M. (1993). One is not born a woman. In H. Abelove, M. A. Barale, & D. M. Halperin (Eds.), *The Lesbian and Gay studies reader* (pp. 103–109). Routledge.
Woolf, V. (1929/2010). *A room of one's own*. Penguin Classics.
WCED. (1987). *Our common future*. Oxford University Press.
Yang, E. C. L., Khoo-Lattimore, C., & Arcodia, C. (2017). A systematic literature review of risk and gender research in tourism. *Tourism Management, 58*, 89–100. https://doi.org/10.1016/j.tourman.2016.10.011
Yudina, O., & Grimwood, B. S. R. (2016). Situating the wildlife spectacle: Ecofeminism, representation, and polar bear tourism. *Journal of Sustainable Tourism, 24*(5), 715–734. https://doi.org/10.1080/09669582.2015.1083996
Zhang, C. X., Kimbu, N. A., Lin, P., & Ngoasong, M. Z. (2020). Guanxi influences on women intrapreneurship. *Tourism Management, 81*, 104137. https://doi.org/10.1016/j.tourman.2020.104137

Part I

Troubling Gender and Tourism Sustainability

LGBTIQ + identities in tourism and leisure research: a systematic qualitative literature review

Faith Ong, Oscar Vorobjovas-Pinta and Clifford Lewis

ABSTRACT

Tourism research on the LGBTIQ + communities has grown over the years, entering mainstream discussions as a segment of interest. This growing focus reflects greater societal acceptance and acknowledgement of the systemic inequalities that challenge their rights. The landscape of current scholarship, though important to academic literature, policy and practice, has not been explored. On this premise, and under the umbrella of social sustainability, a systematic qualitative review of scholarship on the LGBTIQ + community and tourism was conducted with Q1- and Q2-ranked travel and tourism journals (Scimago Journal & Country Rank) as a basis. Articles were analysed to identify the sampling parameters and their topic foci. The findings suggest the literature focuses on sexually diverse groups (gays and lesbians) who are open about their identity, with limited consideration given to bisexual or gender diverse travellers (intersex and transgender). The topics and language used have also evolved in recent years, transforming from earlier fixations on the sexual, to the exploration of other experiences related to the LGBTIQ + communities. This research reflects on this evolution, the implications for the broader queer communities, and proposes a research agenda for more robust inquiry concerning LGBTIQ + travel and leisure.

Introduction

In recent decades, the lesbian, gay, bisexual, transgender, intersex, queer, and other sexually and gender diverse (LGBTIQ+) communities have received significant attention concerning their liberties in the global community. While this has culminated in the right of same-sex marriage being recognised in many countries during the past decade, particularly in the Global North (Ford & Markwell, 2017; Monterrubio, 2019), it has yet to translate into true equality for the LGBTIQ + communities. Similarly, the study of tourism, leisure, hospitality, and events (referred to hereafter as 'tourism and leisure') has often been viewed through a heteronormative lens (Johnston, 2001; Vorobjovas-Pinta & Hardy, 2020), but the growing interest in LGBTIQ + communities as consumers and integral members of society has opened a fertile yet relatively unexplored area for research.

The marginalisation experienced by the LGBTIQ + communities is well documented. Despite homosexuality being depathologised in 1990 by the World Health Organization (Vorobjovas-

Pinta & Hardy, 2016), LGBTIQ+people are still criminalised in 73 jurisdictions across the world. Even in countries with generally tolerant climates, a degree of prejudice still exists within specific areas based on their socio-demographic composition and geographic rurality (Gottschalk & Newton, 2009; Pini et al., 2013). This prejudice can result in minority stress, which Hughes (2002) explains as physical and mental stress caused by stigmatisation and abuse based on membership of a marginal group. Such stress is associated with reduced wellbeing and mental health outcomes and a higher incidence of depression and suicide (Kaniuka et al., 2019). It is within this context that leisure and tourism can provide a means for temporary escape and create environments where the individual can experience, build, and express their identity (e.g., Hughes, 1997; Markwell, 1998). Delivering these benefits necessitates a nuanced understanding of the identities represented within the LGBTIQ+acronym as distinct consumers and participants of leisure and tourism (Vorobjovas-Pinta & Hardy, 2016), however, as researchers in this area, we have found a tendency in the literature to focus on gay and lesbian individuals. It is here where research can play a role by exploring this nuance through a critical lens.

In this paper, we take the position that for research to make meaningful contributions to academic literature, policy and practice focusing on LGBTIQ+tourism and leisure, we need to begin by reviewing the current state of knowledge to identify boundaries and gaps. The need for this review is echoed in the Second Global Report on LGBT Tourism by the United Nations World Tourism Organization, which recognises the diversity within the LGBT communities (UNWTO, 2017). The report warns against assuming LGBT communities are homogenous and notes how specific groups may be more disadvantaged compared to others – for instance, transgender people often experience greater economic discrimination (UNWTO, 2017). At the same time, we are reminded that social inclusion, peace and understanding are catalysts in advancing the universal 2030 Agenda for Sustainable Development and underpin the 17 Sustainable Development Goals (SDGs) (United Nations, 2014). Although the SDGs do not discuss LGBTIQ+inclusivity directly, they address several facets of this notion. For instance, Goals 5 and 10 target gender equality and inequalities across communities, respectively; while Goal 11 talks about inclusively improving urban planning and management; Goal 16 discusses human rights and the need for safe neighbourhoods, and finally Goal 3 explicitly notes that "Multi-sectoral, rights-based and gender-sensitive approaches are essential to address inequalities and to build good health for all". Following the mission of ensuring that "no one is left behind" (United Nations, 2014, p. 9), these goals demonstrate the centrality of inclusivity to the development of sustainable communities where all people enjoy peace and prosperity (United Nations, 2020). As researchers, this call makes it imperative that we critically examine our disciplines to ensure robust scholarship that considers diverse perspectives and provides a voice for underrepresented minorities to enhance their inclusion, and consequently wellbeing, within their communities.

In line with McCabe's (2019) call for greater diversity in constructing a "Tourism for all", herein we present a systematic qualitative literature review on the LGBTIQ+communities in tourism and leisure. Specifically, this review interrogates how the LGBTIQ+communities have been examined in the literature focusing on the methods and samples, as well as the foci and topics of that examination. This enables us to establish the current state of literature in this area and propose an agenda for future research that will have significant impacts on our understanding of LGBTIQ+involvement in tourism and leisure. In doing so, we position this paper under the pillar of social sustainability that advocates for equal opportunity and human rights underpinned by considerations of individual and social wellbeing (Dangi & Jamal, 2016). In the context of this paper, we argue that such rights and opportunities can only be recognised if we understand the voices that have been hitherto underrepresented in the literature and use those voices to guide the development of policy and practice. Boluk et al. (2019) explain that sustainable tourism requires the exploration and deconstruction of power and privilege. Accordingly, in this paper, we challenge the heteronormativity of tourism enquiry (Johnston, 2001; Vorobjovas-Pinta & Hardy, 2020), and call for a greater understanding of the diversities encapsulated under the

LGBTIQ+umbrella. In doing so, we position leisure research as a medium for social change and explain the role researchers can play (Mair & Reid, 2007) by investigating how the LGBTIQ+communities could more holistically benefit from the wellbeing afforded by leisure and tourism. By understanding these benefits and how they are encouraged, these can play an educative role to encourage the leisure and tourism industries in addressing the needs of the LGBTIQ+communities.

We begin first with a discussion of the methods used to gather and analyse data for this research paper, followed by the results, and finally, a critical discussion of the state of literature accompanied by suggestions for future research.

Methods

A systematic literature search with an explicit search strategy was used to explore the LGBTIQ+identities in tourism and leisure research. The search included specific inclusion/exclusion criteria (Gingerich & Peterson, 2013; Yang et al., 2017; Yang & Ong, 2020). The literature search was conducted using Q1 and Q2 journals from the 'Hospitality, Leisure and Hospitality Management' category in Scimago (2019), a worldwide journal ranking tool that serves as an alternative to the Web of Science metrics (Hall, 2011). To narrow the list, only journals with the following terms in their titles were included: 'tourism', 'leisure', 'travel', 'hospitality', 'vacation', and 'events'. Within the resulting set of journals, searches were performed to identify articles with specific key terms in the title, keywords, or abstracts without restriction on the year range. The search terms were: 'LGB*', 'GLB*', 'Gay', 'Lesbian', 'Homosexual', 'Bisexual', 'Transgender', 'Transsexual', 'Transvestite', and 'Queer'. Wildcards (*) were used in light of the dynamic and evolving collection of genders and sexualities represented in the spectrum, ranging from LGB only (representing Lesbian, Gay and Bisexual) to LGBTQIA+ (representing Lesbian, Gay, Bisexual, Transgender, Queer/Questioning, Intersex, Asexual and "+" denoting others). Care was also taken to include terms that are currently obsolete – for instance, although the term 'Transvestite' has now become less acceptable, it was included in the search terms to ensure research from years past would not be excluded. As one of the objectives of this study was to examine the methods and topics of study in the areas of tourism and leisure over time, no restrictions were imposed on the publication year.

The articles identified through the review were analysed using both Leximancer analyses of language as well as a manual qualitative approach. This enabled both temporal and thematic analyses of articles – allowing the researchers to explore themes related to the sample, methods, and purpose of the research; and through the use of Leximancer, the evolution of the literature. The use of Leximancer complements the traditional qualitative approach by identifying underlying themes and concepts and acting as a means of triangulation, helping to deliver robust findings.

Leximancer

Leximancer 4.5, a program-driven natural language processing software, was used to analyse the articles identified. Leximancer uses unsupervised machine-learning and algorithms to analyse the data (Cheng & Edwards, 2019; Smith & Humphreys, 2006; Wilk et al., 2019), bypassing preconceptions and expectation bias while supplementing manual coding results. For this part of the study, only the abstracts of the articles were used as they summarise the key components of the paper. This helped to capture the evolution of defining themes while avoiding the language noise that results from analysing full papers. Mair and Reid (2007) used a similar line of reasoning in their interrogation of leisure research.

Leximancer enables the analysis of large qualitative datasets and has been validated in diverse research contexts and fields, including tourism (e.g., Cheng & Edwards, 2019; Haynes et al., 2019; Spasojevic et al., 2018). Consistent with its role in this research, Leximancer was used for discovering major initial themes in exploratory research through semantic information extraction (Dann, 2010), the results of which are highly reproducible and reliable (Cheng et al., 2018). It was also used to improve reviewing efficiency in systematic literature reviews while identifying, classifying, and summarising data for fast and effective evidence synthesis (Haynes et al., 2019). Two types of results generated by Leximancer are used in this paper: a colour-coded concept map and a quadrant report. These are explained further in the Findings.

Qualitative analysis

Concurrently, qualitative analysis of the sample was conducted manually by two members of the research team to provide a depth to the insight generated (Gingerich & Peterson, 2013). First, content analysis was conducted on full articles within the sample, extracting relevant information about the focus, methods, approaches, and sampling techniques used. Thereon, the objectives of the articles were examined and coded through a thematic analysis process (Saldaña, 2016). This involved conducting an initial review of the sample articles to generate codes, using an open coding process. Once no new codes emerged from open-coding the articles, the codes generated were reviewed by the authors and combined into higher-order categories based on similarity. Through this process, and consistent with Tribe (2010), it became apparent that the articles could also be classified based on their foci into being either critical or business in orientation. Accordingly, following Saldaña (2016), the sample articles were re-reviewed, and the codes and foci identified were assigned to each paper using a team-coding approach. This involved discussing the objective and perspective of the paper and respectively negotiating the code or foci assigned.

Limitations

The limitations of this paper are acknowledged. First, only Scimago Q1 and Q2 journals with tourism, hospitality, or leisure in their title were considered for this review. Accordingly, articles on the LGBTIQ + segment published in Geography or Sexuality focused journals, or those ranked as Q3 or Q4 by Scimago have been omitted. Second, given the search strategy used to find articles, it is possible that articles that considered the various gender and sexual identities but did not identify them in the title, abstract, or keywords, could have been disregarded from consideration. It is, however, likely that those articles would have referred to the LGBTIQ + communities as opposed to being focused on them.

Findings

Our study identified 39 Scimago Q1 and Q2 journals containing the terms tourism, leisure, travel, hospitality, vacation, and events in their titles. Out of the 39 journals, 23 (59%) included research on LGBTIQ + communities in tourism and leisure. The review identified 94 journal articles. Of these, over a third were published in leisure journals (n = 37). *Leisure Studies*, with 17 articles, published the largest number of articles in our sample, followed by *Annals of Leisure Research* (n = 7) and *Leisure Sciences* (n = 7). Figure 1 illustrates the distribution of articles across different journals. The dominance of the leisure journals is unsurprising as LGBTIQ + leisure geographies encompass the nature and extent of LGBTIQ + recreation space development and are linked to tourism consumption.

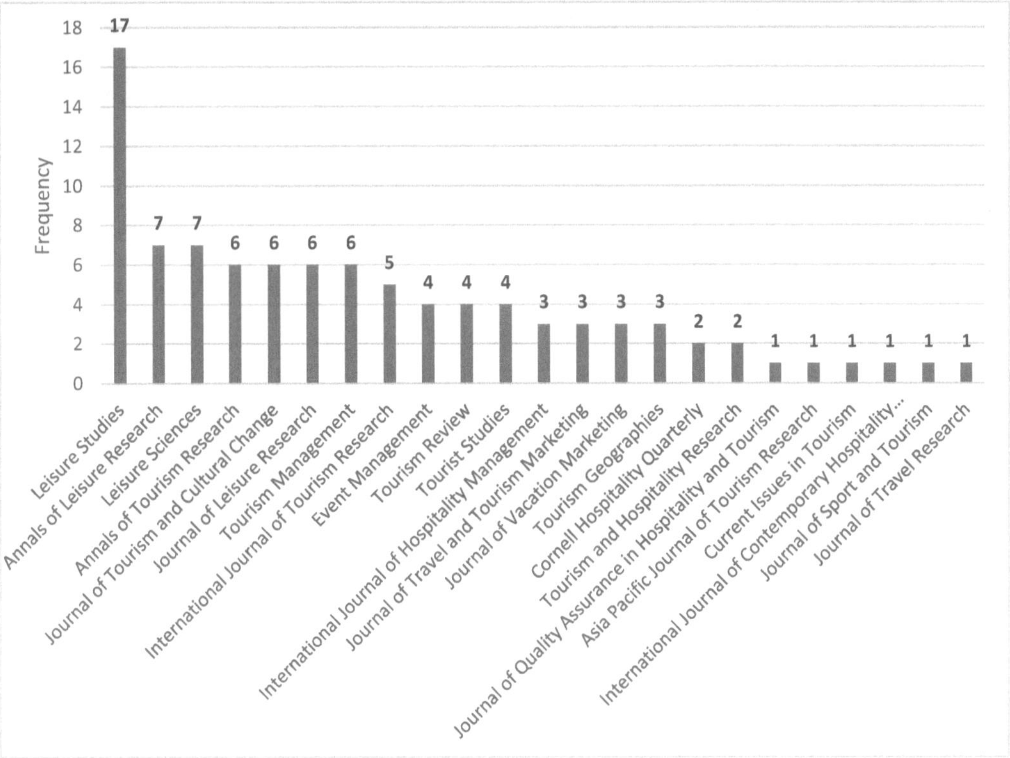

Figure 1. Number of published articles per journal in our sample.

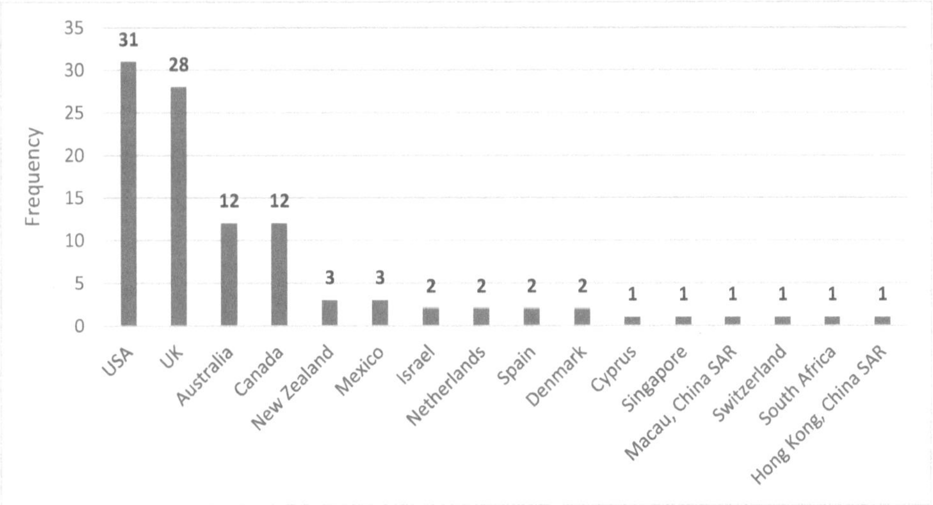

Figure 2. Geographical distribution of the authors publishing on LGBTIQ+ issues in tourism and leisure.

Figure 2 shows the geographical distribution of authors in our sample, which shows a clear concentration around the English-speaking Global North: 31 in the United States, 28 in the United Kingdom, 12 each in Australia and Canada, and three in New Zealand. This is not surprising, considering this study has only used journals that publish in English. Non-English-speaking countries with two articles or more include Mexico (n = 3), as well as Denmark, Israel, Spain and the Netherlands.

Table 1. Number of journal articles by five-year period for Leximancer analysis.

Period	Number of journal articles
1985-1989	1 (not used in analysis due to low count)
1990-1994	0 (not used in analysis due to lack of data)
1995-1999	8
2000-2004	9
2005-2009	11
2010-2014	25
2015-2019	40

Leximancer analysis

This study adopted a two-stage analysis approach using Leximancer: the 'all-in-one' and 'one-in-one' analyses. For the Leximancer analysis, only 93 out of the 94 articles were included with Cohen (1988) excluded as it was published in 1988 and would not have yielded significant findings.

The 'all-in-one' analysis was used for a comparative overview in the shift of the themes and concepts used over time. The analysis was created as one project and captured all 93 article abstracts. The abstracts were split into five folders corresponding to their respective five-year period (quinquennial) (see numbers in Table 1) to facilitate effective comparative analysis. Relatedly, the 'one-in-one' analysis was used to interrogate the unique characteristics of each quinquennial, with each quinquennial captured as a separate project. This was done to address the issue of the different sample size in each folder because "if one data source contributes much more data, then this particular source will also dominate the automatic selection of concept" (Cheng & Edwards, 2019, p. 39). As such, the 2015–2019 folder with 40 abstracts would have overpowered the 1995–1999 folder with only eight abstracts.

All-in-one approach

The 'all-in-one' approach visualised the prevalent themes of the 93 articles published over 25 years. As illustrated in Figure 3, 10 main themes were identified and visualised for the five-yearly tags: travel, gay, sexual, discrimination, space, leisure, media, community, events, and homophobia.

The **gay** theme was most prominent. This is not surprising as most of the articles used 'gay' as an overarching term to denote the nature of the studies. Interestingly, concepts that clustered together include 'lesbian' and 'bisexual', suggesting that some articles either tended to use the terms interchangeably (e.g., Chawansky, 2016; Therkelsen et al., 2013) or covered multiple sexual identities (e.g., Markwell & Tomsen, 2010; Ro et al., 2013). Although the term 'gay' colloquially refers to a homosexual individual – not ascribed to particular genders (Committee on Lesbian & Gay Concerns, 1991), academic parlance has predominantly used it only to reference gay males.

The **leisure** theme encompassed concepts such as 'leisure', 'study', 'experiences', 'social', 'identity', and 'analysis'. Leisure here could be epitomised as a context in which LGBTIQ + people are negotiating their understanding of themselves (Kivel & Kleiber, 2000), especially in sport (Davidson, 2014) and religion (Barbosa & Liechty, 2018).

The **space** theme is the third most prominent theme. Space here pertains to gay spaces or LGBTIQ + safe havens, where people can escape from heteronormative strictures of everyday life (Vorobjovas-Pinta, 2018). Articles exploring the phenomenon of 'space' focused on the consumption of gay spaces, whether through socio-cultural touchstones (Pritchard et al., 2002), the celebration of pride and culture (Caudwell, 2018) or a collective affinity and co-creation (Vorobjovas-Pinta, 2018). The prevalence of such concepts as 'research', 'interviews' and 'findings' indicate the highly empirical nature of the studies in this theme.

The **events** theme included concepts such as 'participants' and 'queer' and was close to the theme of 'homophobia', exploring LGBTIQ + participants' perceptions and experiences of hostility.

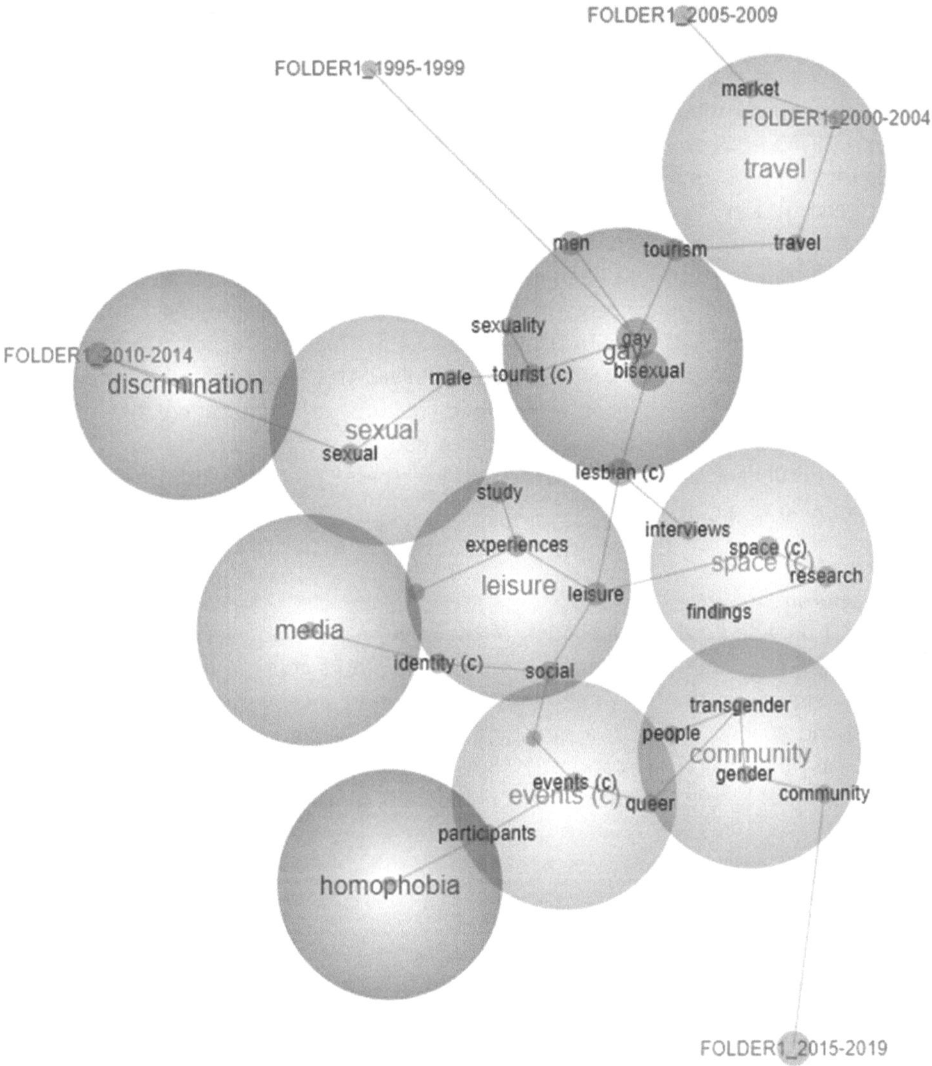

Figure 3. The Leximancer 'all-in-one' concept map.

Events have also been depicted as mediators of social change, calling for the eradication of discriminatory practices (Ong & Goh, 2018). While these concepts also emerge in our qualitative analysis, the notion of 'homophobia', which was closely linked to a paper on discrimination in men's professional sport and fan behaviour (Caudwell, 2011), was not a specific topic of investigation and was therefore not identified through the qualitative examination.

The **community** theme encompassed concepts such as 'transgender', 'gender' and 'people'. The predominance of this theme suggests LGBTIQ+ people have often been grouped together in false conceptualisations of a homogenous market (Vorobjovas-Pinta & Hardy, 2016) – a misconception which has been debunked by the diversity within the community (Therkelsen et al., 2013). The **sexual** theme was dominated by concepts like 'sexual' and 'male'. While some articles focused on the sexual aspects of (gay) male travellers (Mendoza, 2013), others examined sexual harassment either in the work environment or while on holiday (Ineson et al., 2013). This theme is closely positioned to the **discrimination** theme, where articles sought to research and understand the underlying sources of stigma and discrimination based on sexuality (Jacobson & Samdahl, 1998).

The **travel** theme contained concepts 'travel' and 'market'. Early research into gay consumers suggested that the gay travel market did not exist or was not viable; concluding that sexual orientation should not constitute a market segment (Fugate, 1993; Vorobjovas-Pinta & Hardy, 2016). Lastly, the **media** theme has emerged in proximity to the theme of 'leisure', focusing on portrayals of LGBTIQ + voices in mainstream media (e.g., Berbary & Johnson, 2017; Greey, 2018).

The quadrant report and 'one-in-one' analysis results were used to understand how LGBTIQ + research has evolved over time. As illustrated in the quadrant report in Figure 4, the close congregation of the concepts illustrate semantically significant relationships. This implies that LGBTIQ + tourism literature has been relatively homogenous over time in terms of the topics covered. The congregation of concepts in Quadrants 1 and 2 indicates these themes are not unique to their quinquennials. The breakaway cluster in Quadrant 3 points to concepts that occur infrequently but are unique to their quinquennials. This is dominated by the 2015–2019 category. These concepts predominantly emerged in 2015–2019, indicating that our sample's research into LGBTIQ + issues began diversifying but had not reached critical mass. The only concept that often occurs and is unique to Quadrant 4 is 'space', predominantly about the research on the consumption of gay spaces.

One-in-one approach

This section will provide a brief examination of each five-year period to explore the evolution of LGBTIQ + literature over time. Each period is presented with its concept map. Additionally, Table 2 provides an overview of the top three ranked compound concepts for each quinquennial. The ranked compound concept report is a quantitative analysis depicting the most prominent compound concepts (i.e. pairs of concepts), offering additional insight into trends of the LGBTIQ + tourism and leisure literature. These compound concepts are discussed below relevant to each period.

1995–1999

Research published from 1995 to 1999 (Figure 5) came after a decade of fear fuelled by the AIDS epidemic and before an era of societal change (Vorobjovas-Pinta & Hardy, 2016). The compound concepts 'male & sexual', 'gay & male', and 'male & experience' reflected this period and the under-explored dimension of sexuality, with gay men at the centre of the research (e.g., Hughes, 1997; Pritchard et al., 1998). It has been argued that travel provides homosexual men with an opportunity to construct their identity (e.g., Hughes, 1997); nonetheless, the sexual health aspects and its relation to travel were still prominent (e.g., Clift & Forrest, 1999). The 1995–1999 concept map (Figure 5) shows that research emphasised the role tourism and leisure played in escaping or resisting discrimination (e.g., Jacobson & Samdahl, 1998; Laffin, 1999).

2000–2004

The period 2000–2004 experienced similar output quantities as the previous quinquennial. However, there was a notable shift towards the compound concepts of 'travel market', 'travel & tourism', and 'market & tourism'. The literature acknowledged the growing interest in the LGBTIQ + tourism market, as illustrated by Hughes' (2003) and Visser's (2003) discussion of the advantages of promoting gay-friendly destinations. On the other hand, the 2000–2004 concept map (Figure 6) pertains to destination avoidance, which is evoked by the perceptions of risk and discrimination (e.g., Hughes, 2002). Other studies continued to perpetuate the misconceptions of the gay travel segment being somewhat richer and more recession-proof than the 'straight' market (e.g., Ivy, 2001), which was met by challenges in the same era (Carpenter, 2004).

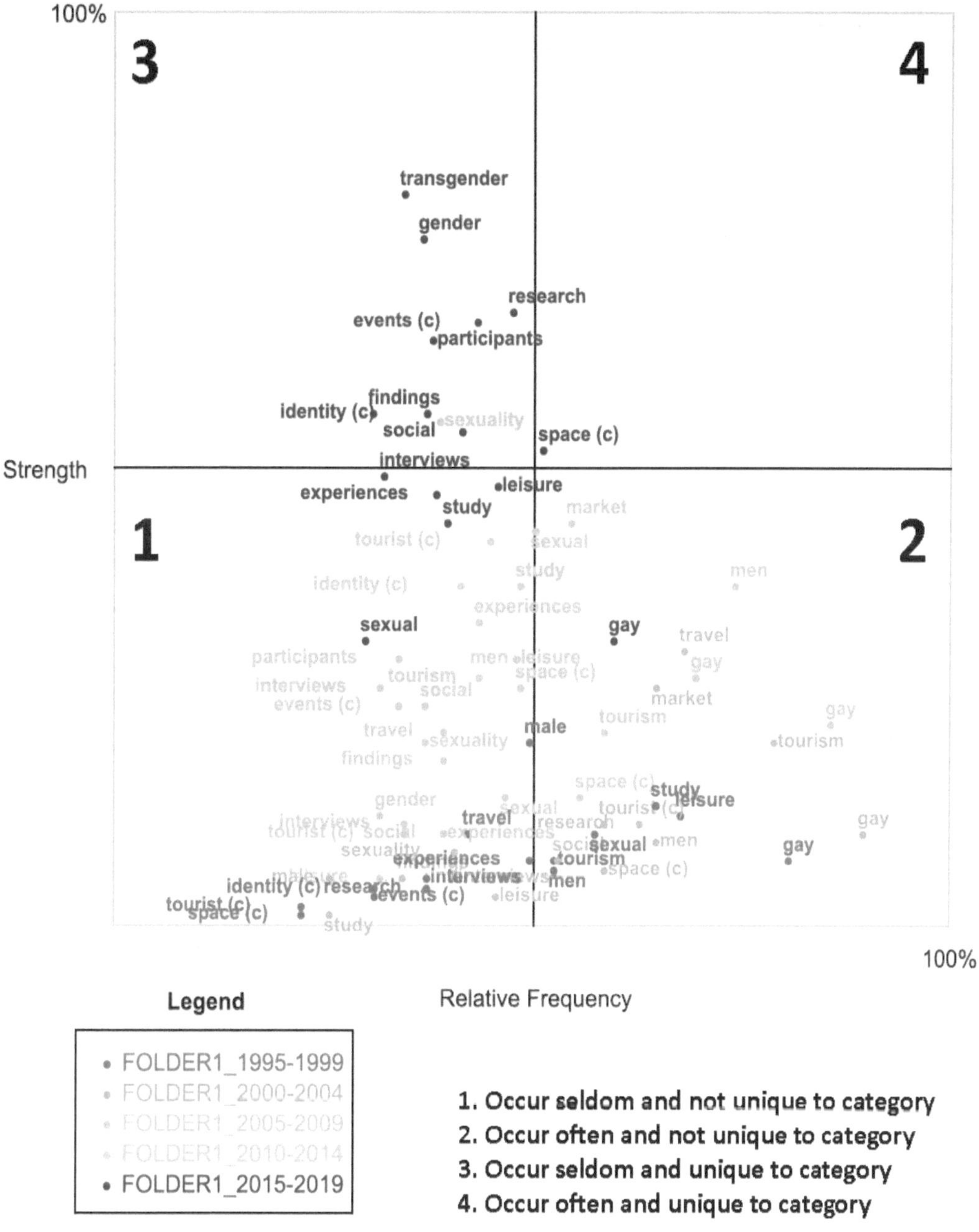

Figure 4. The Leximancer 'all-in-one' quadrant report.

2005–2009

2005–2009 pivoted further in terms of the identities researched. The top three compound concepts were 'sexuality & gender', 'market & tourism', and 'market & travel', reflecting the rise of research into lesbian travellers and their unique travel preferences (e.g., Hughes, 2007). The theoretical focus was on explorations of gender and sexuality and their relationship to leisure

Table 2. Top three leximancer-generated compound concepts for each five-year period.

Compound Concept	Relative Frequency (%)	Strength (%)	Prominence
1995-1999			
Male & sexual	8	35	53.3
Male & study	4	30	24.4
Male & experience	3	29	18.3
2000-2004			
Travel & market	10	40	100.7
Travel & tourism	16	59	67.5
Market & tourism	11	35	48.3
2005-2009			
Sexuality & gender	3	42	26.9
Market & tourism	6	50	26
Market & travel	2	27	25.3
2010-2014			
Tourist & male	2	78	13.3
Sexuality & space	3	55	9.1
Experiences & male	1	71	7.8
2015-2019			
Transgender & gender	1	85	12.6
Gender & identity	2	100	8.4
Transgender & events	1	75	6.9

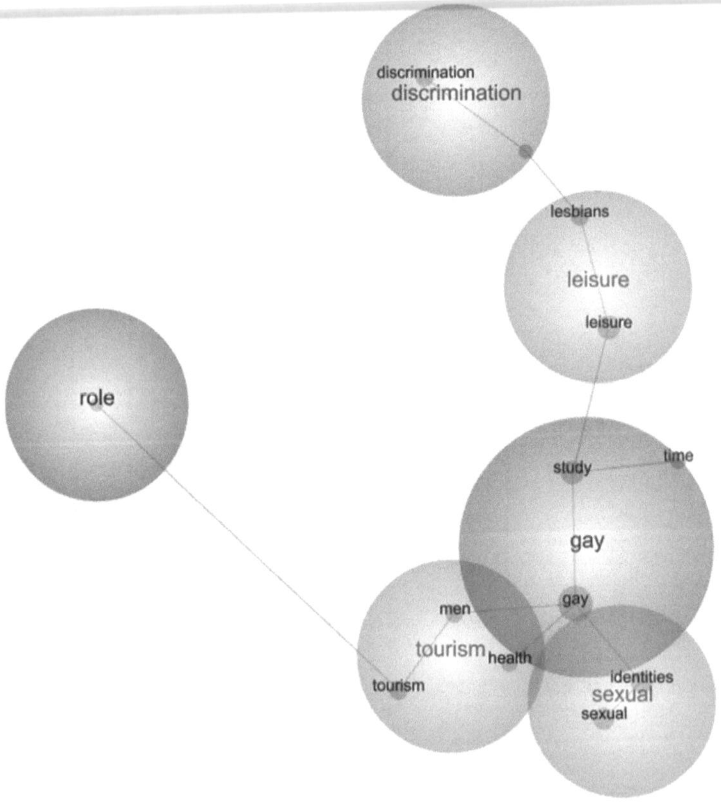

Figure 5. The Leximancer 'one-in-one' concept map: 1995–1999.

spaces, such as how the power dynamics of leisure spaces were informed by gender and sexuality (e.g., Johnson & Samdahl, 2005). This is illustrated in Figure 7, portraying the relationship between a tourism market and festival spaces (e.g., Browne, 2009).

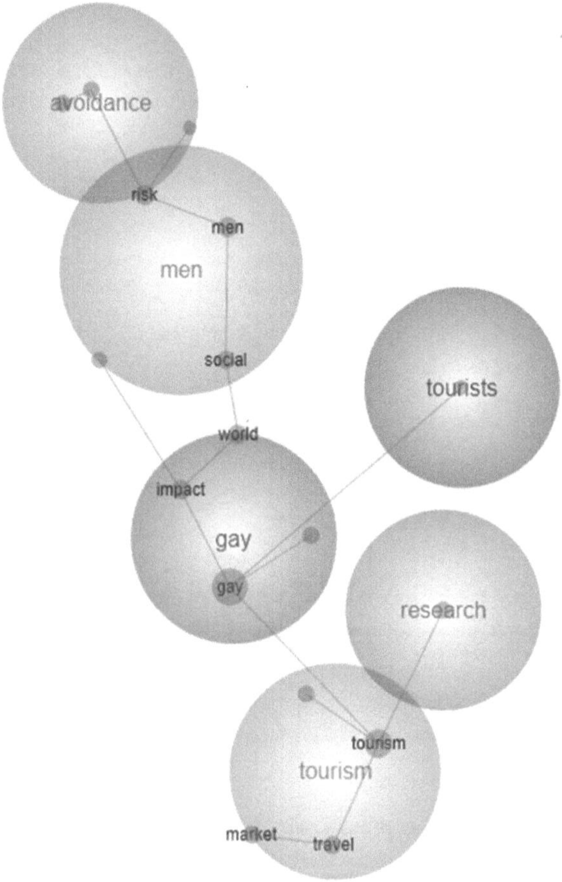

Figure 6. The Leximancer 'one-in-one' concept map: 2000–2004.

2010–2014

Research published in 2010–2014 explored similar compound concepts as in the previous quinquennials: 'tourist & male', 'sexuality & space', and 'experiences & male'. The epistemological and ontological negotiations of space and sexuality remained an important issue in this period (e.g., Browne & Bakshi, 2011). An additional focus here (Figure 8), however, was placed on LGBTIQ + travellers as consumers exploring more subtle tourist spaces and gazes, as well as the explorations of identity through the consumption of media (e.g., Johnson & Dunlap, 2011).

2015–2019

Lastly, 2015–2019 experienced an increase not only in research outputs but also in the diversification of topics. This is revealed through the concept compounds: 'transgender & gender', 'gender & identity' and 'transgender & events'. There was a notable shift from the consumeristic perspectives to more nuanced understandings of research subjects (participants). As illustrated in the 2015–2019 concept map (Figure 9), the emphasis was placed on gender and the understanding of the variety of identities in diverse contexts such as sport (Elling-Machartzki, 2017), events (Caudwell, 2018), hospitality (Vorobjovas-Pinta, 2018) and tourism (Monterrubio, 2019).

Qualitative analysis findings

In addition to the Leximancer findings, an in-depth qualitative analysis of the sample was undertaken. To assist with this, the articles identified through the review were first classified based on

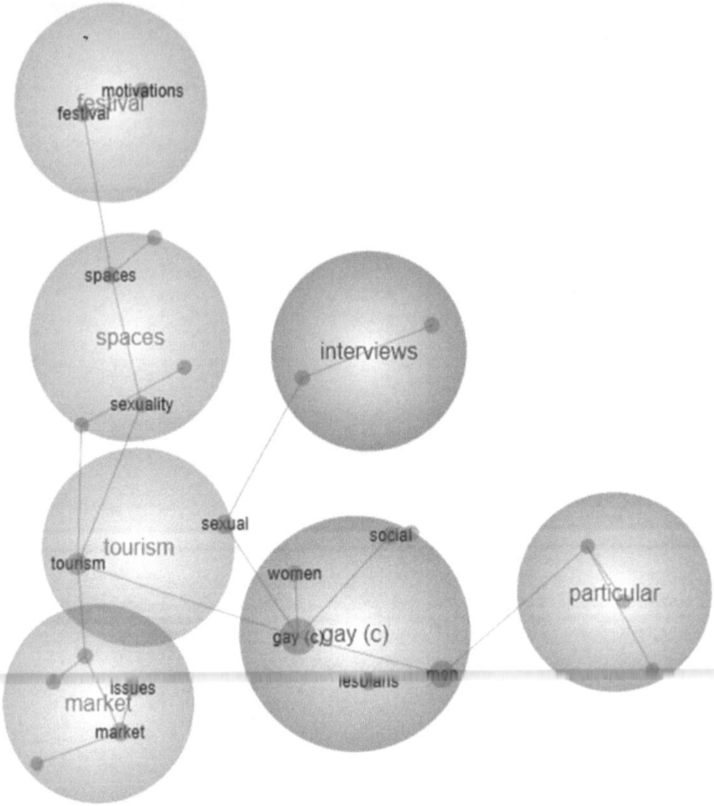

Figure 7. The Leximancer 'one-in-one' concept map: 2005–2009.

the sample, recruitment strategy, methodologies used, and topic foci. These are now discussed with reference to specific articles as a form of evidence, with the caveat that these references are intended as supporting examples and are not exhaustive.

Sample group focus

An understanding of the samples examined is essential to define the boundaries of LGBTIQ+ research in tourism and leisure. Out of the 94 articles reviewed, 31% (n = 29) did not focus specifically on human participants as the element of research – that is, the direct object that possesses the information sought (Malhotra, 2015). These studies were either theoretical (e.g., Hughes, 1997; Sykes & Hamzeh, 2018); or were conducted using auto-ethnographic approaches focusing on the researcher's observations and experience of a phenomenon, (e.g., Caudwell, 2018; Faiman-Silva, 2009); or case study approaches relying on other published material (e.g., Davidson & McDonald, 2018; Ong & Goh, 2018). Another 14 studies focused on sampling stakeholders who were either event organisers (e.g., Binnie & Klesse, 2011; Ford & Markwell, 2017) or others who were not specifically gender or sexually diverse but would be able to provide a perspective on a phenomenon (e.g., Hughes et al., 2010; Paat et al., 2019; Trussell, 2017).

Of the remaining 51 articles, a dominant focus on gay males was observed with more than half of these (61%) explicitly including gay participants and only 31% focusing on lesbian participants. This confirms our impetus for this research: that there was a strong focus on gay men and lesbians in the literature. Noticeably, gender identities were underrepresented (n = 6). The sample articles typically focused on examining single-gender or sexual identities (n = 37 concentrate only on one gender or sexual identity) with only 29% examining more than one. Table 3

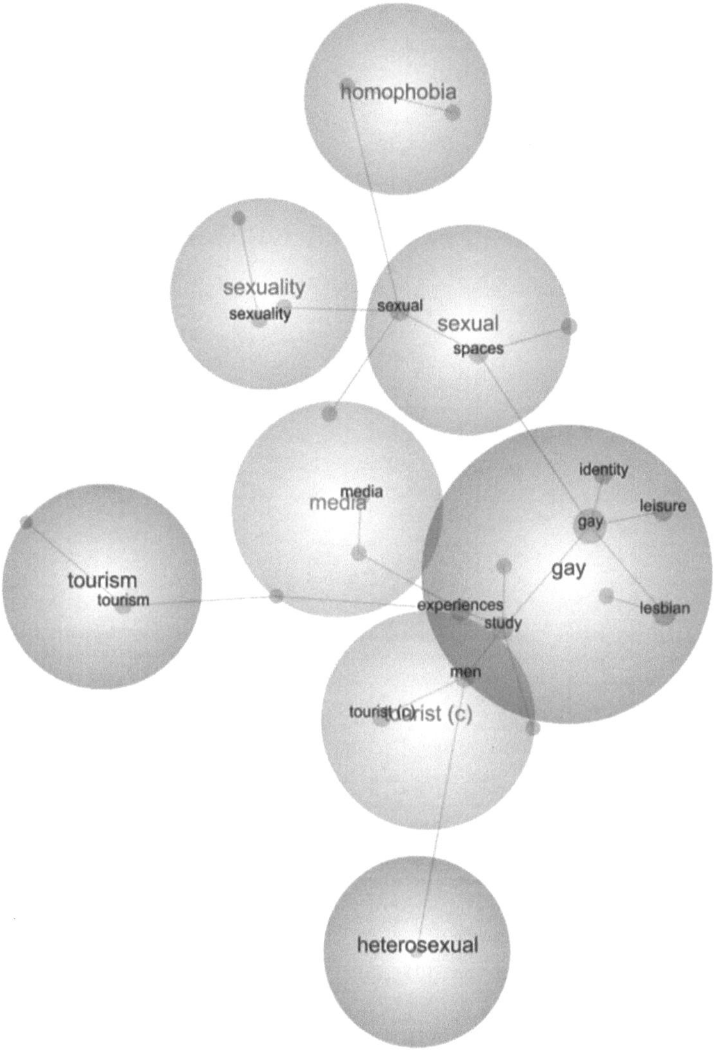

Figure 8. The Leximancer 'one-in-one' concept map: 2010–2014.

summarises the focus of existing research that included gender or sexually diverse participants and the number of groups considered. It is worth noting that studies of transsexual and genderqueer (noted in this study as drag performers) segments are a more recent addition to the literature with the first studies appearing in 2013 looking at drag performers (Barnett & Johnson, 2013) and transsexuals in wilderness (Meyer & Borrie, 2013) and the remaining in the last three years (i.e. Berbary & Johnson, 2017; Olson & Reddy-Best, 2019) beginning to explore specific leisure experiences such as sport (Elling-Machartzki, 2017); this paralleling the growing prominence of these segments within society.

Sample recruitment

The approach used to recruit participants is also relevant as an indication of the boundaries of existing research. Studies have primarily adopted non-probabilistic sampling approaches, relying initially on purposive or convenience sampling techniques. This is often done because of the

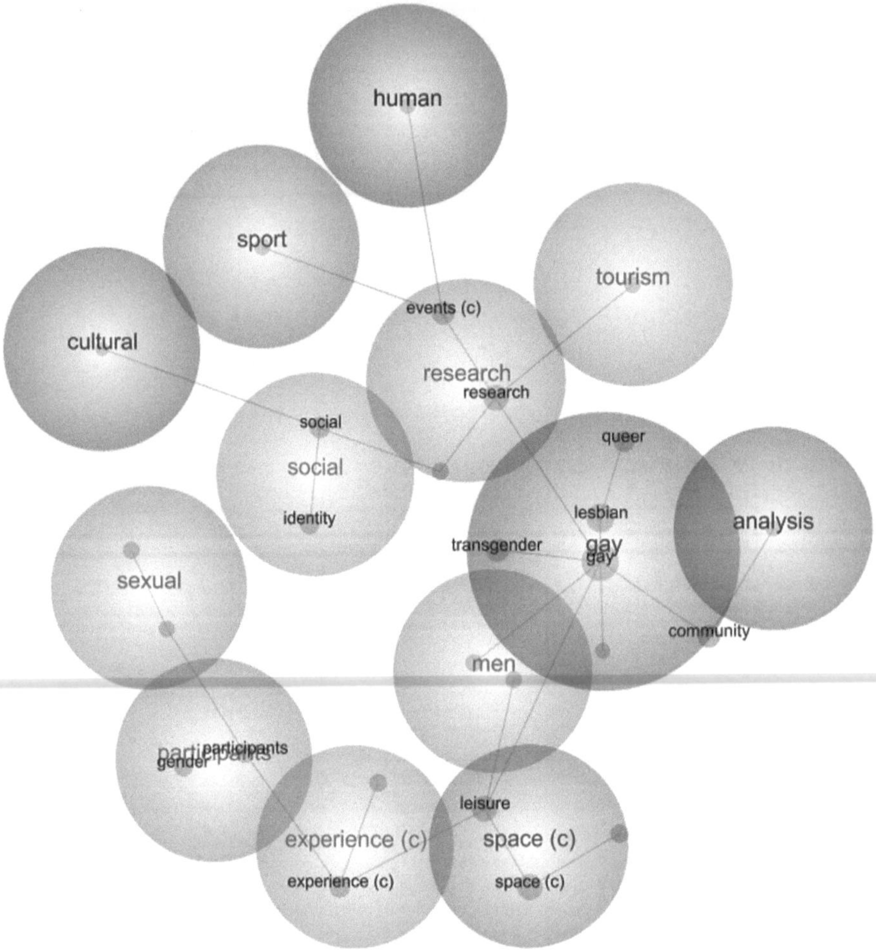

Figure 9. The Leximancer 'one-in-one' concept map: 2015–2019.

Table 3. Gender and sexuality focus of participants in articles from our sample.

One group	Two groups	Three or more groups
Lesbian n = 8	Lesbian + Gay n = 6	LGBT (non-descriptive) n = 6
Gay n = 24	Lesbian + Transsexual n = 1	Lesbian + Gay + Bisexual n = 1
Transsexual n = 2		
Gender Queer n = 3		

topic, which can be contentious, thus requiring a more targeted and personal approach to recruitment as a way of fostering trust with the participant. As such, the articles examined recruited participants based on participation or attendance at LGBTIQ + events and spaces, participation in online or offline LGBTIQ + groups, or personal associations through groups and referrals (e.g., Hughes & Deutsch, 2010; Wong & Tolkach, 2017), allowing researchers to collect a detailed description based on the phenomenon of interest (Etikan et al., 2016). Snowballing techniques have also boosted participation in the research with Pritchard et al. (2000) noting that such an approach is essential given the sensitivity of the topic, which could result in low levels of participation. Snowball sampling was also used in the only study that reported including participants who were not out about their sexual identity (Pritchard et al., 2000). This shows how sampling approaches can be adapted to explore groups, such as those questioning or not-out, who would otherwise be hard to reach.

Table 4. Distribution of sample across methods.

Method	n=	% of total	Approach
Qualitative	41	80%	Interviews, observations, photograph analysis, ethnography, surveys, document analysis
Quantitative	9	18%	Online and offline surveys
Mixed	1	2%	Surveys and interviews

Some studies have attempted to cast a wider net to get a broader representation of the identities they were studying. For instance, Clift and Forrest (1999) used a dual process to recruit participants by approaching event attendees while also distributing their survey instrument as part of a gay magazine. And within the context of sports management, Symons et al. (2017) recruited from LGBTIQ + focused sports groups and venues, but also turned their focus to mainstream sports clubs to recruit "non-community attached LGBT people" (p. 472). Given the increasing use of online research panels, some quantitative studies have adopted such panels to source their required sample (e.g., Olson & Park, 2019; Ro & Olson, 2020), enabling a broader recruitment strategy beyond just those engaging with LGBTIQ + media or groups.

Sample research methods

Of the 51 studies where primary data were collected from gender and sexually diverse individuals, 80% adopted qualitative approaches, generally comprising either in-depth interviews alone; or participant observations combined with in-depth interviews. Qualitative studies have typically used multiple approaches to collect data, integrating, observations, note-taking, and interviews (e.g., Blichfeldt et al., 2013; Jones & McCarthy, 2010); allowing for an emic understanding of the phenomenon being investigated. Only 18% used quantitative survey methods to understand the segment as a market of travel and leisure consumers (e.g., Berezan et al., 2015; Ro & Olson, 2020).

Table 4 summarises this distribution, which reflects the relatively exploratory nature of LGBTIQ + inquiry. An explanation put forward for this qualitative focus is the constantly evolving nature of the social environment wherein LGBTIQ + individuals have experienced increasing acceptance and in turn, participated more openly in tourism and leisure. Compounded by a better understanding of newly defined sexual and gender identities, and an acknowledgement of the influence of these identities on tourism and leisure, these societal trends have provided researchers with a green-field for diverse investigation within a changing environment. This has potential for depth of investigation to accompany the breadth of studies across topic areas. It is also worth noting that the articles in our sample overwhelmingly adopted a point-in-time approach to studying the LGBTIQ + segment – with only Elling-Machartzki (2017) as an exception – suggesting there is an opportunity to explore how tourism and leisure interests have evolved.

Sample foci

Through the review process, it became apparent that the research was approached from a predominantly critical or business perspective. Most of the studies (52%) adopted a **critical** stance to examine the role of identity from an individual and space perspective and explore how tourism and leisure constructs maintain or challenge dominant identities. In doing so, these studies align with Ren et al. (2010), who position critical inquiry as research which questions the dominant narrative by acknowledging diverse perspectives. Literature within this domain, for instance, questioned the role of leisure and tourism in constructing the individual's identity (e.g., Hughes, 1997; Kivel & Kleiber, 2000; Markwell, 1998); challenged the heteronormative gendering and sexualisation of space (e.g., Berbary & Johnson, 2017; Pritchard et al., 2002); examined transformation of a place's gendered and sexual identity (Boyd, 2011) and how sexual identities can coexist (Johnson, 2008). It has explored the gender and sexual politics of hospitality (Binnie & Klesse,

Table 5. Topics emerging from the qualitative analysis.

Topics	Number of articles
Individual experiences	61
Consumer behaviour and process	44
Events	27
Supply perspective	7
Methodologies	3

2011), sport (Elling-Machartzki, 2017), events (e.g., Lamond, 2018; Ong & Goh, 2018), and Pride (de Jong, 2017); and called for perspectives that depart from the colonial discourse of investigation (Sykes & Hamzeh, 2018). These studies call to question the roles of researchers in understanding the phenomenon and the researched, and further our understanding of the interplay between the space or experience and the individual.

The LGBTIQ + communities were also examined from a **business** perspective (28%). These articles take a problem-solution view (Ren et al., 2010) and define the communities as markets, understand their wants as consumers, and examined the development of services for this market. For instance, some articles profiled or defined the market and examined their needs (e.g., Berezan et al., 2015; Olson & Park, 2019; Ro et al., 2017), while others explored how travel decisions were made, underlying motivations, and the resulting traveller experiences (e.g., Clift & Forrest, 1999; Poria, 2006; Poria & Taylor, 2002; Pritchard et al., 1998). Studies also explored how destinations may be marketed as 'gay' (Hughes, 2002), or repositioned as a gay destination (Melián-González et al., 2011). A final set of articles also attempted to divide the market into segments based on demographic descriptors – like the older gay segment (Hughes & Deutsch, 2010; Olson & Park, 2019), the Asian market (Coetzee et al., 2019; Wong & Tolkach, 2017), and families (Lucena et al., 2015); and examine how the identities differ from each other (Therkelsen et al., 2013).

Sample topics

The articles identified were also analysed and classified based on their topic of investigation. Five broad topics emerged as key areas of focus within the sample, as shown in Table 5, noting that some articles were coded against multiple topics.

Perhaps owing to the relative novelty of these communities as a research subject in tourism and leisure, **individual experiences** emerged as the most frequently researched topic. This area encompassed several articles with an identity and experience-related purpose. Sexual identity often presented and with a strong emphasis on tourism and leisure as forms of identity expression, particularly in the dichotomous relationship between heteronormative spaces and non-normative sexual and gender identities. These articles explored the notion of masculinity in gay bars (Johnson, 2008), delved into the minds of drag kings and queens (Barnett & Johnson, 2013; Berbary & Johnson, 2017), and expounded on the use of technologies within gay spaces (Vorobjovas-Pinta & Dalla-Fontana, 2019). Tourism and leisure as identity-forming experiences were explored concerning individual experience, especially about those who were creating nascent homosexual identities (Kivel & Kleiber, 2000), as well as those who used such occasions to test the waters with their families (Trussell, 2017). Several articles also explored experience from a positive or negative perspective – that is those that reinforce identity versus those that discriminate against one's identity (Jacobson & Samdahl, 1998; Symons et al., 2017). Space and place identity also emerged as an essential aspect of the experience, with research exploring gay spaces and their implication (Poria & Taylor, 2002; Vorobjovas-Pinta, 2018), as well as tracing the evolution of destinations to chart their transformation into gay-friendly spaces (Boyd, 2011; Faiman-Silva, 2009). These studies predominantly focused on how space facilitated the experience, development, and expression of identity.

Consumer behaviour and process also featured prominently within the topics coded. Of these, a market segmentation perspective on behaviour dominated, with articles looking at drawing boundaries around market characteristics (Hughes, 2005; Pritchard et al., 1998; Ro et al., 2017), niche market descriptions (Bauer & Giles, 2020; Pitts, 1999), and motivations to travel (Clift & Forrest, 1999; Lucena et al., 2015). While 15 articles were coded against tourist behaviour and decision-making under this topic, when examined through the lens of the consumer decision process (Kotler & Keller, 2012), most of the articles fell under the *evaluation of alternatives* phase, where they determined tourism and leisure choices by primarily considering their LGBTIQ+ identity before making decisions (Caldwell et al., 1998; Wong & Tolkach, 2017).

Events were coded separately, given their dominance in the literature and the multiple perspectives this theme encompassed. While the social outcomes of participation in events were examined (Faiman-Silva, 2009), their impact on individual experiences was also significant (e.g., Binnie & Klesse, 2011; Jones & McCarthy, 2010). The notion of gay space also garnered attention, as the transgressive potential of Pride and other LGBTIQ+-focused events has paralleled the visibility of the LGBTIQ+ communities (e.g., Ford & Markwell, 2017; Jarvis, 2018; Lamond, 2018; Markwell & Waitt, 2009). Of particular concern was the potential for political advocacy through both Pride-related events (e.g., Caudwell, 2018; de Jong, 2017; Greey, 2018; Johnston, 2001; Ong & Goh, 2018) as well as other mega-events such as the Sochi Winter Olympics (e.g., Davidson & McDonald, 2018; Van Rheenen, 2014).

Some articles took a **supply perspective** in examining the spatial distribution of gay-friendly destinations globally (Ivy, 2001), exploring services primarily provided for/by the LGBTIQ+ communities (e.g., Berezan et al., 2015; Collins, 2007; Tan, 2014). Finally, **methodologies** in researching the LGBTIQ+ communities were also a focus for some articles (Laffin, 1999; Vorobjovas-Pinta & Robards, 2017), with these studies putting forward perspectives on how the LGBTIQ+ communities should be studied given the sensitivity of the matter.

Overall, the findings from the in-depth qualitative analysis presented support those generated through Leximancer. For instance, the one-in-one Leximancer analysis results point to a progression of research participant identities, from a mono-market emphasis on gay males to the more recent diversification into lesbian and queer identities. Further, as can be observed, the context has also diversified from a fixation on the sexual to exploring the intersectionality of experiences and identities related to the LGBTIQ+ communities. We expect this diversification to continue as LGBTIQ+ individuals become more strongly integrated into society and enjoy more and different leisure and tourism experiences.

Implications and future directions

The paper so far has reviewed existing literature on the LGBTIQ+ segment on tourism and leisure. The implications of this review are now discussed to provide direction for future research. What is clear from the analysis is that this topic encompasses much potential for exploration, particularly into the complexities of LGBTIQ+ consumption of tourism and leisure, the associated experiences, and their impacts beyond singular encounters. Research can help challenge hegemonic perspectives, and by doing so ensure the benefits of tourism and leisure are experienced by all regardless of their gender or sexual identity and as per the sentiment of the SDGs, "no one is left behind" (United Nations, 2014, p. 9). The inclusion of diverse perspectives about gender and sexual identities is a warranted expectation in education policies. As such, the potential for greater impact of this research lies in incorporating these hegemonic challenges into tourism education, enlisting the industry's partnership to ensure broader benefits that extend into the LGBTIQ+ communities. Such a process would arguably encourage a systemic transition to a more inclusive tourism industry.

Research that has been conducted with a sample of LGBTIQ+ individuals focuses on those who are out or within the Global North, where there is a greater acceptance of those who are sexually or gender diverse. This is consistent with the prominence of Western perspectives in research in general. However, it shines a spotlight on the paucity of research studies focused on LGBTIQ+ leisure and tourism within contexts where LGBTIQ+ behaviours are legally or socially punishable. Within that context, both the researcher and the researched may jeopardise personal freedoms by conducting or participating in such research. Given the ability of tourism and leisure to create a liminal environment where identity can be constructed, experienced, and solidified (Hughes, 1997; Markwell, 1998), those who are not out, questioning, or living in oppressive social climates may benefit the most from these experiences. However, their voices remain relatively silent and little is understood about them and their consumption behaviours. Indeed, such explorations could help foster opportunities for co-existence so that "all people enjoy peace and prosperity" (United Nations, 2020). Accordingly, we need to reflect on how we empower these individuals to participate in research in a way that is respectful and safe for both them and the researcher.

The current homogenisation of LGBTIQ+ communities also necessitates detail in exploring intra-community relations as well as intersectionality between traditional power relations and LGBTIQ+ communities' relations. The exploration of gender diversity has grown in recent years – potentially paralleling the hitherto limited but growing social discourse on non-binary gender identities. While some of the articles in this study considered the notion of genderqueer through an examination of drag kings and queens, limited attention has been paid to those who do not identify as cis-gendered. Arguably, reaching such a sample may require the use of quotas or purposive sampling methods as applied by Olson and Reddy-Best (2019), as they may not be readily apparent or available in a significant sample size using online research panels as found by Hahm et al. (2018). It should be noted that transgender individuals may not necessarily identify themselves as transgender but instead adopt their new gender as their identity. This has implications for how they may be recruited for research and aligns with the SDGs' Goal 3, which calls for more gender-sensitive approaches to address inclusivity (United Nations, 2014). In this context, gender extends beyond traditional binary notions with a focus on empowering individuals by referencing them through how they see themselves. Similarly, there is a palpable absence of bisexual and transgender voices in tourism research (Southall & Fallon, 2011). At the same time, research has primarily examined individual gender and sexual identities. By examining multiple gender or sexual identities, research can explain the diversity encapsulated under the LGBTIQ+ umbrella by comparing the distinct needs and motivations of each identity. Such research would align with the Second Global Report on LGBT Tourism by the United Nations World Tourism Organization, which calls for individualisation of the identities covered under the LGBTIQ+ acronym. Ensuring equity would also encourage a nuanced understanding of each identity to develop tourism experiences or policies that enable the different identities to benefit from the advantages of tourism (Dangi & Jamal, 2016; Vorobjovas-Pinta & Hardy, 2020).

Sample recruitment within the LGBTIQ+ context has been challenging due to limited social acceptance of LGBTIQ+ individuals, and fear that participation may result in a forced outing for individuals (Pritchard et al., 2000). This is particularly the case for subgroups that intersect traditionally heteronormative roles as Bauer and Giles (2020) noted when studying gay fathers. Despite challenges with sampling LGBTIQ+ individuals, if research is to provide a voice for the underrepresented, and assist with achieving the SDGs that discuss inclusivity, greater effort is required to understand these sub-segments. Arguably, given the anonymity and reach in participation afforded by the internet, the use of online research panels may provide access to a broader representation of LGBTIQ+ individuals with different lived experiences of their gender or sexual identity.

Regarding research methods, this review found that LGBTIQ+ research in tourism and leisure tends towards being qualitative and exploratory, typically using traditional qualitative methods like in-depth interviews, participant observation, case studies, and focus groups. In this context, the use of online research methods may provide opportunities for researchers to reach and study samples of those who may wish to remain anonymous. Olson and Reddy-Best (2019) called for research to use more in-the-moment approaches. This could include the use of participant diaries facilitated by digital applications, such as blogs and vlogs, to avoid relying on the participant's memory. With the growing use of social media, blogs and vlogs, as a means of recording one's travel experience, researchers may understand the LGBTIQ+traveller in a more natural environment, based on their posts (Lewis, 2016). And while the transformative and transgressive potential of events has been explored, they face a surfeit of empirical research to substantiate the theoretical assumptions that have underpinned previous research. Additionally, given the limited research that has adopted a longitudinal perspective, opportunities are available to explore behaviours and help explain how greater acceptance impacts the experiences of sexually and gender diverse individuals.

Conclusion

This paper presented an in-depth analysis of how the LGBTIQ+communities have been presented and examined in the tourism and leisure literature. In our work, we placed a particular focus on the methods, samples, as well as foci and topics of examination to make meaningful contributions to academic literature, policy and practice in the area of LGBTIQ+tourism and leisure. Significantly, this research identifies an opportunity to diversify the focus of the research. Existing literature appears focused on providing a critical perspective. While this is important as it questions the norm and paves the way for the future (Ren et al., 2010), there is scope to provide a stronger practical understanding of tourism and leisure behaviours, and to explore how these experiences can be designed to contribute towards creating safe experiences while travelling, as per Goal 16 (United Nations, 2014). Such diversification also can act as an impetus for dialogue, and, importantly, education for industry, policymakers and societies at large. While LGBTIQ+individuals want to be treated like other travel groups, there is a need to experience travel and leisure in a holistic sense, encapsulating the nuances of their sexuality or gender identity (Ro et al., 2013). To this end, studies that quantitatively segment the market based on motivations, lifestyles or psychographics may be appropriate, in addition to those currently in the literature. Ultimately, understanding the needs and desired experiences relevant to LGBTIQ+identities is essential.

Disclosure statement

No potential conflict of interest was reported by the authors.

ORCID

Faith Ong http://orcid.org/0000-0002-0001-4774
Oscar Vorobjovas-Pinta http://orcid.org/0000-0002-2666-9453
Clifford Lewis http://orcid.org/0000-0003-1151-6689

References

Barbosa, C., & Liechty, T. (2018). Exploring leisure constraints among lesbian women who attend a straight-friendly church. *Journal of Leisure Research*, *49*(2), 91–108. https://doi.org/10.1080/00222216.2018.1477679

Barnett, J. T., & Johnson, C. W. (2013). We are all royalty: Narrative comparison of a drag queen and king. *Journal of Leisure Research*, *45*(5), 677–694. https://doi.org/10.18666/jlr-2013-v45-i5-4369

Bauer, M. E., & Giles, A. R. (2020). Where are all the gay fathers?: Reflections on recruiting gay fathers as participants in leisure research. *Leisure Studies*, *39*(3), 454–459. https://doi.org/10.1080/02614367.2019.1684979

Berbary, L. A., & Johnson, C. W. (2017). En/activist drag: Kings reflect on queerness, queens, and questionable masculinities. *Leisure Sciences*, *39*(4), 305–318.

Berezan, O., Raab, C., Krishen, A. S., & Love, C. (2015). Loyalty runs deeper than thread count: An exploratory study of gay guest preferences and hotelier perceptions. *Journal of Travel & Tourism Marketing*, *32*(8), 1034–1050.

Binnie, J., & Klesse, C. (2011). Because it was a bit like going to an adventure park': The politics of hospitality in transnational lesbian, gay, bisexual, transgender and queer activist networks. *Tourist Studies*, *11*(2), 157–174. https://doi.org/10.1177/1468797611424954

Blichfeldt, B. S., Chor, J., & Milan, N. B. (2013). Zoos, sanctuaries and turfs: Enactments and uses of gay spaces during the holidays. *International Journal of Tourism Research*, *15*(5), 473–483. https://doi.org/10.1002/jtr.1890

Boluk, K. A., Cavaliere, C. T., & Higgins-Desbiolles, F. (2019). A critical framework for interrogating the United Nations Sustainable Development Goals 2030 Agenda in tourism. *Journal of Sustainable Tourism*, *27*(7), 847–864.

Boyd, N. A. (2011). San Francisco's Castro district: From gay liberation to tourist destination. *Journal of Tourism and Cultural Change*, *9*(3), 237–248. https://doi.org/10.1080/14766825.2011.620122

Browne, K. (2009). Naked and dirty: Rethinking (not) attending festivals. *Journal of Tourism and Cultural Change*, *7*(2), 115–132. https://doi.org/10.1080/14766820903033666

Browne, K., & Bakshi, L. (2011). We are here to party? Lesbian, gay, bisexual and trans leisurescapes beyond commercial gay scenes. *Leisure Studies*, *30*(2), 179–196. https://doi.org/10.1080/02614367.2010.506651

Caldwell, L. L., Kivel, B. D., Smith, E. A., & Hayes, D. (1998). The leisure context of adolescents who are lesbian, gay male, bisexual and questioning their sexual identities: An exploratory study. *Journal of Leisure Research*, *30*(3), 341–355.

Carpenter, C. (2004). New evidence on gay and lesbian household incomes. *Contemporary Economic Policy*, *22*(1), 78–94. https://doi.org/10.1093/cep/byh007

Caudwell, J. (2011). Does your boyfriend know you're here?' The spatiality of homophobia in men's football culture in the UK. *Leisure Studies*, *30*(2), 123–138. https://doi.org/10.1080/02614367.2010.541481

Caudwell, J. (2018). Configuring human rights at EuroPride 2015. *Leisure Studies*, *37*(1), 49–63. https://doi.org/10.1080/02614367.2017.1383505

Chawansky, M. (2016). Be who you are and be proud: Brittney Griner, intersectional invisibility and digital possibilities for lesbian sporting celebrity. *Leisure Studies*, *35*(6), 771–782. https://doi.org/10.1080/02614367.2015.1128476

Cheng, M., & Edwards, D. (2019). A comparative automated content analysis approach on the review of the sharing economy discourse in tourism and hospitality. *Current Issues in Tourism*, *22*(1), 35–49. https://doi.org/10.1080/13683500.2017.1361908

Cheng, M., Edwards, D., Darcy, S., & Redfern, K. (2018). A tri-method approach to a review of adventure tourism literature: Bibliometric analysis, content analysis, and a quantitative systematic literature review. *Journal of Hospitality & Tourism Research*, *42*(6), 997–1020.

Clift, S., & Forrest, S. (1999). Gay men and tourism: Destinations and holiday motivations. *Tourism Management*, *20*(5), 615–625. https://doi.org/10.1016/S0261-5177(99)00032-1

Coetzee, W. J., Liu, X. N., & Filep, C. V. (2019). Transformative potential of events – the case of gay ski week in Queenstown, New Zealand. *Tourism Review*, *74*(5), 1081–1090. https://doi.org/10.1108/TR-10-2018-0148

Cohen, E. (1988). Tourism and AIDS in Thailand. *Annals of Tourism Research, 15*(4), 467–486. https://doi.org/10.1016/0160-7383(88)90044-8

Collins, D. (2007). When sex work isn't 'work': Hospitality, gay life, and the production of desiring labor. *Tourist Studies, 7*(2), 115–139. https://doi.org/10.1177/1468797607083498

Committee on Lesbian and Gay Concerns. (1991). Avoiding heterosexual bias in language. *American Psychologist, 46*(9), 973–974.

Dangi, T. B., & Jamal, T. (2016). An integrated approach to "sustainable community-based tourism. Sustainability, *8*(5), 475. https://doi.org/10.3390/su8050475

Dann, S. (2010). Redefining social marketing with contemporary commercial marketing definitions. *Journal of Business Research, 63*(2), 147–153. https://doi.org/10.1016/j.jbusres.2009.02.013

Davidson, J. (2014). Racism against the abnormal? The twentieth century Gay Games, biopower and the emergence of homonational sport. *Leisure Studies, 33*(4), 357–378. https://doi.org/10.1080/02614367.2012.723731

Davidson, J., & McDonald, M. G. (2018). Rethinking human rights: The 2014 Sochi Winter Olympics, LGBT protections and the limits of cosmopolitanism. *Leisure Studies, 37*(1), 64–76. https://doi.org/10.1080/02614367.2017.1310284

de Jong, A. (2017). Unpacking Pride's commodification through the encounter. *Annals of Tourism Research, 63*, 128–139. https://doi.org/10.1016/j.annals.2017.01.010

Elling-Machartzki, A. (2017). Extraordinary body-self narratives: Sport and physical activity in the lives of transgender people. *Leisure Studies, 36*(2), 256–268. https://doi.org/10.1080/02614367.2015.1128474

Etikan, I., Musa, S. A., & Alkassim, R. S. (2016). Comparison of convenience sampling and purposive sampling. *American Journal of Theoretical and Applied Statistics, 5*(1), 1–4. https://doi.org/10.11648/j.ajtas.20160501.11

Faiman-Silva, S. L. (2009). Provincetown queer: Paradoxes of 'identity, space, and place. *Journal of Tourism and Cultural Change, 7*(3), 203–220. https://doi.org/10.1080/14766820903267363

Ford, A., & Markwell, K. (2017). Special events and social reform: The case of the Sydney Gay and Lesbian Mardi Gras parade and the Australian marriage equality movement. *Event Management, 21*(6), 683–695. https://doi.org/10.3727/152599517X15073047237214

Fugate, D. L. (1993). Evaluating the US male homosexual and lesbian population as a viable target market segment. Journal of Consumer Marketing, *10*(4), 46–57. https://doi.org/10.1108/07363769310047392

Gingerich, W. J., & Peterson, L. T. (2013). Effectiveness of solution-focused brief therapy: A systematic qualitative review of controlled outcome studies. *Research on Social Work Practice, 23*(3), 266–283.

Gottschalk, L., & Newton, J. (2009). Rural homophobia: Not really gay. *Gay and Lesbian Issues and Psychology Review, 5*(3), 153.

Greey, A. (2018). Queer inclusion precludes (Black) queer disruption: Media analysis of the black lives matter Toronto sit in during Toronto Pride 2016. *Leisure Studies, 37*(6), 662–676. https://doi.org/10.1080/02614367.2018.1468475

Hahm, J., Ro, H., & Olson, E. D. (2018). Sense of belonging to a lesbian, gay, bisexual, and transgender event: The examination of affective bond and collective self-esteem. *Journal of Travel & Tourism Marketing, 35*(2), 244–256.

Hall, C. M. (2011). Publish and perish? Bibliometric analysis, journal ranking and the assessment if research quality in tourism. *Tourism Management, 32*, 16–27.

Haynes, E., Garside, R., Green, J., Kelly, M. P., Thomas, J., & Guell, C. (2019). Semiautomated text analytics for qualitative data synthesis. *Research Synthesis Methods, 10*(3), 452–464. https://doi.org/10.1002/jrsm.1361

Hughes, H. L. (1997). Holidays and homosexual identity. *Tourism Management, 18*(1), 3–7. https://doi.org/10.1016/S0261-5177(96)00093-3

Hughes, H. L. (2002). Gay men's holiday destination choice: A case of risk and avoidance. *International Journal of Tourism Research, 4*(4), 299–312. https://doi.org/10.1002/jtr.382

Hughes, H. L. (2003). Marketing gay tourism in Manchester: New market for urban tourism or destruction of 'gay space'? *Journal of Vacation Marketing, 9*(2), 152–163. https://doi.org/10.1177/135676670300900204

Hughes, H. L. (2005). A gay tourism market: Reality or illusion, benefit or burden? *Journal of Quality Assurance in Hospitality & Tourism, 5*(2–4), 57–74.

Hughes, H. L. (2007). Lesbians as tourists: Poor relations of a poor relation. *Tourism and Hospitality Research, 7*(1), 17–26.

Hughes, H. L., & Deutsch, R. (2010). Holidays of older gay men: Age or sexual orientation as decisive factors? *Tourism Management, 31*(4), 454–463. https://doi.org/10.1016/j.tourman.2009.04.012

Hughes, H. L., Monterrubio, J. C., & Miller, A. (2010). Gay' tourists and host community attitudes. *International Journal of Tourism Research, 12*(6), 774–786. https://doi.org/10.1002/jtr.792

Ineson, E. M., Yap, M. H. T., & Whiting, G. (2013). Sexual discrimination and harassment in the hospitality industry. *International Journal of Hospitality Management, 35*, 1–9. https://doi.org/10.1016/j.ijhm.2013.04.012

Ivy, R. L. (2001). Geographical variation in alternative tourism and recreation establishments. *Tourism Geographies, 3*(3), 338–355. https://doi.org/10.1080/14616680110055448

Jacobson, S., & Samdahl, D. M. (1998). Leisure in the lives of old lesbians: Experiences with and responses to discrimination. *Journal of Leisure Research, 30*(2), 233–255.

Jarvis, N. (2018). The transgressive potential of the 2014 Cleveland/Akron Gay Games legacies. *Event Management, 22*(6), 981–995. https://doi.org/10.3727/152599518X15346132863210

Johnson, C. W. (2008). Don't call him a cowboy": Masculinity, cowboy drag, and a costume change. *Journal of Leisure Research, 40*(3), 385–403. https://doi.org/10.1080/00222216.2008.11950146

Johnson, C. W., & Dunlap, R. (2011). They were not drag queens, they were playboy models and bodybuilders': Media, masculinities and gay sexual identity. *Annals of Leisure Research, 14*(2–3), 209–223. https://doi.org/10.1080/11745398.2011.615716

Johnson, C. W., & Samdahl, D. M. (2005). The night they took over": Misogyny in a country-western gay bar. *Leisure Sciences, 27*(4), 331–348. https://doi.org/10.1080/01490400590962443

Johnston, L. (2001). (Other) bodies and tourism studies. *Annals of Tourism Research, 28*(1), 180–201. https://doi.org/10.1016/S0160-7383(00)00012-8

Jones, L., & McCarthy, M. (2010). Mapping the landscape of gay men's football. *Leisure Studies, 29*(2), 161–173. https://doi.org/10.1080/02614360903261487

Kaniuka, A., Pugh, K. C., Jordan, M., Brooks, B., Dodd, J., Mann, A. K., ... Hirsch, J. K. (2019). Stigma and suicide risk among the LGBTQ population: Are anxiety and depression to blame and can connectedness to the LGBTQ community help? *Journal of Gay & Lesbian Mental Health, 23*(2), 205–220.

Kivel, B. D., & Kleiber, D. A. (2000). Leisure in the identity formation of lesbian/gay youth: Personal, but not social. *Leisure Sciences, 22*(4), 215–232.

Kotler, P., & Keller, K. L. (2012). *Marketing management* (14th ed.). Pearson Education.

Laffin, T. (1999). A methodology for researching a sensitive issue: Gay men, sexuality discrimination and the hospitality industry. *Tourism and Hospitality Research, 1*(3), 243–251. https://doi.org/10.1177/146735849900100305

Lamond, I. R. (2018). The challenge of articulating human rights at an LGBT 'mega-event': A personal reflection on Sao Paulo Pride 2017. *Leisure Studies, 37*(1), 36–48. https://doi.org/10.1080/02614367.2017.1419370

Lewis, C. (2016). *Why Australia is for later? A study of why young Australians rather travel overseas when young* [Paper presentation]. Paper Presented at the Council for Australian University Tourism and Hospitality Education (CAUTHE), Sydney.

Lucena, R., Jarvis, N., & Weeden, C. (2015). A review of gay and lesbian parented families' travel motivations and destination choices: Gaps in research and future directions. *Annals of Leisure Research, 18*(2), 272–289. https://doi.org/10.1080/11745398.2015.1040038

Mair, H., & Reid, D. G. (2007). Leisure research and social change: A millennial state of the art. Leisure/Loisir, *31*(2), 501–522. https://doi.org/10.1080/14927713.2007.9651393

Malhotra, N. K. (2015). *Essentials of marketing research: A hands-on orientation.* : Pearson Essex.

Markwell, K. (1998). Space and place in gay men's leisure. *Annals of Leisure Research, 1*(1), 19–36. https://doi.org/10.1080/11745398.1998.10600864

Markwell, K., & Tomsen, S. (2010). Safety and hostility at special events: Lessons from Australian gay and lesbian festivals. *Event Management, 14*(3), 225–238. https://doi.org/10.3727/152599510X12825895093678

Markwell, K., & Waitt, G. (2009). Festivals, space and sexuality: Gay pride in Australia. *Tourism Geographies, 11*(2), 143–168. https://doi.org/10.1080/14616680902827092

McCabe, S. (2019). Tourism for all?" Considering social tourism: A perspective paper. *Tourism Review, 75*(1), 61–64. https://doi.org/10.1108/TR-06-2019-0264

Melián-González, A., Moreno-Gil, S., & Araña, J. E. (2011). Gay tourism in a sun and beach destination. *Tourism Management, 32*(5), 1027–1037. https://doi.org/10.1016/j.tourman.2010.08.015

Mendoza, C. (2013). Beyond sex tourism: Gay tourists and male sex workers in Puerto Vallarta (Western Mexico). *International Journal of Tourism Research, 15*(2), 122–137. https://doi.org/10.1002/jtr.1865

Meyer, A. M., & Borrie, W. T. (2013). Engendering wilderness: Body, belonging, and refuge. *Journal of Leisure Research, 45*(3), 295–323. https://doi.org/10.18666/jlr-2013-v45-i3-3153

Monterrubio, C. (2019). Tourism and male homosexual identities: Directions for sociocultural research. *Tourism Review, 74*(5), 1058–1069. https://doi.org/10.1108/TR-08-2017-0125

Olson, E., & Park, H. J. (2019). The impact of age on gay consumers' reaction to the physical and social servicescape in gay bars. *International Journal of Contemporary Hospitality Management, 31*(9), 3683–3701. https://doi.org/10.1108/IJCHM-12-2018-0999

Olson, E., & Reddy-Best, K. (2019). Pre-topsurgery, the body scanning machine would most likely error": Transgender and gender nonconforming travel and tourism experiences. *Tourism Management, 70*, 250–261. https://doi.org/10.1016/j.tourman.2018.08.024

Ong, F., & Goh, S. (2018). Pink is the new gray: Events as agents of social change. *Event Management, 22*(6), 965–979. https://doi.org/10.3727/152599518X15346132863292

Paat, Y. F., Torres, L. R., Morales, D. X., Srinivasan, S. M., & Sanchez, S. (2019). Sensation seeking and impulsivity as predictors of high-risk sexual behaviours among international travellers. *Current Issues in Tourism*, 1–17. https://doi.org/10.1080/13683500.2019.1666808

Pini, B., Mayes, R., & Boyer, K. (2013). Scary" heterosexualities in a rural Australian mining town. *Journal of Rural Studies, 32*, 168–176. https://doi.org/10.1016/j.jrurstud.2013.06.002

Pitts, B. G. (1999). Sports tourism and niche markets: Identification and analysis of the growing lesbian and gay sports tourism industry. *Journal of Vacation Marketing, 5*(1), 31–50. https://doi.org/10.1177/135676679900500104

Poria, Y. (2006). Assessing gay men and lesbian women's hotel experiences: An exploratory study of sexual orientation in the travel industry. *Journal of Travel Research, 44*(3), 327–334. https://doi.org/10.1177/0047287505279110

Poria, Y., & Taylor, A. (2002). I am not afraid to be gay when I'm on the net': Minimising social risk for lesbian and gay consumers when using the internet. *Journal of Travel & Tourism Marketing, 11*(2–3), 127–142.

Pritchard, A., Morgan, N. J., Sedgely, D., & Jenkins, A. (1998). Reaching out to the gay tourist: Opportunities and threats in an emerging market segment. *Tourism Management, 19*(3), 273–282. https://doi.org/10.1016/S0261-5177(98)80016-2

Pritchard, A., Morgan, N. J., & Sedgley, D. (2002). In search of lesbian space? The experience of Manchester's gay village. *Leisure Studies, 21*(2), 105–123. https://doi.org/10.1080/02614360110121551

Pritchard, A., Morgan, N. J., Sedgley, D., Khan, E., & Jenkins, A. (2000). Sexuality and holiday choices: Conversations with gay and lesbian tourists. *Leisure Studies, 19*(4), 267–282. https://doi.org/10.1080/02614360050118832

Ren, C., Pritchard, A., & Morgan, N. (2010). Constructing tourism research: A critical inquiry. *Annals of Tourism Research, 37*(4), 885–904. https://doi.org/10.1016/j.annals.2009.11.006

Ro, H., Choi, Y., & Olson, E. (2013). Service recovery evaluations: GLBT versus hetero customers. *International Journal of Hospitality Management, 33*, 366–375. https://doi.org/10.1016/j.ijhm.2012.10.007

Ro, H., & Olson, E. (2020). Gay and lesbian customers' perceived discrimination and identity management. *International Journal of Hospitality Management, 84*, 102319.

Ro, H., Olson, E., & Choi, Y. (2017). An exploratory study of gay travelers: Socio-demographic analysis. *Tourism Review, 72*(1), 15–27.

Saldaña, J. (2016). *The coding manual for qualitative researchers* (3rd ed.). Sage.

Scimago. (2019). *Journal rankings on tourism, leisure and hospitality management*. https://www.scimagojr.com/journalrank.php?area=1400&category=1409

Smith, A. E., & Humphreys, M. S. (2006). Evaluation of unsupervised semantic mapping of natural language with Leximancer concept mapping. *Behavior Research Methods, 68*(2), 262–279.

Southall, C., & Fallon, P. (2011). LGBT tourism. In P. Robinson, S. Heitmann, & P. Dieke (Eds.), *Research themes for tourism* (pp. 218–232). CABI.

Spasojevic, B., Lohmann, G., & Scott, N. (2018). Air transport and tourism—A systematic literature review (2000–2014). *Current Issues in Tourism, 21*(9), 975–997. https://doi.org/10.1080/13683500.2017.1334762

Sykes, H., & Hamzeh, M. (2018). Anti-colonial critiques of sport mega-events. *Leisure Studies, 37*(6), 735–746. https://doi.org/10.1080/02614367.2018.1532449

Symons, C. M., O'Sullivan, G. A., & Polman, R. (2017). The impacts of discriminatory experiences on lesbian, gay and bisexual people in sport. *Annals of Leisure Research, 20*(4), 467–489.

Tan, Q. H. (2014). Orientalist obsessions: Fabricating hyper-reality and performing hyper-femininity in Thailand's kathoey tourism. *Annals of Leisure Research, 17*(2), 145–160. https://doi.org/10.1080/11745398.2014.906312

Therkelsen, A., Blichfeldt, B. S., Chor, J., & Ballegaard, N. (2013). I am very straight in my gay life': Approaching an understanding of lesbian tourists' identity construction. *Journal of Vacation Marketing, 19*(4), 317–327. https://doi.org/10.1177/1356766712474449

Tribe, J. (2010). Tribes, territories and networks in the tourism academy. *Annals of Tourism Research, 37*(1), 7–33. https://doi.org/10.1016/j.annals.2009.05.001

Trussell, D. E. (2017). Parents' leisure, LGB young people and "when we were coming out. *Leisure Sciences, 39*(1), 42–58. https://doi.org/10.1080/01490400.2016.1151844

United Nations. (2014). *Report of the open working group of the general assembly on sustainable development goals: A/68/970*.

United Nations. (2020). *Sustainable development goals*. https://www.undp.org/content/undp/en/home/sustainable-development-goals.html

UNWTO. (2017). *Second global report on LGBT tourism*. https://genctraveller.files.wordpress.com/2017/05/unwtoiglta_globalreport_lgbttourism_lw.pdf

Van Rheenen, D. (2014). A skunk at the garden party: The Sochi Olympics, state-sponsored homophobia and prospects for human rights through mega sporting events. *Journal of Sport & Tourism, 19*(2), 127–144.

Visser, G. (2003). Gay men, tourism and urban space: Reflections on Africa's 'gay capital. *Tourism Geographies, 5*(2), 168–137. https://doi.org/10.1080/1461668032000068261

Vorobjovas-Pinta, O. (2018). Gay neo-tribes: Exploration of travel behaviour and space. *Annals of Tourism Research, 72*, 1–10. https://doi.org/10.1016/j.annals.2018.05.008

Vorobjovas-Pinta, O., & Dalla-Fontana, I. J. (2019). The strange case of dating apps at a gay resort: Hyper-local and virtual-physical leisure. *Tourism Review, 74*(5), 1070–1080. https://doi.org/10.1108/TR-03-2017-0035

Vorobjovas-Pinta, O., & Hardy, A. (2020). Resisting marginalisation and reconstituting space through LGBTQI+. events. *Journal of Sustainable Tourism*. DOI: 10.1080/09669582.2020.1769638

Vorobjovas-Pinta, O., & Robards, B. (2017). The shared An insider ethnographic account of a gay resort. *Tourist Studies, 17*(4), 369–387.

Vorobjovas-Pinta, O., & Hardy, A. (2016). The evolution of gay travel research. *International Journal of Tourism Research, 18*(4), 409–416.

Wilk, V., Soutar, G. N., & Harrigan, P. (2019). Tackling social media data analysis: Comparing and contrasting QSR NVivo and Leximancer. *Qualitative Market Research: An International Journal, 22*(2), 94–113. https://doi.org/10.1108/QMR-01-2017-0021

Wong, C. C. L., & Tolkach, D. (2017). Travel preferences of Asian gay men. *Asia Pacific Journal of Tourism Research, 22*(6), 579–591. https://doi.org/10.1080/10941665.2017.1308396

Yang, E. C. L., Khoo-Lattimore, C., & Arcodia, C. (2017). A systematic literature review of risk and gender research in tourism. *Tourism Management, 58*, 89–100. https://doi.org/10.1016/j.tourman.2016.10.011

Yang, E. C. L., & Ong, F. (2020). Redefining Asian tourism. *Tourism Management Perspectives, 34*, 100667. https://doi.org/10.1016/j.tmp.2020.100667

Gendering sustainability's contradictions: between change and continuity

Hazel Tucker

ABSTRACT

Understanding sustainability to be multifaceted and inherently contradictory, this article draws on long-term ethnographic research in the Cappadocia region of central Turkey to explore the gendered manifestations of various tourism-related 'sustainability fluxes'. In particular, I consider the ways in which both change and continuity are legitimized through sustainability and tourism and so manifest in contradictory ways to shape women's lives and practices. I identify four key sustainability fluxes as especially relevant in this tourism context. Firstly, women's empowerment through tourism work and increased gender equality through social and educational reform are discussed in relation to their enabling and legitimizing of change. Secondly, and reacting against these changes, are fears over sustainable food security plus nostalgic concerns over cultural sustainability. These latter sustainability fluxes engage with tourism and heritagization processes to revalue certain gendered food-production practices, and thereby legitimize continuity tying women's lives and practices to a logic of sameness. Contributing a feminist anthropology approach and guided by post-structuralist feminist thought, the article develops nuanced understanding not only of how these contradictory legitmizations "do things" in shaping women's lives and practices, but also of where ambivalences lay between acceptance and resistance to both change and continuity.

Introduction

Much research at the nexus of tourism, gender and sustainability has espoused the potential benefits of tourism development to enable women to make gains in economic independence and empowerment, thus enhancing the lives and status of women, particularly in rural areas (Gibson, 2001; Gentry, 2007). Additionally, tourism is argued to offer possibilities for patriarchal structures and power relationships to be renegotiated and even resisted (e.g. Scheyvens, 2000; Tucker, 2007; Swain & Wallentin, 2008). At the same time, however, scholars have found that tourism has the potential in some settings to reinforce gendered inequalities, further entrenching gender roles and unequal gender relations (Ferguson, 2009; Vizcaino-Suārez, 2018). Such contradictions in the possibilities afforded by tourism to achieve gender equality and to better the lives of women were highlighted by Tucker and Boonabaana (2012). In their *Journal of Sustainable Tourism* article, Tucker and Boonabaana (2012) called for research into gender and tourism sustainability to take a nuanced approach to understanding the fluid and contradictory ways in which women and men are positioned, and position themselves, in relation to tourism.

The present article picks up on this point to focus on the particular ways in which contradictory ideas regarding sustainability, tourism and gender manifest as lived tensions in touristic milieu. In so doing, "sustainability" is viewed as a mix of competing and contradictory circulating discourses (Sen, 2017), each creating imperatives embedded with their own particular 'regimes of value'. My starting premise here is that the imperatives and values of different 'faces' of sustainability can indicate apparently opposing processes. On the one hand, sustainability often connotes continuity, in its meaning that something is able to be sustained, maintained, or continued. On the other hand, certain sustainability issues suggest and promote change, in the sense of improvement, growth, and in the case of women and sustainability, gender equality and empowerment (Camargo et al., 2016; see Ferguson, 2018 for a discussion of the gender equality global policy context). These opposing indications, that is, "the inertia of 'sustainability' and the dynamism of 'development'" (Weaver et al., 2015, p. 281), are a broadly recognised yet often overlooked contradiction inherent within tourism and sustainability discourses. In this article, by drawing on my long-term ethnographic research on tourism and gender in the Cappadocia region of central Turkey, I will consider how competing imperatives and values engaged through, on the one hand, sustainability legitimizing change (for women), and on the other hand sustainability legitimizing continuity (for women), are played out in relation to tourism. I thus set out to explore how tensions between such contradictory indications of sustainability have particular gendered manifestations in tourism contexts, and how they work to shape women's lives in contradictory ways.

Extant scholarship and practice related to tourism and sustainability frequently either ignores, or recognizes and attempts to find resolutions to the contradictory and paradoxical elements which are pervasive within tourism and sustainability discourses (see, for example, Hughes et al. (2015). Other scholarship, particularly that researched and written with more of a feminist and poststructuralist leaning, has sought to welcome tourism's inherent contradictory and disorderly elements (see Veijola et al., 2014), embracing rather than rejecting the ways in which tourism is relational, unstable and messy (see Jóhannesson et al., 2016). The latter is the approach I adopt here, in line also with feminist anthropology scholarship which has "long been aware of the richness of exploring contradictions in daily, subjective experience as an entry point for accessing gendered meanings, identities and power inequalities" (O'Shaughnessy & Krogman, 2011, p. 137).

In viewing sustainability as a discursive site of power, I consider the ways in which conflicting tourism and sustainability legitimizations regarding 'change' and 'continuity' act as what Jiménez-Esquinas (2017) calls 'affective fluxes' to manifest as inherent contradictions for women's lives and practices. As I will go on to discuss, in Göreme, a prominent touristic town in Cappadocia, various 'sustainability fluxes' have enabled women's paid work and entrepreneurship in tourism, as well as affording young women a university education and the ability to join a profession. On the other hand, certain sustainability fluxes react against these changes and work instead to legitimise continuity and to tie women's lives and practices to a logic of sameness. Taking the concept of sustainability to be multifaceted, in this article I set out to explore the complex and contradictory relationship between gender, tourism and the various 'faces' of sustainability. Rather than ignoring such contradictions, or looking for how they may be resolved, my approach is to accept contradictions with curiosity and to consider how they manifest as tensions for women in their daily lives. This mode of engagement with contradictions, then, is aimed at furthering our understanding of gendered responses and negotiations regarding the multiplicity of tourism and related sustainability fluxes.

"Sustainability", tourism, gender and contradiction

Whilst not exhaustive, the United Nations 17 Sustainable Development Goals (SDGs) are illustrative of the complex and multifaceted nature of the notion of "sustainability". Despite the United Nation's attestations that these wide-ranging SDGs "are integrated and indivisible and balance the three dimensions of sustainable development: the economic, social and environmental" (United Nations, 2015, p. 2), Alarcón and Cole (2019, p. 904) have endeavoured to particularly "highlight the importance of SDG 5, gender equality, for the development of tourism and its relationship to the other SDGs". Further to this point, there is much that may be considered contradictory and incongruent in the relationship between tourism, sustainability and gender. For instance, a significant number of gender and development policies and projects to date have been criticised for their "instrumentalist" approach, with the result that they lead to "women working for development" rather than development "working for women" (Chant, 2006, p. 102). This is little wonder when the concept of "sustainable development", as proffered by the World Commission on Environment and Development (WCED) in 1987, is arguably "entrenched in Enlightenment-driven notions of progress enabled by rapid innovations and growth of scientific knowledge, instrumentality, and technological rationality" (Camargo et al., 2016, p. 75). As Cole (2018a, p.138) argues, too often "development policies lack gender mainstreaming and decisions are being made without the voices of women".

Another point of incongruence in the relationship between sustainability and gender is the criticism proffered by feminist postcolonial scholars that the notion of women's empowerment, as presented in SDG 5 for instance, is based predominantly on western researchers' ideas of how gender relations ought to be. Moreover, these ideas are frequently incongruent with local values and practices (Aghazamani et al., 2020; Chant, 2006; Kabeer, 1999; Singh, 2007). As Singh argues, "For many women, adopting the goals of Western feminism would mean losing their families, the primary source of support for the fulfilment of their social, psychological and economic needs" (2007, p. 104). Cole (2018b, p. 4) concurs with this point in saying that "context is critical and what empowers one woman might not empower another: there are no one-size-fits-all recipes for empowerment". Tucker and Boonabaana (2012) have similarly argued that many attempts to "empower" women through tourism might not only be considered culturally inappropriate or even naïve, but they often fail to meet their aims precisely because they do not fully take into account relevant local complexities, which themselves invariably are messy and contradictory. As Cole puts it, "empowering experiences in one area of a woman's life do not automatically translate into greater capacity to exercise agency and transform power relations in another part of her life" (2018b', p. 4).

Other contradictions apparent in the relationship between gender, tourism and sustainability include tourism's ability both to offer possibilities for the renegotiation of patriarchal structures and power relationships, whilst in many cases also serving to further entrench gender inequality (Scheyvens, 2000; Tucker, 2007; Swain & Wallentin, 2008). For example, Ferguson (2009) has argued that women's employment in tourism is often concentrated in low-paid, part-time and seasonal areas such as cleaning and hospitality work. On the other hand, many researchers have found that the economic opportunities offered by tourism can be beneficial for women and improve their status within and outside of the household in certain conditions, particularly in the case of small and micro-enterprises, and family-run businesses (Gentry, 2007; Gibson, 2001; Scheyvens, 2000). It is mass tourism, with predominantly large-scale business, as opposed to smaller-scale 'alternative' forms of tourism, which is often considered to be the main culprit when it comes to having the most pronounced 'non-sustainability' effects. On this, in Hughes et al.'s (2015) book on resolving the paradoxes in the practice of sustainable tourism, it is argued that an amalgamation of key elements of mass tourism and smaller-scale alternative tourism is the answer to achieving sustainability principles, although, as is often the case in such discussions, gender equality or women's empowerment is not explicitly mentioned.

Hughes et al.'s (2015) main point in their suggesting that an amalgamation – of mass and alternative forms of tourism - could resolve the apparently contradictory impulses of sustainable tourism is that an amalgamation would accommodate both the inertia implied in notions of 'sustainability' and the growth and change implied by the term 'sustainable development'. Whilst attempting to accommodate, or merge, these two seemingly opposing or contradictory processes may help to achieve some sustainability outcomes, it is questionable as to ultimately how useful it is to attempt to resolve contradictions and paradoxes in tourism, or indeed to portray them as *resolvable*. Rather, it may be more insightful to look at how the (inherent and inevitable) contradictions are lived and experienced. Doing that may help to explain some of the resistances to sustainability imperatives. What is of interest, then, is the way in which different sustainability imperatives, connected both directly and indirectly with tourism, collide and clash, sometimes undermining each other, or perhaps at times cancelling each other out. As O'Shaughnessy and Krogman (2011, p. 140) argue, albeit in relation to resource extraction industries rather than tourism, focusing on how material and discursive contradictions "manifest in real consequences for men and women within their daily life" can enable a deeper, more nuanced scholarship on gender and sustainability. Likewise, Sen (2017) deems important what she terms "everyday sustainability", denoting the practices and processes through which women experience and negotiate competing sustainability and justice regimes. In line with this, my aim here is to draw attention to the problematic and contradictory nature of "everyday sustainability" in its relation to tourism and gender. Towards this aim, I pose the following questions: If relationships between sustainability and tourism legitimize change on the one hand, whilst at the same time legitimizing continuity and sameness on the other, how are these contradictions manifested and negotiated in their shaping of people's, and in particular women's, lives? Furthermore, what might this engagement with sustainability's inherent contradictions tell us about gendered acceptances of and resistances to various sustainability principles?

In order to pursue these questions I am guided by feminist anthropology, which not only recognizes the importance of local complexities but also assumes the "ongoing performative nature of gender difference" (Lamphere, 2006, p. xvi). Moreover, whilst gender relations in Cappadocia assert a definite women/men binary (see Tucker, 2003), in my endeavour to build nuanced understanding of the way women's lives are constructed and positioned in relation to tourism and sustainability, I follow the poststructural feminist view that gender roles and relations are fluid and negotiable. As Dowling (2011, p. 6) explains, poststructural feminism "moves beyond theorising "women" as an essentialist category inscribed as victims". Rather, poststructural theory "provides a lens through which to view the potential for the reworking, disruption, contestation, transgression and transformation of the dominant codes and behaviours of society such that change is possible" (Aitchison, 2005, p. 217). However, at the same time as using such "poststructural insights", Aitchison advises also keeping in sight the "material constraints in women's everyday lives" (ibid., p. 219). An example of this approach is Jiménez-Esquinas (2017) work, which analyses quotidian performances in order to look at how certain gendered tourism processes and heritage regimes act as affective fluxes on women's lives, practices and bodies. Jiménez-Esquinas (2017) discusses how affective fluxes are "forces" which "do things" and also "make you do". This approach, Jiménez-Esquinas argues, allows analysis of how different power fluxes, "such as gender stereotypes applied to rural women, the colonial panacea of tourism and heritage regimes", (ibid., p. 314) affect people's lives and practices by "imposing their sense of what is valuable and important and what current generations must care about" (Jiménez-Esquinas, 2017, p. 315). Simultaneously, understanding of the ways in which these stereotypes are used, contested and reinterpreted in the context of daily encounters is developed, so that Jiménez-Esquinas (2017) achieves an "embodied diagnosis", going beyond simple diagnostics of power to understand quotidian performances as a complex mix of both pride and resentment.

In line with this approach, my long-term research in Cappadocia has allowed me to develop understanding of how, over time, tourism and sustainability fluxes operating at multiple levels construct various, often contradictory, legitimizations and regimes of value. Like Jiménez-Esquinas (2017), I too am keen to avoid overly simplistic diagnostics of power and rather to look at how various affective fluxes, or forces, may be lived as ambivalences and negotiations of the in-between. As I will go on to discuss in relation to my Cappadocia research, tourism development, combined with global ideas related to cultural sustainability and heritagization, has led to the (re)valuing of certain agricultural and food production activities as cultural traditions in the local region. Simultaneously, broader processes such as international agreements and global funding organizations exerting pressure to enact social reform for gender equality in Turkey have reconfigured notions of what is possible, especially for girls and women. What is of particular interest here is how the various global and local forces/fluxes combine to (re)construct notions of what is appropriate or do-able. Moreover, in this local context, how do contradictory gendered tourism and sustainability power fluxes, operating at different levels, both combine and undermine each other to manifest as ambivalences in their shaping of women's lives and practices in Cappadocia? Before going on to discuss these contradictory fluxes, I will first introduce the methodology and setting of this research.

Long-term research in Göreme, Cappadocia

This research is part of a long-term ethnographic study, which I began in 1995, looking at the socio-cultural changes brought about by tourism in the rural township of Göreme in the Cappadocia region of central Turkey. With the area having been designated a World Heritage Site in 1985, since that time Göreme's tourism has grown from a small handful of guest-houses and restaurants to now having over 300 hotels, 130 restaurants, numerous tour and activity operators including several hot-air ballooning operations, plus numerous retail shops and stands. Göreme is a small town with a population of approximately 2000 people, and tourism entrepreneurship and employment have become the major source of income there. Continuing for 25 years now, my research in Göreme has developed into a longitudinal study researching the continuing tourism growth and its ongoing implications for society and culture in the area (see, for example, Tucker, 2003, 2007, 2009, 2010, 2011, 2014, 2017).

Following an iterative, interpretative approach, with fieldwork guided by the anthropological methods of participant observation and in-depth interviewing, a primary focus of my research has been the ongoing dynamics of the relationship between tourism development and gender relations in this traditionally conservative patriarchal society. Prior to the development of tourism, the Göreme community largely depended on subsistence farming, although some men found employment in nearby towns and a number of families took advantage of migration programs to northern Europe. During the development of tourism in Göreme in the 1980s–90s, men became the tourism entrepreneurs and gained tourism employment, while women largely remained occupied with garden and farming work for household consumption and with other domestic work (Tucker, 2003). During the past two decades, however, Göreme women have gradually begun to seek employment in tourism businesses, and a few women have established micro-enterprise activity (see Tucker, 2007; Tucker and Boonabaana, 2012). In addition, many of Göreme's younger women are now attending university and pursuing careers other than tourism and in other locations around the country.

In line with reflexivity in feminist anthropology scholarship, I explicitly place myself as the researcher and author in this work. In so doing I acknowledge that "the knowing self is partial in all its guises, never finished, whole, simply there and original" (Haraway, 2003, p. 31). Making repeated fieldwork visits to Göreme over two and a half decades, the research is inevitably a product of the relationships I have built with people in Göreme over the years, and in many of those relationships my initial primary positions of researcher and tourist have transitioned to that of long-term friend. My longitudinal research has also enabled me to gain a sense of the changes and continuities occurring over that time. As the years have gone by I have developed an understanding of how tourism growth is inevitably an ongoing entanglement of local and global processes (Tucker, 2010). In this article, I draw especially on my ethnographic material regarding the gendered discursive contradictions relating to tourism and sustainability. In particular, I use material gathered from purposeful semi-structured interviews I conducted between

2005 and 2019 with twenty-five Göreme women, most of whom were either working in or running business in tourism. Some of the interviews were conducted with small groups of women who worked in the same hotel business. The interviews were mostly conducted in Turkish and later translated and transcribed with the help of a bilingual research assistant in order to ensure comprehension of nuanced meaning.

Whilst these interviews form a wealth of rich material with a purposeful gender and change focus, the bulk of my material is in the form of approximately twenty field note diaries in which I have recorded observations, encounters and the great many informal interviews and casual conversations I have had during fieldwork visits to Göreme. For this particular paper, I have trawled through these field note diaries, as well as the interview transcripts, in order to derive themes concerned with women's positioning in relation to various matters of "sustainability". Whilst Göreme's tourism growth has occurred mostly without directly targeted "sustainability" initiatives, through this process of field note diary and interview material trawling, I identified four broader interrelated "sustainability fluxes" at play which "do things" in shaping Göreme women's lives and practices. These are: women's empowerment through tourism work; gender equality through social and educational reform; sustainable food security; and cultural sustainability. As I will go on to discuss, whilst the first two of these fluxes legitimize change, the latter two legitimize cultural continuity. I will go on next to discuss how the first of these key sustainability fluxes shape women's lives by legitimizing change.

Sustainability legitimizing socio-cultural change

Whilst, in Göreme, there has been significant economic and social change over the past three decades due largely to the growth of tourism, this change has generally not involved any targeted government or nongovernmental organisation-funded "sustainable tourism development" interventions, such as community-based tourism initiatives involving training programmes for women and youth which have occurred elsewhere in Turkey (Tucker, 2017). Rather, Göreme's tourism development has been established predominantly from within the community. Nonetheless, in areas other than tourism national level political, economic and social policy reform has effected significant social change at the local level. These national level processes, often urged by pressures exerted by global-level organisations and funding bodies, include policies and projects which strive towards the principles summarized within UN SDG#5 which are gender equality and empowerment of all women and girls. For example, Turkey's active interest in European Union membership might be considered a key motivation for the country's signing and ratification in 1985 of the Convention on the Elimination of All Forms of Discrimination against Women (CEDAW) (Müftüler-Baç, 2012). Turkey's later adoption of the EU accession criteria during the early 2000s exerted further pressure to enact gender equality reforms.

While some of these national reforms have resulted in substantial changes, there is considerable regional variation across Turkey, with high levels of women's illiteracy and low levels of women's accessing the workforce still prevalent in some areas, especially in central and eastern regions of the country (Durakbaşa & Karapehlivan, 2018). Cappadocia is in the central part of Turkey, and within Cappadocia Göreme is a small rural town, or large village, with what can be described as a conservative Muslim patriarchal society. Most of the elderly women and many of the middle-aged women I have met over the years of my conducting fieldwork have been illiterate, and most have spent their life since puberty undertaking reproductive work in the domestic sphere. The opportunities and prospective lives of those women's granddaughters are substantially different, however, and whilst the older generations still revere notions of *kapalı* (meaning covered or closed-in) as the proper way for women to be, their daughters and granddaughters are tending to live a more *açık* (open) life, with *açık* denoting the ability to go out and move around, their hair uncovered, and able to leave Göreme to attend school and university.

Illustrating the sense of this change starting to occur, in a 2005 interview with four women conducted in the hotel where they worked, I was told "The way people think has changed. Young people have changed. The phrase *'Ev'de otur'* (Stay at home) is finished for young people, completely finished… Education is really important here now. Now the girls go to school. Before they used to learn only how to work in the fields. They do not learn that now. Now they only want to study" [CCS Interview, 2005].

An encounter I had during fieldwork in 2019 further highlighted the extent of the generational change. The encounter was with two women in their early twenties and it took place in the small jewelry shop owned by one of the women. The other woman was her friend and neighbor who was back in Göreme during the university summer break; she was soon to enter her final year of studying business and accountancy at a university in the northwest of Turkey. I had known both of the women's mothers since my first fieldwork period in the late 1990s, as they lived in the neighborhood where I spent much of my time. It was the women, young and old, of that neighborhood who had taught me about the moral codes of being *kapalı* and about the importance of those codes in how they lived their lives (see Tucker, 2003). In the jewelry shop the young women and I laughed together about how we had likely 'met' when they were babies during my earlier fieldwork. An extract from my field note diary reads:

> "Both were wearing tight jeans and neither were wearing a headscarf… Deniz was planning later to visit a friend to wish them a happy birthday and would take a cake she had earlier popped into Nevşehir to buy. I commented that twenty years ago, when I did my earlier fieldwork, I would never have known of a young woman from Göreme taking the bus into Nevşehir, dressed in jeans, to buy a birthday cake. They chorused: "No, we were *kapalı* back then!" [Field note diary#1, 2019].

Also, twenty years ago young women from the neighborhood would not have gone away to university. Nor would they have set up a shop to sell jewelry to tourists. As I will go on next to outline, the significant change which has taken place concerning what is appropriate or otherwise for women to do can be linked both directly and indirectly to tourism, and also to national level reforms related at a broad level to the "sustainability" imperatives of gender equality and women's empowerment. To discuss these different aspects of legitimized change, I will first outline changes to women's accessing paid work or business opportunities in tourism.

Tourism and women's work

Today the majority of Göreme households are engaged in tourism-related work or entrepreneurial activity, although as already mentioned, in this conservative patriarchal society Göreme women have tended to have limited direct access to tourism income and formal work outside of the home. This is because, in accordance with Islamic codes and practice, the familial ideology in Göreme society assigns women to the domestic sphere via a gendered spatial separation upheld by the principles of shame and honour (*namus*) (Elmas, 2007; Tucker, 2007). This was explained to me by one woman as follows: "My husband won't allow me to enter public places, so I couldn't be a waitress, he wouldn't allow it. Also I don't have the confidence." (HW Interview 1, 2005). During Göreme's early tourism development, the honour and shame principles together with the associated being *kapalı* meant that women were only able to access the tourism economy indirectly through the earnings of their husband or other male household members.

In more recent years, however, there has been an increase in women's paid employment in local tourism small businesses as well as in women's micro-scale entrepreneurial activity associated with tourism. In Tucker (2007), I argued that a variety of social, political and economic realms necessarily converged to create an environment in which the codes of shame and honour associated with Göreme women's work outside of the domestic sphere were able to slowly lessen. Firstly, during the early 2000s European Union-funded programmes put on a variety of courses for women in the Göreme municipal building. These included literacy and computer skills, and English language and numeracy skills for women who were interested in tourism employment. These courses not only provided the skills at which they are aimed, but they also gave women legitimacy to 'be' in the town centre. These EU programmes thus enacted the initial steps of what Puwar (2004) describes as "space invasion", an illuminating point of focus in that:

> It sheds light on how spaces have been formed through what has been constructed out. And it is intriguing because it is a moment of change. It disturbs the status quo, while at the same time bearing the weight of the sedimented past (2004, p. 1).

Much of the growth in women's working in tourism which ensued from these initial "space invasions" has tended to be cleaning and laundry work in hotels and guesthouses. During the past two decades many of these establishments have grown to become larger, more formal and compartmentalised 'boutique hotels'. This compartmentalisation has created a "back stage" space where women are now frequently employed. In a group interview with hotel workers, one woman explained: "Here it is safe; we feel comfortable working here" (KBK Interview, 2005). A few women, largely those who are younger, unmarried and have attained some language and service skills, have graduated through to the more public areas, such as into hotel reception or restaurant work. While, initially, these women were often the subject of gossip, many persisted in their employment, thus gradually shifting gendered norms (Tucker, 2007).

Women's seeking employment was also driven by economic necessity within the household or family business. For example, some small businesses were only able to be sustained if the owners' wife or mother at times worked in the business, thereby negating the need to employ paid help. Some women have also attempted to earn a little money from their home by selling handicrafts to passing tourists. Having their cave-house as an attraction and selling handicrafts enables these women to engage with the tourism economy whilst staying in the domestic sphere. For approximately a decade now, also, the town council has provided wooden stalls on a central road for women to display and sell their handwork. While these women complain at the lack of sales they make, they also acknowledged when I spoke with them that having the council stalls provided for them rent-free allows them to make a little bit of money of their own, plus it affords them a legitimacy to be in the center of town where they can socialize with each other and also with passers by, including tourists. Another handicraft shop owned by two sisters has been successfully operating for some years. The owners told me in an interview that when they first dared to open the shop on the main street in town, they felt extremely shy and hid behind the shop counter whenever a local man passed by the front of the shop (AC Interview, 2015). Now they manage the shop confidently and proudly proclaimed themselves to be the first Göreme women "to do business in the centre" (ibid.). After these initial "space invaders" had begun shifting gender norms, young women such as the one whose jewelry shop the encounter described above occurred in could much more easily follow.

Other female entrepreneurs have included a woman who manages her son's small restaurant on the main street and also runs cooking classes from her house for tourists. Another woman accessed a national level 'women-in-business start-up' fund to open a restaurant. In such cases, the ability for women to contribute to the household income has been significant, and this has afforded the women both an increased independence and a heightened sense of self-confidence and self-worth. One woman I interviewed said: "Of course I have more independence because I don't have to ask for money from my husband. Apart from earning money, though, working in tourism affects our head and ideas" (CCS Interview, 2005). This increased ability for women to work and to do business in the central and public spaces of Göreme thus not only represents a significant shift towards "equalising" of the traditionally highly gendered ability to access and participate in the tourism economy, but these change factors, certainly in some women's lives, might also be considered processes of "empowerment".

Social/educational reform

As well as the significant shift that has occurred in normative notions of gender identities and roles enabling women to be in, and to work in, touristic spaces, another significant change, both enabled by and further enabling this spatial shift, is a marked increase in girls and young women entering higher education and completing qualifications. As introduced earlier, this increase can partly be attributed to national social and education reforms that, linked to the modern Turkish Republic's gender equality ideals and urged by European and other international funding organizations and sustainable development agreements, have led to a considerable increase in the number of girls gaining access to all levels of education (Durakbaşa & Karapehlivan, 2018; Müftüler-Baç, 2012). For example, the introduction in 2012 of twelve years of compulsory schooling, including high school attendance for four years, meant that across Turkey the schooling gender gap has been closing. In addition, higher levels of female students have attended University in recent years, with "highly favourable percentages of women among the highly educated professionals which hint at gender equality" (Durakbaşa & Karapehlivan, 2018, p. 75). However, as mentioned above, there is considerable regional variation, with Central and East Anatolia regions being slower in closing the gender gap in completion of secondary and tertiary education. The Göreme and Cappadocia area is in Central Anatolia where in the year 2016–17 only 66.8% of girls attended secondary school (Durakbaşa & Karapehlivan, 2018).

A particularly sharp increase in Göreme girls and young women accessing secondary and tertiary education in recent years is therefore something of an anomaly in the regional context, and this may be linked at least in part to the particular way tourism changes at the local level have interacted with these broader gender equality reforms. This is firstly in that tourism earnings have enabled many Göreme families to meet the financial costs of sending their children, sometimes away from the area so incurring extra living costs, to access education, including private high schools, colleges and universities. As well as gender equality reforms, the Turkish education system has been subject to extreme privatisation policies in recent years (Durakbaşa & Karapehlivan, 2018), and a large number of private schools and tertiary education institutions have been established, including in the Cappadocia region.

Alongside helping to meet financial costs of education, tourism has also afforded increased international links for many Göreme families. For example, marriages between Göreme (mainly) men and foreigners has meant that often rather conservative families have gained in-laws from and in other countries around the world and this has, to varying extents, introduced other societies' gender norms. As well as marriages between Göreme people and foreigners, friendships have created strong links with people in other countries. In some instances, long-term friendships formed between hoteliers and tourists have resulted in these overseas friends becoming mentors, sponsors and hosts for the Göreme businessmen's daughters and sons to attend secondary or tertiary education in the foreign country. That some of the more successful hoteliers and other businessmen in Göreme have followed these 'social capital' opportunities has resulted in setting new precedents which others then wish to emulate. This was illustrated by a young woman I interviewed who had studied to be a clinical psychologist at a university elsewhere in Turkey during which she had completed an internship in northern Europe as well as spending a summer staying with a family friend in Canada to improve her English. She told me that, whilst her grandmother had been vehemently against allowing her to travel overseas, her father had become open-minded because of tourism and also from seeing other families "such as our neighbours or my dad's friends sending their daughters to university, and he saw that it was okay, it can be possible… and so things changed" [LH,H Interview, 2019].

During the past decade, while the tendency for Göreme youngsters to attend college and university has increased overall, what is most notable is that the rate of secondary and tertiary education among Göreme girls has reportedly overtaken that of boys in recent years, with boys these days being more inclined than girls to drop out of school early. This is believed to be due to boys and young men viewing tourism as a viable, enjoyable and lucrative way to make a living. Furthermore, success in tourism business is believed generally not to require any particular education or qualification. If a young man has grown up hanging around his father's or other relative's tourism business, he is likely to have picked up some English or another language, and also to have gained a reasonable understanding of tourists' needs and wants. Keen to work full-time in tourism and sooner or later either entering their father's tourism business or generating their own business, young men are tending overall to leave school earlier than girls.

In the context of national educational reforms designed to bring greater gender equality in access to education and completion of qualifications, then, it is somewhat remarkable that the previous unequal education attainment between boys and girls in Göreme has now become reversed. These days, more girls and young women endeavour to qualify to enter university, and acquire scholarships to help them meet the costs to do so, than their male counterparts. As with the example of Deniz whom I met in the jewelry shop encounter outlined above, in recent years during my fieldwork visits to Göreme I have met several young women who are home for the summer university holidays, hanging out with their friends, sometimes helping their mother prepare food and preserves and perhaps, although less frequently, joining their mother in the fields to harvest produce. They tell me how they are studying to be a nurse, teacher, accountant, architect and so on. Younger teenage girls still completing high school told me of their ambitions to become lawyers or doctors. Whilst national scholarship schemes are available, many Göreme families, because of their relative wealth from tourism, are now able to aspire to sending their daughters to university without having to rely on scholarships. Overall, the increased wealth and international connections, as well as a decreased imperative for women to be *kapalı*, has led to a significantly increased likelihood for young women, even more than young men, to leave Göreme and embark on a professional career.

In Göreme, global and local sustainability fluxes have combined to (re)construct gendered notions of what is appropriate or do-able, especially for women. Moreover, the particular ways in which the sustainability fluxes have worked to legitimize significant change in Göreme have both directly and indirectly related to tourism, although the change outcome which is more directly linked to tourism, that is women's 'empowerment' through tourism work, appears to be less influential in leading to the sustainability goals of gender equality and empowerment of women than is the matter of tourism affording girls and young women access to higher education. Importantly, to refer again to Puwar's (2004) notion of "space invaders", both types of sustainability flux have involved some level of space invasion on the part of Göreme women, and this inevitably incurs risk, for example, risk to family honour, risk of gossip, risk of ostracism, and so on. It seems that this risk "bears the weight of the sedimented past" (Puwar, 2004, p. 1); as an elderly woman told me, "When the young ones work in tourism, they lose their morality" (Mms Interview, 2009). Ultimately, though, both of the gender equality sustainability fluxes, whilst affecting and being affected by local tourism development, can be said to have played a hand in legitimizing significant change for women in Göreme.

Sustainability legitimizing socio-cultural continuity

Whilst gender equality sustainability fluxes have resulted in significant change for women in Göreme, there are also sustainability fluxes that assert and legitimize socio-cultural continuity. These fluxes can be categorized, firstly, as growing concerns related to a lack of sustainable food security, and secondly, an increased nostalgia prompting "cultural sustainability" regimes of value. As I will now go on to discuss, both of these aspects or 'faces' of sustainability act in various ways to shape the lives and practices of women in Göreme.

Sustainable food security

Today, as already said, the majority of Göreme households are engaged in tourism-related work or entrepreneurial activity, and while as a result an increasing number of families have sold their agricultural land, many families have retained at least some of their land in order to continue some farming activities alongside their tourism business activities. As Bryceson (2000, p. 312) points out, retaining land as a fallback is a way of "safeguarding peasant survival in the face of adversity". By continuing with some farming activities, these households continue to maintain some food production for their own consumption, thus maintaining a reasonably stable level of food security. At first glance, the matter of food security is not necessarily a gender-related sustainability flux, and indeed Ambelu et al. (2018), who have recently described the interrelationship between sustainable tourism development and sustainable food security as complex and multifaceted, did not include a gender dimension explicitly in their discussion. However, in Göreme certainly, for reasons outlined above and below, the relationship between tourism and food security is indeed highly gendered.

Due to its capacity to provide alternative livelihood and increased household income opportunities, tourism is generally considered to positively contribute towards food security through increasing households' ability to purchase food, as long as food is accessible and available to buy (ibid.). In Göreme, an increased ability to purchase food using tourism-derived income is evidenced by two supermarket chains opening shops in the town in recent years. However, while making a wider variety of foodstuff available, the increased opportunity to purchase food from supermarkets leads simultaneously to a decrease in local agricultural and food production. This is because an imbalance is created regarding the effort involved, for example in growing and processing wheat for *bulgur* or flour versus buying such products which today are readily and affordably available in shops.

Furthermore, Ambelu et al. (2018) point out that tourism can also harm local agricultural and food production by directly 'competing' over labour and land. In Göreme, tourism does indeed compete with local agricultural practice and food production in both of these respects, with many families selling their fields (for instance, to hot-air balloon companies to use as launch sites), and even when families do hold on to their land, working on it and keeping it productive has lessened in recent years because tourism competes over labour. Importantly, this 'competition' is highly gendered in Göreme, in that it is largely women's agricultural and food production labour with which tourism is in competition. For some families, women's tourism work directly counters the ability, and the will, to engage in agricultural labour. As one woman who works as a housekeeper in a hotel complained to me in late summer, "I work from 8am until 5 pm and have not had a day off for two months. My back is sore… How could I go and work in the garden on top of that?" [Field note diary#2, 2019]. Elsewhere, this kind of competition is what creates the double burden much discussed in relation to women's entering the tourism or another paid workforce (Cole, 2018b).

It is in such situations that, as Ambelu et al. (2018, p. 1756) point out, tourism could be considered to "jeopardize the continuity of farming and thereby food security". In Göreme, the decline in farming is also attributed to increased access to education for girls, as illustrated earlier in an interview quote; "Now the girls go to school. Before they used to learn only how to work in the fields. They do not learn that now. Now they only want to study" [CCS Interview, 2005]. Prompted by an overall decrease in agricultural and food production activities, increased fears over sustainable food security, mixed up with nostalgia which I will go on to discuss shortly, are developing. Whilst a lack of sufficient volume of food is not such a worry now as it used to be some decades ago before the tourism economy developed, the increased purchasing of ready-made foods has led to concerns about the relative health value of convenience foods versus the traditional organic food locally grown and made. One hotelier, who has attempted to resurrect pigeon coops so as to use their droppings for organic fertilizer on his fields, said that when everyone used to eat only the organic locally grown foods, they rarely got sick; "Now people have to go to the doctor all of the time, and that's why this is an important tradition which I want to save" [Field note diary#2, 2019]. Resurrecting pigeon coops is one of many measures taken by the Göreme community to address their fears concerning sustainable food security.

Other measures have more directly engaged tourism, with tourism entrepreneurs actively encouraging the continuation of farming and local food production activities by using local foods in their tourism businesses. Such tourism activities are broadly understood to contribute to the food sector sustainability of the local community by not only encouraging farmers to continue or even increase production, but also by helping to retain tourism earnings in the host destination and helping distribute tourism benefits more broadly amongst the local community (Ambelu et al., 2018; Everett, 2016; Sims, 2009; Torres, 2002). Many hoteliers and restaurant owners in Göreme have, during the past 10 to 15 years especially, increasingly promoted local food production by buying vegetables and fruit from local families, and by using locally made preserves and other specialities on breakfast buffets and restaurant menus. Sometimes a hotelier or restaurant owner will ask his mother and other female relatives to make preserves and other local food items for his restaurant. Others pay women in their neighbourhood to make food products for their business. Frequently, these moves are made explicitly in order to help maintain such practices by providing a small income to those women still willing to undertake such food production activities. They thereby not only distribute tourism benefits, but also sustain a locally-embedded food security system through maintaining a 'fallback' position for when tourism business drops off (which occurs periodically due to, for example, terrorism, political instability or, as currently, a pandemic). Such moves also maintain the practices and knowledge associated with producing organic and healthful products. Importantly, though, because these activities and associated know-how are largely considered women's domain, the food security sustainability flux affects women's lives and bodies by acting to legitimize the maintenance of women's traditional food production knowledge and practices.

Cultural sustainability

The affective flux of nostalgia and an associated "cultural sustainability" imperative add further to the food security sustainability flux in acting to legitimize gendered continuity and sameness in this tourism milieu. As an urging of the continuation of certain aspects of culture "in a linear time perspective" (Soini & Birkeland, 2014, p. 216), the notion of cultural sustainability is frequently tied up with processes of heritagization in tourism contexts. Heritagization, as Rowlands (2002, p. 110) points out, always "implies a threat of loss and the need to preserve or conserve against an inevitable sense of deprivation". In Göreme, a prominent reason for such a perceived threat of loss has been outlined above in relation to the decline in agricultural and food production activities. While this decline is likely felt most strongly as a threat of loss for the Göreme people, since they experience it most directly, it can also engage tourists, both domestic and international, in processes of aestheticized nostalgia.

The use of local food products in restaurants and hotels is a practice broadly recognized as good for tourism sustainability itself (Hall et al., 2003), in that it can enhance tourist experience by promoting ideas of aestheticized "localness". Furthermore, local food production can be used to create new tourism products and experiences. An example of this in Göreme is a hotelier offering an "Organic Breakfast Tour" whereby tourists are taken to a hidden valley garden and fed an expansive array of organic breakfast foods made and served by local women [Field note diary#3, 2015; #2, 2019]. The same hotelier also offers cooking classes taught by the same women who make the breakfast, and, during the autumn harvest season, he offers a "*pekmez*-making" experience [ibid.]. This involves tourists harvesting grapes, crushing them, and then helping boil the grape-juice to produce the local food *pekmez*, or grape molasses. These recently developed tourism products creatively use local cultural practices in order to enhance tourist experience, and in doing so they engage aestheticization processes which serve to revalue *pekmez*-making, along with other local foodstuffs and preserves such as tomato paste, a local speciality tomato and chili pepper sauce, pickles and a local type of flat-bread. This revaluing is itself an affective flux which acts to shape women's lives in particular ways, most poignantly by urging the continuation of these activities identified as "traditional" and "local" activities.

Some of these food production tasks take several days to complete and involve such intense labour that groups of extended family often come together to share the task. One family who are well-known for their especially good *pekmez* have increased their production in recent years because several restaurants purchase large quantities of it from them ahead of time. This family thus calls upon both male and female extended family members to help. Most groupings these days comprise only female family members, however, and during fieldwork I have often joined and helped different groups while they make their *pekmez*, bread and other preserves. One widowed neighbor I joined was helped by her two daughters, one of whom now lives in a nearby city - and so *pekmez*-making had called her back for a while to work again with her mother. Another widow I helped was otherwise making *pekmez* on her own, except for an occasional visit from her son who came over from his restaurant now and then to check on his mother's progress; he had already included a *pekmez* desert on his menu specials board for that evening. I was unable to tell whether he might have been urged to help if I had not been there instead to help his mother [Field note diary#2, 2019]. During an earlier fieldwork visit, a *pekmez*-making group I visited included a young woman who told me that she was studying to become a nurse at Kayseri University. Like the restauranteur, she too exuded a sense of ambivalence, telling me: "The semester starts on Monday and I would have liked to have gone back early except that I still have work to do here in the village" [Field note diary#1, 2010]. Her mother responded telling her this problem would not occur in the future as she would stop making *pekmez* because it was too hard. Another elderly woman in the group then joined in the conversation by saying that she loves making *pekmez* "because it is natural (*doğal*) work", she said [ibid.].

The nursing student also told me that earlier in the day a TV film crew from Istanbul had come by and filmed them. Such scenes would likely ignite nostalgic interest among urbane Turks from other parts of the country, perhaps viewing such food production practices as a part of Turkey's "cultural heritage" which should be sustained. Indeed, these cultural practices have recently started to appear in ethnology museums in the Cappadocia region and, as Roy (2004) suggests: "When formalized as heritage, nostalgia … produces legitimacy through aestheticization". Roy (2004) argues that nostalgia as heritage follows two processes: firstly, it creates "icons that revalorize and revitalize"; and secondly, the aestheticization involved in this process hides, or renders invisible, "the brutal mechanics of capitalist valuation" (Roy, 2004, p. 65). In Göreme, this process of aestheticized, heritagized nostalgia is beginning to act as an affective flux to legitimize the continuation of certain practices largely undertaken by women. Moreover, the aestheticized nostalgia renders somewhat invisible "the brutal mechanics" of the particular power fluxes that are acting to shape women's lives by urging these cultural practices to be revitalized and sustained. The tourism aestheticization is only partly hidden, however, as shown to me when a group of *pekmez*-making women made fun of a tourist by demanding "Pay $10!" when he took photographs of them. When the Japanese man started to take money from his pocket to pay, the women laughed and assured him, with the help of my translation, that they were only joking [Field note diary#2, 2019].

So while heritagized nostalgia in Göreme's tourism in various ways acts to tie women's daily lives, bodies and activities "to a logic of sameness" (Gedalof, 2003, p. 92), it does not go without some level of awareness nor resistance. Indeed, the conversation outlined above among *pekmez*-making women over the practice's "hard" versus "natural" work expressed a certain ambivalence among women concerning the practice. Ambivalence regarding this sustainability flux was also repeatedly illustrated by the hotelier who runs the organic breakfast and *pekmez*-making experiences always becoming melancholic in conversation with me while reflecting on what he saw as the demise of Göreme culture, and his own potential role as a tourism entrepreneur to revitalize that culture [Field note diaries, 2009, 2010, 2015, 2019]. The nostalgic yearning, or as Rosaldo (1989, p. 107) put it, the "mourning for what one has destroyed" is likely entwined with the above-mentioned fears about sustainable food security, so that it becomes difficult to determine which particular affective flux is acting at any point in time. Both the food security sustainability flux and the cultural sustainability flux are a reaction to, but simultaneously work to resist, the changes brought about by other tourism and sustainability fluxes. Furthermore, whilst such aestheticization and revalorization processes work with a logic of sameness to legitimize continuity of certain gendered practices, they are sustainability fluxes which inevitably incur change themselves. The various sustainability fluxes thus work to undermine and destabilize each other in the ways in which they demand constant negotiation in relation to each other.

Concluding discussion

In her advocating for what she terms a "rematerialized poststructural gender project in tourism studies", Aitchison (2009, p. 639) calls for more critical and comprehensive exploration of both the workings and the *re*workings of gender-power relations in tourism. This would necessarily include materialist analyses of the sites and processes of power which "construct, legitimate, and reproduce gender relations in tourism" (ibid.), as well as consideration of the power relations at play in contesting and reworking gender relations. In line with this, and picking up on Jiménez-Esquinas (2017) discussion of affective power fluxes as "forces" which "do things" by imposing their sense of what is valuable and important, I have identified in this article four key "sustainability fluxes" operating as both material and discursive sites of power. Moreover, since poststructuralist feminism recognizes and values the lived experiences of women in order to understand their positioning and negotiations within, in this case, the broader, multifaceted sustainability and tourism nexus, what becomes pertinent within this view is how sustainability power fluxes operate "at different scales from global to the local and from the world stage to the body" (Aitchison, 2009, p. 639). In other words, looking at "what women do, and do not do, and why and how this occurs" is important to understanding how the gender power relations of sustainability operate and are manifested in women's daily lives (Dowling, 2011, p. 7).

Contributing to this understanding my long-term research in Göreme has tracked over time how various contradictory sustainability fluxes, operating at multiple levels and related both directly and indirectly with tourism, collide and clash in their affecting what is and what is not considered valuable, appropriate and do-able, both at different points in time and by different generations. Two of the sustainability fluxes discussed, those of women's tourism-related work and gender-equality education reforms, align with broader, global sustainability principles related to women's empowerment (e.g. through tourism work and entrepreneurship) and gender equality (e.g. in access to education). My longitudinal analysis has allowed tracking of how these global sustainability principles "do things" by prompting, legitimizing and affording acceptance of socio-cultural change at the local level. Indeed, there is no doubt that a significant shift has occurred in Göreme society's normative notions of gender identities and roles, and hence what is deemed do-able for girls and women.

Exactly how far these changes really go in achieving broader gender-related sustainability goals is somewhat questionable, however, with broader contradictions in this regard already established in the tourism, gender and sustainability literature, as outlined above. For example, in line with Ferguson's (2009) suggestion regarding the tendency for women's employment in tourism to be concentrated in lower-paid seasonal jobs, Göreme women's earnings are in most cases low, and within the work places they are entering, women tend to gain the least remunerative jobs. In addition, many women also experience the 'double burden' characteristic of women's workloads due to their household-centred reproductive and food production duties continuing, and competing, when tourism-related work is taken up. In Göreme, as has been found in other parts of the world, women's tourism employment appears to further entrench gender roles and stereotypes regarding the 'type' of work women are capable of doing (Ferguson, 2009; Tucker & Boonabaana, 2012; Vizcaino-Suārez, 2018). As Cole rightly points out, there is not an automatic correlation between tourism employment and empowerment: "Employing women or supporting them to become entrepreneurs does not necessarily change the gendered dynamics in which they are embedded" (Cole, 2018a, p. 133). What is perhaps a more significant sustainability flux enacting change in Göreme, then, is the legitimized increase in girls and young women entering higher education and completing qualifications. Indeed, there is significant difference in the life opportunities for young Göreme women today compared with those their grandmothers, and even their mothers, had.

However, while gender equality sustainability fluxes have resulted in significant change for women in Göreme, other sustainability fluxes, which align with Jiménez-Esquinas (2017) notion of 'affective fluxes', revalorize and legitimize continuity of certain practices. Although this revalorization is itself undoubtedly a change process and the practices themselves incur inevitable change, these sustainability fluxes prompt "cultural sustainability" regimes of value, thus legitimizing continuity and sameness in particular gendered practices and ways of life. Moveover, engaging fear (over perceived lack of food security) and nostalgia (for what has been, or is being, "destroyed"), these affective sustainability fluxes have arisen at least in part as a result of the changes legitimized by the first two sustainability fluxes, those of women's increased involvement in the tourism economy and gender-equality social and education policies, the effects of which are also further compounded by Göreme's tourism. Whilst being a response to changes brought about by the first two sustainability fluxes, the latter two affective sustainability fluxes and their associated legitimizations at the same time "do things" to shape lives in ways which counter, or resist, the first two sustainability fluxes discussed and their associated legitimizations, thus leading at times to apparent ambivalence.

I have previously argued in relation to my longitudinal research in Göreme that all sociocultural change is likely to be shrouded in ambivalence (Tucker, 2010). Bhabha's (1994) notion of ambivalence is pertinent in relation to gender and sustainability too, not only because of its acknowledgement of the continuous pushes and pulls of continuity and change exerted through sustainability fluxes, but also because it acknowledges the possibility of simultaneous acceptance of and resistance to different sustainability fluxes operating at various levels. Indeed, what has become apparent from this present discussion of women's positioning in relation to four separate but intertwined sustainability fluxes, some of which are powerfully affective, is that ambivalence likely exists in the notion of "sustainability" itself. So while sustainability fluxes might usefully be considered discursive sites of power, a poststructural framework necessitates always looking at what else is going on in terms of their material manifestations, always allowing for both complicity and resistance. As we have seen here, since some sustainability fluxes legitimize change whilst others legitimize continuity and sameness, when present together, these different fluxes are likely to pull this way and push that way, thus manifesting as tensions and necessary negotiations in the ways in which they shape women's lives and practices.

In Göreme for instance, women who resist the moral imperative to be *kapalı* by space invading the central streets might choose simultaneously to comply with expected gender roles and relations within their family in order to negotiate appropriate maintenance of their perceived morality. Furthermore, while grandmothers are resistant to their granddaughters' resistances, fathers may be complicit in their resistance. All such negotiations display ambivalence also in the ways in which they variously bear the weight of the sedimented past. Likewise, ambivalence surrounding women's labour in *pekmez*-making shows itself in a hotelier's reflective and troubled attempts to revive what he fears being lost, as well as in two elderly women's negotiations of the pros and cons of that form of backbreaking work. Just as Jiménez-Esquinas (2017) identified ambivalence among women performing bobbin lace in Galicia, Spain as they both adopt and reject the stereotype of picturesque backward women, similar stereotypes are both acknowledged and resisted by Göreme *pekmez*-makers when they jokily demand payment from a tourist for taking a photograph and then refuse the money once he complies.

So whilst the concept of "sustainable development" may be entrenched in Enlightenment-driven types of thinking (Camargo et al., 2016), in following a poststructuralist feminist approach to understand how the positioning of women within the broader sustainability and tourism nexus is negotiated, the importance of recognizing sustainability to be multifaceted and relational comes to the fore. Moreover, by also paying attention to sustainability's gendered material manifestations it has become clear that sustainability is, as feminist anthropologist Anna Tsing (2015, p. 33) might suggest, "recalcitrant to the kind of "summing up" that has become the hallmark of modern knowledge". Rather than assuming that the concept and processes of sustainability could be lived and experienced as coherent and straightforward, and that they could thereby be available for simple power diagnostics, the analysis here has highlighted the importance of identifying and accepting the many incongruities and contradictions in the relationship between gender, tourism and sustainability. Hence, it is precisely through looking at how lives are shaped by sustainability fluxes in such thoroughly contradictory ways that we might be able to develop a better understanding of where ambivalences lay in between acceptance of and resistance to those sustainability fluxes in touristic contexts.

Disclosure statement

No potential conflict of interest was reported by the author.

References

Aghazamani, Y., Kerstetter, D., & Allison, P. (2020). Women's perceptions of empowerment in Ramsar, a tourism destination in northern Iran. *Women's Studies International Forum*, *79*, 1–10.

Aitchison, C. (2005). Feminist and gender perspectives in tourism studies: The socio-cultural nexus of critical and cultural theories. *Tourist Studies*, *5*(3), 207–224.

Aitchison, C. (2009). Gender and tourism discourses: Advancing the gender project in tourism studies. In T. Jamal & M. Robinson (Eds.), *Sage handbook of tourism studies* (pp. 631–644). Sage.

Alarcón, D. M., & Cole, S. (2019). No sustainability for tourism without gender equality. *Journal of Sustainable Tourism*, *27*(7), 903–917.

Ambelu, G., Lovelock, B., & Tucker, H. (2018). Empty bowls: Conceptualising the role of tourism in contributing to sustainable rural food security. *Journal of Sustainable Tourism*, *26*(10), 1749–1765.

Bhabha, H. (1994). *The location of culture*. Routledge.

Bryceson, D. (2000). Disappearing peasantries? Rural labour redundancy in the neo-liberal era and beyond. In D. Bryceson, C. Kay, & J. Moooij (Eds.), *Disappearing peasantries?: Rural labour in Africa, Asia and Latin America* (pp. 299–326). ITDG Publishing.

Camargo, B., Jamal, T., & Wilson, E. (2016). Toward a critical ecofeminist research paradigm for sustainable tourism. In A. M. Munar & T. Jamal (Eds.), *Tourism research paradigms: Critical and emergent knowledges* (pp. 73–86). Emerald.

Chant, S. (2006). Contribution of a gender perspective to the analysis of poverty. In J. S. Jaquette & G. Summerfield (Eds.), *Women and gender equity in development theory and practice: Institutions, resources and mobilisation* (pp. 87–106). Duke University Press.

Cole, S. (2018b). Introduction: Gender equality and tourism: Beyond empowerment. In S. Cole (Ed.), *Gender equality and tourism: Beyond empowerment* (pp. 1–11). CAB International.

Cole, S. (2018a). Conclusions: Beyond empowerment. In S. Cole (Ed.), *Gender equality and tourism: Beyond empowerment.* (pp. 132–142).CAB International.

Dowling, J. (2011). *Just' a fisherman's wife: A post-structural feminist exposé of Australian commercial fishing women's contributions and knowledge, 'sustainability' and 'crisis.* Cambrisge Scholars Publishing.

Durakbaşa, A., & Karapehlivan, F. (2018). Progress and pitfalls in women's education in Turkey (1839–2017). *Encounters in Theory and History of Education, 19*, 70–89.

Elmas, S. (2007). Gender and tourism development: A case study of the Cappadocia region of Turkey. In A. Pritchard (Ed.), *Tourism and gender: Embodiment, sensuality and experience* (pp. 302–314). CAB International.

Everett, S. (2016). *Food and drink tourism: Principles and practice.* Sage Publications.

Ferguson, L. (2018). Gender equality and tourism: The global policy context. In S. Cole (Ed.), *Gender equality and tourism: Beyond empowerment* (pp. 14–22). CAB International.

Ferguson, L. (2009). *Analysing the gender dimensions of tourism as a development strategy* (Working Paper No. pp03/09). ICEI, Universidad Complutense de Madrid.

Gedalof, I. (2003). Taking (a) place: Female embodiment and the re-grounding of community. In S. Ahmed, C. Castaneda, A. M. Fortier, & M. Sheller (Eds.), *Uprooting/regroundings: Questions of home and migration* (pp. 91–114). Berg.

Gentry, K. M. (2007). Belizean women and tourism work: Opportunity or impediment? *Annals of Tourism Research, 34*(2), 477–496.

Gibson, H. (2001). Gender in tourism: Theoretical perspectives. In Y. Apostolopoulos, S. Sonmez, & D. Timothy (Eds.), *Women as producers and consumers of tourism in developing regions* (pp. 19–43). Praeger.

Hall, C. M., Mitchell, R., & Sharples, L. (2003). Consuming places: The role of food, wine and tourism in regional development. In C. M. Hall, L. Sharples, R. Mitchell, N. Macionis, & B. Cambourne (Eds.), *Food tourism around the world: Development management and markets* (pp. 25–59). Elsevier.

Haraway, D. (2003). Situated knowledges: The science question in feminism and the privilege of partial perspective. In Y. S. Lincoln & N. K. Denzin (Eds.), *Turning points in qualitative research: Tying knots in a handkerchief* (pp. 21–46), Altamira Press.

Hughes, M., Weaver, D., & Pforr, C. (Eds.). (2015). *The practice of sustainable tourism: Resolving the paradox.* Routledge.

Jiménez-Esquinas, G. (2017). "This is not only about culture": On tourism, gender stereotypes and other affective fluxes. *Journal of Sustainable Tourism, 25*(3), 311–326.

Johannesson, G., Ren, C., & van der Duim, R. (Eds.). (2015). *Tourism encounters and controversies: Ontological politics of tourism development.* Ashgate.

Kabeer, N. (1999). Resources, agency, achievements: Reflections on the measurement of women's empowerment. *Development and Change, 30*(3), 435–464.

Lamphere, L. (2006). Foreword: takng stock – The transformation of feminist theoretizing in anthropology. In P. L. Geller & M. K. Stockett (Eds.), *Feminist anthropology: Past, present, and future* (pp. viiii–vixvi). University of Pennsylvania Press.

Müftüler-Baç, M. (2012). *Gender equality in Turkey.* European Parliament Directorate General for Internal Affairs.

O'Shaughnessy, S., & Krogman, N. (2011). Gender as contradiction: From dichotomies to diversity in natural resource extraction. *Journal of Rural Studies, 27*(2), 134–143.

Puwar, N. (2004). Space Invaders: Race, Gender and Bodies Out of Place. Oxford: Berg.

Rosaldo, R. (1989). Imperialist nostalgia. Representations, *26*, 107–122.

Rowlands, M. (2002). Heritage and cultural property. In V. Buchlı (Ed.), *The material culture reader* (pp. 105–114). Berg.

Roy, A. (2004). Nostalgias of the modern. In N. AlSayyad (Ed.), *The end of tradition?* (pp. 63–86), Routledge.

Scheyvens, R. (2000). Promoting women's empowerment through involvement in ecotourism: Experiences from the third world. *Journal of Sustainable Tourism, 8*(3), 232–249.

Sen, D. (2017). *Everyday sustainability: Gender justice and fair trade tea in Darjeeling.* State University of New York Press.

Sims, R. (2009). Food, place and authenticity: Local food and the sustainable tourism experience. *Journal of Sustainable Tourism, 17*(3), 321–336.

Singh, S. (2007). Deconstructing "gender and development" for "identities of women". *International Journal of Social Welfare, 16*(2), 100–109.

Soini, K., & Birkeland, I. (2014). Exploring the scientific discourse on cultural sustainability. Geoforum, *51*, 213–223.

Swain, R. B., & Wallentin, F. Y. (2008). *Economic or non-economic factors – what empowers women?* Unpublished working paper, Department of Economics, Uppsala University.

Torres, R. (2002). Toward a better understanding of tourism and agriculture linkages in the Yucatan: Tourist food consumption and preferences. *Tourism Geographies, 4*(3), 282–306.

Tsing, A. L. (2015). *The mushroom at the end of the world: On the possibility of life in capitalist ruins.* Princeton University Press.

Tucker, H. (2003). *Living with tourism: Negotiating identities in a Turkish village.* London.

Tucker, H. (2007). Undoing shame: Tourism and women's work in Turkey. *Journal of Tourism and Cultural Change, 5*(2), 87–105.
Tucker, H. (2009). The cave homes of Göreme: Performing tourism hospitality in gendered space. In P. Lynch, A. McIntosh, & H. Tucker (Eds.), *Commercial homes in tourism: An international perspective* (pp. 127–137). Routledge.
Tucker, H. (2010). Peasant-entrepreneurs: A longitudinal ethnography. *Annals of Tourism Research , 37*(4), 927–946.
Tucker, H. (2011). Success and access to knowledge in the tourist-local encounter: Confrontations with the unexpected in a Turkish community. In D. Theodossopoulos & J. Skinner (Eds.), *Great expectations: Imagination, anticipation, and enchantment in tourism* (pp. 27–39). Berghahn Books.
Tucker, H. (2014). Mind the gap: Opening up spaces of multiple moralities in tourism encounters. In M. Mostafanezhad & K. Hannam (Eds.), *Moral encounters in tourism* (pp. 199–208). Ashgate.
Tucker, H. (2017). Community-based tourism as sustainable development. In I. Egresi (Ed.), *Alternative tourism in Turkey: Role and potential development and sustainability* (pp. 335–347). Springer.
Tucker, H., & Boonabaana, B. (2012). A critical analysis of tourism, gender and poverty reduction. *Journal of Sustainable Tourism, 20*(3), 437–456.
United Nations. (2015). *Transforming our world: The 2030 Agenda for Sustainable Development. A/RES/70/1. Resolution adopted by the General Assembly on 25 September 2015.* United Nations.
Veijola, S., Molz, J. G., Pyyhtinen, O., Hockert, E., & Grit, A. (Eds.). (2014). *Disruptive tourism and its untidy guests: Alternative ontologies for future hospitalities.* Palgrave Macmillan.
Vizcaino-Suārez, P. (2018). Tourism as empowerment: Women artisan's experiences in Central Mexico. In S. Cole (Ed.), *Gender equality and tourism: Beyond empowerment* (pp. 46–54). CAB International.
Weaver, D., Hughes, M., & Pforr, C. (2015). Paradox as a pervasive characteristic of sustainable tourism: Challenges, opportunities and trade-offs. In M. Hughes, D. Weaver, & C. Pforr (Eds.), *The practice of sustainable tourism: Resolving the paradox* (pp. 281–290). Routledge.

Tourists' water conservation behavior in hotels: the role of gender

Ana B. Casado-Díaz, Franco Sancho-Esper, Carla Rodriguez-Sanchez and Ricardo Sellers-Rubio

ABSTRACT

Hotels and destination managers are increasingly expressing concern about the impacts of climate change and sustainable water use, especially in crowded and water-scarce destinations. The aim of this study is to examine gender differences in hotel guests' reported water conservation behavior (WCB) when on vacation. The study examines several factors that can potentially affect these gender differences, namely attitudes toward water conservation, normative and hedonic motives, destination problem awareness, and destination attachment. Data from a sample of 680 hotel guests reveal significant gender differences, with specific factors affecting the WCB of guests of each gender. Attitudes exert a positive influence on guests' WCB. This influence is greater for women than for men. Normative motives also positively influence attitudes, although this effect is greater for men than for women. Conversely, hedonic motives negatively influence water conservation attitudes, and this effect is greater for women than for men. Lastly, destination problem awareness positively influences normative motives, while destination attachment negatively influences hedonic motives. No gender effect is found for these relationships. The implications for research and practice in sustainable tourism and pro-environmental behavior are presented.

Introduction

Water is a key resource in the tourism industry, since hotel facilities and tourists depend heavily on water for direct and indirect use (Gössling et al., 2019). Considerable academic research has focused on this topic in the last decade. An increasing number of studies has examined water-saving measures by hotels, such as reducing water consumption (e.g. through water-efficient appliances, low-flow showerheads, and towel reuse initiatives) and recycling water (Gössling et al., 2019). However, further research is needed to better understand the underlying factors behind guests' water conservation behavior. Only then can more specific strategies be developed to reduce water overuse and encourage more sustainable behaviors (Hadjikakou et al., 2015).

The research objectives of this study are twofold. First, building on environmental psychology theories such as goal-framing theory (Lindenberg & Steg, 2007), this study examines the importance of cognitive and affective determinants of water conservation attitudes. The determinants considered in this study are normative and hedonic motives and contextual factors of the destination (destination place attachment and destination problem awareness). Most studies of water conservation behavior or intention have considered either social or cognitive drivers (e.g. Han & Hyun, 2018b, 2018c). However, affective motives can influence the extent to which individuals interact with their surroundings, thus influencing their pro-environmental behavior (Steg et al., 2014). Moreover, research on guests' pro-environmental behavior in hotels has typically measured general intentions to conserve water through items such as "I am willing to conserve water in a lodging context when traveling" (e.g. Han & Hyun, 2018b; Untaru et al., 2016). This approach prevents the extrapolation of findings to specific behaviors such as in-room water conservation behaviors. To overcome this limitation, the present study examines reported water conservation behavior and measures this behavior by analyzing *specific* in-room activities (e.g. "I've turned off the tap while brushing my teeth"). Therefore, hotel managers can use the findings of the present study to implement specific measures to solve specific problems. They can thus incentivize guests' reduction of in-room water use and prevent misuse. This issue is critical in regions with water scarcity problems. These problems affect many tourist destinations around the world, but they are particularly acute in the Mediterranean region.

Second, when examining the sociodemographic and psychological aspects of individuals' environmental concern and behavior, research has shown that women seem to have greater concern for the environment than men (e.g. Dietz et al., 2002; Dzialo, 2017; Zelezny et al., 2000). Research also suggests that women tend to engage more frequently in general pro-environmental behavior (PEB) than men (Dzialo, 2017) and that several individual and contextual factors can influence this PEB (Knight, 2019). The most common explanations are based on socialization theory (Zelezny et al., 2000), evolutionary psychology (Saad & Gill, 2000), and ecofeminism (Dzialo, 2017). In the context of the hotel industry, evidence of gender effects is mixed. For example, some studies of guests' willingness to pay for green initiatives by hotels have shown that women are more likely to pay a premium (e.g. Han et al., 2011), while others have indicated the opposite (e.g. Kang et al., 2012). This difference in behavior might be influenced by many factors, including context, location, culture, and perceived environmental degradation (Knight, 2019). Therefore, research in different contexts is needed to fully understand gender differences in PEB (Briscoe et al., 2019). This study addresses this gap and examines the key moderating role of gender in the context of water conservation behavior in hotels. Nevertheless, the novelty of the research does not circumscribe to examining the effect of gender on specific relationships between variables, but rather on the whole model. The rationale is that the expected differences between female and male tourists should apply to the proposed relationships because the findings in the consumer behavior literature show that gender affects the relationships between affective, cognitive, and attitudinal variables (Han & Hyun, 2018a).

Theoretical framework

The relationship between water conservation attitudes and behavior and the role of gender

Attitude is the most frequently used concept in sustainable tourism to explain environmentally responsible behavior based on the attitude–behavior model (e.g. Gao et al., 2017). Research on attitudes in relation to tourists' sustainable choices has primarily focused on general environmental attitudes and attitudes toward ecotourism or sustainable tourism (Passafaro, 2020). In a hotel context, empirical studies have considered attitudes toward staying at, visiting, and revisiting green hotels (e.g. Chen & Tung, 2014). To a lesser extent, studies have also examined attitudes toward green practices in the lodging industry (Manaktola & Jauhari, 2007) and attitudes toward performing eco-friendly behaviors at a hotel, such as towel reuse (Han et al., 2018) and water conservation (Untaru et al., 2016). This study focuses on the latter approach by examining the association between actual guests' attitudes toward water-saving practices and these guests' water conservation behavior. Regarding hotel guests' behaviors, the present study focuses on hotel guests' reported in-room water conservation behavior when on vacation. In-room water use includes behaviors such as showering, flushing the toilet, and brushing teeth, which are directly related to water use. However, in-room water use also covers bed linen changes and towel replacement. In short, in-room water conservation behavior includes any individual effort to save water consumed through any such use (Marandu et al., 2010).

The literature stresses the positive and significant influence of attitudes on individuals' intentions to conserve water in both household settings (e.g. Dolnicar et al., 2012) and hotel contexts (e.g. Untaru et al., 2016). Assuming that intention is a strong antecedent of behavior, a similar relationship between attitudes and behaviors is to be expected (Gao et al., 2017). Thus, an individual with a favorable attitude toward water conservation should act accordingly. For example, Marandu et al. (2010) reported a positive relationship between attitudes toward water conservation and water conservation behavior in a household context. Therefore, the following hypothesis is proposed:

> *H1a.* Attitudes toward water conservation are positively associated with guests' reported in-room water conservation behavior.

Gender has been studied as a possible moderator of the relationship between attitudes and behaviors (Ajzen & Fishbein, 1980). Studies of the moderating role of gender in environment-related contexts suggest that when environmental attitudes are strong, green purchasing behaviors are greater among women than among men (e.g. Dagher et al., 2015). Hasnain et al. (2020) also found a positive relationship between environmental attitudes and greater green buying intention among women than among men. In contrast, Chekima et al. (2016) did not find support for such a moderating effect. These inconclusive results highlight the need for further research on this topic to improve the current understanding of gender differences in tourism behaviors (Vicente-Molina et al., 2018).

In the hotel industry, research has shown that gender may be used to segment hotel guests by their PEB and attitudes, with studies revealing that female guests behave in a more environmentally friendly way than male visitors (Dolnicar & Leisch, 2008). However, in the literature on hotel water conservation behavior, this perspective tends to be omitted from the models (e.g. Untaru et al., 2016). Only recently have researchers started considering the moderating role of gender in this context, although not in relation to attitudes (Han & Hyun, 2018b; Han et al., 2020). Despite a lack of evidence of gender moderation in the attitude–behavior relationship in the hotel context, research has shown a gender effect when the two variables are examined independently. More specifically, women have been found to have greater (positive) attitudes toward water conservation than men (Lipchin et al., 2005). Women are also more inclined than men to engage in water conservation behaviors when traveling (Gabarda-Mallorquí et al., 2018). According to some authors, women seem more likely to engage in pro-environmental behaviors such as water conservation because these behaviors are linked to their private sphere (Dietz et al., 2002; Vicente-Molina et al., 2018). As stated by Vicente-Molina et al. (2018, p. 91), "gender differences can be expected to affect explanatory factors and, through them, pro-environmental behavior." Based on the previous findings, the following hypothesis is proposed:

> *H1b.* The relationship between attitudes toward water conservation and guests' reported in-room water conservation behavior is stronger for women than for men.

Cognitive and affective components of attitudes toward water conservation behavior

Normative and hedonic motives for water conservation behavior

The most widely used and well-known theories in environmental research, such as the theory of planned behavior (TPB; Ajzen, 1985) and the norm-activation model (NAM; Schwartz, 1977), are based on the assumption that people are rational in their decision-making processes and actions. Accordingly, cognitive determinants (e.g. environmental beliefs, knowledge, and norms) are important motivators for individuals who engage in PEB. Nevertheless, in recent decades, scholars have advocated the inclusion of affective factors and emotions to extend these theories (Steg & Vlek, 2009). For instance, Eagly and Chaiken (2007) argued that people's attitudes are typically based on both affective and cognitive components. Thus, recent studies of environmental consumer behavior have sought to explain evaluations of an attitude object or event (e.g. recycling) in terms of people's cognition/thinking and feeling-based evaluations (e.g. Rodriguez-Sanchez et al., 2018).

This idea is also supported by goal-framing theory in environmental psychology (Lindenberg & Steg, 2007). According to this theory, individuals behave subject to three overarching goal frames: the normative goal (to act appropriately), the gain goal (to protect or improve one's resources), and the hedonic goal (to feel better right now). While normative and gain goals activate cognitive and instrumental motives such as moral norms or saving money, hedonic goals activate one or more sub-goals to improve the way one feels in each situation (e.g. avoiding effort or seeking direct pleasure). The prevalence of a goal in a decision depends on the predominant values of the decision maker and on external cues (e.g. the context) that activate different values (Steg et al., 2014). In the case of in-room water conservation behavior, normative and hedonic goals seem to play a key role (Han & Hyun, 2018b; Miao & Wei, 2013).

Concerning normative motives, a multidisciplinary stream of research indicates that morality plays an influential role in attitude formation and that moral appeals can be a powerful tool for persuasion (Feinberg & Willer, 2013). Perceiving environmental issues in moral terms is associated with the strength and valence of individuals' environmental attitudes (Feinberg & Willer, 2013). Cognitive factors such as ascription of responsibility and awareness of environmental problems are prerequisites for the development of moral norms, which in turn influence environmental decision making (Bamberg & Moser, 2007). Unsurprisingly, therefore, past studies have found that one important reason why travelers behave pro-environmentally is that they feel it is the moral or right thing to do (Han & Hyun, 2018c; Kiatkawsin & Han, 2017). Based on this idea, the following hypothesis is proposed:

H2a. Guests' normative motives regarding responsible water use are positively associated with attitudes toward water conservation.

According to goal-framing theory, hedonic goals should be stronger than normative goals because they relate to the most basic need, namely satisfaction (Lindenberg & Steg, 2007). Hedonic goal-framing suggests that affect has the strongest impact on people's attitudes and behavior (Huijts et al., 2010). Hedonic motives are rooted in hedonic values, which are primarily focused on improving one's feelings and reducing effort (Steg et al., 2014). Therefore, hedonic values are negatively linked to environmental beliefs, attitudes, and preferences because they are likely to make people concentrate on the possible hedonic aspects of environmental behavior such as pleasure, discomfort, or effort (Steg et al., 2014). In the lodging context, research has shown that hedonic motives are more salient than moral motives when guests are on vacation (e.g. Miao & Wei, 2013, 2016). For example, several studies have shown that the likelihood that an individual behaves environmentally in a hotel is predominantly a negative function of motives such as personal comfort and enjoyment (Miao & Wei, 2013, 2016; Wang et al., 2018). Based on this line of reasoning, the following hypothesis is proposed:

H3a. Guests' hedonic motives are negatively associated with attitudes toward water conservation.

The effect of destination context on normative and hedonic motives for water conservation behavior

Scholars have emphasized the need for further research on the role of contextual or situational factors in the attitude–behavior relationship, as well as motivators to behave pro-environmentally when traveling (Passafaro, 2020). In response to these calls, this study examines the effect of destination context (i.e. destination problem awareness and destination attachment) on normative and hedonic motives.

Awareness of consequences refers to "whether someone is aware of the negative consequences for others or for other things on values when not acting prosocially" (De Groot & Steg, 2009, p. 426). This awareness might influence a tourist's sense of obligation to behave pro-environmentally through personal norms. The NAM theory describes the role of problem awareness, normative motives (personal norms), and ascription of responsibility in explaining prosocial intentions or behavior (Schwartz, 1977). Scholars have examined the associations between these variables, where problem awareness, ascribed responsibility, and personal norms are related to pro-environmental intentions or behavior in different ways (e.g. De Groot & Steg, 2009).

Regarding destination problem awareness, it might be expected that if tourists are aware of environmental problems in a destination, they feel obliged to adapt their behavior to reduce the negative impact of their actions (Han & Hyun, 2018c). This study tests the relationship between problem awareness and personal norms, in line with previous studies in the hotel context (Chen & Tung, 2014; Han et al., 2015). Chen and Tung (2014) found that hotel guests perceive a stronger moral obligation when they are highly concerned about the environment and its harmful impact. Similarly, Han et al. (2015) showed a direct link between problem awareness and moral norms when explaining hotel customers' pro-environmental intention to revisit. Therefore, the following hypothesis is proposed:

H4a. Guests' destination problem awareness is positively associated with normative motives.

Place attachment can be defined as "an affective bond to a particular geographical area and the meaning attributed to that bond" (Morgan, 2010, p. 12). In tourism, place attachment is often described as "destination attachment" (e.g. Yuksel et al., 2010), which refers to the link between tourists and the place or destination where they spend their vacations. Applying goal-framing theory to the uniquely hedonic context of tourism (Juvan & Dolnicar, 2014) and vacations (Otto & Ritchie, 1996), hedonic motives are expected to strongly drive tourists' behavior. With tourists' pro-environmental behavior and intention, if tourists have a strong attachment to a destination, they might be willing to protect it (Gifford & Nilsson, 2014). In fact, several authors have reported an association between place attachment and PEB (e.g. Raymond et al., 2011; Tonge et al., 2015). Similarly, when tourists are strongly attached to a destination, they are more inclined to make sacrifices for the environment, even at the expense of immediate self-interest, effort, or cost (Davis et al., 2011). Thus, a negative relationship between destination attachment and hedonic motives is expected, and a stronger attachment to the destination should lessen the importance of hedonic motives (perceived effort and cost associated with saving water during the stay). Accordingly, the following hypothesis is proposed:

H5a. Guests' destination attachment is negatively associated with hedonic motives.

Cognitive and affective components of attitudes toward water conservation behavior: the role of gender

Several attempts to explain environmental moral obligation using theories such as the NAM and value-belief-norm theory (VBN; Stern, 2000) have revealed significant gender differences (e.g. Nordfjaern & Rundmo, 2019; Stern et al., 1993). These studies have consistently shown that women have a greater moral obligation than men to behave pro-environmentally. Studies have also shown that women tend to be more concerned and knowledgeable than men about the environment and associated problems (e.g. Tindall et al., 2003; Zelezny et al., 2000). Women also report stronger pro-environmental views about environmental issues than men (Xiao & McCright, 2015). However, other studies have revealed the opposite, indicating that men are more concerned about environmental issues and have greater knowledge than women (e.g. Shen & Saijo, 2008). Women seem to express greater levels of concern about site-specific and local environmental problems (e.g. water scarcity) than men (Davidson & Freudenburg, 1996). By contrast, with broader issues such as preserving wildlife and biodiversity, gender differences are weak or non-existent (Hayes, 2001). These results show that individual findings may be contingent on the specific issue at hand.

Gender differences have been explained using gender socialization theory (Zelezny et al., 2000). Gender socialization theory posits that women have a greater moral obligation to act pro-environmentally than men as a result of gender-based socialization processes (Lee et al., 2013). Early childhood socialization renders women more sensitive to the feelings and needs of others. Thus, they are more concerned about the environment than men and are more motivated to work for the environment. On the contrary, men are more focused on economic issues, are the main earners, and see themselves as more detached from the natural world than women (Zelezny et al., 2000).

In the context of sustainable tourism (nature-based destination), Meng and Uysal (2008) showed that women place more importance than men on most destination attributes. Unsurprisingly, therefore, environmental studies in the hospitality industry have evidenced that women show greater levels of moral obligation and problem awareness than men when they are traveling. For instance, Han and Hyun (2018a) found that women had a greater sense of obligation toward eco-friendly behaviors while traveling, which translated into higher levels of eco-friendly purchasing and recycling behavior. Similarly, in a study of university students, Alonso-Almeida (2013) found that women were more prone to engaging in environmental practices in tourism management because they were more aware of environmental problems in the tourism industry. Regarding water conservation behavior, very few studies have examined the influence of gender on internal perceptions and responses such as moral obligation and environmental awareness. These studies have yielded inconsistent results. For instance, Han and Hyun (2018b) found that the linkage between personal norms and water conservation intentions in a hotel context is not moderated by gender. In contrast, Han et al. (2020) empirically found that the relationship between environmental awareness and the intention to conserve water, energy, or local resources when traveling is stronger for young women than for men. Given the scant literature on hotels and gender, and drawing on the environmental and sustainable tourism literature cited above, the following hypotheses are proposed:

> *H2b.* The relationship between normative motives and attitudes toward in-room water conservation behavior is stronger for women than for men.

> *H4b.* The relationship between destination problem awareness and normative motives is stronger for women than for men.

Hedonic values have been used in environmental research because they have been found to be important predictors of PEB (Steg et al., 2014). Hedonic values, along with egoistic values, shape self-enhancement values, which reflect a concern for one's own interests. Such values are contrary to self-transcendence values, which signal a concern for collective interests (Steg et al., 2014). Research has shown that men exhibit greater levels of self-enhancement than women, resulting in lower PEB among men. However, the evidence of this effect warrants further examination (Mobley & Kilbourne, 2013). For example, Balundé et al. (2019) recently failed to find significant gender differences in the relationship between hedonic values and PEB. Moreover, in a tourism context, Han et al. (2018) found that the relationship between willingness to make sacrifices for the environment when traveling and intentions to reduce waste and recycle do not differ significantly by gender.

The situation seems to be different with respect to the consumption of products and services. The consumer behavior literature hypothesizes that hedonic value may be derived from the

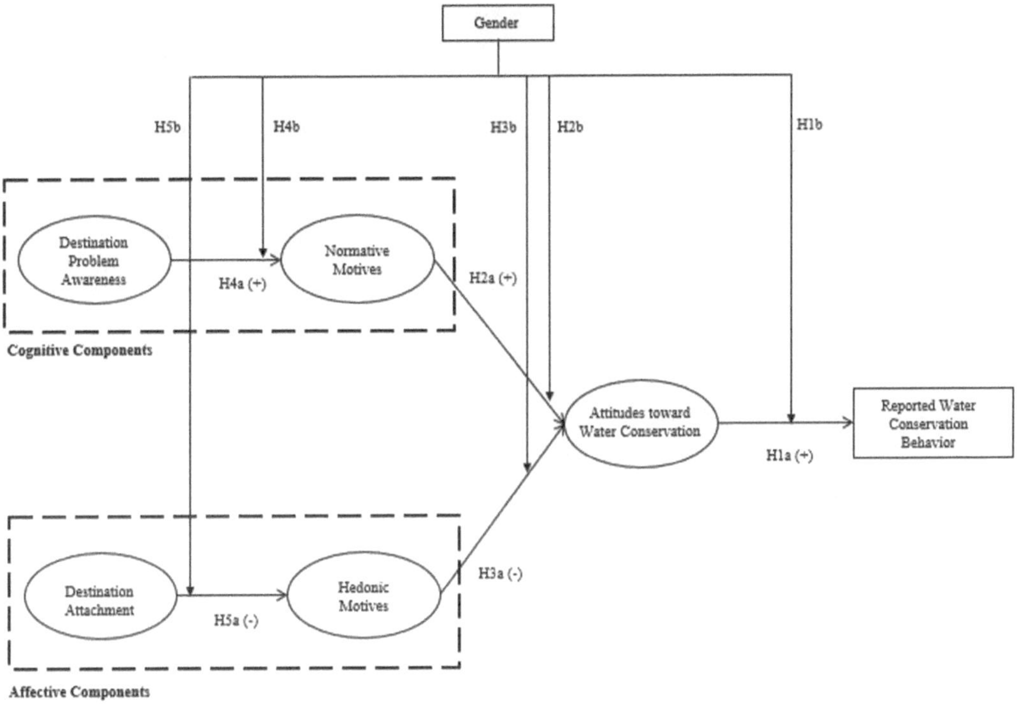

Figure 1. Proposed theoretical model of guests' reported in-room water conservation behavior.

immediate pleasure or joy of consumption, and studies have consistently shown that the effect of hedonic value on attitudes and behavior is stronger for women than for men (Jackson et al., 2011; Yang & Lee, 2010). For instance, Gatersleben et al. (2014) analyzed consumer identities and environmental variables such as pro-environmental attitudes and ecological purchases. They found that women are more likely to identify themselves with a hedonistic consumer identity than men, but they did not find gender differences in conscious consumer identity. The limited evidence suggests that women might be more hedonistic than men when traveling (consuming the tourist experience) and when using water in their hotel rooms. For example, an in-room water conservation action such as towel reuse means some loss of comfort, reduces the hedonic value of the accommodation, and lessens the joy of using fresh, dry towels every day. This greater hedonistic tendency may mean that the negative relationship between hedonic motives and attitudes to save water in hotels is stronger for women than for men. Finally, the relationship between place attachment and hedonic motives may also be stronger for women than for men. Several authors have affirmed that gender plays a critical role in place attachment, with women developing more attachment and stronger affective connections to a place (Hidalgo & Hernández, 2001; Rollero & De Picolli, 2010). However, other authors have found that men and women show the same levels of place attachment (Brown et al., 2003). Within the tourism context, Ramkissoon and Mavondo (2015) showed that although gender has no significant main effect on any dimension of place attachment, it moderates the link between visitors' pro-environmental behavioral intention and certain dimensions of place attachment (i.e. place identity, place affect, and place dependence). Based on the scant literature and the above ideas, the following hypotheses are proposed:

> *H3b.* The relationship between hedonic motives and attitudes toward in-room water conservation behavior is stronger for women than for men.

Table 1. Sample profile.

Criteria	Levels	Full sample	(%)	Female	(%)	Male	(%)
Age	18–29	66	9.7	29	8.5	37	11.1
	30–44	118	17.3	67	19.6	51	15.3
	45–65	317	46.5	165	48.4	151	45.2
	66 or more	175	25.7	80	23.5	95	28.4
Education	No studies	44	6.5	24	5.7	20	6.0
	Primary education	96	14.1	53	12.6	43	12.8
	Secondary education	324	47.6	154	36.5	169	50.3
	University/college	215	31.6	111	26.3	104	31.0
Income (net monthly)	No income	36	5.3	21	6.2	15	4.5
	€300 or less	2	.3	2	.6	0	.0
	€301–€600	5	.7	3	.9	2	.6
	€601–€900	14	2.1	9	2.7	5	1.5
	€901–€1200	48	7.0	27	8.0	21	6.3
	€1201–€1800	176	25.8	85	25.1	90	26.8
	€1801–€2400	240	35.2	117	34.5	123	36.6
	€2401–€3000	107	15.7	51	15.0	56	16.7
	€3001–€4500	28	4.1	14	4.1	14	4.2
	€4501 or more	20	2.9	10	2.9	10	3.0
Country of residence	Spain	303	44.5	149	43.4	153	45.4
	UK	256	37.6	134	39.1	122	36.2
	Other	67	9.8	32	9.3	35	10.4
	Ireland	33	4.8	17	5.0	16	4.7
	Germany	14	2.1	6	1.7	8	2.4
	Russia	6	.9	4	1.2	2	.6
	France	2	.3	1	.3	1	.3

$n_{total} = 680$, $n_{female} = 337$ (49.5%) and $n_{male} = 343$ (50.5%).

H5b. The relationship between destination attachment and hedonic motives is stronger for women than for men.

The hypotheses are summarized in Figure 1.

Method

Sample and data collection

A face-to-face survey (CAPI with Qualtrics software®) was used to gather data at the hotel premises. Although initial contact with respondents was in person, individuals self-administered the questionnaire using a tablet. The interviewer was on hand to provide support if necessary. This form of data collection, together with the choice of question order so as not to disclose the purpose of the study earlier than necessary, was used to reduce social desirability bias (Juvan & Dolnicar, 2017). An external market research institute conducted the fieldwork using actual tourists in six four-star hotels in Benidorm between July and August 2019 (summer season). Benidorm is an ideal study setting for two reasons. First, it is one of the most important "sun and sand" tourist destinations in the Mediterranean (Ivars-Baidal et al., 2013), welcoming more than 11 million tourists in 2018 (INE, 2019). Second, it has a high density of hotels, which means that its tourism activity exerts extra pressure on water demand (Casares-Blanco et al., 2019). Although Benidorm officially has around 67,000 inhabitants, the average number of water consumers in the summer season rises to above 100,000 (Sánchez-Galiano et al., 2017). Based on previous studies, which highlight the importance of hotel quality on guests' water consumption (e.g. Rico et al., 2020), this study centers on understanding the water conservation behavior of guests in four-star hotels (mid-to-high level). A final sample of 680 guests from 758 responses (89.70%) was used. This sampling procedure met previously established quotas (GVA, 2016) and ensured that respondents had stayed at the hotel for at least three nights. Table 1 shows the respondent profile.

Table 2. Descriptive statistics.

	Destination problem awareness (DPA)	Normative motives (NM)	Destination attachment (DA)	Hedonic motives (HM)	Attitudes toward water conservation (ATT)	Reported water conservation behavior (RWCB)
DPA	–	.33**	.22**	−.34**	.32**	.25**
NM		–	.15**	−.34*	.39**	.31**
DA			–	.06	.04	−.05
HM				–	−.53**	−.61**
ATT					–	.42**
RWCB						–
Range	1–7	1–7	1–7	1–7	1–7	1.78–6.78
Mean (SD) for whole sample	4.54 (1.29)	4.45 (1.29)	4.99 (1.34)	3.63 (1.45)	5.51 (1.01)	3.99 (.82)
Mean (SD) for women	4.52 (1.26)	4.38 (1.31)	5.08 (1.31)	3.78 (1.43)	5.40 (1.02)	3.94 (.77)
Mean (SD) for men	4.56 (1.32)	4.52 (1.29)	4.90 (1.37)	3.48 (1.46)	5.61 (1.00)	4.03 (.86)
t test for equality of means	−.35	−1.45	1.69	2.62**	−2.66**	−1.50

n = 680 individuals. The aggregate variables were calculated as the arithmetic mean of the items of each factor.
**$p < .01$.
*$p < .05$.

Survey instruments

The scales used to operationalize the variables in the model were all at the same behavior-specific level, related to water consumption and conservation (Steg & De Groot, 2010). The antecedents of reported water conservation behavior (RWCB) were selected and adapted from the literature. These antecedents are attitudes toward water conservation (ATT), hedonic motives (HM), normative motives (NM), destination problem awareness (DPA), and destination attachment (DA). All constructs were measured using multiple items on a scale ranging from 1 (*strongly disagree*) to 7 (*strongly agree*). RWCB was measured using the mean (aggregate index) of nine items from the literature (see Appendix). The responses to the statements related to in-room water conservation actions ranged from 1 (*never*) to 7 (*all the time*). These statements on actual hotel guests' reported in-room water conservation behavior were preferred to items on generic PEB to reduce self-reported bias (Kormos & Gifford, 2014). Table 2 shows the main descriptive statistics of the variables in the model.

Data analysis

The data were analyzed using IBM-SPSS v26 and EQS 6.2. A two-step estimation process was used to estimate the model (Anderson & Gerbing, 1988). Confirmatory factor analysis (CFA) was performed to assess reliability and validity of the measurement model. Once these psychometric features had been confirmed for the full sample, the measurement invariance between genders was assessed (Hair et al., 2006). A sequence of multigroup CFA estimations was used to assess the loose cross-validation, configural invariance, and metric invariance (Steenkamp & Baumgartner, 1998). Finally, multigroup structural equation modeling (SEM) was used to test the proposed hypotheses. This model was considered the optimal model to analyze the moderating role of a discrete variable (i.e. gender) jointly in all relationships of the model (Byrne, 2013). This model also utilized the factor structure of the data (latent constructs and observable variables).

Results

CFA of the whole sample of guests was performed to confirm the dimensional structure and assess convergent validity. Overall, the model showed acceptable fit (see Table 3). Convergent validity was checked using several criteria (Jöreskog & Sörbom, 1993). All items satisfied both the weak convergence criterion (the factor loadings of all items were statistically significant) and the strong convergence criterion (all factor loadings were greater than 0.5). Convergent validity was also confirmed because all average variance extracted (AVE) scores were greater than 0.5 (Fornell & Larcker, 1981; Table 3). All composite reliability indexes (CRI) were above 0.65, thus confirming construct reliability (Steenkamp & Geyskens, 2006).

Discriminant validity was confirmed by two findings: (i) none of the 95% confidence intervals of the correlations between each pair of factors included unity (Anderson & Gerbing, 1988) and (ii) all AVE scores were greater than the squared between-construct correlations (Wiertz & de Ruyter, 2007; see Table 4).

Analysis of the invariance of the measurement instrument between genders consisted of a sequence of CFA estimations (Steenkamp & Baumgartner, 1998; see Table 5).

First, the independent CFA estimations for women and men supported loose cross-validation because both had good overall fit. Second, configural invariance was assessed by imposing the same structure of latent factors between both groups (equal form) on the CFA. Configural invariance was supported by the results because the restricted model had good overall fit, and all latent variables were specified by the same indicators. Third, metric invariance was assessed by imposing equality restrictions of the factor loadings of items for both groups on the CFA. This invariance was assessed by analyzing the incremental fit of these nested models compared to

Table 3. Reliability and convergent validity.

Construct & items	Std. loading	Robust t	CRI	AVE
Attitudes toward water conservation				
ATT1	.80	38.33*	.89	.62
ATT2	.81	41.16*		
ATT3	.82	51.92*		
ATT4	.78	35.65*		
ATT5	.71	30.55*		
Normative motives				
NM1	.69	23.53*	.72	.50
NM2	.57	18.02*		
NM3	.78	31.02*		
Hedonic motives				
HM1	.91	73.72*	.93	.65
HM2	.88	64.42*		
HM3	.80	54.46*		
HM4	.81	44.16*		
HM5	.73	36.35*		
HM6	.76	38.20*		
HM7	.71	37.98*		
Destination problem awareness				
DPA1	.51	14.56*	.74	.50
DPA2	.68	25.69*		
DPA3	.87	34.24*		
Destination attachment				
DA1	.93	71.05*	.98	.84
DA2	.94	93.80*		
DA3	.95	113.88*		
DA4	.93	65.29*		
DA5	.88	58.92*		
DA6	.86	50.99*		
DA7	.94	90.17*		
DA8	.91	54.58*		
S-B χ^2 (289 df) = 1085.458 ($p < .01$), BBNFI = .933; BBNNFI = .943 , CFI = .950 , RMSEA = .066 [.062–.070]				

n = 680 individuals. CRI = composite reliability index; AVE = average variance extracted.
*p < .01.

equal form estimation (Putnick & Bornstein, 2016). The results supported metric invariance because (i) ΔCFI = .002 (less than the recommended cutoff of 0.01) and (ii) ΔSRMS = .001 (less than the recommended cutoff of .030; Chen, 2007). Given the large sample size, these indicators were preferred to $\Delta\chi^2$, as recommended by Cheung and Rensvold (2002).

Multigroup structural equation modelling (SEM) was performed to test the hypotheses (see Table 6). The sample estimation and multigroup estimation have acceptable overall fit.[1]

All proposed relationships in the model are significant and have the expected sign. All three estimations (full sample, subsample of women, and subsample of men) indicate that attitudes are positively and significantly related to reported water conservation behavior (RWCB), supporting H1a. Moreover, the association between attitudes and RWCB is greater for women than for men ($\beta_{women} = .76$, $\beta_{men} = .57$, $\Delta\chi^2$ [1] = 27.07, $p < .05$), supporting H1b.

Regarding the antecedents of attitudes, normative motives and attitudes are positively and significantly related, supporting H2a. Unexpectedly, however, the relationship between norms and attitudes is stronger for men than for women ($\beta_{women} = .25$, $\beta_{men} = .44$, $\Delta\chi^2$ [1] = 3.88, $p < .05$), so the results do not support H2a. In contrast, hedonic motives seem to be significantly negatively associated with attitudes (supporting H3a) for all three estimations. In terms of gender, hedonic motives to save water are greater for women than for men ($\beta_{women} = -.79$, $\beta_{men} = -.53$, $\Delta\chi^2$ [1] = 51.80, $p < .05$), supporting H3b.

Regarding perceived situational factors, guests' perceived destination problem awareness is significantly associated with greater normative motives to conserve water, supporting H4a. In

Table 4. Discriminant validity.

	DPA	MN	DA	HM	ATT
Destination problem awareness (DPA)	**.49**	.26	.13	.23	.21
Normative motives (NM)	[.43, .59]	**.46**	.05	.37	.40
Destination attachment (DA)	[.29, .44]	[.14, .29]	**.84**	.01	.01
Hedonic motives (HM)	[−.55, −.42]	[−.68, −.55]	[.06, .06]	**.65**	.55
Attitudes toward water conservation (ATT)	[.39, .52]	[.57, .70]	[.01, .12]	[−.78, −.70]	**.62**

n = 680 individuals. Values on the diagonal correspond to the average variance extracted. Values above the diagonal correspond to the shared variances (squared correlations). Values below the diagonal correspond to the 95% confidence intervals for the estimated factor correlations.

Table 5. Invariance testing.

	χ^2	Robust χ^2	df	$\Delta\chi^2$	Δdf	RMSEA (90% CI)	SRMR	CFI	TLI (NNFI)
Single group (loose cross-validation):									
Women	748.75	689.58	289			.066 [.059, −.072]	.16	.96	.96
Men	751.31	681.35	289			.066 [.059, −.072]	.16	.96	.96
Measurement invariance:									
Equal form (configural invariance)	1500.06	1370.86	578			.066 [.061, −.070]	.16	.96	.96
Equal factor loading (metric invariance)	1593.35	1441.12	604	68.66*	26	.066 [.062, −.070]	.16	.96	.96

*p < .01.

Table 6. Hypothesis testing: multigroup estimation.

		Total sample (1)		Multigroup model (2)						
				Women		Men				
Hyp.	Relationship	Std. load	Robust t	Std. load	Robust t	Std. load	Robust t	$\Delta\chi^2$ (Δdf = 1)	p	Gender effect
H1	ATT -> RWCB	.63*	24.62	.76	26.82*	.57	17.33*	27.07	.01	Difference
H2	NM -> ATT	.34*	10.09	.25	6.96*	.44	11.23*	3.88	.05	Difference
H3	HM -> ATT	−.69*	25.61	−.79	−24.04*	−.53	−17.39*	51.80	.01	Difference
H4	DPA -> NM	.63*	11.17	.70	10.90*	.70	12.38*	1.27	.26	Equality
H5	DA -> HM	−.39*	−13.48	−.49	−13.13*	−.50	−14.66*	1.98	.16	Equality

(1) Total sample (n = 680): SBχ^2(df = 319) = 1690.12; BBNNFI = .91; CFI = .92; RMSEA = .084 [.080, −.088]
R^2(RWCB) = .391; R^2(ATT) = .679; R^2(NM) = .397; R^2(HM) = .152
(2) Multigroup (n = 680): SBχ^2(df = 319) = 2156.76; BBNNFI = .93; CFI = .93; RMSEA = .087 [.082, −.091]
Group 1 (women), Group 2 (men): R^2(RWCB) = .572, .324; R^2(ATT) = .790, .618; R^2(NM) = .490, .493; R^2(HM) = .237, .252

*p < .01.

this case, gender does not seem to play a moderating role (β_{women} = .70, β_{men} = .70, $\Delta\chi^2$ [1] = 1.27, p > .1), so the results do not support H4b. Lastly, guests' destination attachment is negatively and significantly related to hedonic motives for conserving water, supporting H5a. As in the previous case, gender does not seem to play a moderating role (β_{women} = −.49, β_{men} = −.50, $\Delta\chi^2$ [1] = 1.98, p > .1), so the results do not support H5b.

Discussion and conclusions

This study enhances the current understanding of hotel guests' attitudes and reported in-room water conservation behavior by considering moral and hedonic motives, perceived situational factors (destination problem awareness and attachment), and gender. First, the results show that guests' attitudes toward saving water on vacation are crucial to explain guests' reported in-room water conservation behavior. This finding supports the literature on sustainable tourism, according to which attitudes are considered a prerequisite of PEB and the intention to behave pro-environmentally (e.g. Han et al., 2018; Untaru et al., 2016). Furthermore, the relationship between attitudes and reported water conservation is stronger for women than for men, in line with previous studies (Lipchin et al., 2005). This finding suggests that when women have stronger attitudes toward saving water, they are more likely to behave pro-environmentally than men. Consequently, the attitude–behavior gap is smaller for women than for men in a hotel context.

Second, the findings indicate that guests' hedonic motives to conserve water in a hotel when on vacation are stronger antecedents of attitudes than normative motives. This interesting finding implies that the relative importance of a goal in a decision depends not only on its value for the decision maker but also on the context where the decision is made (Steg et al., 2014). In a vacation context, hedonic motives seem to be more salient than moral motives. A possible explanation may be that hedonic motives are less rational and involve people's desire to feel comfortable (Lindenberg & Steg, 2007). In addition to this predominance of hedonic motives, gender also moderates the relationship between motivators and attitudes toward water conservation, as suggested in previous studies (e.g. Lee et al., 2013). Unexpectedly, moral obligation (moral motives) seems to be a weaker predictor of attitudes toward conserving water for women than for men. This result contradicts the findings of other environmental studies (e.g. Joireman et al., 2001; Nordfjaern & Rundmo, 2019) and differs from those of studies in the hotel sector showing no significant gender differences (e.g. Han & Hyun, 2018b). One probable reason for this finding is the study setting. Benidorm is a "sun and sand" mass tourism destination. Tourists usually visit Benidorm to relax and escape from their daily routines. In such a context, tourists' sense of responsibility may be susceptible to temporary "suspension" (Miao & Wei, 2016). For example, Dolnicar et al. (2017) reported that even highly environmentally aware tourists (*eco-centric* tourists) are not willing to behave pro-environmentally when they are on vacation. It is therefore likely that women, who usually bear the brunt of family obligations, experience a greater degree of suspension of obligations (including environmental obligations) than men. Regarding the moderating role of gender on the relationship between hedonic motives and attitudes, the negative impact of hedonic motives on attitudes toward saving water is stronger among women than men. This finding, which is consistent with previous studies (e.g. Jackson et al., 2011; Yang & Lee, 2010), implies that women are more hedonistic water consumers than men. Women are less inclined to sacrifice their own personal comfort for the sake of the environment, probably because of the intrinsically hedonic nature of tourism. This difference is observed despite the absence of gender differences in previous studies in the hotel sector (Han et al., 2018).

Third, concerning the perceived situational factors included in the model (i.e. destination problem awareness and destination attachment), the results suggest that guests' perceptions of a water scarcity problem in the destination are strongly related to guests' moral obligations. These results are in line with previous research on the hotel industry supporting the positive relationship between awareness of the consequences of a specific behavior and moral norms (e.g. Chen & Tung, 2014; Han et al., 2015). The results also imply that the weaker a guest's attachment to a specific destination is, the greater that guest's hedonic motivation regarding water conservation will be. This finding is aligned with previous studies suggesting that tourists with a strong attachment to a destination are more likely to make sacrifices for the environment, even at the expense of immediate self-interest, effort, or cost (Davis et al., 2011).

Finally, the results show no gender differences in the relationships between perceived situational factors and motives. In fact, the results also show no statistical differences between the mean values of DPA and DA for men and women (see Table 2). These findings somewhat contradict past research showing that women are more aware of local environmental problems (e.g. Davidson & Freudenburg, 1996) and develop stronger attachment and affective connections to a place than men (Hidalgo & Hernández, 2001; Rollero & De Picolli, 2010). A possible explanation for this discrepancy may again be associated with the type of destination. Benidorm is not a nature-related destination with fragile social and ecological systems. These conditions are some of the prerequisites for building a strong connection with a place from a pro-environmental perspective (Silva et al., 2013).

In terms of theory, this study makes an important contribution by being one of the few studies to empirically show the superiority of hedonic motives over moral motives in explaining individuals' in-room water conservation behavior. Furthermore, the bulk of framing studies in sustainable tourism have predicted water conservation behavior or intention using only social or cognitive drivers (e.g. Han & Hyun, 2018b; Han et al., 2020). This study also included affective variables (hedonic motives and destination place attachment) to better characterize this behavior. Accordingly, this study is one of the first to explain in-room water conservation behavior using data collected from actual hotel guests while they were staying at the hotel. The data were gathered by interviewing individuals in a real consumption context. This fieldwork approach enabled a reduction in attribution bias with respect to recollections about past consumption because guests' feelings were being experienced at the time of data collection (Haggag et al., 2019). Similarly, this fieldwork approach meant that individuals could be asked about a specific tourist destination (Benidorm). Therefore, the data captured the importance of the destination in tourists' pro-environmental decisions. Additionally, no previous study has focused on analyzing gender differences in relation to water conservation behavior at hotels. Documenting the moderating effect of gender on in-room water conservation decision making fills a sizeable gap in the literature. The findings therefore contribute to research on guests' responsible behaviors and water consumption by providing a fresh perspective for future studies.

Practical implications

These findings have several practical implications. First, the positive relationship between guests' attitudes and behavior highlights the importance of developing strategies that foster tourists' positive attitudes toward the environment. These strategies could be implemented by hotel managers and other stakeholders that would benefit from a more sustainable use of water resources. Such stakeholders include destination management organizations (DMOs), hotel associations, and regional tourism organizations. One of the most widely applied measures consists of communication strategies to encourage positive attitudes toward the environment during a guest's stay. However, this strategy could likewise be applied from the moment at which tourists begin to plan and book their holidays. Many international tourists book their holidays in advance. For example, summer holidays are planned the previous winter. Therefore, these stakeholders have time to develop tourists' pro-environmental attitudes before these tourists actually arrive at their hotels. The results also show that the positive relationship between attitudes and behavior is stronger for women than men but that water conservation activities are actually quite rare among respondents (Mean = 3.99, SD = .82). In fact, previous research has shown that even in the presence of strong pro-environmental attitudes, consumers do not always behave in an environmentally friendly manner (Juvan & Dolnicar, 2014). This finding could be explained by the tradeoffs that often accompany this behavior (e.g. less comfort or convenience). Therefore, hotel managers seeking to decrease in-room water use should intensify their efforts to encourage guests' engagement in water conservation behaviors. They can do so by enhancing the attitudes of guests and by implementing strategies to encourage cooperation during their stay, especially among male customers. Here, nudges might be useful to promote the desired behavior. A nudge consists of any message, idea, or device that influences people to do the right thing without compromising their freedom of choice. Thus, hotel managers should provide clear and unambiguous messages that help tourists consciously adapt their habits and routine behaviors. If hotel managers do not remind tourists of the need to behave sustainably, then guests might simply forget or even neglect to perform this kind of behavior.

The findings also suggest that these strategies should emphasize normative motives (moral obligation) as opposed to hedonic motives because the latter have a negative impact on attitudes toward saving water. Tourists might feel less obliged to behave in an environmentally friendly way if the amount of effort required to do so makes it inconvenient to save water or if it represents more effort than it is worth. Thus, strategies to save water that require guests' collaboration must be as simple and easy to use as possible (e.g. dual flush buttons on toilets). This approach can convince guests that their level of comfort and well-being will not be compromised.

Moreover, given that the effect of normative motives on attitudes is greater for men and that the effect of hedonic motives on attitudes is greater for women, the proposed strategy could actually have a greater effect on men than on women. A more effective strategy for female guests could include offering rewards for conserving water to mitigate the perceived effort and inconvenience associated with saving water when on vacation (hedonic motives).

Given the positive and significant effect of destination problem awareness on normative motives and the negative and significant effect of destination attachment on hedonic motives, it seems reasonable to assume that messages containing information on specific water-related destination problems could increase the effectiveness of communication strategies. Similarly, hotels could try to increase guests' pro-environmental attachment to the destination (destination loyalty), and not only to their facilities (brand loyalty), to indirectly promote in-room water conservation behaviors. Destination problem awareness and destination attachment could also be enhanced by DMOs. These organizations usually focus their communication strategies on perceptions of functional and hedonic dimensions. However, they should also raise tourists' awareness by providing information on the environmental problems in the destination (i.e. water scarcity) and should increase tourists' attachment to the destination. Providing a positive tourist experience is a critical way of strengthening place attachment (Hosany et al., 2017).

Finally, hotel managers could also try to personalize their in-room messages according to their guests' gender. While some rooms might be occupied by both male and female guests, it is very likely that other rooms will be occupied only by men or only by women. Using big data techniques, hotel managers might be able to anticipate these situations to develop different in-room strategies depending on guests' gender. A combination of destination information on water scarcity and messages focusing on the moral obligation to save water could effectively increase men's attitudes toward water conservation and thus men's water conservation behavior (e.g. "Water availability is a serious issue in Benidorm. We trust you to help us by saving as much water as possible during your stay."). For women, it might be more effective to highlight destination attachment and the negative environmental consequences of behaving hedonistically while staying at the hotel (e.g. "We know you love Benidorm, but it needs a little help from its friends. Please save water during your stay.").

Limitations and future research

The current study also has some limitations that should be addressed in future research. First, the data were gathered from a sample of hotel guests in Benidorm and cannot be generalized to all hotel customers or lodging contexts. Future research should validate these results for other types of lodging facilities (e.g. home sharing, vacation rentals, and campsites), other hotel categories (e.g. 1–3-star, luxury, or exclusive hotels), and other destinations with different water scarcity levels (e.g. Nordic countries and the Caribbean). Second, further research is needed to better understand gender differences by, for instance, comparing different tourism destinations with different affective images and personal attachment levels (e.g. natural protected areas, urban destinations, and rural settings). Third, future research should examine the existence of omitted variables that could improve the explanatory power of the proposed model (e.g. attribution of control). Finally, the proposed model is based on guests' self-reported water conservation behavior while staying at a hotel. Although self-reported behavior measures represent a more accurate indicator than, for example, behavioral intentions, it could be of interest to consider using measures of actual in-room water use and objective measures.

Note

1. As per the literature on the limitations of goodness-of-fit indicators (Chen et al., 2008), overall model fit should be assessed by examining several measures of fit. Although the RMSEA and its confidence interval are on the limit of being considered acceptable (McCallum et al., 1996), other indicators such as BBNNFI and CFI are above the acceptable threshold.

Disclosure statement

No potential conflict of interest was reported by the authors.

Funding

This work was partially supported by the Spanish Ministry of Science, Innovation and Universities under research project INTETUR (RTI2018-099467-B-I00) and Emerging Project grant of the University of Alicante (GRE17-15).

ORCID

Ana B. Casado-Díaz http://orcid.org/0000-0002-9960-0078
Franco Sancho-Esper http://orcid.org/0000-0002-6073-1935
Carla Rodriguez-Sanchez http://orcid.org/0000-0002-5801-6801
Ricardo Sellers-Rubio http://orcid.org/0000-0002-9108-1904

References

Ajzen, I. (1985). From intentions to actions: A theory of planned behavior. In J. Kuhl & J. Beckman (Eds.), *Action control: From cognition to behavior* (pp. 11–39). Springer.

Ajzen, I., & Fishbein, M. (1980). *Understanding attitudes and predicting social behaviour*. Prentice-Hall.

Alonso-Almeida, M. M. (2013). Environmental management in tourism: Students' perceptions and managerial practice in restaurants from a gender perspective. *Journal of Cleaner Production*, *60*, 201–207.

Anderson, J. C., & Gerbing, D. W. (1988). Structural equation modeling in practice: A review and recommended two-step approach. *Psychological Bulletin*, *103*(3), 411–423. https://doi.org/10.1037/0033-2909.103.3.411

Balundé, A., Perlaviciute, G., & Steg, L. (2019). The relationship between people's environmental considerations and pro-environmental behavior in Lithuania. *Frontiers in Psychology*, *10*, 2319.

Bamberg, S., & Moser, G. (2007). Twenty years after Hines, Hungerford and Tomera: A new meta-analysis of psycho-social determinants of pro-environmental behavior. *Journal of Environmental Psychology*, *27*(1), 14–25. [Database] https://doi.org/10.1016/j.jenvp.2006.12.002

Briscoe, M. D., Givens, J. E., Hazboun, S. O., & Krannich, R. S. (2019). At home, in public, and in between: Gender differences in public, private and transportation pro-environmental behaviors in the US Intermountain West. *Environmental Sociology*, *5*(4), 374–392.

Brown, B., Perkins, D., & Brown, G. (2003). Place attachment in a revitalizing neighbourhood: Individual and block levels of analysis. *Journal of Environmental Psychology*, *23*(3), 259–271. https://doi.org/10.1016/S0272-4944(02)00117-2

Byrne, B. M. (2013). *Structural equation modeling with EQS: Basic concepts, applications, and programming*. Routledge.

Casares-Blanco, J., Fernández-Aracil, P., & Ortuño-Padilla, A. (2019). Built environment and tourism as road safety determinants in Benidorm (Spain). *European Planning Studies*, *27*(7), 1314–1328.

Chekima, B., Wafa, S. A., Wafa, S. K., Igau, O. A., Chekima, S., & Sondoh, S. L. (2016). Examining green consumerism motivational drivers: Does premium price and demographics matter to green purchasing? *Journal of Cleaner Production, 112*, 3436–3450. https://doi.org/10.1016/j.jclepro.2015.09.102

Chen, F., Curran, P. J., Bollen, K. A., Kirby, J., & Paxton, P. (2008). An empirical evaluation of the use of fixed cutoff points in RMSEA test statistic in structural equation models. *Sociological Methods & Research, 36*(4), 462–494.

Chen, F. F. (2007). Sensitivity of goodness of fit indexes to lack of measurement invariance. *Structural Equation Modeling, 14*(3), 464–504.

Chen, M. F., & Tung, P. J. (2014). Developing an extended theory of planned behavior model to predict consumers' intention to visit green hotels. *International Journal of Hospitality Management, 36*, 221–230. https://doi.org/10.1016/j.ijhm.2013.09.006

Cheung, G. W., & Rensvold, R. B. (2002). Evaluating goodness-of-fit indexes for testing measurement invariance. *Structural Equation Modeling: A Multidisciplinary Journal, 9*(2), 233–255. https://doi.org/10.1207/S15328007SEM0902_5

Dagher, G. K., Itani, O., & Kassar, A. N. (2015). The impact of environment concern and attitude on green purchasing behavior: Gender as the moderator. *Contemporary Management Research, 11*(2), 179–206. https://doi.org/10.7903/cmr.13625

Davidson, D. J., & Freudenburg, W. R. (1996). Gender and environmental risk concerns: A review and analysis of available research. *Environment and Behavior, 28*(3), 302–339. https://doi.org/10.1177/0013916596283003

Davis, J. L., Le, B., & Coy, A. E. (2011). Building a model of commitment to the natural environment to predict ecological behavior and willingness to sacrifice. *Journal of Environmental Psychology, 31*(3), 257–265. https://doi.org/10.1016/j.jenvp.2011.01.004

De Groot, J. I. M., & Steg, L. (2009). Morality and prosocial behavior: The role of awareness, responsibility, and norms in the Norm Activation Model. *The Journal of Social Psychology, 149*(4), 425–449. https://doi.org/10.3200/SOCP.149.4.425-449

Dietz, T., Kalof, L., & Stern, P. C. (2002). Gender, values, and environmentalism. *Social Science Quarterly, 83*(1), 353–364. https://doi.org/10.1111/1540-6237.00088

Dolnicar, S., Cvelbar, L. K., & Grün, B. (2017). Do pro-environmental appeals trigger pro-environmental behavior in hotel guests? *Journal of Travel Research, 56*(8), 988–997. https://doi.org/10.1177/0047287516678089

Dolnicar, S., & Leisch, F. (2008). An investigation of tourists' patterns of obligation to protect the environment. *Journal of Travel Research, 46*, 381–391.

Dolnicar, S., Hurlimann, A., & Grün, B. (2012). Water conservation behavior in Australia. *Journal of Environmental Management, 105*, 44–52. https://doi.org/10.1016/j.jenvman.2012.03.042

Dzialo, L. (2017). The feminization of environmental responsibility: A quantitative, cross-national analysis. *Environmental Sociology, 3*(4), 427–437.

Eagly, A. H., & Chaiken, S. (2007). The advantages of an inclusive definition of attitude. *Social Cognition, 25*(5), 582–602. https://doi.org/10.1521/soco.2007.25.5.582

Feinberg, M., & Willer, R. (2013). The moral roots of environmental attitudes. *Psychological Science, 24*(1), 56–62. https://doi.org/10.1177/0956797612449177

Fornell, C., & Larcker, D. (1981). Structural equation models with unobserved variables and measurement error. *Journal of Marketing Research, 18*, 39–50.

Gabarda-Mallorquí, A., Fraguell, R. M., & Ribas, A. (2018). Exploring environmental awareness and behavior among guests at hotels that apply water-saving measures. Sustainability, *10*(5), 1305. https://doi.org/10.3390/su10051305

Gao, J., Huang, Z., & Zhang, C. (2017). Tourists' perceptions of responsibility: An application of norm-activation theory. *Journal of Sustainable Tourism, 25*(2), 276–291. https://doi.org/10.1080/09669582.2016.1202954

Gatersleben, B., Murtagh, N., & Abrahamse, W. (2014). Values, identity and proenvironmental behaviour. *Contemporary Social Science, 9*(4), 374–392. https://doi.org/10.1080/21582041.2012.682086

Gifford, R., & Nilsson, A. (2014). Personal and social factors that influence pro-environmental concern and behaviour: A review. *International Journal of Psychology: Journal International de Psychologie, 49*(3), 141–157. https://doi.org/10.1002/ijop.12034

Gössling, S., Araña, J. E., & Aguiar-Quintana, J. T. (2019). Towel reuse in hotels: Importance of normative appeal designs. *Tourism Management, 70*, 273–283. https://doi.org/10.1016/j.tourman.2018.08.027

GVA. (2016). *Profile of the tourist who visits the Valencian Community Results 2016* [in Spanish]. https://xurl.es/1or4m

Hadjikakou, M., Miller, G., Chenoweth, J., Druckman, A., & Zoumides, C. (2015). A comprehensive framework for comparing water use intensity across different tourist types. *Journal of Sustainable Tourism, 23*(10), 1445–1467. https://doi.org/10.1080/09669582.2015.1044753

Haggag, K., Pope, D. G., Bryant-Lees, K. B., & Bos, M. W. (2019). Attribution bias in consumer choice. *The Review of Economic Studies, 86*(5), 2136–2183. https://doi.org/10.1093/restud/rdy054

Hair, J., Anderson, R., Tatham, R., & Black, W. (2006). *Multivariate data analysis* (6th ed.). Prentice Hall.

Han, H. (2015). Travelers' pro-environmental behavior in a green lodging context: Converging value-belief-norm theory and the theory of planned behavior. *Tourism Management, 47*, 164–177.

Han, H., Chua, B. L., & Hyun, S. S. (2020). Eliciting customers' waste reduction and water saving behaviors at a hotel. *International Journal of Hospitality Management*, *87*, 102386. https://doi.org/10.1016/j.ijhm.2019.102386

Han, H., Hsu, L. T., Lee, J. S., & Sheu, C. (2011). Are lodging customers ready to go green? An examination of attitudes, demographics, and eco-friendly intentions. *International Journal of Hospitality Management*, *30*(2), 345–355. https://doi.org/10.1016/j.ijhm.2010.07.008

Han, H., Hwang, J., Kim, J., & Jung, H. (2015). Guests' pro-environmental decision-making process: Broadening the norm activation framework in a lodging context. *International Journal of Hospitality Management*, *47*, 96–107. https://doi.org/10.1016/j.ijhm.2015.03.013

Han, H., Lee, M. J., & Kim, W. (2018). Promoting towel reuse behavior in guests: A water conservation management and environmental policy in the hotel industry. *Business Strategy and the Environment*, *27*(8), 1302–1312. https://doi.org/10.1002/bse.2179

Han, H., & Hyun, S. S. (2018a). College youth travelers' eco-purchase behavior and recycling activity while traveling: An examination of gender difference. *Journal of Travel & Tourism Marketing*, *35*(6), 740–754.

Han, H., & Hyun, S. S. (2018b). Eliciting customer green decisions related to water saving at hotels: Impact of customer characteristics. *Journal of Sustainable Tourism*, *26*(8), 1437–1452.

Han, H., & Hyun, S. S. (2018c). What influences water conservation and towel reuse practices of hotel guests? *Tourism Management*, *64*, 87–97.

Han, H., Yu, J., Kim, H. C., & Kim, W. (2018). Impact of social/personal norms and willingness to sacrifice on young vacationers' pro-environmental intentions for waste reduction and recycling. *Journal of Sustainable Tourism*, *26*(12), 2117–2133. https://doi.org/10.1080/09669582.2018.1538229

Hasnain, A., Raza, S. H., & Qureshi, U. S. (2020). The impact of personal and cultural factors on green buying intentions with mediating roles of environmental attitude and eco-labels as well as gender as a moderator. *South Asian Journal of Management Sciences*, *14*(1), 1–27. https://doi.org/10.21621/sajms.2020141.01

Hayes, B. C. (2001). Gender, scientific knowledge, and attitudes toward the environment: A cross-national analysis. *Political Research Quarterly*, *54*(3), 657–671.

Hidalgo, M. C., & Hernández, B. (2001). Place attachment: Conceptual and empirical questions. *Journal of Environmental Psychology*, *21*(3), 273–281. https://doi.org/10.1006/jevp.2001.0221

Hosany, S., Prayag, G., Van Der Veen, R., Huang, S., & Deesilatham, S. (2017). Mediating effects of place attachment and satisfaction on the relationship between tourists' emotions and intention to recommend. *Journal of Travel Research*, *56*(8), 1079–1093. https://doi.org/10.1177/0047287516678088

Huijts, N., Molin, E., Steg, L. (2010, November). *Understanding the public acceptance of hydrogen technologies in transport: A conceptual framework*. 11th TRAIL Congress. http://citeseerx.ist.psu.edu/viewdoc/download?doi=10.1.1.462.597&rep=rep1&type=pdf

INE. (2019). *Hotel occupation national survey* [in Spanish]. Spanish National Institute of Statistics (INE). Madrid. https://www.ine.es/dynt3/inebase/index.htm?padre=238&capsel=238

Ivars-Baidal, J. A., Rodríguez-Sánchez, I., & Vera-Rebollo, J. F. (2013). The evolution of mass tourism destinations: New approaches beyond deterministic models in Benidorm (Spain). *Tourism Management*, *34*, 184–195.

Jackson, V., Stoel, L., & Brantley, A. (2011). Mall attributes and shopping value: Differences by gender and generational cohort. *Journal of Retailing and Consumer Services*, *18*(1), 1–9. https://doi.org/10.1016/j.jretconser.2010.08.002

Joireman, J. A., Lasane, T. P., Bennett, J., Richards, D., & Solaimani, S. (2001). Integrating social value orientation and the consideration of future consequences within the extended norm activation model of proenvironmental behaviour. *The British Journal of Social Psychology*, *40*(Pt 1), 133–155. https://doi.org/10.1348/014466601164731

Jöreskog, K., & Sörbom, D. (1993). *LISREL 8 structural equation modeling with the SIMPLIS command language*. Scientific Software International.

Juvan, E., & Dolnicar, S. (2014). The attitude-behaviour gap in sustainable tourism. *Annals of Tourism Research*, *48*, 76–95. https://doi.org/10.1016/j.annals.2014.05.012

Juvan, E., & Dolnicar, S. (2017). Drivers of pro-environmental tourist behaviours are not universal. *Journal of Cleaner Production*, *166*, 879–890. https://doi.org/10.1016/j.jclepro.2017.08.087

Kang, K. H., Stein, L., Heo, C. Y., & Lee, S. (2012). Consumers' willingness to pay for green initiatives of the hotel industry. *International Journal of Hospitality Management*, *31*(2), 564–572. https://doi.org/10.1016/j.ijhm.2011.08.001

Kiatkawsin, K., & Han, H. (2017). Young travelers' intention to behave pro-environmentally: Merging the value-belief-norm theory and the expectancy theory. *Tourism Management*, *59*, 76–88.

Knight, K. W. (2019). Explaining cross-national variation in the climate change concern gender gap: A research note. *The Social Science Journal*, *56*(4), 627–632.

Kormos, C., & Gifford, R. (2014). The validity of self-report measures of proenvironmental behavior: A meta-analytic review. *Journal of Environmental Psychology*, *40*, 359–371. https://doi.org/10.1016/j.jenvp.2014.09.003

Lee, E., Park, N. K., & Han, J. H. (2013). Gender difference in environmental attitude and behaviors in adoption of energy-efficient lighting at home. *Journal of Sustainable Development*, *6*(9), 36. https://doi.org/10.5539/jsd.v6n9p36

Lindenberg, S., & Steg, L. (2007). Normative, gain and hedonic goal frames guiding environmental behavior. *Journal of Social Issues*, *63*(1), 117–137. https://doi.org/10.1111/j.1540-4560.2007.00499.x

Lipchin, C. D., Antonius, R., Roishmawi, K., Afanah, A., Orthofer, R., & Trottier, J. (2005). Public perceptions and attitudes towards the declining water level of the Dead Sea basin: A multi-cultural analysis. In S. Schoenfeld (Ed.), *Palestinian and Israeli environmental narratives*. Centre for International and Security Studies, York University.

MacCallum, R. C., Browne, M. W., & Sugawara, H. M. (1996). Power analysis and determination of sample size for covariance structure modeling. *Psychological Methods*, *1*(2), 130–149. https://doi.org/10.1037/1082-989X.1.2.130

Manaktola, K., & Jauhari, V. (2007). Exploring consumer attitude and behavior towards green practices in the lodging industry in India. *International Journal of Contemporary Hospitality Management*, *19*(5), 364–377. https://doi.org/10.1108/09596110710757534

Marandu, E. E., Moeti, N., & Joseph, H. (2010). Predicting residential water conservation using the theory of reasoned action. *Journal of Communication*, *1*(2), 87–100. https://doi.org/10.1080/0976691X.2010.11884774

Martínez-Ibarra, E. (2015). Climate, water and tourism: Causes and effects of droughts associated with urban development and tourism in Benidorm (Spain). *International Journal of Biometeorology*, *59*(5), 487–501. https://doi.org/10.1007/s00484-014-0851-3

Meng, F., & Uysal, M. (2008). Effects of gender differences on perceptions of destination attributes, motivations, and travel values: An examination of a nature-based resort destination. *Journal of Sustainable Tourism*, *16*(4), 445–466. https://doi.org/10.1080/09669580802154231

Miao, L., & Wei, W. (2013). Consumers' pro-environmental behavior and the underlying motivations: A comparison between household and hotel settings. *International Journal of Hospitality Management*, *32*, 102–112. https://doi.org/10.1016/j.ijhm.2012.04.008

Miao, L., & Wei, W. (2016). Consumers' pro-environmental behavior and its determinants in the lodging segment. *Journal of Hospitality & Tourism Research*, *40*(3), 319–338.

Mobley, C., & Kilbourne, W. (2013). Gender differences in pro-environmental intentions: A cross-national perspective on the influence of self-enhancement values and views on technology. *Sociological Inquiry*, *83*(2), 310–332. https://doi.org/10.1111/j.1475-682X.2012.00431.x

Morgan, P. (2010). Towards a developmental theory of place attachment. *Journal of Environmental Psychology*, *30*(1), 11–22.

Nordfjaern, T., & Rundmo, T. (2019). Acceptance of disincentives to driving and pro-environmental transport intentions: The role of value structure, environmental beliefs and norm activation. Transportation, *46*(6), 2381–2396.

Nordlund, A. M., & Garvill, J. (2003). Effects of values, problem awareness, and personal norm on willingness to reduce personal car use. *Journal of Environmental Psychology*, *23*(4), 339–347. https://doi.org/10.1016/S0272-4944(03)00037-9

Otto, J. E., & Ritchie, J. B. (1996). The service experience in tourism. *Tourism Management*, *17*(3), 165–174. https://doi.org/10.1016/0261-5177(96)00003-9

Passafaro, P. (2020). Attitudes and tourists' sustainable behavior: An overview of the literature and discussion of some theoretical and methodological issues. *Journal of Travel Research*, *59*(4), 579–601.

Prayag, G., & Lee, C. (2019). Tourist motivation and place attachment: The mediating effects of service interactions with hotel employees. *Journal of Travel & Tourism Marketing*, *36*(1), 90–106.

Putnick, D. L., & Bornstein, M. H. (2016). Measurement invariance conventions and reporting: The state of the art and future directions for psychological research. *Developmental Review*, *41*, 71–90. https://doi.org/10.1016/j.dr.2016.06.004

Ramkissoon, H., & Mavondo, F. T. (2015). The satisfaction–place attachment relationship: Potential mediators and moderators. *Journal of Business Research*, *68*(12), 2593–2602. https://doi.org/10.1016/j.jbusres.2015.05.002

Raymond, C., Brown, G., & Robinson, G. (2011). The influence of place attachment, and moral and normative concerns on the conservation of native vegetation: A test of two behavioural models. *Journal of Environmental Psychology*, *31*(4), 323–335. https://doi.org/10.1016/j.jenvp.2011.08.006

Rico, A., Olcina, J., Baños, C., Garcia, X., & Sauri, D. (2020). Declining water consumption in the hotel industry of mass tourism resorts: Contrasting evidence for Benidorm, Spain. *Current Issues in Tourism*, *23*(6), 770–783. https://doi.org/10.1080/13683500.2019.1589431

Rodriguez-Sanchez, C., Schuitema, G., Claudy, M., & Sancho-Esper, F. (2018). How trust and emotions influence policy acceptance: The case of the Irish water charges. *British Journal of Social Psychology*, *57*(3), 610–629.

Rollero, C., & De Piccoli, N. (2010). Place attachment, identification and environment perception: An empirical study. *Journal of Environmental Psychology*, *30*(2), 198–205. https://doi.org/10.1016/j.jenvp.2009.12.003

Saad, G., & Gill, T. (2000). Application of evolutionary psychology in marketing. *Psychology and Marketing*, *17*(12), 1005–1034. https://doi.org/10.1002/1520-6793(200012)17:12<1005::AID-MAR1>3.0.CO;2-H

Sánchez-Galiano, J. C., Martí-Ciriquián, P., & Fernández-Aracil, P. (2017). Temporary population estimates of mass tourism destinations: The case of Benidorm. *Tourism Management*, *62*, 234–240. https://doi.org/10.1016/j.tourman.2017.04.012

Schwartz, S. H. (1977). Normative influences on altruism. In L. Berkowitz (Ed.), *Advances in experimental social psychology* (Vol. 10, pp. 221–279). Academic Press.

Shen, J., & Saijo, T. (2008). Reexamining the relations between socio-demographic characteristics and individual environmental concern: Evidence from Shanghai data. *Journal of Environmental Psychology, 28*(1), 42–50. https://doi.org/10.1016/j.jenvp.2007.10.003

Silva, C., Kastenholz, E., & Abrantes, J. L. (2013). Place-attachment, destination image and impacts of tourism in mountain destinations. Anatolia, *24*(1), 17–29. https://doi.org/10.1080/13032917.2012.762312

Steenkamp, J. B. E., & Baumgartner, H. (1998). Assessing measurement invariance in cross-national consumer research. *Journal of Consumer Research, 25*(1), 78–90. https://doi.org/10.1086/209528

Steenkamp, J. B. E., & Geyskens, I. (2006). How country characteristics affect the perceived value of web sites. *Journal of Marketing, 70*(3), 136–150. https://doi.org/10.1509/jmkg.70.3.136

Steg, L., Bolderdijk, J. W., Keizer, K., & Perlaviciute, G. (2014). An integrated framework for encouraging pro-environmental behaviour: The role of values, situational factors and goals. *Journal of Environmental Psychology, 38*, 104–115. https://doi.org/10.1016/j.jenvp.2014.01.002

Steg, L., & De Groot, J. (2010). Explaining prosocial intentions: Testing causal relationships in the norm activation model. *The British Journal of Social Psychology, 49*(Pt 4), 725–743. https://doi.org/10.1348/014466609X477745

Steg, L., & Vlek, C. (2009). Encouraging pro-environmental behaviour: An integrative review and research agenda. *Journal of Environmental Psychology, 29*(3), 309–317. https://doi.org/10.1016/j.jenvp.2008.10.004

Stern, P. (2000). Towards a coherent theory of environmentally significant behavior. *Journal of Social Issues, 56*(3), 407–424.

Stern, P., Dietz, T., & Kalof, L. (1993). Value orientations, gender, and environmental concern. *Environment and Behavior, 25*(5), 322–348. https://doi.org/10.1177/0013916593255002

Tindall, D. B., Davies, S., & Mauboules, C. (2003). Activism and conservation behavior in an environmental movement: The contradictory effects of gender. *Society & Natural Resources, 16*(10), 909–932.

Tonge, J., Ryan, M. M., Moore, S. A., & Beckley, L. E. (2015). The effect of place attachment on pro-environment behavioral intentions of visitors to coastal natural area tourist destinations. *Journal of Travel Research, 54*(6), 730–743. https://doi.org/10.1177/0047287514533010

Untaru, E. N., Ispas, A., Candrea, A. N., Luca, M., & Epuran, G. (2016). Predictors of individuals' intention to conserve water in a lodging context: The application of an extended theory of reasoned action. *International Journal of Hospitality Management, 59*, 50–59. https://doi.org/10.1016/j.ijhm.2016.09.001

Vicente-Molina, M. A., Fernández-Sainz, A., & Izagirre-Olaizola, J. (2018). Does gender make a difference in pro-environmental behavior? The case of the Basque Country University students. *Journal of Cleaner Production, 176*, 89–98. https://doi.org/10.1016/j.jclepro.2017.12.079

Wang, W., Wu, J., Wu, M. Y., & Pearce, P. L. (2018). Shaping tourists' green behavior: The hosts' efforts at rural Chinese B&Bs. *Journal of Destination Marketing & Management, 9*, 194–203.

Wiertz, C., & de Ruyter, K. (2007). Beyond the call of duty: Why customers participate in firm-hosted online communities. *Organization Studies, 28*(3), 347–378. https://doi.org/10.1177/0170840607076003

Xiao, C., & McCright, A. M. (2015). Gender differences in environmental concern: Revisiting the institutional trust hypothesis in the USA. *Environment and Behavior, 47*(1), 17–37. https://doi.org/10.1177/0013916513491571

Yang, K., & Lee, H. J. (2010). Gender differences in using mobile data services: Utilitarian and hedonic value approaches. *Journal of Research in Interactive Marketing*, *4*(2), 142–156. https://doi.org/10.1108/17505931011051678

Yuksel, A., Yuksel, F., & Bilim, Y. (2010). Destination attachment: Effects on customer satisfaction and cognitive, affective and conative loyalty. *Tourism Management, 31*(2), 274–284. https://doi.org/10.1016/j.tourman.2009.03.007

Zelezny, L. C., Chua, P. P., & Aldrich, C. (2000). Elaborating on gender differences in environmentalism. *Journal of Social Issues, 56*(3), 443–457. [Database] https://doi.org/10.1111/0022-4537.00177

APPENDIX

Constructs, items and main sources

Construct & items	Sources
REPORTED WATER CONSERVATION BEHAVIOR RWCB1. I've turned off the tap while brushing my teeth. RWCB 2. I've let the water run until it was the right temperature (R). RWCB 3. I've had a shower more than once a day (R). RWCB 4. I've turned off the shower while lathering up. RWCB 5. I've had a long shower when a shorter one would have done (R). RWCB 6. I've flushed the toilet every time I've used it (R). RWCB 7. I've chosen between the small and large buttons when flushing the toilet. RWCB 8. I've used the same towel(s) more than one day. RWCB 9. I've used the same bed sheets more than one day.	Gabarda-Mallorquí et al. (2018); Miao & Wei (2013, 2016)
ATTITUDES TOWARDS WATER CONSERVATION ATT1. For me, saving water in hotels when I'm travelling is: 1 (very bad) - 7 (very good) ATT2. For me, saving water in hotels when I'm travelling is: 1 (Not important) - 7 (very important). ATT3. For me, saving water in hotels when I'm travelling is: 1 (Very unpleasant) - 7 (very pleasant) ATT4. For me, saving water in hotels when I'm travelling is: 1 (Very useless) - 7 (very useful) ATT5. For me, saving water in hotels when I'm travelling is: 1 (Very harmful) - 7 (very beneficial)	Fisbein & Ajzen (1975), Han (2015)
NORMATIVE MOTIVES TO CONSERVE WATER NM1. I feel guilty when I use a lot of water on holiday. NM2. I feel obliged to do things to save water (e.g., reusing towels) when I stay at a hotel. NM3. Because of my own values/principles, I feel that I should use water responsibly when I stay at a hotel, regardless of what other people do.	Gatersleben, Murtagh, & Abrahamse (2014), Han & Hyun (2018c)
HEDONIC MOTIVES TO CONSERVE WATER HM1. It's too much effort to save water during my stay. HM2. The amount of effort required makes it inconvenient to do things to save water. HM3. I have too many other things to do (during my trip) to think about saving water. HM4. Saving water during my stay is more effort than it's worth. HM5. My own comfort is more important to me. HM6. Because of the amount of daily activity during my trip, I often forget saving water. HM7. My lifestyle (the quality of my hotel experience) would change for the worse if I focused on saving water	Miao & Wei (2013, 2016)
DESTINATION PROBLEM AWARENESS DPA1. I'm aware that there have been problems with water availability in Benidorm. DPA2. I believe that Tourism makes Benidorm a heavy consumer of water. DPA3. I feel that tourists' water consumption is a serious problem for Benidorm city.	Martínez-Ibarra (2015), Nordlund & Garvill (2003)
DESTINATION ATTACHMENT DA1. Benidorm is a very special destination to me. DA2. I identify strongly with this destination. DA3. Going on holiday in Benidorm means a lot to me. DA4. I am very attached to this destination. DA5. No other place can provide the same holiday experience as Benidorm. DA6. Benidorm is the best place for what I like to do on holiday. DA7. Going on holiday here is more important to me than going on holiday anywhere else. DA8. I would not change Benidorm for any other destination to do what I do when I'm on holiday here.	Hosany et al. (2017), Prayag & Lee (2019)

An examination of critical determinants of carbon offsetting attitudes: the role of gender

Gregory Denton, Hengxuan Chi and Dogan Gursoy

ABSTRACT
Utilizing two studies, determinants of travelers' carbon offsetting attitudes and the role that gender plays in cognitive appraisal and attitude formation are examined by exploring the interactions between knowledge, credibility, and trust and their relative impacts on behavioral intentions. Findings suggest that objective knowledge, subjective knowledge, trust, and credibility impact attitudes in dramatically different degrees. Females are found to be higher in objective knowledge, credibility of climate science, and carbon offsetting attitudes, whereas males are found to be higher in subjective knowledge. Knowledge is found to be not only an imperfect predictor of carbon offsetting attitudes but can negatively influence attitudes through indirect effects on trust and perceived credibility. Furthermore, travelers who process information heuristically (primarily males) will have significantly different responses to stimuli than travelers who process information systematically (primarily females). Implications for message framing and possible interventions to increase participation levels in voluntary carbon offset programs are discussed.

Introduction

Climate change has been considered a looming global crisis and a threat to the future of humanity for over a generation. In the three decades following the U.N. Framework Convention on Climate Change and the first "World Scientists' Warning to Humanity" (Kendall, 2000) carbon emissions from tourism have remained on an upward trajectory. (Lenzen et al., 2018) Tourism is viewed as a major contributor to climate change (Becken, 2004; Kim et al., 2019) and is receiving increasing attention as a "polluting" industry. Both scholars and environmental organizations have emphasized the importance of changing tourists' behaviors to be more environmentally friendly (Arana et al., 2013; Gössling & Scott, 2018). However, studies have shown that despite pro-environmental beliefs, tourists' actual behaviors suggest a lack of commitment (Juvan & Dolnicar, 2014; Weaver, 2007). Furthermore, gaps exist between tourists' beliefs and their attitudes and behaviors (Arana et al., 2013; Becken, 2004; Choi & Ritchie, 2014; Gössling et al., 2009; Gössling & Scott, 2018; Mair, 2011; Segerstedt & Grote, 2016).

Although the concept of voluntary carbon offsetting (VCO) has been criticized as a form of "buying forgiveness" (Mair, 2011), carbon sequestration has the potential to contribute meaningfully to the tourism industry's objective of reduced net carbon emissions (Chazdon & Brancalion, 2019). A significant proportion of travelers express willingness to offset tourism carbon emissions (Scott et al., 2015) making VCO a potentially effective adjunct to other carbon reduction strategies.

A number of factors have been identified that contribute to low VCO participation rates, including tourists' knowledge about climate change and carbon offsetting (Cohen & Higham, 2011; Segerstedt & Grote, 2016), perceived credibility of climate science (Hartman et al., 2017), and trust in carbon offsetting programs and the companies that offer them (Dhanda & Hartman, 2011; Kim & Kim, 2014). Additional factors identified in the literature include assigning responsibility for resolving climate change issues to others (Gössling, 2009) or the sense that one person's effort will not make a difference (Gössling & Peeters, 2007). However, a factor that is largely missing from studies on sustainable attitudes is the influence of gender in the cognitive processing of environmental information. Although past research has addressed relationships between gender and environment from the perspectives of gender equity (Dankelman & Jansen, 2010) and pro-environmental attitudes (Economou & Halkos, 2020), few studies have considered methods of increasing VCO participation based upon the different cognitive processes of males and females, with largely insignificant or mixed findings (Dolnicar & Leisch, 2008; Juvan & Dolnicar, 2017). Missing from the literature are assessments of how environmental attitudes are influenced differently between genders and how the differences in cognitive processing of information between females and males might lead each gender to change their attitudes differently. This research gap is significant because, if indeed females and males have different psychological attributes and underlying mechanisms influencing their environmental attitudes, unlocking those differences may be the key to finally effecting changes in environmental behaviors.

This study will address the gap in previous research by investigating the moderating role of gender on tourists' attitudes toward addressing climate change (specifically, on attitudes related to offsetting tourism-related carbon emissions) and exploring how cognition and information processing affect attitudes and carbon offsetting decisions. More specifically, this study will examine how the cognitive appraisal process produces different environmental attitudes for females than males as a result of different levels of knowledge, trust in carbon offsetting companies and perceived credibility of climate science. The effects of interactions between those attributes and environmental attitudes will be explored to provide examine whether different information processing strategies influence attitudes in diverging ways. Findings of this study will contribute to the current knowledge by identifying whether gender plays a significant role in determining environmental attitudes and illustrating whether male and female attitudes toward carbon offsetting are affected by different stimuli. Implications for industry are made and gender-specific strategies for improving attitudes toward carbon offsetting are recommended.

Literature review

Theoretical background

The theoretical model of this study is built on Lazarus' (1991) cognitive appraisal theory (CAT), which suggests that an individual's response to a stimulus is determined by emotions generated through a two-stage cognitive appraisal process (Kuo & Wu, 2012). The primary stage involves quick evaluation about a stimulus based on pre-existing knowledge, whereas the secondary stage comprises more deliberate assessments which are strongly influenced by the primary appraisal (Kuo & Wu, 2012). As mental states evolve from psychological appraisals that are unique to each individual, people are likely to have disparate responses to the same situations depending on their past experiences (Choi & Choi, 2019) and the degree to which the situation conforms or conflicts with their personal beliefs (Cai et al., 2018). For an overview of CAT please see Roseman (1984) or Lazarus (1991).

Tourism scholars recognize that both tangible and intangible attributes influence tourists' appraisal of their travel experiences (Bagozzi et al., 1999; Choi & Choi, 2019), thus making CAT an appropriate theoretical framework. CAT has also been applied previously to studying environmental risk perceptions, recognizing that consumers process environmental risks with dual affective and cognitive approaches (Keller et al., 2012). CAT has also been utilized to explain behaviors for complex stimuli in hospitality research, including acceptance of artificial intelligence devices (Gursoy et al., 2019), residents' support toward events based upon their trust in government (Ouyang et al., 2017), and assessment of disruptive restaurant behavior (Cai et al., 2018).

The role of gender on environmental attitudes

The influence of gender on environmentalism has been studied for decades by sociologists (Dankelman & Jansen, 2010; McCright, 2010; Zelezny et al., 2000) and gender is widely considered an important determinant in attitudes toward the environment (McKercher et al., 2011) but it is still considered to have received little attention from scholars (Scannell & Gifford, 2013; Terry, 2009). Several theories have been introduced to explain gender differences in pro-environmental attitudes, including socialization theory, structural theory, value-belief-norm theory, social roles thesis, and the institutional trust hypothesis (Economou & Halkos, 2020; McKercher et al., 2011; Strapko et al., 2016). The result of this stream of research is a combination of robust patterns but inconclusive results that predominantly treat gender as a control variable in studies rather than a primary predictor of environmental attitudes (McCright, 2010).

Early studies identified largely consistent differences between genders in environmental attitudes but could not coalesce around meaningful interpretations of the underlying factors (Bord & O'Connor, 1997; Xiao & McCright, 2012). Females are generally shown to exhibit greater levels of environmental concern and to engage in more environmentally-oriented behavior (Zelezny et al., 2000). However, gender differences in environmental attitudes have been attributed by some researchers to perceived vulnerability and risk tolerance (Bord & O'Connor, 1997; Brody, 1984; Finucane et al., 2000) socioeconomic factors and political ideology (Davidson & Haan, 2012; Economou & Halkos, 2020; Finucane et al., 2000; McCright & Dunlap, 2011; Strapko et al., 2016), or have been found to be insignificant (Dietz et al., 2007; Heath & Gifford, 2006; Strapko et al., 2016; Whitmarsh & O'Neill, 2010). The preponderance of findings are that the environmental attitudes of males and females are significantly different (McKercher et al., 2011; Strapko et al., 2016; Xiao & McCright, 2012; Zelezny et al., 2000) but it is not clear what might be causing this significant difference. Studies argue that males and females are likely to process information differently (McCright, 2010; Scannell & Gifford, 2013), which can result in significant differences in the level of knowledge and the type of knowledge each gender value and retain, in their trust levels and their perception of the credibility, all of which can influence attitudes. Since it is not clear what might be causing these significant differences, it is important to investigate the cause of differences in male and female attitudes toward carbon offsetting.

Psychological barriers to sustainability

Tourists express concern for the environment (Gössling & Scott, 2018) but little interest in changing their travel behaviors (Cohen & Higham, 2011; Mair, 2011), prompting tourism researchers to explore psychological factors as barriers to sustainable behaviors. The role of personal variables such as attitudes, beliefs, norms and values have been documented in past research, but the conclusions regarding the influence of these variables on environmental behavior have not been consistent (Vicente-Molina et al., 2018).

Knowledge

A number of studies have identified tourists' lack of knowledge about tourism's impact on climate change as an impediment to environmental behavior by tourists (Hares et al., 2010; Juvan & Dolnicar, 2017; White et al., 2019). Consumer knowledge has been found to have a strong impact on decision-making across a broad range of environmental behaviors (Milfont et al., 2012; O'Connor et al., 1999), although the relationship between knowledge and environmental behavior has been shown to be inconsistent (Juvan & Dolnicar, 2017; Marquart-Pyatt et al., 2011).

A significant gap has been identified in the carbon offsetting literature, with large numbers of tourists unaware of their carbon emission levels or VCO options. Although some research suggests that addressing the knowledge gap would help stimulate more pro-environmental behaviors (Becken, 2004; Choi & Ritchie, 2014; Gössling et al., 2009; Hares et al., 2010; Segerstedt & Grote, 2016), the link between knowledge and environmental behavior is not clear (Vicente-Molina et al., 2018). Somestudies suggest that there is no significant link between knowledge and environmental behavior (Laroche et al., 2001) or that increased levels of knowledge actually correspond with reduced levels of concern for the environment (McCright, 2010) while others suggest that increasing the levels of knowledge by providing information can make tourists behave more environment friendly (Dolnicar, 2020). Because of these contradictory findings reported in previous studies about the role of knowledge and how knowledge is processed, thus, requires further exploration.

Studies of social cognition have revealed that different types of knowledge and different knowledge processing strategies have widely disparate influences over consumer behavior (Carlson et al., 2009; Ellen, 1994). Studies differentiate between objective knowledge (accurate information about a subject) and subjective knowledge (self-beliefs about one's relative level of knowledge) and have shown that each form of knowledge influences consumer behaviors independently (Carlson et al., 2009; Ellen, 1994). As an example, Malka et al. (2009) found complex interactions between subjective knowledge, credibility of climate science, and beliefs in global warming that suggest that the relationship between knowledge about global warming and concern for the planet is only significant for people who trust scientists (Malka et al., 2009; Milfont et al., 2012).

Researchers have also studied the impact of knowledge processing strategies on consumer behavior. The Heuristic and Systematic Processing Model is a widely-applied theory which identifies two distinct pathways employed to process information. Systematic processing is described as comprehensive and analytical, whereas heuristic processing is more inferential and relatively effortless (Hlee et al., 2018; Meyers-Levy & Maheswaran, 2004). Subjective information has been shown to be processed heuristically and is believed to have weaker persuasive power, whereas objective information is processed systematically and has been found to have more persuasibility (Chaiken et al., 1989). A person who has a higher-level of subjective knowledge may be due to the exposure of subjective information (e.g., other people's opinions regarding climate change). In contrast, a person who has a higher-level of objective knowledge is mainly caused by the processing of objective information (e.g., a precise number of how the level of carbon emission changed). Thus, each form of knowledge is hypothesized to have different impacts on consumer attitudes toward carbon offsetting.

Studies of the interactions between knowledge and trust and between knowledge and credibility of climate science reveal complex interrelationships (Denton et al., 2020). Malka et al. (2009) found that the association between subjective knowledge and concern over the environment was uniformly positive among people who trust scientists, but studies evaluating objective knowledge are limited. The interactions between both subjective and objective knowledge and both company trust and credibility of climate science are expected to be similarly complex, with the strongest correlations occurring between subjective knowledge - heuristic information processing and objective knowledge - systematic information processing.

It is hypothesized that interactions between knowledge type and information processing will result in differences in tourists' attitudes toward carbon offsetting. Tourists higher in objective knowledge who employ systematic information processing strategies are more likely to form opinions based upon scientific research and academic sources because they gather information comprehensively and employ more careful reasoning by virtue of their systematic processing. These tourists are expected to assess facts rather than relying upon opinions or conjecture, which would lead them to place greater credibility in climate science. Systematic information processors are also more likely to be aware of past "greenwashing" transgressions and other deceptive or irresponsible corporate behaviors and therefore be less inclined to place trust in carbon offset providers.

H1: Higher levels of objective knowledge will correspond with lower levels of trust in carbon offsetting companies

H2: Higher levels of objective knowledge will correspond with higher levels of credibility of climate science

Conversely, tourists who have higher levels of subjective knowledge are more likely to utilize heuristic information processing strategies because of their self-assessed prior knowledge (Meyers-Levy & Maheswaran, 1991). This processing strategy is more prone to superficial assessments, including reliance upon unsubstantiated claims, non-scientific opinions, and media sources that may have pro-corporate bias (Dietz et al., 2007; Meyers-Levy & Maheswaran, 2004). These heuristic information processors are hypothesized to be influenced to a greater degree by the opinions of others (including climate-deniers) as well as advertisements and company claims about environmentalism. This would lead heuristic processors to place greater trust in companies while also questioning the credibility of climate science due to the influence of climate deniers and other non-scientific opinion-makers.

H3: Higher levels of subjective knowledge will correspond with higher levels of trust in carbon offsetting companies

H4: Higher levels of subjective knowledge will correspond with lower levels of credibility of climate science

Gender differences in knowledge

Despite a significant body of literature indicating that science and math education facilitates gender inequality in favor of males (McCright, 2010), Scannell and Gifford (2013) identified that women are more likely than men to gather information on global warming to make their assessments. Males indicate higher levels of self-assessed knowledge compared to females (Bord & O'Connor, 1997), but females actually possess higher levels of actual knowledge than males (McCright, 2010). The propensity for females to underestimate their knowledge (and for males to overestimate it) is indicative of gender differences in subjective and objective knowledge.

H5: Males and females have significantly different levels of subjective knowledge regarding climate change and carbon offsetting

H6: Males and females have significantly different levels of objective knowledge regarding climate change and carbon offsetting

Trust in company

In this study, trust in company refers to trust in tourism operators who offer carbon-offsetting programs. Although greater knowledge is considered necessary to increasing VCO participation, knowledge by itself is not considered sufficient to engender needed changes (Milfont et al., 2012). Trust, or the accepting of vulnerability based upon positive expectations of the intentions or behavior of another (Rousseau et al., 1998), has also been shown to determine customers' willingness to engage in sustainability (Ponnapureddy et al., 2017; Rahman et al., 2015). Trust involves consumers not just placing trust in a single actor but in an industry or group of actors (Doney & Cannon, 1997; Nunkoo & Gursoy, 2016), and trusting both their benevolence and their performance (Doney & Cannon, 1997; Lee et al., 2014). Trust between the stakeholders of a potential exchange has also been shown to influence decision-making outcomes (Blau, 1964; Ouyang et al., 2017), particularly in situations with high levels of uncertainty or incomplete knowledge (Dietz et al., 2007). Perceived untrustworthiness of companies has been shown to reduce customers' perception of control and lower their cooperation intentions (Seetanah & Sannassee, 2015).Trust is considered particularly salient to potential exchanges involving hospitality companies due to past instances of "greenwashing" (Ponnapureddy et al., 2017; Rahman et al., 2015) undermining trust in companies' motives for environmental initiatives.

H7: Trust in companies is positively correlated with attitudes toward carbon offsetting

Gender differences in trust

Studies have identified a strong correlation between company trust and risk perceptions, but also point to decisions to "trust the experts" often being made heuristically rather than systematically (Trumbo & McComas, 2003). Women have been found to be more critical of reported science and technology findings and, thus, less willing to accept the accuracy of those findings based on their face value compared to men (Jacobs & Simpkins, 2006; McCright, 2010). Women are more likely to form their opinions after systematic processing of the information while men are more likely to utilize heuristic processing and to make assessments based on a small number of heuristic cues. Thus, this study hypothesizes that there is a gender difference in levels of trust placed in carbon offsetting companies, with females placing lower levels of trust than males due to their reliance upon systematic information processing rather than heuristic assessments of company claims.

H8: Males and females have significantly different levels of trust in companies engaged in carbon offsetting

Credibility of climate science

Credibility of climate science affects how customers process information and their predisposition to act upon the information (Arora et al., 2006; Kim & Kim, 2014). Tourists considering engaging in carbon offsetting must also make an assessment of the overall credibility of the climate science relating to carbon emissions and potential global warming. Climate science is a controversial and heavily-debated topic (Gössling et al., 2015; McCright et al., 2013) and one where the consumers are compelled to rely upon the expertise of others due to the complexity and ambiguity of the information involved (Hartman et al., 2017). Compounding the issue for the past two decades have been disparate claims made by groups either espousing the risks of climate change or claiming the concerns are overblown, with each group espousing their own versions of scientific evidence (McCright et al., 2013).

Identified gaps in the perceived credibility of climate science (Gössling et al., 2015; Kim & Kim, 2014) are particularly relevant in the context of tourists' disparate knowledge levels and information processing strategies – objective, systematic information processors are hypothesized to rely more upon scientific sources whereas subjective, heuristic information processors are more likely to be influenced by non-scientific opinion leaders.

H9: Credibility of climate science is positively correlated with attitudes toward carbon offsetting

Gender differences in credibility of climate science

Studies have established that women are more likely than men to believe in global warming (Bord & O'Connor, 1997; Joireman & Liu, 2014; Malka et al., 2009; McCright, 2010; Tranter, 2011),

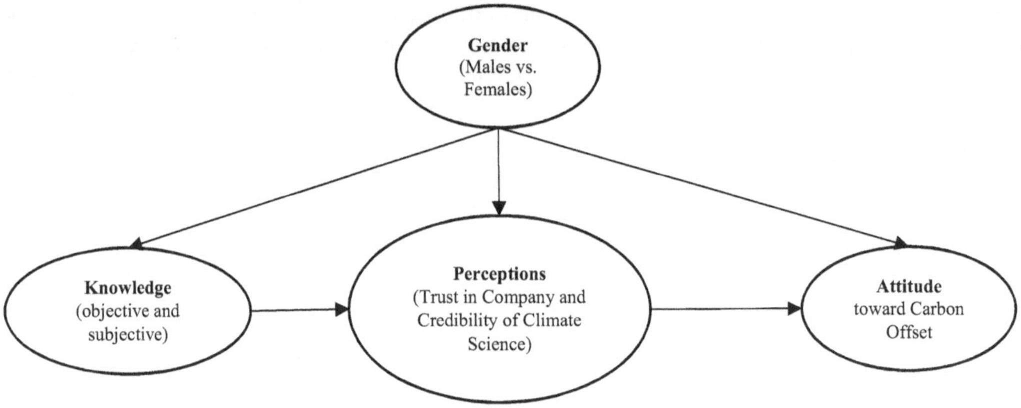

Figure 1. Proposed conceptual model for cross-gender difference.

and that men are more likely to be climate deniers (Dunlap & McCright, 2011; Norgaard, 2011). Thus, this study hypothesizes that the differences in credibility of climate science are at least partially attributed to gender differences in information collection and processing. The well-established disjuncture between the views expressed by members of the national policy elite and the scientific consensus (Dietz et al., 2007; McCright & Dunlap, 2000) will lead to different outcomes for systematic versus heuristic information processors.

H10: Males and females have significantly different levels of credibility of climate science

Gender differences in attitudes toward carbon offsetting

Although few studies have compared attitudes toward carbon offsetting based upon gender, a significant body of literature has established that females have more environmental concern than men (Gifford et al., 1982) and are more likely to engage in behaviors to mitigate global warming (Bord & O'Connor, 1997; Dietz et al., 2007; Meier & Christen, 2012; O'Connor et al., 1999; Park & Vedlitz, 2013) and support climate change mitigation policies (Maibach et al., 2011; McCright et al., 2013). Thus, this study hypothesizes that the gender gap in attitudes toward carbon offsetting will mirror those of other environmental issues, which would be consistent with the cognitive appraisal process and the other hypotheses contained herein.

H11: Males and females have significantly different overall attitudes toward carbon offsetting

Based on the preceding discussion, the proposed conceptual model presented in Figure 1 is developed. The model suggests that individuals' level of objective and subjective knowledge is critical determinants of their trust in companies and their perceptions of credibility of climate science, which in turn influence their attitudes toward carbon offsetting.

Study 1: Examining the proposed conceptual model

Methodology

Data collection

The purpose of Study 1 was to validate the proposed cognitive appraisal model. First, 93 college students (47 female, 46 male) from a university located in North America were recruited to take part in the pilot study. The students were first asked to read descriptions of various carbon offsetting programs, and were then asked to answer a number of survey questions. Once they had completed the survey questions, they were asked to rate how clear the item was to ensure the survey was comprehensible. After refining the questionnaire based on the results the pilot study,

Study 1 recruited a customer panel via Amazon Mechanical Turk (Mturk). These customers (231 females, 231 males) were all from the United States. Participants who completed the survey and passed the attention checks received a small amount of monetary incentive ($0.50).

Measurement instruments

As presented in Appendix A, *subjective knowledge* was measured by a three-item scale developed by Ellen (1994). Three items that were adopted from studies by Lang and Hallman (2005) were used to measure *trust in company. Perceived credibility of climate science* was measured by six items that were adopted from Hartman et al. (2017). *Attitudes toward carbon offsetting* were measured using the four-item scale developed by Hsu and Huang. (Appendix A). All items were measured on a 5-point Likert scale. Respondents' *objective knowledge toward carbon offsetting* was measured utilizing an eight-question quiz, following the approach used in previous studies (Huy Tuu et al., 2011; Pieniak et al., 2010). This method was used by several previous studies and is believed to be able to accurately measure objective knowledge (Huy Tuu et al., 2011; Pieniak et al., 2010). The quiz score was standardized for the data analysis and the reliability was fixed at eighty-five percent to assume fifteen percent of residual variance in objective knowledge (Huy Tuu et al., 2011; Jöreskog & Sörbom, 1996).

Results

Pilot study

In the pilot study, a confirmatory factor analysis was performed to validate the properties of the measurement instrument. In this analysis, common method bias, factor loadings, Cronbach's Alphas, convergent validities, and discriminant validities were examined. In addition, the score of understandability for each measurement item was evaluated. Based on these results, the measurement instruments were refined. Results of the pilot study indicated that all the items were understandable and had desired measurement property. Based on these results, all the survey items retained.

Main study

In the main study, the demographic profile (Appendix B) of respondents was examined, and the data normality was investigated. A confirmatory factor analysis was then performed to validate the measurement model. Finally, the validity of the structural model was assessed through a covariance-based structural equation modeling (CB-SEM) with a maximum likelihood estimation. CB-SEM aims to estimate the difference between the observed covariance matrix and the estimated covariance matrix based on the theoretical model by providing useful model fit indexes (Hair et al., 2011). Furthermore, compared to other SEM approaches, the CB-SEM is considered to be more appropriate for model testing (Hair et al., 2017) (Table 1).

Before conducting the CFA analysis, researchers examined the distribution of the dataset for skewness and kurtosis. Afterwards, a Harman's single-factor test was conducted by loading all items into a single exploratory factor. Results of the Harman's single-factor test indicated that 41.05% of the total variance was explained by a single factor, suggesting that common method bias was not an issue. The results of CFA (Table 2) demonstrated that factor loadings of measurement items were all greater than 0.60, factors' Cronbach's Alphas were all greater than 0.80, and AVEs were all higher than 0.50. In addition, the factors' squared roots AVEs were all greater than their associated factor correlations (Table 3). These provided satisfactory evidence for the reliability and both convergent and discriminant validities of the measurement scale. Furthermore, model fit indices ($\chi^2 = 199.30$, df = 110, χ^2 to df ratio = 1.91, CFI = 0.99, TLI = 0.98, RMSEA = 0.04, SRMR = 0.03) point to a well-fitted measurement model.

Table 1. Demographic profile.

	Demographic distribution % (n = 462)
Age	
18-25	14.7
26-34	31.2
35-44	22.9
44-54	15.1
55-64	12.1
65 or over	3.9
Marital Status	
Single	37.2
Married	45.6
Live together	8.1
Divorced	0.7
Widowed	8.5
Occupation	
Student	6.7
Professional	49.4
Managerial	14.1
Sales	9.5
Homemaker	7.4
Other	13.0
Education	
Less than high school	0.7
High school graduate	6.9
College Degree	75.7
Professional degree	14.3
Doctorate degree	2.3
Annual Income	
Under $10,000	5.8
$10,000-$29,999	18.0
$30,000-$49,999	22.3
$50,000-$69,999	22.1
$70,000 – $99,999	15.6
$100,000 and above	16.2

Table 2. Measurement scale properties.

Constructs/items	Item loadings	Cronbach's alphas	AVE
Subjective Knowledge (SK)		0.89	0.74
SK1	0.88		
SK2	0.84		
SK3	0.86		
Trust in Company (TC)		0.89	0.72
TC1	0.84		
TC2	0.85		
TC3	0.86		
Credibility (CR)		0.96	0.79
CR1	0.91		
CR2	0.86		
CR3	0.88		
CR4	0.92		
CR5	0.89		
CR6	0.86		
Attitude (A)		0.93	0.78
A1	0.93		
A2	0.86		
A3	0.89		
A4	0.85		
Objective Knowledge (OK)		0.85	0.85
OK1	0.92		

Notes: $\chi^2 = 199.30$, $df = 110$, CFI = 0.99, TLI = 0.98, RMSEA = 0.04, SRMR = 0.03.

Table 3. Correlations between Factor and Square Root of AVEs.

Factors	SK	TC	CR	A	OK
Subjective Knowledge (SK)	(0.860)				
Trust in company (TC)	**0.282**	(0.849)			
Credibility (CR)	**-0.115**	**-0.423**	(0.889)		
Attitude toward carbon offset (A)	**-0.174**	−0.093	**0.535**	(0.883)	
Objective knowledge (OK)	**0.218**	**-0.140**	**0.476**	**0.544**	(0.922)

Notes: Numbers in parentheses are square roots of AVE; Correlations in bold are significant at $p < 0.5$ level.

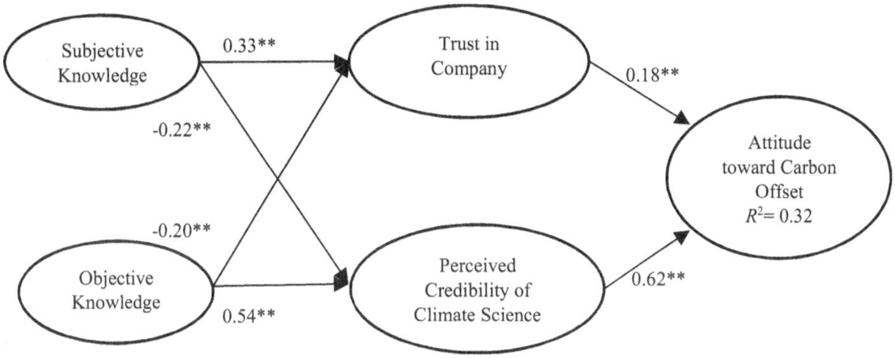

Figure 2. Results of structural equation modeling analysis.
Notes:
Model-fit: $\chi^2 = 261.27$, $df = 112$, CFI = 0.98, TLI = 0.97, RMSEA = 0.05, SRMR = 0.06.
**: significant at the 0.05 level

The results of SEM analysis (Figure 2) revealed a well-fitted structural model (χ^2= 261.27, df = 112, χ^2 to df ratio = 2.33, CFI = 0.98, TLI = 0.97, RMSEA = 0.05, SRMR = 0.06), suggesting that the proposed cognitive appraisal model can explain a significant proportion of attitudes toward carbon offsetting (R^2= .32, p < .001). Furthermore, the results indicated that objective knowledge negatively predicts trust (β= −.20, p <.001) and positively predicts credibility of climate science (β= .54, p <.001). Thus, H1 and H2 were supported. Conversely, subjective knowledge positively predicts trust (β= .33, p <.001) and negatively predicts credibility of climate science (β= −.22, p <.001). Therefore, H3 and H4 were supported. In addition, the results suggested that both trust (β= .18, p <.001) and credibility (β= .62, p <.001) have a significant impact on attitude, supporting H7 and H9.

Discussion of study 1 findings

Study 1 confirms that subjective and objective knowledge have significant but contradictory impacts on attitudes toward carbon offsetting, and further validates the different roles of trust and credibility in the cognitive appraisal process. Trust has a positive mediation effect on the relationship between subjective knowledge and attitudes, whereby the attitudes of tourists with high subjective knowledge are enhanced by the higher levels of trust that they place in carbon offsetting companies, but lower levels of credibility of climate science more than offset the effect of trust, resulting in lower overall attitudes toward carbon offsetting as levels of subjective knowledge increase. For tourists high in objective knowledge the effects are significantly different, with reduced levels of company trust being more than offset by increasing credibility of climate science, resulting in overall higher attitudes toward carbon offsetting.

The reported split-group effect between subjective knowledge and objective knowledge is a significant finding. By better understanding the effects of knowledge and information processing strategies, researchers and industry practitioners can devise messaging strategies (e.g., messages with different levels of objective or subjective information) to more effectively influence attitudes and avoid messaging that inadvertently lowers attitudes through unintended conditional interactions.

Study 2: Exploring gender differences

Methodology

Study 2 replicated the methodology of Study 1 by presenting questions to a panel of respondents recruited from Amazon MTurk in order assess male and female respondents' objective knowledge, subjective knowledge, trust, credibility of climate science, and attitude. Objective knowledge about carbon emissions was measured in Study 2 using nine questions that were taken from the U.S. Department of Energy Climate IQ Quiz (https://www.energy.gov/articles/quiz-test-your-climate-change-iq). A tenth question was dropped from the IQ Quiz because it was pertained to per capita carbon emission levels between U.S. states and was considered unrelated to the research topic. A series of independent sample t-tests was performed to investigate whether there were any significant differences in objective knowledge, subjective knowledge, trust, credibility of climate science, and attitude between male and female respondents.

Results

Four hundred and ninety participants took part in Study 2 (245 female, 245 male) by completing the survey and passing attention checks. The demographic statistics (Appendix B) suggest there was no significant deviation in terms of age, marital status, occupation, education, and family income between genders. Similar to the approach used in Study 1, data normality, measurement model properties including factor loadings, AVEs and Cronbach's Alphas, and overall measurement model fit ($\chi^2 = 162.33$, df = 110, χ^2 to df ratio = 1.48, CFI = 0.99, TLI = 0.98, RMSEA = 0.03, SRMR = 0.03) were examined to ensure the reliability of the data. All of the model fit indicators suggested that the measurement model was appropriate.

T-tests

The results of t-tests (Table 4, Figure 3) demonstrated statistically significant differences in factor mean scores across gender groups. Study 2 found that the factor mean of subjective knowledge was higher for males ($\bar{x}_m = 3.47$, $\bar{x}_f = 3.24$, $t = 2.89$, $p = .004$) while the factor mean of objective knowledge was significantly higher for females ($\bar{x}_f = 2.90$, $\bar{x}_m = 2.71$, $t = -2.68$, $p = .008$). Females also achieved significantly higher scores in perceived credibility of climate science ($\bar{x}_f = 3.52$, $\bar{x}_m = 3.21$, $t = -2.94$, $p = .003$), and attitudes toward carbon offsetting ($\bar{x}_f = 3.94$, $\bar{x}_m = 3.59$, $t = -4.61$, $p < .001$). Therefore, support was provided for hypotheses H5, H6, H10, and H11. Interestingly, no significant main effect of gender was found on trust ($\bar{x}_f = 3.22$, $\bar{x}_m = 3.16$, $t = -0.65$, $p = .52$), thus failing to support hypothesis H8. These findings were confirmed by tests of the pilot study and Study 1 datasets, which resulted in findings that were consistent using different samples and different measures of objective knowledge.

The non-significant findings of the main effect in trust across gender was unexpected. To further explore the differences in trust across genders, this study examined the moderation effect of gender in the relationship between trust, credibility and attitude using floodlight analysis.

Table 4. Results of Independent Samples T-tests.

	Pilot Study (n = 93) Student Sample				**Study 1 (n = 462)** Online Sample				**Study 2 (n = 490)** Online Sample				
Factors	$\bar{X}_{male}$	$\bar{X}_{female}$	*t*	*p*	$\bar{X}_{male}$	$\bar{X}_{female}$	*t*	*p*	$\bar{X}_{male}$	$\bar{X}_{female}$	*t*	*p*	Hypothesis
SK	3.36	2.94	2.07	**.04****	3.20	3.02	1.77	**.08***	3.47	3.24	2.89	**.004****	**H7**
OK	3.20	3.56	−2.64	**.01****	3.70	3.97	−3.96	**<.001****	2.71	2.90	−2.68	**.008****	**H8**
TC	2.40	2.66	−1.58	.12	2.67	2.65	0.27	.79	3.16	3.22	−0.65	.52	H9
CR	2.67	3.01	−2.29	**.03****	3.25	3.63	−3.30	**.001****	3.21	3.52	−2.94	**.003****	**H10**
A	3.74	4.11	−2.05	**.04****	3.78	4.05	−2.94	**.003****	3.59	3.94	−4.61	**<.001****	**H11**

Notes:.
1. Objective Knowledge (OK), Subjective Knowledge (SK), Trust in Company (TC), Perceived Credibility of Climate Science (CR), Attitude toward carbon offset (A).
2. Objective knowledge was tested using two different quiz across study 1 and 2.
3. *: *t* statistic is significant at the 0.1 level.
4. **: *t* statistic is significant at the 0.05 level.

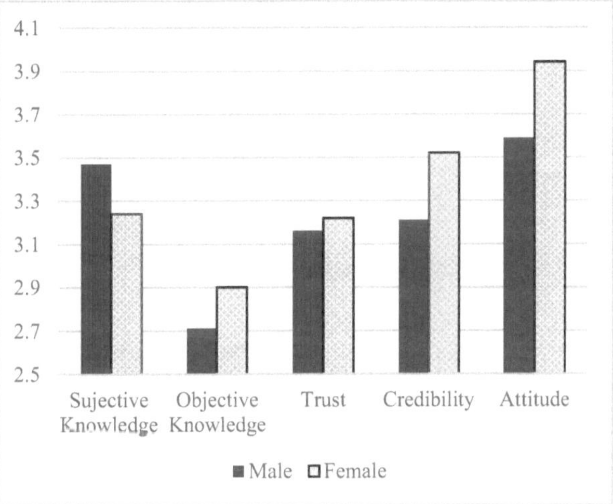

Figure 3. Mean differences across genders. PLEASE CORRECT SPELLING OF 'SUBJECTIVE' IN FIGURE 3...THE PROGRAM DOES NOT ALLOW ME TO EDIT THE FIGURE.....

Moderation test

The results (Figure 4) of the moderation analyses provided evidence of an interaction between gender and trust and the outcome variable of attitude such that the influence of trust on attitudes was lower for females at lower levels of trust (*b*= −0.20, *p*= .005, 95% CI (-0.33 to −0.06)). This creates a suppression effect between trust and attitude for females which was statistically significant when trust scores were less than or equal to 3.98. Among respondents with low levels of initial trust, an increased level of trust predicted significantly greater increases in attitudes for males than for females, whereas the effect of increases in trust was not different for males and females with high levels of initial trust.

Indirect effect

To further examine the moderating effects of gender differences, this study conducted an indirect effect test using 10,000 bootstrapped samples (Hayes, 2018). The results indicated that there was a significant negative indirect effect of gender on attitude through subjective knowledge and trust (Gender ➔ SK➔ TC➔ A; 95% CI (-0.037 to −0.005)) and a significant positive indirect effect of gender on attitude through objective knowledge and credibility (Gender ➔ OK➔ CR➔ A; 95% CI (0.001 to 0.015)). This indicates that female attitudes may be increased by the

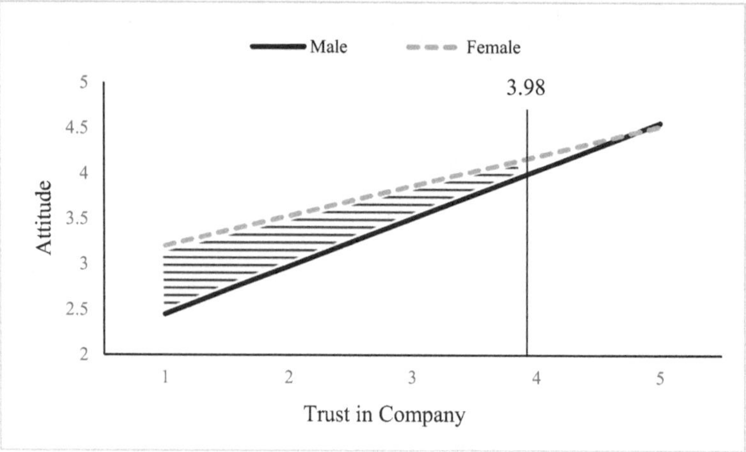

Figure 4. Results of floodlight analysis.
Note. Gender difference is significant ($p < .05$) within the shaded (Johnson-Neyman) regions.

pathway from objective knowledge to perceived credibility of climate science whereas male attitudes are decreased indirectly via the negative influence of subjective knowledge on trust.

Gender differences in environmental attitudes

Study two provides strong evidence of gender differences not only in environmental attitudes but in the psychological attributes that contribute to the formulation of environmental attitudes. Females show higher levels of objective knowledge, which is consistent with systematic information processing. Objective knowledge is associated with higher levels of credibility of climate science and lower levels of trust in carbon offsetting companies, which may be attributed to more comprehensive information gathering and reliance upon more scientific information sources. By contrast, males show higher levels of subjective knowledge which is consistent with heuristic information processing. Subjective knowledge predicts higher levels of trust in carbon offsetting companies but lower levels of perceived credibility of climate science.

Tracing knowledge levels through the cognitive appraisal model presented in Study 1 shows that an increased level of objective knowledge corresponds with increases in attitudes toward carbon offsetting, although the increases are dampened by the negative indirect effect of reduced trust. Conversely, an increased level of subjective knowledge corresponds with lower levels of attitude toward carbon offsetting as a result of the negative indirect effect of lower perceived credibility of climate science overshadowing the positive impact of higher trust.

General discussion

This study contributes to the literature by supporting assertions that gender should be treated as an explanatory variable of environmental attitudes rather than a control variable (McKercher et al., 2011; Zelezny et al., 2000). This is one of the first studies that reveals that the interactions between subjective knowledge and objective knowledge are complex and are different between males and females. Findings further suggest an interaction effect between subjective and objective knowledge on individuals' attitudes towards carbon offsetting. However, the interactions between knowledge are mediated by trust in carbon offsetting companies and credibility of climate science, with opposing consequences on attitudes toward carbon offsetting. Subjective knowledge is shown to have a positive influence on trust but a negative influence on credibility of climate science, suggesting that people who consider themselves to be knowledgeable about the environment are more likely to employ superficial heuristic information processing strategies which leads them to rely upon non-scientific information, opinions, and company appeals. This lower perceived credibility of climate science has a stronger negative impact on attitudes that more than offsets the positive impacts from increased knowledge and trust, suggesting that measures intended to increase knowledge may have a suppressing effect on attitudes toward carbon offsetting.

Conversely, those who are high in objective knowledge and employ systematic information processing strategies show a positive correlation between knowledge and credibility of climate science but a negative impact on trust in carbon offsetting companies. This suggests that tourists who employ comprehensive systematic information processing will rely more upon scientific information sources that increase their attitudes toward climate science but also act to lower their trust in the companies providing carbon offsetting. The negative indirect effect of objective knowledge on attitudes through trust partially attenuates but does not outweigh the combined positive direct effect of knowledge on attitudes and the indirect effect through credibility of climate science.

This study evaluates the role that gender plays in influencing tourists' attitudes toward carbon offsetting as a result of different cognitive processing, building upon the existing body of literature addressing gender-specific roles in environmental management (Dankelman & Jansen, 2010). This study shows that males are higher in subjective knowledge whereas females are higher in objective knowledge, credibility, and attitudes toward carbon offsetting. This finding suggests that females are significantly more likely to follow systematic information processing strategies whereas males are more likely to form attitudes using heuristic information processing. This suggests that a single strategy for modifying consumer behaviors may have positive effects on one gender but negative effects on the other through the diverging impacts of knowledge on attitudes through the prior influence of knowledge on trust and credibility.

Support is provided by this study for a moderating influence of gender on trust. Among respondents with low levels of initial trust, increases in trust predicted significantly greater increases in attitudes of males than of females, whereas the effect of increases in trust was not different for males and females with high levels of initial trust. These findings suggest that trust in company exhibits different influences on attitude across different genders. These results further suggest that increasing trust levels, especially increasing trust level of males, can have a magnifying effect on the impact of knowledge on attitudes toward carbon offsetting. This finding is especially important for enhancing the effect of objective knowledge on attitudes since low trust level was found to dampen the effects of objective knowledge.

The findings of this study have several implications for creating environmental appeals to change tourist behaviors. Understanding the factors that contribute to environmental attitudes is essential before policymakers can design and apply policies intended to change behaviors (Economou & Halkos, 2020). To that end, understanding how cognitive appraisals differ based upon knowledge and gender will help practitioners frame messages so as to achieve desired outcomes and avoid unintended consequences.

Appeals that focus on subjective knowledge would likely be more effective for males than for females due to the congruence with males' heuristic information processing strategies. These appeals could employ opinion leaders and superficial or limited hard information and should focus on presenting carbon offsetting companies as trustworthy to benefit from the positive correlation between subjective knowledge and company trust. These appeals would minimize scientific discussion due to the negative correlation between subjective knowledge and credibility of climate science.

Appeals that focus on objective knowledge would likely be more effective for females employing systematic information processing strategies. These appeals could include more facts and data; however, recognizing the inverse relationship between objective knowledge and company trust, these appeals should de-emphasize the carbon offsetting companies and focus the tourists' attention on data and science.

Conclusion, limitations and future directions

This study investigates determinants of tourists' carbon offsetting attitudes and the role that gender plays in cognitive appraisal and attitude formation. Results suggest that objective knowledge, subjective knowledge, trust, and credibility impact attitudes in dramatically different degrees. Females are found to be higher in objective knowledge, credibility of climate science, and carbon offsetting attitudes, whereas males are found to be higher in subjective knowledge. Knowledge by itself is found to be not only insufficient in predicting carbon offsetting attitudes but can actually reduce attitudes through indirect effects on trust and perceived credibility.

Although this study attempted to improve validity of the findings by employing multiple independent samples, it is not free of limitations. Relying on survey data limits the generalizability of the findings and precludes conclusions of causality. Experimental studies should be conducted under controlled conditions to evaluate the impact of these, and other, psychological attributes on tourists' carbon offsetting attitudes. Furthermore, studies that test the impact of knowledge manipulations on actual carbon offsetting behaviors would provide valuable additional insights.

This study confirms that the relationships between psychological attributes are complex and that cognitive appraisals differ for females and males, but much work remains to be done to identify additional variables and interactions between those variables, and to evaluate how those variables and interactions differ between males and females. A number of scientific fields and theories each contribute hypotheses regarding the causes of gender differences, including evolutionary psychology (Buss & Schmitt, 1993), sociocultural theory (Eagly & Wood, 1999) gender schema/cognitive social theory (Bussey & Bandura, 1999; Hyde, 2014), and biology (Penn, 2003); however, methods of applying these theories to produce tangible change in environmental attitudes and behaviors remains to be fully explored.

To the extent that cognitive appraisal contributes to the formulation of attitudes, it also needs to be explored whether these disparate information processing strategies can be effectively stimulated through message framing and "nudging" (Thaler & Sunstein, 2008; Tversky & Kahneman, 1981). This study suggests that environmental appeal messages that are effective with heuristic males may be counterproductive with systematic females by triggering reductions in trust of carbon offsetting companies, whereas messages that are effective for females might reduce attitudes among males through reduced credibility of climate science – these and other complex interactions should be substantiated by more in-depth experiments.

It also should be noted that this study focuses on explicit attitudes as self-reported by respondents. An interesting field of research developing over recent years explores the divergence between explicit and implicit attitudes and the potential for greater predictability of behaviors through assessments of implicit attitudes (see Beattie & McGuire, 2020). Furthermore, the potential exists for conditional interactions between explicit and implicit attitudes, with compounding effects when the two attitudes are in alignment and suppressing effects when they diverge. Although this topic is beyond the scope of this study, a much more robust understanding of the attitude-behavior gap may be possible through studies that jointly assess explicit and implicit attitudes.

Lastly, this study aims to investigate the cross-gender differences for general U.S. tourists, and respondents showed disproportionately high education levels. Therefore, the observations and conclusions from this study may not generalize to other cultures or samples with different education backgrounds. Additional studies should be conducted to determine whether information processing mechanisms operate similarly among different sample groups across cultural and socioeconomic factors.

Disclosure statement

No potential conflict of interest was reported by the authors.

References

Arana, J., Leon, C., Moreno-Gil, S., & Zubiaurre, A. (2013). A comparison of tourists' valuation of climate change policy using different pricing frames. *Journal of Travel Research, 52*(1), 82–92. https://doi.org/10.1177/0047287512457260

Arora, R., Stoner, C., & Arora, A. (2006). Using framing and credibility to incorporate exercise and fitness in individuals' lifestyle. *Journal of Consumer Marketing, 23*(4), 199–207. https://doi.org/10.1108/07363760610674329

Bagozzi, R., Gopinath, M., & Nyer, P. (1999). The role of emotions in marketing. *Journal of the Academy of Marketing Science, 27*(2), 184–206. https://doi.org/10.1177/0092070399272005

Beattie, G., & McGuire, L. (2020). The modifiability of implicit attitudes to carbon footprints and its implications for carbon choice. *Environment and Behavior, 52*(5), 467–494. https://doi.org/10.1177/0013916518808571

Becken, S. (2004). How tourists and tourism experts perceive climate change and carbon-offsetting schemes. *Journal of Sustainable Tourism, 12*(4), 332–345. https://doi.org/10.1080/09669580408667241

Blau, P. (1964). *Exchange and power in social life*. John Wiley.

Bord, R. J., & O'Connor, R. E. (1997). The gender gap in environmental attitudes: the case of perceived vulnerability to risk. *Social Science Quarterly, 78*(4), 830–840.

Brody, C. J. (1984). Differences by sex in support for nuclear power. *Social Forces, 63*(1), 209–228. https://doi.org/10.2307/2578866

Buss, D. M., & Schmitt, D. P. (1993). Sexual strategies theory: an evolutionary perspective on human mating. *Psychological Review, 100*(2), 204–232. https://doi.org/10.1037/0033-295X.100.2.204

Bussey, K., & Bandura, A. (1999). Social cognitive theory of gender development and differentiation. *Psychological Review, 106*(4), 676–713. https://doi.org/10.1037/0033-295x.106.4.676

Cai, R., Lu, L., & Gursoy, D. (2018). Effect of disruptive customer behaviors on others' overall service experience: an appraisal theory perspective. *Tourism Management, 69*, 330–344. https://doi.org/10.1016/j.tourman.2018.06.013

Carlson, J., Vincent, L., Hardesty, D., & Bearden, W. (2009). Objective and subjective knowledge relationships: a quantitative analysis of consumer research findings. *Journal of Consumer Research, 35*(5), 864–876. https://doi.org/10.1086/593688

Chaiken, S., Liberman, A., & Eagly, A. H. (1989). *Heuristic and systematic information processing within and beyond the persuasion context. Unintended Thought*. Guilford Press.

Chazdon, R., & Brancalion, P. (2019). Restoring forests as a means to many ends. *Science (New York, N.Y.), 365*(6448), 24–25. http://science.sciencemag.org https://doi.org/10.1126/science.aax9539

Choi, H., & Choi, H. C. (2019). Investigating tourists' fun-eliciting process toward tourism destination sites: an application of cognitive appraisal theory. *Journal of Travel Research, 58*(5), 732–744. https://doi.org/10.1177/0047287518776805

Choi, A., & Ritchie, B. (2014). Willingness to pay for flying carbon neutral in Australia: an exploratory study of offsetter profiles. *Journal of Sustainable Tourism, 22*(8), 1236–1256. https://doi.org/10.1080/09669582.2014.894518

Cohen, S., & Higham, J. (2011). Eyes wide shut? UK consumer perceptions on aviation climate impacts and travel decisions to New Zealand. *Current Issues in Tourism, 14*(4), 323–335. https://doi.org/10.1080/13683501003653387

Dankelman, I., & Jansen, W. (2010). Gender, environment and climate change: understanding the linkages. In I. Dankelman (Ed.), *Gender and climate change: An introduction* (pp. 37–62). Earthscan.

Davidson, D. J., & Haan, M. (2012). Gender, political ideology, and climate change beliefs in an extractive industry community. *Population and Environment, 34*(2), 217–234. https://doi.org/10.1007/s11111-011-0156-y

Denton, G., Chi, O. H., & Gursoy, D. (2020). n examination of the gap between carbon offsetting attitudes and behaviors:Role of knowledge, credibility and trust. International Journal of Hospitality Management, Vol 90, 102608

Dhanda, K., & Hartman, L. (2011). The ethics of carbon neutrality: a critical examination of voluntary carbon offset providers. *Journal of Business Ethics, 100*(1), 119–149. https://doi.org/10.1007/s10551-011-0766-4

Dietz, T., Dan, A., & Shwom, R. (2007). Support for climate change policy: social psychological and social structural influences. *Rural Sociology, 72*(2), 185–214. https://doi.org/10.1526/003601107781170026

Dolnicar, S. (2020). Designing for more environmentally friendly tourism. *Annals of Tourism Research, 84*, 102933. https://doi.org/10.1016/j.annals.2020.102933

Dolnicar, S., & Leisch, F. (2008). An investigation of tourists' patterns of obligation to protect the environment. *Journal of Travel Research*, *46*(4), 381–391. https://doi.org/10.1177/0047287507308330

Doney, P., & Cannon, J. (1997). An examination of the nature of trust in buyer-seller relationships. *Journal of Marketing*, *61*(2), 35–51. https://doi.org/10.2307/1251829

Dunlap, R. E., & McCright, A. M. (2011). Climate change denial: sources, actors, and strategies. In C. Lever-Tracy (Ed.), *Routledge handbook of climate change and society* (pp. 240–260). Taylor & Francis.

Eagly, A. H., & Wood, W. (1999). The origins of sex differences in human behavior: evolved dispositions versus social roles. *American Psychologist*, *54*(6), 408–423. https://doi.org/10.1037/0003-066X.54.6.408

Economou, A., & Halkos, G. (2020). The gender environmentalism gap in Germany and the Netherlands. *Social Science Quarterly*, *101*(3), 1038–1055. https://doi.org/10.1111/ssqu.12785

Ellen, P. (1994). Do we know what we need to know? Objective and subjective knowledge effects on pro-environmental behaviors. *Journal of Business Research*, *30*(1), 43–52. https://doi.org/10.1016/0148-2963(94)90067-1

Finucane, M. L., Slovic, P., Mertz, C. K., Flynn, J., & Satterfield, T. A. (2000). Gender, race and perceived risk: the white male effect. *Health, Risk, and Society*, *2*(2), 159–172. https://doi.org/10.1080/713670162

Gifford, R., Hay, R., & Boros, K. (1982). Individual differences in environmental attitudes. *The Journal of Environmental Education*, *14*(2), 19–23. https://doi.org/10.1080/00958964.1983.10801933

Gössling, S. (2009). Carbon neutral destinations: a conceptual analysis. *Journal of Sustainable Tourism*, *17*(1), 17–37. https://doi.org/10.1080/09669580802276018

Gössling, S., Hultman, J., Haglund, L., Källgren, H., & Revahl, M. (2009). Swedish air travelers and voluntary carbon offsets: towards the co-creation of environmental value? *Current Issues in Tourism*, *12*(1), 1–19. https://doi.org/10.1080/13683500802220687

Gössling, S., & Peeters, P. (2007). It does not harm the environment!' an analysis of industry discourses on tourism, air travel and the environment. *Journal of Sustainable Tourism*, *15*(4), 402–417. https://doi.org/10.2167/jost672.0

Gössling, S., & Scott, D. (2018). The decarbonisation impasse: global tourism leaders' views on climate change mitigation. *Journal of Sustainable Tourism*, *26*(12), 2071–2086. https://doi.org/10.1080/09669582.2018.1529770

Gössling, S., Scott, D., & Hall, C. M. (2015). Inter-market variability in CO_2 emission-intensities in tourism: implications for destination marketing and carbon management. *Tourism Management*, *46*, 203–212. https://doi.org/10.1016/j.tourman.2014.06.021

Gursoy, D., Chi, O. H., Lu, L., & Nunkoo, R. (2019). Consumers acceptance of artificially intelligent (AI) device use in service delivery. *International Journal of Information Management*, *49*, 157–169. https://doi.org/10.1016/j.ijinfomgt.2019.03.008

Hair, J., Hult, G., Ringle, C., Sarstedt, M., & Kai, T. (2017). Mirror, mirror on the wall: a comparative evaluation of composite-based structural equation modeling methods. *Journal of the Academy of Marketing Science*, *45*(5), 616–632. https://doi.org/10.1007/s11747-017-0517-x

Hair, J. F., Ringle, C. M., & Sarstedt, M. (2011). PLS-SEM: Indeed a silver bullet. *Journal of Marketing Theory and Practice*, *19*(2), 139–152. https://doi.org/10.2753/MTP1069-6679190202

Hares, A., Dickinson, J., & Wilkes, K. (2010). Climate change and the air travel decisions of UK tourists. *Journal of Transport Geography*, *18*(3), 466–473. https://doi.org/10.1016/j.jtrangeo.2009.06.018

Hartman, R. O., Dieckmann, N. F., Sprenger, A. M., Stastny, B. J., & DeMarree, K. G. (2017). Modeling attitudes toward science: development and validation of the credibility of science scale. *Basic and Applied Social Psychology*, *39*(6), 358–371. https://doi.org/10.1080/01973533.2017.1372284

Hayes, A. F. (2018). *Introduction to mediation, moderation, and conditional process analysis* (2nd ed.). Guilford Press.

Heath, Y., & Gifford, R. (2006). Free- market ideology and environmental degradation: The case of belief in global climate change. *Environment and Behavior*, *38*(1), 48–71. https://doi.org/10.1177/0013916505277998

Hlee, S., Lee, H., & Koo, C. (2018). Hospitality and tourism online review research: a systematic analysis and heuristic-systematic model. *Sustainability*, *10*(4), 1141. https://doi.org/10.3390/su10041141

Huy Tuu, H., Ottar Olsen, S., & Thi Thuy Linh, P. (2011). The moderator effects of perceived risk, objective knowledge and certainty in the satisfaction-loyalty relationship. *Journal of Consumer Marketing*, *28*(5), 363–375. https://doi.org/10.1108/07363761111150017

Hyde, J. S. (2014). Gender similarities and differences. *Annual Review of Psychology*, *65*(1), 373–398. https://doi.org/10.1146/annurev-psych-010213-115057

Jacobs, J. E., & Simpkins, S. D. (2006). *Leaks in the pipeline to math, science, and technology careers*. Jossey-Bass.

Joireman, J., & Liu, R. (2014). Future-oriented women will pay to reduce global warming: Mediation via political orientation, environmental values, and belief in global warming. *Journal of Environmental Psychology*, *40*, 391–400. https://doi.org/10.1016/j.jenvp.2014.09.005

Jöreskog, K. G., & Sörbom, D. (1996). *LISREL 8 user's reference guide*. Scientific Software International.

Juvan, E., & Dolnicar, S. (2014). The attitude-behaviour gap in sustainable tourism. *Annals of Tourism Research*, *48*, 76–95. https://doi.org/10.1016/j.annals.2014.05.012

Juvan, E., & Dolnicar, S. (2017). Drivers of pro-environmental tourist behaviours are not universal. *Journal of Cleaner Production*, *166*, 879–890. https://doi.org/10.1016/j.jclepro.2017.08.087

Keller, C., Bostrom, A., Kuttschreuter, M., Savadori, L., Spence, A., & White, M. (2012). Bringing appraisal theory to environmental risk perception: a review of conceptual approaches of the past 40 years and suggestions for future research. *Journal of Risk Research*, *15*(3), 237–256. https://doi.org/10.1080/13669877.2011.634523

Kendall, HW (2000) Press Release: Announcing World Scientists' Warning to Humanity. In: Kendall H.W. (ed) A Distant Light. Masters of Modern Physics, vol 0, Springer, New York, NY. https://doi.org/10/1007/978-1-4419-8507-1_18

Kim, Y. H., Barber, N., & Kim, D.-K. (2019). Sustainability research in the hotel industry: Past, present, and future. *Journal of Hospitality Marketing & Management*, *28*(5), 576–620. https://doi.org/10.1080/19368623.2019.1533907

Kim, S.-B., & Kim, D.-Y. (2014). The effects of message framing and source credibility on green messages in hotels. *Cornell Hospitality Quarterly*, *55*(1), 64–75. https://doi.org/10.1177/1938965513503400

Kuo, Y.-F., & Wu, C.-M. (2012). Satisfaction and post-purchase intentions with service recovery of online shopping websites: Perspectives on perceived justice and emotions. *International Journal of Information Management*, *32*(2), 127–138. https://doi.org/10.1016/j.ijinfomgt.2011.09.001

Lang, K.T., and Hallman, W.K. (2005). Who does the public trust? The case of genetically modified foods in the United States. *Risk Analysis 25*(5), 1241–1252.

Laroche, M., Bergeron, J., & Barbaro-Forleo, G. (2001). Targeting consumers who are willing to pay more for environmentally friendly products. *Journal of Consumer Marketing*, *18*(6), 503–520. https://doi.org/10.1108/EUM0000000006155

Lazarus, R. (1991). *Emotion and adaptation*. Oxford University Press.

Lee, K., Conklin, M., Cranage, D. A., & Lee, S. (2014). The role of perceived corporate social responsibility on providing healthful foods and nutrition information with health-consciousness as a moderator. *International Journal of Hospitality Management*, *37*, 29–37. https://doi.org/10.1016/j.ijhm.2013.10.005

Lenzen, M., Sun, Y.-Y., Faturay, F., Ting, Y.-P., Geschke, A., & Malik, A. (2018). The carbon footprint of global tourism. *Nature Climate Change*, *8*(6), 522–528. https://doi.org/10.1038/s41558-018-0141-x

Maibach, E., Leiserowitz, A., Roser-Renouf, C., & Mertz, C. K. (2011). Identifying like-minded audiences for global warming public engagement campaigns: An audience segmentation analysis and tool development. *Plos One*, *6*(3), e17571. https://doi.org/10.1371/journal.pone.0017571

Mair, J. (2011). Exploring air travellers' voluntary carbon-offsetting behavior. *Journal of Sustainable Tourism*, *19*(2), 215–230. https://doi.org/10.1080/09669582.2010.517317

Malka, A., Krosnick, J. A., & Langer, G. (2009). The association of knowledge with concern about global warming: Trusted information sources shape public thinking. *Risk Analysis*, *29*(5), 633–647. https://doi.org/10.1111/j.1539-6924.2009.01220.x

Marquart-Pyatt, S., Shwom, R., Dietz, T., Dunlap, R., Kaplowitz, S., McCright, A., & Zahran, S. (2011). Understanding public opinion on climate change: a call for research. *Environment*, *53*(4), 38–42.

McCright, A. M. (2010). The effects of gender on climate change knowledge and concern in the American public. *Population and Environment*, *32*(1), 66–87. https://doi.org/10.1007/s11111-010-0113-1

McCright, A. M., & Dunlap, R. E. (2000). Challenging global warming as a social problem. *Social Problems*, *47*(4), 499–522. https://doi.org/10.1525/sp.2000.47.4.03x0305s

McCright, A. M., & Dunlap, R. E. (2011). Cool dudes: The denial of climate change and polarization in the American public's views of global warming, 2001-2010. *The Sociological Quarterly*, *52*(2), 155–194. https://doi.org/10.1111/j.1533-8525.2011.01198.x

McCright, A. M., Dunlap, R. E., & Xiao, C. (2013). Perceived scientific agreement and support for government action on climate change in the USA. *Climatic Change*, *119*(2), 511–518. https://doi.org/10.1007/s10584-013-0704-9

McKercher, B., Pang, S., & Prideaux, B. (2011). Do gender and nationality affect attitudes towards tourism and the environment? *International Journal of Tourism Research*, *13*(3), 266–300. https://doi.org/10.1002/jtr.816

Meier, T., & Christen, O. (2012). Gender as a factor in an environmental assessment of the consumption of animal and plant-based foods in Germany. *The International Journal of Life Cycle Assessment*, *17*(5), 550–564. https://doi.org/10.1007/s11367-012-0387-x

Meyers-Levy, J., & Maheswaran, D. (1991). Exploring differences in males' and females' processing strategies. *Journal of Consumer Research*, *18*(1), 63–70. https://doi.org/10.1086/209241

Meyers-Levy, J., & Maheswaran, D. (2004). Exploring message framing outcomes when systematic, heuristic, or both types of processing occur. *Journal of Consumer Psychology*, *14*(1–2), 159–167. https://doi.org/10.1207/s15327663jcp1401&2_18

Milfont, T., Wilson, J., & Diniz, P. (2012). Time perspective and environmental engagement: A meta-analysis. *International Journal of Psychology: Journal International de Psychologie*, *47*(5), 325–334. https://doi.org/10.1080/00207594.2011.647029

Norgaard, K. M. (2011). *Living in denial: Climate change, emotions and everyday life*. MIT Press.

Nunkoo, R., & Gursoy, D. (2016). Rethinking the role of power and trust in tourism planning. *Journal of Hospitality Marketing & Management*, *25*(4), 512–522. https://doi.org/10.1080/19368623.2015.1019170

O'Connor, R. E., Bard, R. J., & Fisher, A. (1999). Risk perceptions, general environmental beliefs, and willingness to address climate change. *Risk Analysis*, *19*(3), 461–471. https://doi.org/10.1111/j.1539-6924.1999.tb00421.x

Ouyang, Z., Gursoy, D., & Sharma, B. (2017). Role of trust, emotions and event attachment on residents' attitudes toward tourism. *Tourism Management*, *63*, 426–438. https://doi.org/10.1016/j.tourman.2017.06.026

Park, H., & Vedlitz, A. (2013). Climate hazards and risk status: Explaining climate risk assessment, behavior and policy support. *Sociological Spectrum*, *33*(3), 219–239. https://doi.org/10.1080/02732173.2013.732900

Penn, D. J. (2003). The evolutionary roots of our environmental problems: Toward a Darwinian ecology. *The Quarterly Review of Biology*, *78*(3), 275–301. https://doi.org/10.1086/377051

Pieniak, Z., Aertsens, J., & Verbeke, W. (2010). Subjective and objective knowledge as determinants of organic vegetables consumption. *Food Quality and Preference*, *21*(6), 581–588. https://doi.org/10.1016/j.foodqual.2010.03.004

Ponnapureddy, S., Priskin, J., Ohnmacht, T., Vinzenz, F., & Wirth, W. (2017). The influence of trust perceptions on German tourists' intentions to book a sustainable hotel: a new approach to analyzing marketing information. *Journal of Sustainable Tourism*, *25*(7), 970–988. https://doi.org/10.1080/09669582.2016.1270953

Rahman, I., Park, J., & Chi, C. G.-Q. (2015). Consequences of "greenwashing": Consumers' reactions to hotels' green initiatives. *International Journal of Contemporary Hospitality Management*, *27*(6), 1054–1081. https://doi.org/10.1108/IJCHM-04-2014-0202

Roseman, I. J. (1984). Cognitive determinants of emotion: A structural theory. In P. Shaver (Ed.), *Review of personality and social psychology: emotions, relationships and health* (Vol. 5, pp. 11–36). Beverly Hills.

Rousseau, D., Sitkin, S., Burt, R., & Camerer, C. (1998). Not so different after all: A cross-discipline view of trust. *Academy of Management Review*, *23*(3), 393–404. https://doi.org/10.5465/amr.1998.926617

Scannell, L., & Gifford, R. (2013). Personally relevant climate change: The role of place attachment and local versus global message framing in engagement. *Environment and Behavior*, *45*(1), 60–85. https://doi.org/10.1177/0013916511421196

Scott, D., Gössling, S., Hall, C. M., & Peeters, P. (2015). Can tourism be part of the decarbonized global economy? The costs and risks of alternate carbon reduction policy pathways. *Journal of Sustainable Tourism*, *24*(1), 1–72. https://doi.org/10.1080/09669582.2015.1107080

Seetanah, B., & Sannassee, R. V. (2015). Marketing promotion financing and tourism development. The case of Mauritius. *Journal of Hospitality Marketing & Management*, *24*(2), 202–215. https://doi.org/10.1080/19368623.2014.914359

Segerstedt, A., & Grote, U. (2016). Increasing adoption of voluntary carbon offsets among tourists. *Journal of Sustainable Tourism*, *24*(11), 1541–1554. https://doi.org/10.1080/09669582.2015.1125357

Strapko, N., Hempel, L., MacIlroy, K., & Smith, K. (2016). Gender differences in environmental concern: reevaluating gender socialization. *Society & Natural Resources*, *29*(9), 1015–1031. https://doi.org/10.1080/08941920.2016.1138563

Terry, G. (2009). No climate justice without gender justice: An overview of the issues. *Gender & Development*, *17*(1), 5–18. https://doi.org/10.1080/13552070802696839

Thaler, R., & Sunstein, C. (2008). *Nudge: Improving decisions about health, wealth, and happiness*. Yale University Press.

Tranter, B. (2011). Political divisions over climate change and environmental issues in Australia. *Environmental Politics*, *20*(1), 78–96. https://doi.org/10.1080/09644016.2011.538167

Trumbo, C., & McComas, K. (2003). The function of credibility in information processing for risk perception. *Risk Analysis: An Official Publication of the Society for Risk Analysis*, *23*(2), 343–353. https://doi.org/10.1111/1539-6924.00313

Tversky, A., & Kahneman, D. (1981). The framing of decisions and the psychology of choice. *Science (New York, N.Y.)*, *211*(4481), 453–458. https://doi.org/10.1126/science.7455683

Vicente-Molina, M. A., Fernandez-Sainz, A., & Izagirre-Olaizola, J. (2018). Does gender make a difference in pro-environmental behavior? The case of the Basque Country university students. *Journal of Cleaner Production*, *176*, 89–98. https://doi.org/10.1016/j.jclepro.2017.12.079

Weaver, D. (2007). Towards sustainable mass tourism: paradigm shift of paradigm nudge? *Tourism Recreation Research*, *32*(3), 65–69. https://doi.org/10.1080/02508281.2007.11081541

White, K., Habib, R., & Hardisty, D. (2019). How to SHIFT consumer behaviors to be more sustainable: a literature review and guiding framework. *Journal of Marketing*, *83*(3), 22–49. https://doi.org/10.1177/0022242919825649

Whitmarsh, L., & O'Neill, S. (2010). Green identity, green living? the role of pro-environmental self-identity in determining consistency across diverse pro-environmental behaviors. *Journal of Environmental Psychology*, *30*(3), 305–314. https://doi.org/10.1016/j.jenvp.2010.01.003

Xiao, C., & McCright, A. M. (2012). Explaining gender differences in concern about environmental problems in the United States. *Society & Natural Resources*, *25*(11), 1067–1084. https://doi.org/10.1080/08941920.2011.651191

Zelezny, L., Chua, P.-P., & Aldrich, C. (2000). Elaborating on gender differences in environmentalism. *Journal of Social Issues*, *56*(3), 443–457. https://doi.org/10.1111/0022-4537.00177

Appendix A. Measurement items and properties

Item No.	Item Description	Mean Study 1/ Study 2	SD Study 1/ Study 2	Skewness Study 1/ Study 2	Kurtosis Study 1/ Study 2
Subjective Knowledge (Ellen, 1994)					
SK1	I consider myself to be knowledgeable about the impact of travel on the environment	3.12/ 3.61	1.20/ 0.93	−0.15/ -0.63	−0.98/ -0.19
SK2	Compared to other people, I know a lot about the impact of travel on the environment	3.00/ 3.28	1.21/ 0.98	0.01/ -0.18	−0.97/ -0.51
SK3	People who know me consider me to be knowledgeable about the impact of travel on the environment	2.97/ 3.18	1.26/ 1.04	0.04/ -0.04	−1.04/ -0.70
Trust in Company (Lang & Hallman, 2005; Nunkoo et al., 2012)					
TC1	You can generally trust decisions regarding carbon offset made by tourism companies	2.71/ 3.28	1.05/ 0.97	0.16/ -0.42	−0.62/ -0.06
TC2	You can generally trust the people who run tourism companies to do what is right regarding carbon offset	2.66/ 3.30	1.06/ 1.04	0.15/ -0.33	−0.65/ -0.25
TC3	Tourism companies can be trusted to do what is right regarding carbon offset without our having to constantly check on them	2.58/ 3.00	1.14/ 1.09	0.33/ -0.06	−0.67/ -0.61
Credibility of Climate Science (Hartman et al., 2017)					
CR1	*People trust climate scientists a lot more than they should	3.56/ 3.25	1.35/ 1.29	−0.44/ -0.22	−1.12/ -1.06
CR2	*People don't realize just how flawed a lot of climate science research really is	3.56/ 3.27	1.30/ 1.30	−0.42/ -0.23	−1.03/ -1.08
CR3	*A lot of climate science theories and predictions are dead wrong	3.71/ 3.42	1.31/ 1.26	−0.54/ -0.32	−1.03/ -1.02
CR4	*Sometimes I think we put too much faith in climate science	3.68/ 3.40	1.37/ 1.34	−0.63/ -0.32	−0.97/ -1.17
CR5	*Our society places too much emphasis on climate science	3.67/ 3.48	1.38/ 1.33	−0.55/ -0.40	−1.12/ -1.07
CR6	*I am concerned by the amount of influence that climate scientists have in society	3.63/ 3.39	1.26/ 1.33	−0.42/ -0.28	−1.11/ -1.17
Attitude (Hsu & Huang, 2012)					
A1	I like the idea of offsetting carbon emissions when I travel	4.10/ 3.76	1.04/ 0.94	−1.18/ -0.84	0.98/ 0.77
A2	I have a favorable attitude toward offsetting carbon emissions	4.09/ 3.82	1.05/ 0.91	−1.19/ -0.94	1.01/ 1.14
A3	Offsetting carbon emissions when I travel would be worthwhile	4.06/ 3.72	1.04/ 0.98	−1.20/ -0.81	1.18/ 0.54
A4	Offsetting carbon emissions when I travel would be rewarding	3.97/ 3.68	1.03/ 1.00	−0.91/ -0.71	0.37/ 0.45
Objective Knowledge					
OK	Total score of a 8-question quiz (Study 1) and a 9-question quiz (Study 2)	3.21/ 3.84	0.72/ 0.76	−0.20/ -0.87	0.19/ 0.77

Appendix B. Demographic profile across genders

		Study 1		Study 2	
		Male (n = 231)	Female (n = 231)	Male (n = 245)	Female (n = 245)
Age					
	18-25	13.4	16.0	15.9	17.1
	26-34	32.5	29.9	34.3	29.0
	35-44	22.5	23.4	25.7	28.2
	44-54	17.3	13.0	14.7	13.5
	55-64	10.8	13.4	6.9	9.4
	65 or over	3.5	4.3	2.4	2.9
Marital Status					
	Single	42.4	32.0	41.2	31.4
	Married	45.0	46.3	49.8	49.8
	Live together	7.4	8.7	5.3	11.8
	Divorced	0.4	0.9	2.9	5.7
	Widowed	4.8	12.1	0.8	1.2
Occupation					
	Student	7.4	6.0	9.4	11.4
	Professional	51.1	47.6	50.6	35.1
	Managerial	16.0	12.1	16.7	8.6
	Sales	10.4	8.7	5.3	7.3
	Homemaker	0.9	13.9	2.0	14.7
	Other	14.3	11.7	15.9	22.9
Education					
	Less than high school	0.9	0.4	0.0	1.2
	High school graduate	8.7	5.2	11.4	7.8
	College Degree	74.9	76.6	82.0	86.0
	Professional degree	12.5	16.0	2.9	3.4
	Doctorate degree	3.0	1.7	3.7	1.6
Annual Income					
	Under $10,000	6.1	5.6	14.9	21.5
	$10,000-$29,999	16.9	19.1	13.9	19.2
	$30,000-$49,999	22.5	22.1	19.9	17.1
	$50,000-$69,999	23.4	20.8	29.7	20.4
	$70,000 – $99,999	15.6	15.6	8.2	10.6
	$100,000 and above	15.6	16.9	13.5	11.2

Appendix C. Quiz questions for examining objective knowledge

Study 1

Please indicate whether you believe the following statements to be true or false (*true, false, or don't know*) regarding the impact of travel on the environment and carbon emissions generated by travel.

1. Levels of carbon in the atmosphere are higher than what is considered sustainable.
   (Correct answer: True)
2. Tourism is considered to be a significant contributor to carbon emissions in the atmosphere.
   (Correct answer: True)
3. Levels of carbon in the atmosphere are approximately the same as they have been for thousands of years.
   (Correct answer: False)
4. The current levels of carbon in the atmosphere have been recorded in the past during previous warming periods.
   (Correct answer: False)
5. There is scientific consensus that there will be environmental impacts as a result of excess levels of carbon in the atmosphere.
   (Correct answer: True)

6. Carbon offsetting programs are available where consumers can invest in programs that absorb or reduce carbon in the atmosphere.
   (Correct answer: True)
7. A number of airlines and retail companies use carbon offsetting and/or make offsetting available to their customers.
   (Correct answer: True)
8. A single tree can absorb up to a ton of carbon from the atmosphere.
   (Correct answer: True)

Study 2

Please indicate the correct answer(s).

Questions were adopted from: https://www.energy.gov/articles/quiz-test-your-climate-change-iq

1. How does the greenhouse effect work? (Correct answer: b)

a. Greenhouse gases reflect the sun's energy, causing it to warm the Earth.
b. Greenhouse gases absorb the sun's energy, slowing or preventing heat from escaping into space.
c. Greenhouse gases directly warm oceans and cause dramatic weather.
d. Oceans absorb greenhouse gases, which cause the Earth's temperature to rise.
e. I don't know

2. What is the most potent greenhouse gas? (Correct answer: a)
   a. Fluorinated gases
   b. Nitrous oxide
   c. Carbon dioxide
   d. Methane
   e. I don't know
3. How much have global average temperatures increased in the last century? (Correct answer: d)
   a. 2.1 degrees Fahrenheit
   b. 0.6 degrees Fahrenheit
   c. 4.3 degrees Fahrenheit
   d. 1.4 degrees Fahrenheit
   e. I don't know
4. How much have sea levels risen in the past 100 years? (Correct answer: a)
   a. 7 inches
   b. 2 inches
   c. 5 inches
   d. 16 inches
   e. I don't know
5. How much have carbon dioxide emissions changed in the United States since 2006? (Correct answer: b)
   a. Increased 560 million metric tons
   b. Dropped 480 million metric tons
   c. Increased 230 million metric tons
   d. Dropped 120 million metric tons
   e. I don't know
6. Which of these is considered a critical threshold for carbon dioxide levels in the atmosphere? (Correct answer: d)
   a. 123 parts per million
   b. 250 parts per million
   c. 685 parts per million
   d. 400 parts per million
   e. I don't know
7. What is the biggest source of greenhouse gas emissions in the United States? (Correct answer: c)
   a. Farming, logging and manufacturing
   b. Heating and cooling buildings
   c. Producing electricity
   d. Using transportation
   e. I don't know
8. What does carbon intensity measure? (Correct answer: a)
   a. Carbon dioxide produced per dollar of gross domestic product

  b. Carbon dioxide produced per electrical charge
  c. Carbon dioxide produced per kilowatt hour
  d. Carbon dioxide produced per British thermal unit of energy
  e. I don't know

9. During the 2015 United Nations Climate Change Conference in Paris, how many countries committed to doubling clean energy research and development? (Correct answer: c)
  a. 76
  b. 195
  c. 20
  d. 12
  e. I don't know

Sustainability through the tourism entrepreneurship journey: a gender perspective

Cristina Figueroa-Domecq, Albert Kimbu, Anna de Jong and Allan M. Williams

ABSTRACT

Women's tourism entrepreneurship has been identified as fundamental to meeting the UN's Sustainable Development Goals of both 'gender equality' and 'decent growth and economic growth' but neither entrepreneurship nor sustainability are gender neutral in the tourism industry. Therefore, further research is required into how gender influences sustainable entrepreneurship, providing insights for tourism entrepreneurship policy. In response to a prevalent essentialism in much of the literature, this paper adopts a post-structuralist framework, alongside a mixed-methods approach, to understand the complex role of gender and sustainability at different stages of entrepreneurship. The initial focus is on a survey of 539 tourism students (women and men) which analyses the latent and nascent entrepreneurship stages, while 19 interviews with established tourism entrepreneurs provide further insights into these issues. The analysis focusses especially on the individual characteristics of risk, personal attitudes to entrepreneurship and behavioural control. While broad gender differences are observed, notably in societal perceptions of risk aversion, there is also considerable blurring of the approaches of established entrepreneurs in particular to sustainability and entrepreneurship. If entrepreneurship is to enhance sustainability, policy needs to account for the non-essentialised gendered dimensions that inhibit and enable sustainable tourism entrepreneurship.

Introduction

Women's entrepreneurship in the tourism industry has been identified as fundamental to meeting the UN's Sustainable Development Goals of both 'gender equality' and 'decent growth and economic growth' (UN World Tourism Organization (UNWTO, 2019). Yet research on this topic remains limited despite consensus among researchers about the complex role that gender plays in the different phases of the entrepreneurial process (i.e. pre-start up, start up and growth), as well as in entrepreneurial performance (Ahl & Marlow, 2019; Figueroa-Domecq et al., 2020; Gupta et al., 2009; Minniti et al., 2005). Fundamentally, women are perceived to have less human capital and tend to perceive themselves as less competent in terms of entrepreneurial behaviour (Ploum et al., 2018), reflecting deep gender inequalities in power and status that have become engrained in society and accepted as 'normal and inevitable' (Ahl, 2006; Ahl & Marlow, 2012). Furthermore, the tendency in the entrepreneurship literature to position masculinity as the norm when describing entrepreneurial traits, or using masculine characteristics to describe entrepreneurs, tends to make women invisible and inconspicuous (Marlow & Patton, 2005). There is, therefore, a considerable gap between the aspirations of the UN Sustainable Development Goals and the gendered dimensions of entrepreneurship. This is compounded by the highly gendered nature of sustainability in the working environment (Hesselbarth & Schaltegger, 2014; Ploum et al., 2018). Women working for sustainable development individually, and as members of civil society, play a key role in supporting their communities (Global Volunteers, 2019; UN Women, 2014; UNWTO, 2019) and are generally more socially oriented than men when operating businesses (Betz et al., 1989).

Bringing together the gender and sustainability perspectives indicates the need to understand gendered attitudes towards, and experiences of, the increasing business commitment to more sustainable entrepreneurship (Zahra et al., 2014). In short, how is sustainable entrepreneurship gendered? This question has largely been neglected by tourism scholars who have tended to focus on the barriers encountered by women entrepreneurs that is, on enhancing economic sustainability, whilst ignoring the social and environmental dimensions of entrepreneurship. This neglect is surprising because tourism has been acknowledged as a fertile field for entrepreneurial initiatives, due to a predominance of small firms and relatively low entry barriers (Li, 2008; Nikraftar & Hosseini, 2016; Shaw & Williams, 1999), while tourism's diverse, dynamic and flexible nature (Peeters & Ateljevic, 2009) has led to it being perceived as a potentially empowering sector for women entrepreneurs (UNWTO, 2019; Vujko et al., 2019). Additionally, it is contended that, in the long term, tourism cannot survive if it is not *both* sustainable and entrepreneurial (Crnogaj et al., 2014:377) and, as indicated, these are both gendered.

Consequently, this article aims to explore the application of a gender perspective to understanding the importance given to sustainability in tourism entrepreneurship. This will provide perspectives on how sustainable entrepreneurship and sustainability in entrepreneurship are gendered. In addressing this aim, we adopt a post-structuralist approach, which recognises the limitations of structuralism, particularly its essentialism. Following on from this, our analysis focusses especially on risk, attitudes to entrepreneurship, and behavioural control as providing non-essentialist insights into the gendered nature of sustainable entrepreneurship. Given that entrepreneurialism is an individualised, complex journey, the need to avoid oversimplified generalisations becomes even greater: different influences shape the varied experiences of women and men over the journey.

A mixed method approach is adopted to studying the entrepreneurship journey, drawing on findings from Spain. More specifically, this study proposes a Quantitative Dominant (QUANT-qual) concurrent complementary method design (Truong et al., 2020): QUANT through a survey of latent and nascent entrepreneurs among tourism students in Spain (539), and the impact of sustainability on their entrepreneurial ideas and steps (Study A); and Qual through interviews with 19 established entrepreneurs about their nuanced entrepreneurial experience and the varied importance placed on sustainability (Study B). This design emphasises the complementarity of the results (Dayour et al., 2019; Truong et al., 2020) from the survey among students (QUANT), with the reflections of a group of established entrepreneurs (qual) providing further insights into complexity and greater nuance.

Literature review and theoretical framework

Sustainability is a holistic and integrated approach that considers ecological, social and economic dimensions, recognising that all three must be considered together to find lasting prosperity. It demands ways of living, working and being that enables all human beings to lead healthy, fulfilling, and economically secure lives while reducing pressure on the environment and not endangering the welfare of future generations (Santillo, 2007) Sustainability is seen as factor of success in the working environment and it is increasingly integrated into business processes (Ploum et al., 2018).

Sustainable entrepreneurship is defined in this paper as the recognition, establishment, assessment and exploitation of entrepreneurial opportunities that enable the production of goods and services that contribute to profitability and sustainability of a firm whilst directly and indirectly engendering positive social and environmental impacts for community and its members (Shepherd & Patzelt, 2011). Sustainable entrepreneurship in the tourism sector broadly fits this definition, yet also incorporates a more holistic emphasis (Crnogaj et al., 2014) on the destination scale, alongside the individual and organisational. The literature review first sets out the rationale for considering how attitudes, behavioural control and risk provide a means to look beyond the inherent essentialism of focusing on socio-economic, socio-demographic and institutional considerations. It then considers the value of a post structuralist approach.

Sustainable entrepreneurship, gender and tourism

Entrepreneurship is highly gendered and the relevance of sustainability to entrepreneurs, employees and consumers, is also considered to be gendered. It has been argued that women play a stronger role than men in establishing positive values in terms of social welfare, family and community wellbeing (Kimbu & Ngoasong, 2016), reducing carbon emissions and greenhouse gases (De Silva & Pownall, 2014). Relatedly, firms characterised by gender diverse leadership teams/boards tend to be more effective at pursuing environmentally friendly strategies and have more enhanced sustainability reporting mechanisms (Kassinis et al. 2016). Although this is questioned in some studies (e.g. Glass et al., 2016), the literature generally suggests that women tend to demonstrate a more environmentally conscious responsible attitude than men as entrepreneurs.

In tourism research, Braun's (2010) analysis of attitudes and behaviour towards sustainability among Australian businesses surmised that, women's desire to make their enterprises more sustainable/greener was overwhelmingly motivated by broader ethical concerns. Similarly, Polk's (2003) survey of sustainable transportation systems in Sweden concluded that women were more prepared than men to participate in ecologically benign activities and were consistently more supportive of ecological issues. This is because, as Scheyvens (2000) argues, women develop different knowledge of the environment while their roles within their families and communities often place them in close connection with the natural environment. However, the relationship between gender, sustainability and tourism entrepreneurship is largely unexplored with very few exceptions (see e.g. Dahles et al., 2020; Kokkranikal & Morrison, 2002; Thompson et al., 2018). In particular, the explicit role of gender as a moderating variable/influence for nascent and extant entrepreneurs to engage in sustainable entrepreneurship (Outsios & Farooqi, 2017) has not been explored.

In terms of the determinants of these different levels of interest, the social and environmental contexts are of course important. For example, societal obligations manifested, in family commitments and other caring responsibilities are often ascribed to women (Kimbu & Ngoasong, 2016; Ngoasong & Kimbu, 2019). Marlow (1997) contends that many women go into entrepreneurship and self-employment as a result of balancing family commitments and work. That is, the conditional flexibility conferred by self-employment and entrepreneurship is strongly associated with female (sustainable) entrepreneurs (Outsios & Farooqi, 2017). In summary, within tourism, societal demands impact on women's time commitment and ability to acquire and/or improve the skills needed to grow their businesses and engage in more sustainable activities (Moreno-Alarcón & Cole, 2019; Roomi et al., 2009).

These broad structural considerations do not explain the considerable variation in individual women's experiences of sustainable entrepreneurship. Looking beyond societal norms and expectations (Koe et al., 2014), sustainable entrepreneurs need to perceive entrepreneurship as desirable, have positive perceptions of their capabilities, self-confidence and aspire to be one's own boss (Koe et al., 2014). They also need to be prepared to take risks (Greco & de Jong, 2017) which, although gendered, do not follow any simple gender divisions. The remainder of the literature review considers personal attitude towards sustainable entrepreneurship, behavioural control and risk: these have been little researched in general, and not at all in tourism, in relation to gender and the entrepreneurial journey.

Personal attitude towards entrepreneurship (pae)

The entrepreneurial journey begins with potential entrepreneurs who believe they have the necessary skills, can perceive business opportunities, and wish to exploit these. In contrast, sustainable entrepreneurship also takes into account environmental and societal goals (Cohen & Winn, 2007; Crnogaj et al., 2014, p. 378.; Dean & McMullen, 2007)

Personal attitudes are important in engaging with sustainable entrepreneurship and these are gendered. Men tend to be more interested in entrepreneurship, and "a significant portion of the gender gap in entrepreneurial propensity is explained by subjective perceptions whereas socio-economic variables appear to play a smaller role" (Koellinger et al., 2013, p. 229). Perceptual factors are ergo crucial in explaining how gender differences influence PAE, and how these impact attitudes towards sustainability and enterprise performance. However, a cautious and reflective approach towards business growth and engagement in sustainability is not limited to women. Outsios and Farooqi (2017) note that there is no gendered difference in the attitude and desire of entrepreneurs to growing their businesses; instead, both men and women manifested a positive desire to grow their enterprises but were wary of the impact on the environment and the sustainability of their businesses. This cautiousness stems from ethical issues in dealing with the (in)congruity between sustainability and growth of their businesses as well as the community support they provide (Hallak et al., 2013), rather than gendered perceptions about family commitments, or their roles in business and society. Irrespective of gender, consideration should be given to ethical and strategic challenges that are encountered as a business grows whilst ensuring it remains sustainable.

Behavioural control (BC)

Bamberg and Möser (2007) identified BC, alongside attitudes and personal norms, as essential in driving (sustainable) behaviour. Individuals with a favourable attitude towards a behaviour perceive it as producing desirable outcomes or eliciting positive emotions when manifested (Kraft et al., 2005). Additionally, Ajzen (2002) notes that an individual's belief that they have a high probability to behave in a particular way successfully, as well as persistence in adopting that behaviour, indicates a high perceived BC. Such individuals often take personal responsibility for addressing 'a problem' (De Groot & Steg, 2009). Individuals with favourable attitudes, perceived supportive social norms, perceived behavioural control, and strong personal norms are more inclined to perform, behave and relate to others in a particular way. This is important for tourism entrepreneurs for whom networks (social and professional) are crucial in enabling the sustainability of their businesses, through access to capital and market opportunities (e.g. Kimbu & Ngoasong, 2016; Ngoasong & Kimbu, 2019), knowledge and other resources, as well as hedging against risks and uncertainties (Williams et al., 2020); these networks are strongly gendered especially amongst nascent entrepreneurs (Braun, 2010).

From a gender perspective, tourism entrepreneurship studies have mainly examined issues such as women's agency in challenging gender roles, employment, empowerment and emancipation (Duffy et al., 2015; Ngoasong & Kimbu, 2019), women transitioning from corporate jobs to entrepreneurship (Mattis, 2004), and women intrapreneurship (Zhang et al., 2020). In contrast, they have tended to ignore the influence of behaviour although certain pro-environmental behaviours – including BC - are largely driven by internal personal attributes.

Risk

Several studies (see e.g. Williams et al., 2020, Williams & Baláž, 2015, Hall & Williams, 2019; Zhang et al., 2020;) have explored the influence of risk and uncertainty in tourism entrepreneurship and innovation. In the most detailed study to date, Williams et al. (2020) note that dynamic capabilities determine entrepreneurs' ability to engage effectively with, and manage, the shifting risks and uncertainties their businesses encounter. Furthermore, the relationship between sustainable entrepreneurship and risk arises since this kind of entrepreneurial philosophy can confer opportunities that are sometimes observed in market failures and/or uncertain environments which detract from sustainability, including those that are environmentally relevant (Ateljevic & Doorne, 2000; Dean & McMullen, 2007).

But risk is multifaceted and influences entrepreneurship from different angles. Researchers have postulated that cognitive perception makes a significant contribution to the study of entrepreneurship, and risk is inherent to this (Barbosa et al., 2007). In this regard, according to Ajzen's (1991) Theory of Planned Behaviour, attitudes, subjective norms and perceived behavioural control determine entrepreneurial intentions. At the same time, entrepreneurs' risk perceptions impact on their attitudes towards the control of events (Behavioural Control) (Keh et al., 2002).

Surprisingly, a gender lens is largely absent from this and other studies despite considerable extant evidence that gender variations in approaches to risk are the outcome of divergences in cognitive abilities and in how men and women assemble and process information and solve problems (Barrett,1995; Chung,1998). Women, it has been argued (e.g. Barber & Odean, 2001; Rowe & Wright, 2001), tend to be more risk averse than men, which Kepler and Shane (2007) attribute to their "caring and nurturing" role, reinforcing stereotypical norms. This may influence engagement with crises: for example, Carlsson-Kanyama et al. (2010) in their study of climate-change related risk judgements demonstrated that women, more than men, felt a responsibility to act to limit climate change (being more risk averse).

Sexton and Bowman-Upton (1990) note that women entrepreneurs are more risk averse than men entrepreneurs and consequently, dedicate more time searching for information on how to reduce potential risk and uncertainties. They are also less inclined to forgo potential benefits as they have a greater fear of failure (Camelo-Ordaz et al., 2016), are more loss averse, and more likely to engage in businesses with lower probability of failure and profit (Brush et al., 2009).

Women's risk aversion propensity is also manifested when accessing finance. Raising capital, and managing finance and debt, whether in the start-up or expansion/growth phases, is fraught with risks and uncertainties for both men and women (e.g. Kassinis et al., 2016). However, women owners of sustainable SMEs are more reluctant to take on debt due to concerns about how this will impact on their goals (Outsios & Farooqi, 2017). Instead, they are more likely than men to rely on personal savings, family and friends networks, and/or ploughing back profits into the businesses (Kassinis et al., 2016; Ngoasong & Kimbu, 2019). Marlow and Swail (2014) and Zhang et al. (2020) surmise that women are often positioned as a group that is structurally marginalised and lacking the masculine attributes often mirrored in the normative male entrepreneurial model due to how researchers portray women as "risk averse", coupled with prior experiences, gender socialisation, and cognitive cues.

While PAE, BC and risk significantly broaden, and complicate, our understanding of the gendering of sustainable entrepreneurship, the challenge of essentialising gender differences remains highly germane. We seek to address this via a post structuralist approach.

A post-structuralist theoretical framework

Western feminists have become increasingly engaged with environmental issues and theories since the 1980s (Goebel, 2003). Fundamental to early conceptualisations was the association thought to exist between women and the environment – whereby women were recognised as possessing a form of privileged epistemological understanding of nature compared to men. Women and nature were perceived as being innately interconnected, driven by patriarchal power relations that intersect with other inequalities (such as, culture and class) (Wachholz, 2011: 289). This conceptualisation rendered 'women' as a homogenous, essentialised category, who tended to be characterised by victimhood in inequitable gender systems, and environmental contexts of deterioration (Goebel, 2003). An underlying thread was the premise that Western capitalism was the cause of environmental destruction, and that Westernised ideas were thus incapable in responding to environmental concerns, in contrast to indigenous approaches and ways of living.

Whilst feminists have engaged in deconstructing innate relations between women and the environment since the 1990s, broader tourism management scholarship has consistently identified 'women' as having higher concern and awareness of the environment, compared to 'men' (cf. Betz et al. 1989; Kato, 2019; Kimbu & Ngoasong, 2016; Scheyvens 2000). There is therefore tension within tourism scholarship regarding how to understand gendered relationships to sustainability and the implications for tourism management. There has been particular interest in the role of tourism entrepreneurs as offering insights into the ways tourism and sustainability are conceptualised at the individual scale, by those negotiating everyday choices regarding sustainability (Outsios & Farooqi, 2017).

A post structuralist approach helps to address the complexity of entrepreneurial performance by focusing on how the intersectionality of identity (including gender, age nationality, etc.) shapes the varying pathways of tourism entrepreneurship and informs perceptions and performances relating to sustainability within entrepreneurial journeys. A post structuralist approach allows entrepreneurship to be understood as a complex, multi-layered activity where shifts in ideas occur in different social interactive systems. A detailed analysis of the varying experiences of gendered entrepreneurship allows the identification of what is often made invisible in more 'traditional', structuralist models, whilst also questioning internal bias within the literature that seeks to construct 'sustainability' as an essentialised and fixed category (Kato, 2019). In summary, this paper undertakes a feminist analysis of tourism entrepreneurship and sustainability, drawing on post structuralism to deconstruct the tensions and complexities constituting gender and sustainability.

Sustainable entrepreneurship also needs to be viewed not as an outcome but as a journey, with different stages. Whilst recognising the limitations to generalisation, three main phases tend to be identified in the entrepreneurial journey: latent entrepreneurs, with preferences for self-employment; nascent entrepreneurs already taking steps to start a business; entrepreneurs in new businesses and subsequent consolidation and growth phases (Bönte & Piegeler, 2013; Global Entrepreneurship Monitor, 2019). Post structuralism reminds us these phases are not necessarily uni-linear. Whether and how the relationship between gender, entrepreneurship and sustainability shift across these stages remains largely under-researched.

Methodology

The division of research into the dichotomy of quantitative and qualitative approaches is debatable and viewed by some scholars as "neither meaningful nor productive" (Shah et al., 2018, p.

91). Consequently, mixed methods have become a third major research paradigm ((Creswell and Creswell, 2017)) and is increasingly applied in tourism research (Dayour et al., 2019; Truong et al., 2020).

This study adopts a quantitative dominant mixed methods approach informed by the need to follow latent, nascent and established entrepreneurs in the different phases of the entrepreneurial journey. Through a quantitative survey of university students in tourism and hospitality, the study analyses how gender relates to latent and nascent sustainable entrepreneurship. A qualitative approach is further taken to the study of established entrepreneurs both because it allows exploration of the gendered complexity of sustainability, and because of the difficulties in identifying a sufficiently large sample for a quantitative analysis of this group. This design has a **complementarity** purpose, as the interviews supplement the survey, providing further nuance and complexity in understanding sustainability (Truong et al., 2020).

This analysis follows the entrepreneurial journey, through different samples and methods that adapt to the research methodological limitations in each stage. There are limitations since this involves two contrasting cross-sectoral, rather than a longitudinal, analysis, but it is still – to the best of our knowledge – the first study of different stages of the journey to sustainable entrepreneurship.

Study A. Quantitative analysis – survey

Research question and hypothesis: Following on from the broad aims set out in the introduction, the quantitative analysis addresses four main research questions, with related hypotheses, that will be approached through a gendered lens:

- Research question 1: Are women more interested than men in sustainability issues around entrepreneurship?
- H1. Women are more interested in sustainability in entrepreneurship in tourism (Kimbu & Ngoasong, 2016; Marlow, 1997; Moreno-Alarcón & Cole, 2019; Ngoasong & Kimbu, 2019; Outsios & Farooqi, 2017; Roomi et al., 2009)
  - Research Question 2: How does PAE influence the importance attached to sustainability?
- H2. Perceived attitude towards entrepreneurship (the capacity to identify business opportunities and become an entrepreneur) impacts positively on the importance attached to sustainability in entrepreneurship (Cohen & Winn, 2007; Crnogaj et al., 2014; Dean & McMullen, 2007; Hallak et al., 2013; Koellinger et al., 2013).

- Research Question 3: What is the influence of BC on the importance attached to sustainability?

- H3a. Higher BC impacts positively on the importance attached to sustainability in entrepreneurship (Bamberg & Möser, 2007; Kimbu & Ngoasong, 2016; Ngoasong & Kimbu,2019
- H3b, Higher BC has a positive impact on having a PA (; Ajzen, 2002;De Groot & Steg, 2009).

- Research Question 4: How does risk acceptance influence the importance attached to sustainability and attitudes towards entrepreneurship?

- H4a. The higher the risk acceptance the greater the importance attached to Sustainability in Entrepreneurship (seen as an opportunity) (Ateljevic & Doorne, 2000; Dean & McMullen, 2007).
- H4b. Higher risk acceptance has a positive impact on a positive personal perception of BC (Barbosa et al., 2007; Keh et al., 2002).

- H4c. Higher risk acceptance has a positive impact on a positive personal perception of PAE (Williams et al., 2020, Williams & Baláž, 2015, Hall & Williams, 2019; Zhang et al., 2020).

Research instrument

Data were collected using a structured questionnaire to understand tourism and hospitality university student's perception of entrepreneurship as a career path in Spain. A pilot study was conducted (50 respondents) leading to minor revisions of the research instrument (Dayour et al., 2019). The survey was based on previously tested scales about PAE, perceived support and barriers, entrepreneurial knowledge, perceived BC, networking, the relevance of sustainability and risk management, as well as socio-demographic characteristics. All items are measured using the same 5-point Likert scale, requesting a level of agreement, of certainty or efficiency. For detailed information about the survey, see Appendix 1.

Data collection and procedure

The focus on students is supported by (1) the strong relationship between education and entrepreneurship (Santero-Sanchez et al., 2015); (2) the high percentage of the population aged 18-30 studying at university in Spain (2015, UNESCO Gross Enrolment Ratio (GER) of 86.67%); and (3) access and convenience sampling needs.

The population are undergraduate students in tourism studies in Spain, where there were 26,735 students in tourism in 2016 (Source: S.G. de Coordinación y Seguimiento Universitario. Ministerio de Educación, Cultura y Deporte), implying a minimum sample size of 359 participants. Data collection started in January 2019 and finished in September 2019. 892 completed questionnaires were received from 14 universities drawn from the range of the autonomous regions (Madrid, Andalusia, Catalonia, Canary Islands, Galicia, Cantabria). After removing all incomplete questionnaires, a total of 562 was included in the analysis. As the current study only focuses on students that are already entrepreneurs, already taking steps to become entrepreneurs or potentially interested in becoming entrepreneurs, the 23 respondents not interested in entrepreneurship (4.1%) were excluded. The final sample had 539 respondents, of which 70.3% were women and 29.7% were men, reflecting the over representation of women in this academic area (Chambers et al., 2017; Pritchard & Morgan, 2017). All the participants were residents in Spain: while 78.7% were Spanish, 9.8% were European (not including Spain), and 5.9% from Central and South America.

Data analysis

Descriptive analysis and Structural Equation Models (SEM), with AMOS 26, were performed.

Study B. Qualitative analysis - interviews

Research instrument

Interview design. The semi-structured interviews were based on a general script, but the interviewee had enough flexibility to develop their own speech. The script was based on Risman's (1998) theory "Gender as a Social Structure", and the adaptation of (Segovia-Pérez et al., 2019) to the hospitality industry. The interviews sought to explore the gendered aspects of the entrepreneurial journey, and how this was negotiated by individuals, at an individual, interactional and institutional level. Sustainability being a dominant theme in the ways participants narrated and made sense of such entrepreneurial journeys. All the interviews started with the question about how they had become entrepreneurs. This general and open question lead to more specific questions around the previously mentioned theory, when needed.

Data collection and procedure

Sample. Women's entrepreneurship in tourism has been highly selective, typically being linked to stereotypical women's home and family care work (Peeters & Ateljevic, 2009); e.g. handicrafts (Kwaramba et al., 2012), the Sharing Economy (Juul, 2015), and social entrepreneurship (Kimbu & Ngoasong, 2016). Therefore, in order to avoid stereotyping, the sample sought to incorporate male and female entrepreneurs who have the tools and the skills to challenge the roles they have been traditionally assigned in the industry (Costa et al., 2013). Although snowball sampling was used, multiple entry points drew on the range of experiences in the different tourism sub-sectors: Hospitality (accommodation and restaurant industry); commercialisation (communication and distribution); transportation; events or other tourist services; and high-tech organisations. We also considered geographic location (rural-urban differential), when selecting entrepreneurs. In total, 19 interviews were undertaken. A complete description of the participants is included in Appendix 2.

Data analysis

N-VIVO 12 was used to support a post structuralist thematic analysis (deconstruction) to identify interconnections and explanations associated with specific social-economic contexts. Two researchers undertook and evaluated an analysis of the interview transcripts in two-phases (Segovia-Perez et al. 2019; Hsieh and Shannon, 2005). The initial inductive process (phase 1) identified a long list of issues around tourism entrepreneurship, regardless of gender: motivations, encountered barriers, critical success factors, risk, training, funding and the importance of networking. The inductive process was followed by a deductive process (phase 2) that raised the issues identified as relevant for sustainability support through SEM. These main issues were PAE, CB and risk management. These issues were evaluated from a gender perspective, differentiating interventions from men and women. This analysis allowed the identification of similarities and differences towards the relevance of entrepreneurship and the gendered impact of PAE, BC and risk on the relevance of sustainability.

Findings

Study A. Quantitative analysis – survey

Entrepreneurship is a potential career path for most participating students with only 4.1% of the respondents rejecting this possibility. These results are aligned with other studies of university students (e.g. del Rio et al., 2016), which indicate that higher levels of education have a positive impact on entrepreneurial intention (GEM, 2001-2019).

In the current study, although the numbers are small, women seem a little more negatively biased towards entrepreneurship (4.5% never want to become entrepreneurs, compared to men, 3%). These results were expected since the literature indicates that women have lower entrepreneurial intentions than men (Nowiński et al., 2019; Wilson et al., 2007).

In relation to Research Question 1, the evaluation throughout the entrepreneurship path shows how a very similar percentage of women and men are already entrepreneurs (6.33% and 6.24% respectively), though more men are more likely than women to be taking steps towards entrepreneurship (nascent entrepreneurship) (9.38% and 3.96%, respectively). Regarding latent entrepreneurship, men and women are broadly similar: respectively, 26.88% and 26.65% want to become entrepreneurs in less than three years, and 63% and 57.5% want to do so eventually. In comparison, in 2018, only 17% of men in the European Union (28 countries) and 9% of women were entrepreneurs; it is highly likely that many of the latent entrepreneurs will not complete the entrepreneurial journey.

The general literature (e.g. De Silva & Pownall, 2014; Glass et al., 2016; Kimbu & Ngoasong, 2016) shows how women students tend to give more support to sustainable ideas about entrepreneurship. In relation to the relevance of sustainability among latent and nascent entrepreneurs among university students, the majority of participants in the study agree or strongly agree that firms should take a leading role in the field of sustainability (S4) (89.05%; mean= 4.4 for women and 4.7 for men)., and that firms that care and are environmentally and socially sustainable have advantages in recruiting and retaining qualified employees (S5) (74.21%; mean= 3.97 for women and 3.98 for men). A gender perspective informs that women were more likely to state that the environmental performance of a company will be considered more important by financial institutions in future (S3) (78.85%; mean= 4.15 for women and 3.97 for men), that Corporate Social Responsibility (CSR) should be foundational for companies (S2) (83.67%; mean= 4.30 for women and 4.09 for men) and that environmental problems are one of the biggest challenges for society (S1) (88.13%; mean= 4.49 for women and 4.36 for men). Results show how women are positively biased towards more than half of the variables explaining the importance given to sustainability (S3, S2 and S1) with S3 and S2 being statistically significant (Pearson Chi-Square = 9.201; $p = 0.5$), but do not have significantly higher interest in all aspects of sustainability. Sustainability in entrepreneurship is important for men and women, in different ways.

This dichotomous position towards sustainability - either you are or are not interested in sustainability as an entrepreneur - from a gender perspective is in practice blurred, as suggested by post-structuralist perspectives, as will become evident when considering the findings of the Structural Equation Modelling (SEM).

To support the previous statistical analysis and to understand the construction of sustainable entrepreneurship in university students, and how this is influenced by gender and other gendered variables related to sustainability and entrepreneurship, a SEM was performed. The conceptual model was validated using CFA (Dayour et al., 2019; Joo et al., 2020) and a two-step approach was adopted. A validation of the conceptual model using CFA was followed to ensure the fit of the data in the model as well as test the hypotheses. See Appendix 3 for detailed analysis.

The detailed description of the SEM shows how the importance given to sustainability in entrepreneurship (dependent variable) is page 11). influenced by four variables (independent variable): gender (control variable), the capacity to manage risk (latent variable), BC (latent variable) and PAE (latent variable). The Squared Multiple Correlation shows how the independent variables explain 17% of the dependent variable. The final set of variables (Table A1 in Appendix 3) included in the model are the following:

SUSTAINABILITY: S1 (Environmental problems are one of the biggest challenges for our society); S2 (Corporate social responsibility should be part of the foundations of each company); S3 (The environmental performance of a company will be considered more and more by financial institutions, in the future); S4 (Firms should take a leading role in the field of sustainability).

GENDER: Female/Male

RISK: R1 (I adapt more flexibly to new situations than my friends); R2 (I manage my problems better than my friends); R3 (I would adapt better to changing living conditions than my friends); R4 (I am willing to take more risks than my friends).

PERCEIVED ATTIT. TOWARDS ENTREPRENEURSHIP (PAE): P1 (If I had the opportunity and resources, I'd like to start a firm); P2 (Being an entrepreneur would entail great satisfaction for me); P3 (Among various options, I would rather be an entrepreneur).

PERCEIVED BEHAVIOURAL CONTROL (BC): B1 (Creating and putting into operation a new venture); B2 (Interacting with key people to raise capital to create a new venture); B3 (Recognizing opportunities in the market for new products and/or services); B4 (Negotiating and maintaining favourable relationships with potential investors and banks); B5 (Keeping the new-venture creation process under control); B6 (Defining your business idea and a new business strategy).

The SEM model (see Table A3, Appendix 3) confirms that the interest in sustainability within entrepreneurship was jointly predicted by Gender, and specifically by being a woman ($\beta = 0.177$; $p < 0.000$) (H1 is confirmed). Further to this, the SEM confirms that PAE has a direct and positive relationship ($\beta = 0.153$; $p < 0.000$) (H2 is confirmed) with sustainable entrepreneurial relevance This construct relates to the satisfaction entailed within entrepreneurship and what it means, as well as an entrepreneur's orientation to innovation, society and the environment (Cohen & Winn, 2007; Crnogaj et al., 2014; Dean & McMullen, 2007).

The study sought to explore relations between BC and the importance given to Sustainable Entrepreneurship. The SEM results confirm BC has a positive impact on sustainability ($\beta = 0.108$; $p < 0.006$) (H3a is confirmed) and entrepreneurs' PAE ($\beta = 0.233$; $p < 0.000$) (H3b is confirmed) and (see Table A3, Appendix 3). BC, in this context refers to a wide range of variables that are relevant for innovation and entrepreneurship (Ajzen, 2002; Kraft et al., 2005; Ngoasong & Kimbu, 2019; Kimbu & Ngoasong, 2016): including, for example, creating and putting into operation a new venture; interacting and negotiating with people; recognising new opportunities; and keeping new projects under control.

Finally, the SEM results confirm the relevance of Risk for Sustainable Entrepreneurship. Risk, a highly gendered variable, has a positive impact on the importance attached to sustainability ($\beta = 0.158$; $p < 0.000$) (H4a is confirmed), student's perception of their BC ($\beta = 0.519$; $p < 0.000$) (H4b is confirmed), and PAE ($\beta = 0.188$; $p < 0.005$) (H4c is confirmed) (see Table A3, Appendix 3). These results are aligned with extant research on the capacity to innovate and take risks in entrepreneurship (e.g. Williams et al., 2020, Williams & Baláž, 2015, Hall & Williams et al., 2020). The findings also support Dean and McMullen's (2007) insights regarding how sustainability in entrepreneurship can be seen as a market failure and/or render uncertain environments.

Study B. Qualitative analysis - interviews

While the quantitative and qualitative analyses are not directly comparable, in this section we seek to provide further nuanced insights into the gendered nature of sustainable entrepreneurship by considering the views and experiences of established entrepreneurs.

The qualitative findings presented thus far align with the SEM Hypothesis testing, which sought to examine the relations between variables – with women being identified as being more engaged and motivated by sustainability.

In the entrepreneur's interviews, gender was important in understanding the intersection of sustainability and entrepreneurial performance (Research Question 1). The deductive process indicates that whilst both women and men are interested in sustainability, for women sustainability has a more pivotal role. For example, all women participants integrated environmental and/or social sustainability into the design of their products and services:

> We have created a company with three objectives... to improve the health of people in the city, save the planet and fix rural population (WomEnt_02, woman)

> We want to bring mindfulness to our surroundings and to tourists. We want to help. (WomENt_8, woman)

Sustainability was also understood as a motivating reason for many women to become entrepreneurs:

> It arose from a family need, basically, when my daughters were born because I am very worried about food, and well, also in what is a little healthy, where to buy organic products, that were not at disproportionate prices, and above all not only ecological but also had a certain guarantee of origin. (WomENt_04, woman)

> We really like the bio-construction [of the hotel], here are materials that... for example, we made, and that was important for us. (WomENt_06, woman)

Importantly, sustainability narratives were less evident within the men's accounts. When they did emerge, they tended to be framed within the context of broader structural frameworks (such as the Sustainable Development Goals (SDGs)) and how their business might benefit from alignment with such frameworks. As evident in MenENt_06's narrative:

> These are opportunities that you detect there, right? So, when you see that works at the political level and that anything that carries the SDGs ahead seems like it is already done, the project is already half done, so then, we jump on the bandwagon. (MenENt_06, man)

Social sustainability, moreover, was identified as particularly important for the women entrepreneurs, with several participants being engaged in social development across varying local, national and international scales:

> We buy everything we can from local producers. It is important to help our neighbours. (WomENt_05, woman)

> These are opportunities that you are detecting to give visibility to female entrepreneurship, … if I think I can make a small change, I will. I have to make female entrepreneurship visible so that things are not just about finishing your career and looking for a job. There are more options. (WomENt_09, woman)

Again, the value in supporting social sustainability was also visible in MenENt_02's (man) account: *The town where it is located, and its people are important to us*. Yet, it was notable that within the narratives of the male participants, less time and focus was granted to the role of enterprises within specific social contexts. MenENt_02's account services to represent the more abstract and brief engagement given to concerns about social sustainability, within the context of male entrepreneurial journeys.

A particular area of social concern for women participants was human resource management, with team engagement being key to women's entrepreneurial performance:

> I prioritize a model of self-management where people find their fulfilment by doing the project that… what else… in which they feel most, in which they see the most. I think it's going to be a winning concept because people are going to be happy and going to give their best. (WomEnt_02, woman)

> For me it is very important that my colleagues also find that point of personal development within the company. (WomENt_03, woman)

> In the end I am looking for a long-term collaboration. If the person is good, I want them to retire with me… The most important thing for me is the commitment of that person and to establish a link with them. (WomENt_09, woman)

Whilst again being less evident across the interviews, men also highlighted the importance of social responsibility in entrepreneurial management:

> For me the team is fundamental. I want them to talk, to feel part of this. Your personal development is fundamental to me. (MenENt_07, man)

When evaluating interviews with entrepreneurs in the consolidated and growth phases, the main differences arise around the motivation behind their entrepreneurial ventures and the varying starting points for the entrepreneurial journeys. Their starting point and their PAE are quite diversified (Research Question 2). WomENt_02, MenENt_05, WomENt_09 and MenENt_01, for example, all had a persistent feeling about doing something different, and they had long felt they would become entrepreneurs. Others, like WomENt_01, MenENt_06, MenENt_07, WomENt_03 and WomENt_05 became entrepreneurs as a result of their professional experience as employees; they needed more, and entrepreneurship fulfilled those needs. MenENt_04, WomENt_07, MenENt_03 and WomENt_08 turned to entrepreneurship out of necessity, while others, such as WomENt_04 and MenENt_02, identified market opportunities. The diverse map of motivations that leads these entrepreneurs towards their journey does not appear to be perceived and narrated through a gendered lens and the interviews can't identity any relationship between PAE and the importance given to sustainability.

The finding within the literature that women are more **risk** averse than men (e.g. De la Fuente-Cabrero et al. 2014; Brush et al., 2009) was also evident in our qualitative interviews with entrepreneurs. However, the qualitative narratives also provide more insights into the nuanced relations between women and risk; whereby women were found to accept risk, as long as it is managed conservatively (Research Question 4):

> I believe that [entrepreneurship] is a risk, and from my point of view we have always been much more realistic and cautious in how to do things, how to spend money. In entrepreneurship it is important to start with the minimum risk. (WomENt_03, woman)

> I have always clung to the false stability of a fixed job, every month I get my salary, not much, but I have my salary. (WomENt_08, woman)

Others held much more complex notions of risk, highlighting the fluidity of risk perception and the limitations in attempting to uphold fixed notions of women as necessarily and innately risk averse:

> I give importance to risk. However, in the end that focus is not usually required. (WomENt_09, woman)

Whilst a further set of women entrepreneurs exhibited more masculine discourses in their narration of risk; these serve to question constructions of women as risk averse and contest gendered assumptions within the literature:

> I am not worried about uncertainty, that is, I am not afraid, I am not paralyzed. (WomENt_02, woman)

> To become an entrepreneur was a very deep reflection that I made personally when I decided that we had to look for a Plan B. Yet, the reality is that we were quite unconscious. I want to say that I believe that much of what we do as entrepreneurs is done by illusion. (WomENt_05, woman)

Men, by contrast, conformed to gendered expectations found within the literature, which suggest they are less concerned with risk. The interview narratives indicate that men's positive perception regarding behavioural control (particularly within the context of enterprise objectives and outcomes), assisted in heightening confidence and reducing perceptions of risk:

> To control the risk, I control the objectives, I can control the way we work, I can bring the team I work with, and then that risk exists, of course, I will define it, I will go every time dwarfing more. (MenENt_05, man)

> I am quite unconscious in the sense that I am quite thrown and, and no, and I don't stop to think about the consequences, but I do it with knowledge of the cause. (MenENt_03, man)

> Fear is a sensation that I believe that any entrepreneur has, but that you have to know how to tame it, you have to know how to placate it. How? I believe that, basically, with a blind confidence in yourself and in your project. (MenENt_02, man)

Despite possessing high confidence in relation to behavioural control and risk, many men entrepreneurs were involved in large scale, tech organisations; enterprises that are, arguably understood as highly vulnerable to failure and external influence (Reid & Smith, 2002). Such disconnections between context and risk perception highlight the influence of gender in making sense of risk and behavioural control within entrepreneurial performance.

It is not possible to identify how BC impacts sustainability, within the context of the interviews. It is possible, however, to explore relations between BC and gender (Research Question 3). The women's entrepreneurship narratives appeared to conform to gendered expectations (Braun, 2010; Ngoasong & Kimbu, 2019), that suggest women are less likely to exhibit high levels of behavioural control:

> Women still question whether we have to talk, because much more has always been questioned. (WomENt_03, woman)

> Although we don't know, we have a lot of capacity, we are prepared, we have been gifted to raise children, to give birth. I believe that we have many super powers that men do not have, so we should use them … Just like we manage our house and we are very good managers. (WomENt_08, woman)

Crucially, there is a complex positioning of BC within these qualitative narratives. On one level, the women participants were clear and open in their hesitancies and vulnerabilities within the entrepreneurial context – suggesting low levels of BC. And yet, at the same time, we can see that low levels of BC are not necessarily the perceptions of the women themselves but rather stem from broader social constructions that position women's abilities. WomENt_08, for example, is quick to question these constructions in stating her belief in the 'super powers' of women, whilst WomENt_03 confirms that it is the questioning from others that influences perceptions.

Nevertheless, such perceptions influenced the ways women participants tended to position themselves within the enterprise, taking on high levels of planning and low-profile positions:

> Well, I like to be the one that is often not seen, but it is the one that helps behind the team. (WomENt_09, woman)

> Planning is perhaps more my strength. (WomENt_05, woman)

Such positionings contrast to those within the narratives of some of the men entrepreneurs:

> I am a nonconformist, I do not usually like anything of 'no' for an answer, I ask why not, or why. (MenENt_05, man)

Men too were transparent in sharing concerns with planning and learning, undertaken throughout the entrepreneurial journey – likewise, deconstructing masculine expectations concerning high levels of confident and controlling performance:

> I have learned .... Well, look, assuming I have no fucking idea of anything, because that's the way it is, really. Really. Reading a lot, finding out about things I did not know, by people who could give me their experience, and especially surrounding me with people who knew things that I do not know. (MenENt_05, man)

> Success? Always surround myself with people who know more than me, and in the end that feedback makes you, if you are a bit shabby. (MenENt_03, man)

Conclusions

This article provides three main contributions to the literature on sustainable tourism entrepreneurship. First, in utilising a mixed-methods post-structuralist approach, we uncover some of the circumstances under which men and women engage in sustainability at different points in the entrepreneurial journey, as well as the importance they attach to sustainability in tourism entrepreneurship. The women in this study tended to give considerable importance to specific aspects of sustainability, particularly the social and environmental dimensions, when engaging with tourism entrepreneurship (cf De Silva & Pownall, 2014; Kimbu & Ngoasong, 2016); confirming previous studies (Glass et al., 2016; Walls et al., 2012). Building on these ideas, we found that when men were concerned with sustainability, it tended to be conceptualised within the context of potential economic contributions if they focused on the environment and social dimensions of sustainability.

The research also unpacks some of the complexity in this picture by following up our principal focus on latent and nascent entrepreneurs with a further analysis of interviews with a group of established entrepreneurs. Although for university students, mostly at the latent and nascent stages of entrepreneurship, gender has an impact, this does not follow simple dichotomous lines. Women students are more likely to attach importance to the social and environmental aspects of sustainability, and the risk variable (conceptually, known to be highly gendered) is also significant, but neither behavioural control nor personal attitudes seem to be gendered at this stage of the journey. Amongst more established entrepreneurs, men and women, engage in sustainability differently, and in different areas. Sustainability is an important motivator for women entrepreneurs, strongly linked to the potentials of their entrepreneurship to generate environmental products and inclusive workplace environments, while men are more likely to see sustainability in terms of business opportunities. There are gender differences in approaches to risk, although these are complex, with women commenting on both their concerns, and social constructions of gender differences, while there are considerable variations in the extent to which men express their competence, or perhaps over-confidence, to manage risks and uncertainty. Gender differences are more blurred in terms of behavioural control, and absent in respect of personal attitudes. Therefore, as the findings from our analyses of individuals at different stages of the journey to entrepreneurship indicate, in keeping with post-structuralist perspectives gender differences should not be essentialised.

Second, our research approach contributes to developing a theoretical framework for unpacking and resolving tensions in understanding relations between gender and behaviour in sustainable tourism entrepreneurship. This study highlighted that, even though BC has a positive impact on the entrepreneurs' PAE and on sustainability, societal attitudes which construct, and stereotype women's abilities and capabilities directly influence their propensity to take risk and introduce specific innovations (Ajzen, 2002; Kraft et al., 2005; Williams et al., 2020). Societal attitudes constrain the actions of women participants, meaning they are less likely to engage in 'risky behaviours', negatively influencing their likely engagement with sustainable (tourism) entrepreneurship. By contrast, men entrepreneurs are traditionally seen as being confident risk takers. However, unlike previous studies (cf. De la Fuente-Cabrero et al. 2014; Brush et al., 2009) which indicate women are risk averse (in all circumstances), our findings caution against such generalisation; there were multiple perceptions and performances relating to risk, with some women being tolerant of risk if it could be carefully managed. We suggest that such an awareness, yet tolerance, of risk may be an influential positioning for tourism entrepreneurs of all genders who are seeking to engage with sustainability, but that uncertainty (unknown risk) may be a key discriminator (Williams et al., 2020) which requires further investigation.

Extending these ideas, our third major contribution is revealing the complexity of engaging in sustainable entrepreneurship (in tourism) in which societal perceptions and PAE inhibit women's risk-taking propensities (Marlow & Swail, 2014; Zhang et al., 2020) but encourages pro-environmental and prosocial behaviours. This results in women not accessing and acting on resources and opportunities in a timely manner, constraining them to focus more on social and environmental dimensions of sustainability to the detriment of the economic dimension; thereby impacting on their business performance. Women tourism entrepreneurs, just like men, need to be capacitated to become more independent, confident and able to access resources (Outsios & Farooqi, 2017) that enable them to undertake business activities in ways that consider all three dimensions of sustainability. By examining the different enabling variables (PAE, BC, risk), we have demonstrated the initial conditions, triggering events and self-reinforcing mechanisms (Bidegain Ponte, 2017; Kraft et al., 2005; Moreno-Alarcón & Cole, 2019) needed to start-up and ensure engagement in entrepreneurship by women (and men) and the effects of these on their engagement in sustainability and the sustainability of their businesses.

Our findings have policy implications for fostering gender equality through sustainable entrepreneurship in tourism, including questioning how women tourism entrepreneurs are often associated with traditional caring roles when enacting sustainable entrepreneurship (Figueroa-Domecq et al., 2020) rather than being constructed as valuable, conscious and responsible business choices. Extant tourism policies are also gender neutral and primarily focused on economic sustainability, rendering limited engagement with social and environmental sustainability. This particularly impacts women tourism entrepreneurs who, as we have identified, are more likely to engage in social and environmental sustainability, whilst attaching less significance to economic sustainability compared to men (UNWTO, 2019). Creating enabling environments and opportunities through policy initiatives at the macro and micro levels that specifically recognise gendered distinctions should be prioritised. Public sector bodies can encourage entrepreneurs concerned with environmental and social sustainability by providing training and enabling market access. As Ngoasong and Kimbu (2019, p.55) note, "these interventions can enable women entrepreneurs to move beyond their own entrepreneurial goals to secure buy-in from those who are otherwise resistant or indifferent" to engaging in all three dimensions of sustainability in tourism entrepreneurship. Otherwise, women's sustainable entrepreneurial performances will continue to be undervalued; a construction that may inhibit the capacity for sustainable tourism entrepreneurship to be utilised to help achieve the SDG's and, crucially and paradoxically, counter the SDG of 'gender equality'.

Finally, it is important to re-emphasise that as this study drew on findings from two populations, using different methodologies, they are not directly comparable. Therefore, it would be beneficial, in future studies, to undertake longitudinal research to gain further insights into how individual entrepreneurs' perceptions and approaches to sustainability shift across the entrepreneurial journey. We further suggest that, as revealed within this study, mixed methods approaches are beneficial in rendering both broad scale and nuanced insights into how gender influences sustainability and tourism entrepreneurship. Moreover, a deeper understanding of the intersections of gender and sustainability within the context of tourism entrepreneurship is required, if the medium of entrepreneurship is to be used to achieve the SDGs. For this reason, extending this research beyond the Spanish context will render more complex insights into the multiple ways through which gender influences sustainability perceptions and approaches through tourism entrepreneurship. Further, in recognising the dominance of research on tourism entrepreneurship in the Global North, we argue that future investigation should focus on how sustainability is influenced by gender in Global South contexts.

Disclosure statement

No potential conflict of interest was reported by the author(s).

ORCID

Albert Kimbu http://orcid.org/0000-0001-8505-6705
Anna de Jong http://orcid.org/0000-0003-3776-1607
Allan M. Williams http://orcid.org/0000-0001-6134-3611

References

Hall, C.M., & Williams, A.M. (Eds.). (2019). *Tourism and Innovation*. Routledge.

Ahl, H. (2006). Why research on women entrepreneurs needs new directions. *Entrepreneurship Theory and Practice, 30*(5), 595–621. https://doi.org/10.1111/j.1540-6520.2006.00138.x

Ahl, H., & Marlow, S. (2012). Exploring the dynamics of gender, feminism and entrepreneurship: advancing debate to escape a dead end?. Organization, *19*(5), 543–562. https://doi.org/10.1177/1350508412448695

Ahl, H., & Marlow, S. (2019). Exploring the false promise of entrepreneurship through a postfeminist critique of the enterprise policy discourse in Sweden and the UK. *Human Relations*, 001872671984828–001872671984848. https://doi.org/10.1177/0018726719848480

Ajzen, I. (1991). The theory of planned behavior. *Organizational Behavior and Human Decision Processes, 50*(2), 179–211. https://doi.org/10.1016/0749-5978(91)90020-T

Ajzen, I. (2002). Perceived behavioral control, self-efficacy, locus of control, and the theory of planned behavior 1. *Journal of Applied Social Psychology, 32*(4), 665–683. https://doi.org/10.1111/j.1559-1816.2002.tb00236.x

Ateljevic, I., & Doorne, S. (2000). Staying within the 'fence': Lifestyle entrepreneurship in tourism. *Journal of Sustainable Tourism, 8*(5), 378–392. https://doi.org/10.1080/09669580008667374

Bamberg, S., & Möser, G. (2007). Twenty years after Hines, Hungerford, and Tomera: A new meta-analysis of psycho-social determinants of pro-environmental behaviour. *Journal of Environmental Psychology, 27*(1), 14–25. https://doi.org/10.1016/j.jenvp.2006.12.002

Barber, B. M., & Odean, T. (2001). Boys will be boys: Gender, overconfidence, and common stock investment. *The Quarterly Journal of Economics, 116*(1), 261–292. https://doi.org/10.1162/003355301556400

Barbosa, S. D., Gerhardt, M. W., & Kickul, J. R. (2007). The role of cognitive style and risk preference on entrepreneurial self-efficacy and entrepreneurial intentions. *Journal of Leadership & Organizational Studies, 13*(4), 86–104. https://doi.org/10.1177/10717919070130041001

Barrett, M. (1995). Feminist perspectives on learning for entrepreneurship: The view from small business. In W. D. Bygrave, B. J. Bird, S. Birley, N. C. Churchill, M. Hay, R. H. Keeley, & W. E. Wetzel, Jr (Eds.), *Frontiers of entrepreneurial research* (pp. 323–336). Babson College..

Betz, M., O'Connell, L., & Shepard, J. M. (1989). Gender differences in proclivity for unethical behavior. *Journal of Business Ethics, 8*(5), 321–324. https://doi.org/10.1007/BF00381722

Bidegain Ponte, N. (2017). La Agenda 2030 y la Agenda Regional de Género: sinergias para la igualdad en América Latina y el Caribe. Santiago: United Nations. https://repositorio.cepal.org/bitstream/handle/11362/41016/S1700105A_es.pdf?sequence=7

Boley, B. B., McGehee, N. G., Perdue, R. R., & Long, P. (2014). Empowerment and resident attitudes toward tourism: Strengthening the theoretical foundation through a Weberian lens. *Annals of Tourism Research, 49*, 33–50. https://doi.org/10.1016/j.annals.2014.08.005

Bönte, W., & Piegeler, M. (2013). Gender gap in latent and nascent entrepreneurship: Driven by competitiveness. *Small Business Economics, 41*(4), 961–987. https://doi.org/10.1007/s11187-012-9459-3

Braun, P. (2010). Going green: Women entrepreneurs and the environment. *International Journal of Gender and Entrepreneurship, 2*(3), 245–259. https://doi.org/10.1108/17566261011079233

Brush, C. G., De Bruin, A., & Welter, F. (2009). A gender-aware framework for women's entrepreneurship. *International Journal of Gender and Entrepreneurship, 1*(1), 8–24. https://doi.org/10.1108/17566260910942318

Camelo-Ordaz, C., Diánez-González, J. P., & Ruiz-Navarro, J. (2016). The influence of gender on entrepreneurial intention: The mediating role of perceptual factors. *BRQ Business Research Quarterly, 19*(4), 261–277. https://doi.org/10.1016/j.brq.2016.03.001

Carlsson-Kanyama, A., Julia, I., & Röhr, U. (2010). Unequal representation of women and men in energy company boards and management groups: Are there implications for mitigation?. *Energy Policy, 38*(8), 4737–4740. https://doi.org/10.1016/j.enpol.2010.03.072

Chambers, D., Munar, A. M., Khoo-Lattimore, C., & Biran, A. (2017). Interrogating gender and the tourism academy through epistemological lens. *Anatolia, 28*(4), 501–513. https://doi.org/10.1080/13032917.2017.1370775

Chung, J. T. (1998). Risk reduction in public accounting firms: Are women more effective?. *International Review of Women and Leadership, 4*(1), 39–45.

Cohen, B., & Winn, M. I. (2007). Market imperfections, opportunity and sustainable entrepreneurship. *Journal of Business Venturing, 22*(1), 29–49. https://doi.org/10.1016/j.jbusvent.2004.12.001

Costa, C., Carvalho, I., Caçador, S., & Breda, Z. (2013). Gender and entrepreneurship in tourism: An analysis of tourism graduates' entrepreneurial profile. *Revista Turismo & Desenvolvimento, 2*(17/18), 623–635.

Creswell, J. W., & Creswell, J. D.(2017). Research design: Qualitative, quantitative, and mixed methods approaches. California: Sage publications.

Crnogaj, K., Rebernik, M., Hojnik, B. B., & Gomezelj, D. O. (2014). Building a model of researching the sustainable entrepreneurship in the tourism sector. *Kybernetes, 43*(3/4), 377–393. https://doi.org/10.1108/K-07-2013-0155

Dahles, H., Khieng, S., Verver, M., & Manders, I. (2020). Social entrepreneurship and tourism in Cambodia: Advancing community engagement. *Journal of Sustainable Tourism, 28*(6), 816–833. https://doi.org/10.1080/09669582.2019.1706544

Dayour, F., Park, S., & Kimbu, A. N. (2019). Backpackers' perceived risks towards smartphone usage and risk reduction strategies: A mixed methods study. *Tourism Management, 72*, 52–68. https://doi.org/10.1016/j.tourman.2018.11.003

De Groot, J. I., & Steg, L. (2009). Morality and prosocial behavior: The role of awareness, responsibility, and norms in the norm activation model. *The Journal of Social Psychology, 149*(4), 425–449. https://doi.org/10.3200/SOCP.149.4.425-449

De la Fuente Cabrero, C., Segovia-Perez, M., & Figueroa-Domecq, C. (2014). Implications of financial institution support for women's business projects. *ESIC Market Economic and Business Journal, 45*(3), 515–552.

De Silva, D. G., & Pownall, R. A. J. (2014). Going green: Does it depend on education, gender or income? *Applied Economics, 46*(5), 573–586. https://doi.org/10.1080/00036846.2013.857003

Dean, T. J., & McMullen, J. S. (2007). Toward a theory of sustainable entrepreneurship: Reducing environmental degradation through entrepreneurial action. *Journal of Business Venturing, 22*(1), 50–76. https://doi.org/10.1016/j.jbusvent.2005.09.003

Dean, T. J., & McMullen, J. S. (2007). Toward a theory of sustainable entrepreneurship: Reducing environmental degradation through entrepreneurial action. *Journal of Business Venturing, 22*(1), 50–76. https://doi.org/10.1016/j.jbusvent.2005.09.003

del Rio, M. D. L. C., Peris-Ortiz, M., Álvarez-García, J., & Rueda-Armengot, C. (2016). Entrepreneurial intentions and entrepreneurship education to University students in Portugal. *Technology, Innovation and Education, 2*(1), 7.

Duffy, L. N., Kline, C. S., Mowatt, R. A., & Chancellor, H. C. (2015). Women in tourism: Shifting gender ideology in the DR. *Annals of Tourism Research, 52*, 72–86. https://doi.org/10.1016/j.annals.2015.02.017

Figueroa-Domecq, C., de Jong, A., & Williams, A. M. (2020). Gender, tourism & entrepreneurship: A critical review. *Annals of Tourism Research, 84*, 102980.

Figueroa-Domecq, C., Palomo, J., Flecha-Barrio, M. D., & Segovia-Pérez, M. (2020). Technology double gender gap in tourism business leadership. *Information Technology & Tourism*, 1–32.

Filieri, R., Alguezaui, S., & Mcleay, F. (2015). Why do travelers trust TripAdvisor? Antecedents of trust towards consumer-generated media and its influence on recommendation adoption and word of mouth. *Tourism Management, 51*, 174–185. https://doi.org/10.1016/j.tourman.2015.05.007

Glass, C., Cook, A., & Ingersoll, A. R. (2016). Do women leaders promote sustainability? Analyzing the effect of corporate governance composition on environmental performance. *Business Strategy and the Environment, 25*(7), 495–511. https://doi.org/10.1002/bse.1879

Global Entrepreneurship Monitor (GEM). (2019). *Global Report 2018/2019.* https://www.gemconsortium.org/report/gem-2018-2019-global-report

Global Volunteers. (2019). *The global role of women – caretakers, conscience, farmers, educators and entrepreneurs, global volunteers*. https://globalvolunteers.org.

Goebel, A. (2003). Women and sustainability: What kind of theory do we need?. *Canadian Woman Studies, 23*(1), 77–84.

Greco, A., & de Jong, G. (2017). *Sustainable entrepreneurship: Definitions, themes and research gaps.* Working paper series 1706-CSE, University of Groningen.

Gupta, V. K., Turban, D. B., Wasti, S. A., & Sikdar, S. (2009). The role of gender stereotypes in perceptions of entrepreneurs and intentions to become an entrepreneur. *Entrepreneurship Theory and Practice, 33*(2), 397–417. https://doi.org/10.1111/j.1540-6520.2009.00296.x

Hallak, R., Brown, N., & Lindsay, N. J. (2013). Examining tourism SME owners' place attachment, support for community and business performance: The role of enlightened self-interest model. *Journal of Sustainable Tourism, 21*(5), 658–678. https://doi.org/10.1080/09669582.2012.709861

Hesselbarth, C., & Schaltegger, S. (2014). Education future change agents for sustainability-learnings from the first sustainability management master of business administration. *Journal of Cleaner Production, 62*, 24–36. https://doi.org/10.1016/j.jclepro.2013.03.042

Hsieh, H.-F., & Shannon, S. E. (2005). Three Approaches to Qualitative Content Analysis. *Qualitative Health Research, 15*(9), 1277–1288. https://doi.org/10.1177/1049732305276687

Jaén, I., Moriano, J. A., & Liñán, F. (2013). Personal values and entrepreneurial intentions: An empirical study. *Conceptual Richness and Methodological Diversity in Entrepreneurship Research*, 15

Joo, D., Woosnam, K. M., Strzelecka, M., & Boley, B. B. (2020). Knowledge, empowerment, and action: Testing the empowerment theory in a tourism context. *Journal of Sustainable Tourism*, *28*(1), 69–85. https://doi.org/10.1080/09669582.2019.1675673

Juul, M. (2015). The sharing economy and tourism: Tourist accommodation. *European Parliamentary Research Service*, *26*(5), 2016. Disponível em: http://www.europarl.europa.eu/RegData/etudes/BRIE/2015/568345/EPRS_BRI(2015)568345_EN.pdf. Acesso em,

Kassinis, G., Panayiotou, A., Dimou, A., & Katsifaraki, G. (2016). Gender and Environmental Sustainability: A Longitudinal Analysis. *Corporate Social Responsibility and Environmental Management*, *23*(6), 399–412. https://doi.org/10.1002/csr.1386

Kato, K. (2019). Gender and sustainability–exploring ways of knowing–an ecohumanities perspective. *Journal of Sustainable Tourism*, *27*(7), 939–956. https://doi.org/10.1080/09669582.2019.1614189

Keh, H. T., Der Foo, M., & Lim, B. C. (2002). Opportunity evaluation under risky conditions: The cognitive processes of entrepreneurs. *Entrepreneurship Theory and Practice*, *27*(2), 125–148. https://doi.org/10.1111/1540-8520.00003

Kepler, E., & Shane, S. (2007). *Are male and female entrepreneurs really that different*? Washington, DC: US Small Business Administration, Office of Advocacy.

Kimbu, A. N., & Ngoasong, M. Z. (2016). Women as vectors of social entrepreneurship. *Annals of Tourism Research*, *60*, 63–79. https://doi.org/10.1016/j.annals.2016.06.002

Koe, W. L., Omar, R., & Majid, I. A. (2014). Factors associated with propensity for sustainable entrepreneurship. *Procedia – Social and Behavioral Sciences*, *130*, 65–74. https://doi.org/10.1016/j.sbspro.2014.04.009

Koellinger, P., Minniti, M., & Schade, C. (2013). Gender differences in entrepreneurial propensity. *Oxford Bulletin of Economics and Statistics*, *75*(2), 213–234. https://doi.org/10.1111/j.1468-0084.2011.00689.x

Kokkranikal, J., & Morrison, A. (2002). Entrepreneurship and sustainable tourism: The houseboats of Kerala. *Tourism and Hospitality Research*, *4*(1), 7–20. https://doi.org/10.1177/146735840200400102

Kollmann, T., Stöckmann, C., & Kensbock, J. M. (2017). Fear of failure as a mediator of the relationship between obstacles and nascent entrepreneurial activity—an experimental approach. *Journal of Business Venturing*, *32*(3), 280–301. https://doi.org/10.1016/j.jbusvent.2017.02.002

Kraft, P., Rise, J., Sutton, S., & Røysamb, E. (2005). Perceived difficulty in the theory of planned behaviour: Perceived behavioural control or affective attitude? *The British Journal of Social Psychology*, *44*(Pt 3), 479–496. https://doi.org/10.1348/014466604X17533

Kuckertz, A., & Wagner, M. (2010). The influence of sustainability orientation on entrepreneurial intentions—Investigating the role of business experience. *Journal of Business Venturing*, *25*(5), 524–539. https://doi.org/10.1016/j.jbusvent.2009.09.001

Kwaramba, H. M., Lovett, J. C., Louw, L., & Chipumuro, J. (2012). Emotional confidence levels and success of tourism development for poverty reduction. *Tourism Management*, *33*(4), 885–894. https://doi.org/10.1016/j.tourman.2011.09.010

Landon, A. C., Woosnam, K. M., & Boley, B. B. (2018). Modeling the psychological antecedents to tourists' pro-sustainable behaviors: An application of the value-belief-norm model. *Journal of Sustainable Tourism*, *26*(6), 957–972. https://doi.org/10.1080/09669582.2017.1423320

Li, L. (2008). A review of entrepreneurship research published in the hospitality and tourism management journals. *Tourism Management*, *29*(5), 1013–1022. https://doi.org/10.1016/j.tourman.2008.01.003

Marlow, S. (1997). Self-employed women- new opportunities, old challenges?. *Entrepreneurship & Regional Development*, *9*(3), 199–210. https://doi.org/10.1080/08985629700000011

Marlow, S., & Patton, D. (2005). All credit to men? Entrepreneurship, finance, and gender. *Entrepreneurship Theory and Practice*, *29*(6), 717–735. https://doi.org/10.1111/j.1540-6520.2005.00105.x

Marlow, S., & Swail, J. (2014). Gender, risk and finance: Why can't a woman be more like a man? *Entrepreneurship & Regional Development*, *26*(1-2), 80–96. https://doi.org/10.1080/08985626.2013.860484

Mattis, M. C. (2004). Women entrepreneurs: Out from under the glass ceiling. *Women in Management Review*, *19*(3), 154–163. https://doi.org/10.1108/09649420410529861

Minniti, M., Arenius, P., & Langowitz, N. (2005). *2004 Global entrepreneurship monitor special topic report: Women and entrepreneurship*. Center for Women's Leadership at Babson College.

Moreno-Alarcón, D. A., & Cole, S. (2019). No sustainability for tourism without gender equality. *Journal of Sustainable Tourism*, *27*(7), 903–919. https://doi.org/10.1080/09669582.2019.1588283

Ngoasong, M. Z., & Kimbu, A. N. (2019). Why hurry? The slow process of high growth in women-owned businesses in a resource-scarce context. *Journal of Small Business Management*, *57*(1), 40–58. https://doi.org/10.1111/jsbm.12493

Nikraftar, T., & Hosseini, E. (2016). Factors affecting entrepreneurial opportunities recognition in tourism small and medium sized enterprises. *Tourism Review*, *71*(1), 6–17. https://doi.org/10.1108/TR-09-2015-0042

Nowiński, W., Haddoud, M. Y., Lančarič, D., Egerová, D., & Czeglédi, C. (2019). The impact of entrepreneurship education, entrepreneurial self-efficacy and gender on entrepreneurial intentions of university students in the Visegrad countries. *Studies in Higher Education*, *44*(2), 361–379. https://doi.org/10.1080/03075079.2017.1365359

Outsios, G., & Farooqi, S. A. (2017). Gender in sustainable entrepreneurship: Evidence from the UK. *Gender in Management: An International Journal, 32*(3), 183–202. https://doi.org/10.1108/GM-12-2015-0111

Peeters, L. W. J., & Ateljevic, I. (2009). Women empowerment entrepreneurship nexus in tourism: Processes of social innovation. In J. e. Ateljevic & S. J. Page (Eds.), *Tourism and entrepreneurship: International perspectives* (pp. 75–90). Elsevier.

Ploum, L., Blok, V., Lans, T., & Omta, O. (2018). Toward a validated competence framework for sustainable entrepreneurship. *Organization & Environment, 31*(2), 113–132. https://doi.org/10.1177/1086026617697039

Pritchard, A., & Morgan, N. (2017). Tourism's lost leaders: Analysing gender and performance. *Annals of Tourism Research, 63*, 34–47. https://doi.org/10.1016/j.annals.2016.12.011

Reid, G. C., & Smith, J. A. (2002). How do venture capitalists handle risk in high technology ventures? *Performance Measurement and Management Control, 12*, 361–379.

Roomi, M. A., Harrison, P., & Beaumont-Kerridge, J. (2009). Women-owned small and medium enterprises in England analysis of factors influencing the growth process. *Journal of Small Business and Enterprise Development, 16*(2), 270–288. https://doi.org/10.1108/14626000910956056

Rowe, G., & Wright, G. (2001). Differences in expert and lay judgements of risks: Myth or reality?. *Risk Analysis : Analysis, 21*(2), 341–356. https://doi.org/10.1111/0272-4332.212116

Santero-Sanchez, R., Segovia-Pérez, M., Castro-Nuñez, B., Figueroa-Domecq, C., & Talón-Ballestero, P. (2015). Gender differences in the hospitality industry: A job quality index. *Tourism Management, 51*, 234–246. https://doi.org/10.1016/j.tourman.2015.05.025

Scheyvens, R. (2000). Promoting women's empowerment through involvement in ecotourism: Experiences from the Third World. *Journal of Sustainable Tourism, 8*(3), 232–249. https://doi.org/10.1080/09669580008667360

Segovia-Pérez, M., Figueroa-Domecq, C., Fuentes-Moraleda, L., & Muñoz-Mazón, A. (2019). Incorporating a gender approach in the hospitality industry: Female executives' perceptions. *International Journal of Hospitality Management, 76*, 184

Sexton, D., & Bowman-Upton, N. (1990). Female and male entrepreneurs: Psychological characteristics and their role in gender-related discrimination. *Journal of Business Venturing, 5*(1), 29–36. https://doi.org/10.1016/0883-9026(90)90024-N

Shah, S. S., Shah, A. A., & Khaskhelly, N. (2018). Pragmatism research paradigm: a philosophical framework of advocating methodological pluralism in social science research. *Grassroots, 52*(1), 90

Shaw, G. J., & Williams, A. M. (1999). Tourism Entrepreneurship: The Role of Small Business in British Coastal Resorts. Wirtschaftsfaktor Tourismus, Münster, Selbstverlag des Instituts für Geographie der Westfälischen Wilhelms-Universität Münster, 25–31.

Shepherd, D. A., & Patzelt, H. (2011). The new field of sustainable entrepreneurship: Studying entrepreneurial action linking "what is to be sustained" with "what is to be developed". *Entrepreneurship Theory and Practice, 35*(1), 137–163. https://doi.org/10.1111/j.1540-6520.2010.00426.x

Thompson, B. S., Gillen, J., & Friess, D. A. (2018). Challenging the principles of ecotourism: Insights from entrepreneurs on environmental and economic sustainability in Langkawi, Malaysia. *Journal of Sustainable Tourism, 26*(2), 257–276. https://doi.org/10.1080/09669582.2017.1343338

Truong, D., Liu, R. X., & Yu, J. J. (2020). Mixed methods research in tourism and hospitality journals. *International Journal of Contemporary Hospitality Management, 32*(4), 1563–1579. https://doi.org/10.1108/IJCHM-03-2019-0286

UN Women (2014). World survey on the role of women in development, UN Women, New York (online). www.unwomen.org.

UN World Tourism Organization (2019). *Global Report on Women in Tourism 2019*. WTO.

Vujko, A., Tretiakova, T. N., Petrović, M. D., Radovanović, M., Gajić, T., & Vuković, D. (2019). Women's empowerment through self-employment in tourism. *Annals of Tourism Research, 76*(C), 328–330. https://doi.org/10.1016/j.annals.2018.09.004

Wachholz, S. (2011). Ecofeminism. In Deen K. Chatterjee (ed.). *Encyclopedia of Global Justice.*, 289–290.

Walls, J. L., Berrone, P., & Phan, P. H. (2012). Corporate governance and environmental performance: Is there really a link? *Strategic Management Journal, 33*(8), 885–913. https://doi.org/10.1002/smj.1952

Williams, A. M., & Baláž, V. (2015). Tourism risk and uncertainty: Theoretical reflections. *Journal of Travel Research, 54*(3), 271–287. https://doi.org/10.1177/0047287514523334

Williams, A. M., Rodríguez Sánchez, I., & Škokić, V. (2020). Innovation, risk, and uncertainty: A study of tourism entrepreneurs. *Journal of Travel Research*, https://doi.org/10.1177/0047287519896012

Wilson, F., Kickul, J., & Marlino, D. (2007). Gender, entrepreneurial self-efficacy, and entrepreneurial career intentions: Implications for entrepreneurship education. *Entrepreneurship Theory and Practice, 31*(3), 387–406. https://doi.org/10.1111/j.1540-6520.2007.00179.x

Zahra, S. A., Wright, M., & Abdelgawad, S. G. (2014). Contextualization and the advancement of entrepreneurship Research. *International Small Business Journal: Researching Entrepreneurship, 32*(5), 479–500. https://doi.org/10.1177/0266242613519807

Zampetakis, L. A., Bakatsaki, M., Kafetsios, K., & Moustakis, V. S. (2016). Sex differences in entrepreneurs' business growth intentions: An identity approach. *Journal of Innovation and Entrepreneurship, 5*(1), 29. https://doi.org/10.1186/s13731-016-0057-5

Zhang, C. X., Kimbu, A. N., Lin, P., & Ngoasong, M. Z. (2020). Guanxi influences on women intrapreneurship. *Tourism Management, 81*, 104137. https://doi.org/10.1016/j.tourman.2020.104137

Appendix 1. Questionnaire design

	Source
ENTREPRENEURIAL INTENTION	
Are you, alone or with others, currently the owner of a business you help manage, self-employed, or selling any goods or services to others? Yes / No	GEM (2019), Costa et al. (2013), and Kollmann et al. (2017)
Are you, alone or with others, currently trying to start a new business, including any self-employment or selling any goods or services to others? Yes / No	GEM (2019), Costa et al. (2013), and Kollmann et al. (2017)
You are not an entrepreneur yet, but how well do the following periods describe your potential interest in becoming an entrepreneur in the future?	Authors
SUSTAINABILITY (5 point Likert scale) (Coefficient Alphas = 0.898). *Please value your level of agreement with the following statements on a scale from "Completely disagree" (1) to "Completely agree" (5)*	
Environmental problems are one of the biggest challenges for our society (S1)	Kuckertz and Wagner (2010)
Corporate social responsibility should be part of the foundations of each company (S2)	Kuckertz and Wagner (2010)
The environmental performance of a company will be considered more and more by financial institutions, in the future (S3)	Kuckertz and Wagner (2010)
Firms should take a leading role in the field of sustainability (S4)	Kuckertz and Wagner (2010)
Firms that care and are environmentally and socially sustainable have advantages in recruiting and retaining qualified employees (S5)	Kuckertz and Wagner (2010)
RISK (5 point Likert scale) (Coefficient Alphas = 0.898). *Please compare yourself to your best friends and rate your abilities on a scale from "Certainly Not" (1) to "Certainly Yes"*	
I adapt more flexibly to new situations than my friends (R1)	Williams and Baláž (2015)
I manage my problems better than my friends (R2)	Williams and Baláž (2015)
I would adapt better to changing living conditions than my friends (R3)	Williams and Baláž, (2015)
I am willing to take more risks than my friends (R4)	Williams and Baláž, (2015)
PERCEIVED ATTITUDE TOWARDS ENTREPRENEURSHIP (PAE) (5 point Likert scale) (Coefficient Alphas = 0.848). *Please, answer to the following questions about your perception of entrepreneurship on a scale from "Completely disagree" (1) to "Completely agree" (5)*	
If I had the opportunity and resources, I'd like to start a firm (P1)	Zampetakis et al. (2016)
Being an entrepreneur would entail great satisfaction for me (P2)	Zampetakis et al. (2016)
Among various options, I would rather be an entrepreneur (P3)	Zampetakis et al. (2016)
Being an entrepreneur implies more advantages than disadvantages (P4)	Zampetakis et al. (2016)
PERCEIVED BEHAVIOURAL CONTROL (BC) (5 point Likert scale) (Coefficient Alphas = 0.961). *To what extent are you able to effectively perform the following tasks: on a scale from* "Not effectively at all" *(1) to* "Completely effectively" *(5)*	
Creating and putting into operation a new venture (B1)	(Jaén et al., 2013)
Interacting with key people to raise capital to create a new venture (B2)	Jaén et al (2013)
Recognising opportunities in the market for new products and/or services (B3)	Jaén et al (2013)
Negotiating and maintaining favourable relationships with potential investors and banks (B4)	Jaén et al (2013)
Keeping the new-venture creation process under control (B5)	Jaén et al (2013)
Defining your business idea and a new business strategy (B6)	Jaén et al (2013)
Having a successful business (B7)	Jaén et al (2013)

Table A1. Descriptive and CFA statistics for each item.

	Factor loading	Means	Cronbach's alpha
Sustainability			
S1 (Environmental problems are one of the biggest challenges for our society)	0.747	4.45	0.799
S2 (Corporate social responsibility should be part of the foundations of each company)	0.772	4.24	0.799
S3 (The environmental performance of a company will be considered more and more by financial institutions, in the future)	0.634	4.10	0.799
S4 (Firms should take a leading role in the field of sustainability)	0.687	4.36	0.799
Risk			
R1 (I adapt more flexibly to new situations than my friends)	0.822	3.78	0.747
R2 (I manage my problems better than my friends)	0.551	3.61	0.747
R3 (I would adapt better to changing living conditions than my friends)	0.621	3.67	0.747
R4 (I am willing to take more risks than my friends)	0.641	3.53	0.747
Attitude towards entrepreneurship (PAE)			
P1 (If I had the opportunity and resources, I'd like to start a firm)	0.781	4.20	0.817
P2 (Being an entrepreneur would entail great satisfaction for me)	0.806	4.02	0.817
P3 (Among various options, I would rather be an entrepreneur)	0.736	3.66	0.817
Perceived behavioural control (BC)			
B1 (Creating and putting into operation a new venture)	0.760	3.31	0.858
B2 (Interacting with key people to raise capital to create a new venture)	0.643	3.31	0.858
B3 (Recognising opportunities in the market for new products and/or services)	0.746	3.42	0.858
B4 (Negotiating and maintaining favourable relationships with potential investors and banks)	0.662	3.25	0.858
B5 (Keeping the new-venture creation process under control)	0.723	3.21	0.858
B6 (Defining your business idea and a new business strategy)	0.729	3.44	0.858

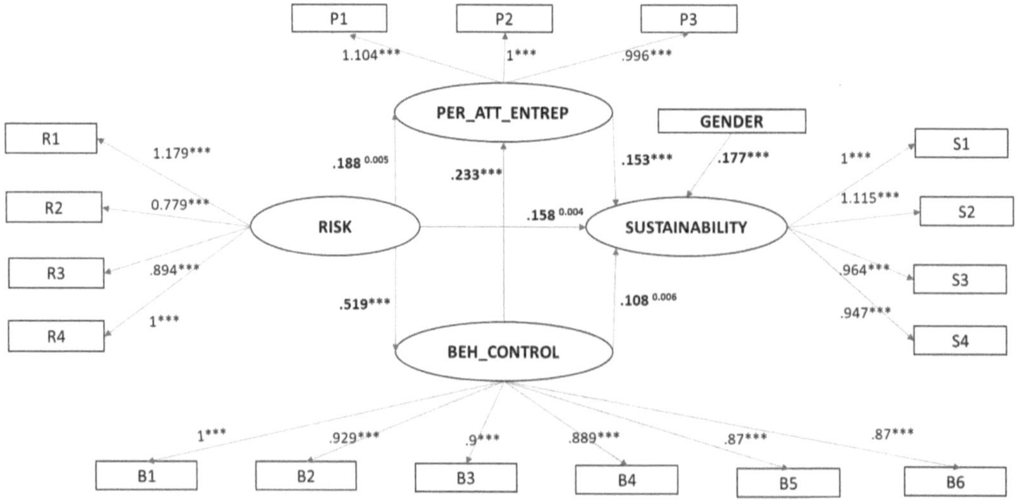

Figure A1. Structural path model.

Appendix 2. Participants in interviews

Num.	Code	Sex	Age	Tourism industry
1	WomENt_01	Women	30–40	Consultancy
2	WomENt_02	Women	40–50	Tourism intermediaries
3	MenENt_01	Men	50–60	Consultancy
4	WomENt_03	Women	30–40	Tourism intermediaries
5	WomENt_04	Women	30–40	Tourism intermediaries
6	MenENt_02	Men	40–50	Hotels and accommodation
7	WomENt_05	Women	40–50	Hotels and accommodation
8	WomENt_06	Women	40–50	Hotels and accommodation
9	MenENt_03	Men	40–50	Consultancy
10	MenENt_04	Men	40–50	Hotels and accommodation
11	WomENt_07	Women	40–50	Hotels and accommodation
12	WomENt_08	Women	40–50	Consultancy
13	MenENt_05	Men	30–40	Tourism intermediaries
14	WomENt_09	Women	30–40	Tourism intermediaries
15	MenENt_06	Men	40–50	Consultancy
16	WomENt_10	Women	30–40	Food and beverage
17	MenENt_07	Men	30–40	Hotels and accommodation
18	MenENt_08	Men	30–40	Transportation
19	MenENt_09	Men	50–60	Hotels and accommodation

Table A2. Convergent and discriminant validity.

	CR	AVE	Risk	Sustain	Per_Att_Entrep	Beh_Control
Risk	0.757	0.444	**0.666**			
Sustainability	0.803	0.507	0.272***	**0.712**		
PAE	0.818	0.601	0.268***	0.286***	**0.775**	
BC	0.86	0.507	0.383***	0.282***	0.336***	**0.712**

Appendix 3

Following Dayour et al. (2019) factor correlations, loadings, model fit indices, as well as validity and reliability were estimated. Four constructs (Sustainability, Risk, PAE, BC) and one control variable (Gender) were included in the proposed model. Items with loadings below 0.5 were removed (Dayour et al., 2019, Boley et al., 2014; Filieri et al., 2015).). Table A1 shows how the included items loaded significantly between 0.551 and 0.806 which is indicative that the interrelationships between items and associated constructs were high. In addition, and for internal consistency of the measurements, the Cronbach's alpha values were checked against the lower limit of 0.7. Values exhibit adequate internal consistency.

The adequacy of the model regarding its ability to mirror variance and covariance of the dataset was validated using the goodness-of-fit in CFA (Duffy et al. 2015; Dayour et al. (2019). Results of the CFA show that fit indices fall between suggested thresholds (PCMIN/DF = 2.570; CFI = 0,941;RMSE = 0.054; NFI = 0.908). According to the results in Table A2 only Risk violates the test of discriminant and convergent validity, with an AVE of 0.444 (44.4%). Although this might be a limitation, other studies (e.g. Boley et al., 2014) confirm the possibility to use this construct with a lower validity when there is strong precedence in the use of those items (e.g. Landon et al., 2018) and adequate factor loading and construct reliability are shown; also, the measurement model, including this construct, demonstrated acceptable fit, while modification indices did not reveal mechanisms for improvement (Landon et al., 2018).

Following the measurement model, the hypothesized relationships among constructs were tested using structural equation modelling (Figure A1). Co-variances were estimated for the four exogenous latent variables, and the outcome variables measuring intent, respectively.

The calculation of the R square show how the independent variables explain 17% of the dependent variable. The model shows a positive impact of the four exogeneous variables (three latent variables and one control variable). Results (Table A3) confirm that the interest in sustainability within entrepreneurship was jointly predicted by Gender, and specifically by being a woman ($\beta = 0.177$; $p < 0.000$) (H1), Risk acceptance ($\beta = 0.158$; $p < 0.000$) (H4a),

Table A3. Hypothesis relationships between constructs and observed relationship from the SEM.

Hypothesis	Hypothesized relationship	Beta coefficient	p	Support to hypothesis
H1	H1. Women are more interested in sustainability in entrepreneurship in tourism; Sustaina <— COgender	0.177	***	YES
H2	H2. Perceived attitude towards entrepreneurship (the capacity to identify business opportunities and become an entrepreneur) impacts positively on the importance attached to sustainability in entrepreneurship; Sustaina <— PAE	0.153		YES
H3a	H3a. Higher BC impacts positively on the importance attached to sustainability in entrepreneurship; Sustaina <— BC	0.108	0.006	YES
H3b	H3b, Higher BC has a positive impact on having a PA; PAE<— BC	0.233	***	YES
H4a	H4a. The higher the risk acceptance the greater the importance attached to Sustainability in Entrepreneurship; Sustaina <— Risk	0.158	***	YES
H4b	H4b. Higher risk acceptance has a positive impact on a positive personal perception of BC; BC <— Risk	0.519	***	YES
H4c	H4c. Higher risk acceptance has a positive impact on a positive personal perception of PAE; PAE <— Risk	0.188	0.005	YES

positive perception about their BC ($\beta = 0.108$; $p < 0.006$) (H3a) and a positive PAE as a career (($\beta = 0.153$; $p < 0.000$) (H2). Hypothesis testing of the relations amongst variables have also been confirmed. A highly gendered variable, Risk, has a positive impact on the student's perception of their BC ($\beta = 0.519$; $p < 0.000$) (H4b) and their PAE ($\beta = 0.188$; $p < 0.005$) (H4c). Furthermore, BC has a positive impact on their PAE ($\beta = 0.233$; $p < 0.000$) (H3b).

Part II

Ideology, Social Organising and the Human-Nature Divide

Are we all in this together? Gender intersectionality and sustainable tourism

Donna Chambers

ABSTRACT

In this paper I provide a critical discussion of gender intersectionality and its relevance for sustainability in tourism, focusing on the intersection between gender and race. I argue that Black women in tourism suffer from a double negation caused by both sexism and racism, but this has received little acknowledgement or critical discussion in studies of sustainable tourism. However, an intersectional approach to gender is vital as it rejects reductionist views of women's experiences in tourism and the attendant power relationships that such an approach (re)produces. I argue that it is through a critical understanding of the importance of an intersectional approach to gender that we can move closer to achieving equity in the development of tourism. My discussion is theoretically underpinned by critical race theory (CRT) and the related logics of Black feminism which I use to highlight the (re)presentation of Black women in tourism. Methodologically I draw on the storytelling technique popularised in CRT to analyse a fictional film – 'Heading South', to explore Black women's elision as agentic beings as well as their (re)presentation as vulnerable and submissive. I argue that such (re)presentations of Black women in tourism popular cultural narratives have material implications for sustainable development.

Introduction

> Many women of color feel obliged to make [a choice] between ethnicity and womanhood; how can they? You never have/are one without the other. The idea of two, illusory separated identities, one ethnic, the other woman (or more precisely female) again partakes in the Euro-American system of dualistic reasoning and its age-old divide-and-conquer tactics (Trinh, 1989, p. 104)

I begin this paper with the above quotation from T. Minh-ha Trinh (1989) who speaks of the impossibility of separating one's gender identity from one's ethnic identity. This pluralist approach to identity is the foundational logic of the concept of intersectionality (Crenshaw, 1989). In this paper I focus on the intersection between gender and race and contend that this has been neglected in studies of sustainable tourism. Gender has been extensively discussed in tourism studies (Ferguson & Alarcón, 2015; Figueroa-Domecq et al., 2020) and there are also several publications which focus on race (Boukhris, 2017; Jamerson, 2016). However, there have been very limited explorations of both gender *and* race despite several tourism scholars advocating for this more pluralist approach (see Alarcón & Cole, 2019; Gibson, 2001; Mooney, 2018, 2020; Pritchard, 2014). Specifically, I argue that in traditional discourses of tourism, Black women's voices have been largely elided (see Small, 2013) although a counter-narrative is slowly emerging (see for example Chambers & Buzinde, 2015; Gill, 2019; Lee, 2017).

Black women in tourism are affected by a double negation – they are frequently erased from both discourses of sexism and racism. Black women in tourism have scarcely been perceived as having agency. Instead, they have primarily been the objects of research and too often (re)presented by white men and women as socio-culturally, economically and politically vulnerable and thus in need of 'empowerment' (Arnfred, 2004; Syed, 2010). Where their voices are 'heard' it is often through the interpretive lens of Western scholars and (re)producers of culture in popular media such as in film. This is particularly troubling in the context of sustainable tourism as it belies the equality and social justice agendas which underpin the very notion of sustainability. Indeed, almost 20 years ago Cohen (2002) advocated for a more critical interrogation of the concept of sustainable tourism to include notions of social equity. Much later, Jamal and Camargo (2014) spoke of the need to discuss sustainable tourism in the context of theories of justice and ethics and argued that how people are represented are important considerations in sustainable tourism. In similar vein, Bright et al. (2021) in an article in a Special Issue of the Journal of Sustainable Tourism on Justice and Tourism, spoke of 'just representation' in reference to memorial tourist sites in Tennessee in America and contended that many of these sites occluded representations of non-white male history. It is evident therefore that issues of 'just representation' are important for sustainable tourism and in this paper, I contend that this issue is pertinent in the context of the (re)presentation of Black women.

In this paper I will therefore provide a critical discussion of gender intersectionality and its importance for sustainable tourism. Such an approach rejects reductionist views of women's experiences in tourism and attendant power relationships that are thereby (re)produced. My discussion is theoretically underpinned by critical race theory and relatedly, Black feminism. Methodologically, the discussion adapts the CRT technique of storytelling which enables me to unpack the way in which Black women are (re)presented in a film on female sex tourism in post-colonial Haiti. I begin with a discussion of critical race theory which provides the overarching theoretical foundation for this work.

Critical race theory

It is generally accepted that race is a powerful socio-cultural construct and important signifier (Morrison, 1992; Harris, 1995). Writing on New Zealand as a tourist state, Werry (2011) provides a cogent description of the complex nature of race:

> Race is a cultural construction or formation: it is a fluid constellation of discourse, belief, affect, physical or linguistic repertoires, sensibilities and signs attributed to or performed by subjects in connection with a claim regarding biology, and in relation to specific social, political, and economic interests (2011, pp. xxviii–xxiv).

Race is not an immutable, objective fact but is multifaceted, shifting, and unstable. Despite this understanding of race as a socio-cultural rather than biological (innate) phenomenon, people are nevertheless ascribed into racial categorisations involving strict binary oppositions, the most predominant being that between those racialised as white (Western) and those racialised as Black (the 'others'), with the former assuming a superior role. The notion of blackness has historically been associated with several negative stereotypes and these are always juxtaposed against positive notions of whiteness. As such blackness, according to Morrison (1992), acts as a foil against which whiteness is defined. Understanding the ways in which blackness (and necessarily whiteness) has been socio-culturally constructed and its material effects on the ways that we experience our contemporary worlds has been an issue which has long preoccupied scholars across a variety of disciplines including in the social sciences. One way in which scholars have sought to understand the problematics associated with the construction of race is through critical race theory.

Critical race theory (CRT) emerged in the unique socio-cultural, politico-legal historical context of the United States of the 1970s and perceived racism not as an exception in American society and polity, but as a normal and natural part of it that required immediate action (Delgado & Stefancic, 2000). CRT was seen as an integral ideological framework which articulated the way in which racism was a system of oppression that was deeply embedded in American society (Parker & Lynn, 2002). Those who subscribed to CRT sought to dismantle the very systems and structures of the American socio-legal framework which underpinned racial inequality and oppression. However, it is important to note that CRT is not a cohesive set of theories (Bell,1995) but those who subscribe to this approach do share a number of principles, including but not limited to the following: racism is inherent in our societies; racism is intersectional and must be understood also in the context of other inequalities such as gender, class and sexuality; race is a social construct which is historically contingent; traditionally dominant ideologies that subscribe to objectivity and colour blindness must be challenged; there must be a commitment to social justice that is both liberating and transformational; adherence to the central role of marginalised voices and a concomitant methodological approach that draws on the notion of 'cultural intuition' which includes storytelling and counter-storytelling; and finally the embrace of trans disciplinarity (Carbado & Roithmayr, 2014; Solorzano & Yosso, 2001).

Gloria Anzaldúa (1990) cited in Solorzano and Yosso (2001) captured CRT cogently in that for Black people:

> Because we are not allowed to enter discourse, because we are often disqualified and excluded from it, because what passes for theory these days is forbidden territory for us, it is vital that we occupy theorizing space, that we not allow white men and women solely to occupy it. By bringing in our own approaches and methodologies, we transform that theorising space (p. 488).

CRT is a product of the particular social and discursive system within which it was created. Nevertheless, it would be true to say that some of its underlying precepts can be applied in other geographical contexts and other academic disciplines where there are multiple problematics associated with race such as in sustainable tourism. So far, CRT has been extensively used in the field of education (Ladson-Billings, 1998; Parker & Lynn, 2002; Solorzano & Yosso, 2001) and to a lesser extent in the field of sport (Anderson & McCormack, 2010; Burdsey, 2011; Hylton, 2011, 2018).

CRT in tourism research

CRT has received little attention in sustainable tourism studies although race is an important marker of difference in tourism. It is this difference which has stimulated Western tourism flows to countries and cultures of the Global South. Werry (2011) argues that tourism 'operates as a significant technology of racial governmentality [and also makes of its] subjects ambivalent collaborators in the racial project of the state' (p. xxviii). Undoubtedly, critical theoretical discussions of the racialised, marginalised 'other' is underserved in tourism scholarship. Indeed, in my exegesis of the mainstream and related tourism literature I have found very few explicit references to CRT as a way of understanding the racialisation of Black bodies and cultures in tourism.

There are a few exceptions – Henry (2019) argues for the use of CRT to understand 'the cultural politics of racism in the volunteer tourism encounter' (p. 326). He went on to suggest that volunteer tourism is a 'manifestation of contemporary whiteness' (Henry, 2019, p. 327); Jamerson (2016) draws on theories of racial formation and racial neoliberalism to highlight the link between culture and economics in tourism and suggests that racial difference has value and is an important asset in tourism; Dillette et al. (2019) use CRT explicitly to reveal the critical role that understanding the Black travel experience can play in learning about race in tourism; Bright et al. (2021) also draw on the logics of CRT to examine roadside markers in Tennessee in terms of whether they depict 'just representations' of the past. Torabian and Miller (2017), while not explicitly citing CRT are nevertheless inspired by its precepts in their discussion of the impact of race and nationality on freedom of movement in tourism. Similar influences of CRT can be found in the writings of Duffy et al. (2019) who analyse how racial violence in the United States influences Black Travel and Dillette (2021) who uses the theory of double consciousness, originally coined by early CRT scholar W.E.B DuBois to understand African Americans' experiences of roots tourism in their travels to Ghana.

There are then limited studies which apply CRT explicitly. I would suggest however that CRT is part of the canon of critical theory and there have been several publications in the history of tourism scholarship which have drawn on this broad theoretical approach (although not explicitly cited as such). For example, Mellinger (1994) examined photographic postcards of African Americans from the South between 1893 and 1917 and concluded that in these images Black people were 'positioned in a racist regime of representation that constructed subjectivities for those depicted, and identities for their viewers' (p. 756). More recent research in tourism has embraced postcolonial perspectives and tend to focus on tourism as neo-colonialism or as a new form of imperialism in which former colonial peoples are exploited culturally, economically, and sexually by Western tourists (Akama et al., 2011; Amoamo & Thompson, 2010; Hall & Tucker, 2004; Hoppe, 2010; Osagie & Buzinde, 2011).

A distinct body of critical scholarship emerged approximately 15 years ago and has been deemed as 'critical tourism studies'. Critical tourism scholars subscribe to a values driven approach to research and action, disrupting hegemonic discourses and practices in tourism in order to achieve transformations in terms of planetary justice and sustainability (Ateljevic et al., 2013; Boluk et al., 2019; Cole & Morgan, 2010; Devine & Ojeda, 2017; Higgins-Desbiolles, 2008; Lyon & Hunter-Jones, 2019). Critical tourism scholars are concerned with dismantling a range of inequalities including those associated with gender and race. Many of the works in critical tourism speak of the way in which indigenous (primarily non-white) cultures are exploited in tourism but surprisingly little have provided any in-depth theorisations of the concept of race and its material effects in tourism. Jamerson suggests that:

> Critical tourism scholars may benefit from an increased racial awareness in their work towards providing a counternarrative to strictly business-based tourism research. Meanwhile, race scholars might benefit from an increased understanding of ways racial difference operates within tourism, as it is a major site of negotiations of Otherness (2016, p. 1038)

There are some exceptions where tourism scholars have provided reflexive accounts of their own racial and ethnic identities and how this has been influenced by the colonial encounter (Chambers & Buzinde, 2015; Wijesinghe, 2020). I suggest that CRT provides an appropriate organising framework for an examination of race and its centrality to tourism studies. Further, an important precept of CRT is that race must also be understood in relation to its intersections to other socio-cultural constructs such as gender and in the following section I will discuss Black feminism, inspired by CRT, and which highlights the intersection between gender and race.

Black feminism

Race of course has intersections with other identity categories such as class, sexuality, and gender. However, it is the intersection with gender, namely in terms of women, where the discourse on intersectionality is, arguably, most fully developed. Specifically, Black feminism is very much affiliated with CRT and Kimblerlé Crenshaw, is attributed with popularising the term 'intersectionality'. Crenshaw observed the marginalisation of African American women in both the legal and the public spheres and for her, intersectionality refers to the way in which race, class and sexual subordination are intimately intertwined (Crenshaw, 1989). Within the sociology literature it is perhaps well established that 'any analysis of women that ignores race will be incomplete' and might refer only to issues that affect white women (Browne & Misra, 2003, p. 487). In similar fashion, analyses of racial inequality that elides the lives of Black women is glaringly inadequate. The issue is not simply one of adding gender to analyses of race and vice versa but understanding that both need to be integrated in such a way that a fuller and richer appreciation of the challenges and opportunities faced by all the groups involved is achieved (Browne & Misra, 2003; Hill Collins, 1990). According to Hill Collins (1990, p. 555) 'replacing additive models of oppression with interlocking ones creates possibilities for new paradigms'.

In tourism studies there is increasing recognition that an intersectional approach to gender provides a more nuanced and richer means of understanding the issues affecting women in tourism (Chambers & Rakic, 2018; Cole & Ferguson, 2015; Gao & Kerstetter, 2016; Hutchings et al., 2020; Mooney, 2018; Mooney et al., 2017). However, most of these publications do not focus on the intersection between gender and race and while a few reflect on issues to do with ethnicity (Khoo-Lattimore et al., 2019; Patil, 2011), they do not provide theoretical analyses drawing on insights from Black feminist literature where much of the work on gender intersectionality resides. Bott (2018) who examines 'tribal' tourism in Vietnam using critical discourse analyses does speak about the power of racialising discourses and draws on Said's Orientalism but does not make any reference to Black feminist work.

CRT and Black feminism have been critiqued due to their apparent creation of dualisms between concepts such as black/white, Western/non-Western, dominant/oppressed, man/woman. For example, Sen,(2006) rejects notions of the singularity of cultures and identities as an unhelpful obsession with Western supremacy and Western domination. Suleri (1992) speaks of the rhetoric of 'us' and 'them' as being 'devastating' and as beleaguering issues of identity formation in our contemporary world while Trinh (1989) argues for the rejection of a homogenisation of categories such as 'Third World women' and 'postcolonial women' instead calling for a non-binary understanding of difference. Mohanty (1984) presents a cogent argument in speaking about the way in which some Western feminist writers have framed the 'Third World woman' as a category of analysis which objectifies them. She argues that such objectification is dangerous and needs to be challenged.

Aziz (1997, p.74) argues for a 'feminism of difference' that acknowledges the 'actual historical differences in colonialism, imperialism, racism and representation – and how these are appropriated'. Anthias and Yuval-Davis (1990) also suggest that while there are struggles that affect all postcolonial women, some struggles are specific to particular ethnic groups. Trinh (1989) goes so far as to say that such dualisms reflect the logics of domination (such as patriarchy and colonialism) and to continue to use these is to submit to these logics. This is a view shared by Black feminist Hill Collins (1990) who suggests that a simple binary between black/white represents Eurocentric masculinist dualistic thinking which necessarily results in relationships where one side is privileged and the other denigrated.

While these are plausible arguments, it is also important to recognise the positive aspects of collective identities, their political role in advancing equality and in fostering a sense of belonging amongst people. For example, Black feminists have played a key role in exposing the multiple oppressions faced by Black women while acknowledging the heterogeneity in the very notion of 'Black women'. In fact, in Black British feminism the term is used as a political statement and a shared consciousness experienced by all those women from the British postcolonial diaspora who have been silenced and objectified in discourses of 'British whiteness' (Mirza, 1997). In this paper my reference to 'Black women' focuses on women of African descent in a particular postcolonial context as a way of making a necessarily political argument about the power of touristic representations and their material effects on equity in tourism practices. In the following section I discuss the methodological underpinnings of this study.

Methodology

Crenshaw et al. (1995) argued that there is no systematic methodological or doctrinal approach adopted by all CRT scholars. For her part, Ladson-Billings (1998, p. 9) contends that CRT is:

> an important intellectual and social tool for deconstruction, reconstruction and construction: deconstruction of oppressive structures and discourses, reconstruction of human agency, and construction of equitable and socially just relations of power (p. 9).

In CRT legitimate sources of 'data' include poetry, parables, stories, counter-stories, fiction, and historical narratives. Delgado (2000) argues that 'stories, parables, chronicles and narratives are powerful means for destroying mindsets' (p. 61). This destructive function of storytelling is a means by which those traditionally subjugated can find a 'way out of the trap of unjustified exclusion. They can help us understand when it is time to reallocate power' (Delgado, 2000, p. 61). Critiques of CRT's methodological approach include that its arguments are too negative and leave very limited room for hope for the liberation of Black people; that it constantly draws on the 'race card' which implies an economical use of the 'truth' (Delgado & Stefancic, 2000); that it is not sufficiently analytical and traditional in its approach; and that is does not produce a racially unique form of scholarship (Bell, 1995).

I suggest that many of these criticisms are consistent with the debate between scientism and interpretivism, between objectivism and subjectivism with the latter seen as being less rigorous, valid and 'truthful'. In fact, storytelling is seen by qualitative researchers as a legitimate means of knowledge production. Qualitative researchers recognise that telling stories is an integral part of the human condition. Storytelling is seen within the parameters of narrative research which is important for critical theoretical scholars as it can open up understanding to different aspects of our social world (Fraser, 2004). Narrative researchers reject the notion of researcher objectivity and do not attempt to 'sanitize research by appealing to scientific facts and linear trajectories' (Fraser, 2004, p. 183). Recognising that there are no certainties in research, what storytelling produces are tentative, often iterative and fluid ideas which represent a plausible way of interpreting social phenomena. In using CRT and Black feminism to understand the (re)presentations of Black women in tourism I am providing my own interpretation as a Black woman, from a postcolonial island in the Caribbean. My intersectional subject positions as both Black and female has provided me with invaluable insights for the analysis of a fictional film about white female sex tourism, to (re)tell a story about tourism, race, and gender so that it brings Black women to the fore.

Ladson-Billings (2009) utilised film as a 'storytelling vehicle' (p. 89) in her CRT study of the representation of Black women teachers and the way in which they had been denigrated in this media. According to Ladson-Billings (2009) what makes this use of film as storytelling credible as a CRT strategy is the way in which the central message of the filmmaker is relegated to the margins while issues of race and gender become pivotal to understandings of the film. According to Yosso (2002), who also draws on CRT, entertainment media including film can perpetuate racism as well as gender oppression and critical analyses of film can effect social change. Yosso (2002) further suggests that an analysis of film through the lens of CRT can serve as an important 'consciousness raising' (p. 54) strategy.

The film analysed in this paper is about female sex tourism. The link between sex and tourism is well established in tourism studies (Carr et al., 2016; Clift & Carter, 2000; Opperman, 1999; Ryan & Hall, 2001) and in sustainable tourism in particular (Bandyopadhyay & Nascimento, 2010; Kibicho, 2005; Yan et al., 2018). Sex tourism is a complex phenomenon and goes beyond the rather simplistic definition of 'tourism for commercial sex purposes' (see Oppermann (1999) for a more extensive discussion of the various dimensions of sex tourism). This paper is not focused on sex tourism *per se* as this has already been covered extensively elsewhere as previously indicated. What is important to note here is that none of these extant studies have drawn explicitly on CRT or black feminism as theoretical framings nor have they used narrative analysis/storytelling. Most have also been focused on the sex tourist as a Western (white) male and fewer have acknowledged the centrality of Western (white) women as sex tourists (but see Herold et al., 2001; Jeffreys, 2003; Pruitt & LaFont, 1995: Sanchez Taylor, 2006). Jamerson (2016) has suggested that sex tourism provides a noteworthy research area which can highlight racial dynamics in tourism. Sex tourism thus provides the backdrop to the current discussion which interrogates the way in which Black women have been (re)presented in film but I do not focus on the phenomenon of sex tourism itself.

My discussion focuses on the film Heading South (*Ver le Sud*) (running time approximately 104 minutes). The film was first shown at the Venice Film Festival in 2005 and then on general release in 2006 to primarily white audiences in Europe and America. I have selected this film as it is one of only two full length fiction films (dramas) that I am aware of that have focused on the power of whiteness particularly as this relates to women (the other film is Paradise Love which I have discussed elsewhere – Chambers, 2021). I have examined Heading South in a previous publication through the prism of CRT (see Chambers, 2021). In this earlier work I focused on the experiences of the white female sex tourists and their Black male 'lovers'. However, in this paper I bring the Black women in the film to the fore and discuss the ways in which they have been (re)presented.

Heading South is directed by a white Frenchman Lauren Cantet and I suggest that as such, the story of the film is filtered through a white, heterosexual male gaze. The ubiquitous concept of the 'tourist gaze' coined by John Urry (1990), simply put, refers to the power and authority that Western tourists have over the people and places that they visit. Of the myriad adaptations of this concept in tourism studies, that by Pritchard and Morgan (2000) who argued that tourism representations are gendered and that the 'language of tourism ... is scripted for a male heterosexual audience' (p. 886) is pertinent for this discussion. The 'male gaze' is authoritative because men, as the 'master subject' have the power to construct the media and create the images and identities. In this scenario women are often portrayed as one-dimensional (Pritchard & Morgan, 2000). It is therefore apposite to interrogate how Black women (who suffer from double jeopardy due to racism and sexism) are (re)presented in this film. The film is set in the context of the political turmoil of late 1970s postcolonial Haiti and tells the story of three middle aged North American white female tourists who visit the island for sex with local young men or beach boys – Ellen (played by Charlotte Rampling), Brenda (played by Karen Young) and Sue (played by Louise Portal). Other central characters in the film are beach boy Legba (played by Mènothy Cesar) and hotel waiter and driver, Albert (played by Lys Ambroise).

Representations of Black women in 'Heading South'

Black women are not central characters in the film and are often (re)presented as foils for the key Black male protagonists. There is little if any interaction between the Black women in the film and the white female sex tourists. Interestingly the film's director claimed that the film was more political than it might first have appeared (CineEuropa, 2006, 22 January). Yet while Black women are no doubt central to an understanding of the political context of the tumultuous environment of late 1970s Haiti, their characters are underdeveloped in contrast to the white female sex tourists who are the central characters and with whom the audience are encouraged to relate. In a review of the film (Holden, 2006), it is stated that the film is 'too sophisticated to demonise these women whose relationships with their young lovers are more tender and nourishing than overtly crass'. The question is, how are the Black women who appear in the film (re)presented and what does this say about tourism practices in the postcolonial world? In examining the film, I have drawn out two main themes which articulate the ways in Black women are (re)presented – *Black female vulnerability* and *Black female submission.*

Black female vulnerability

The film opens with a well-dressed Black man in an airport building (we later learn that his name is Albert) where he has arrived to collect a white female tourist. While writing the name of the tourist (Brenda) on a small chalkboard he is approached by a smartly dressed Black woman who interrupts him and asks whether she can have a word. Albert, apparently unenthused by this interruption, indicates that he does not have much time to listen to what she has to say – he is initially quite dismissive albeit respectful. The woman recounts a story seemingly symptomatic of the poverty and danger that exists for the Black women of Haiti:

> Woman: Let me tell you this. Once we had a nice life. My husband was a sanitary inspector in the Public Health Department. But one day and I don't know why, he was handcuffed in his office, I never saw him again. Now I don't have a penny. I can't pay my rent, or my daughter's tuition. Sir, I may be ugly, but my daughter is like a gift from God. She's 15. She's as beautiful as she is sweet. She wants to be a nurse.
>
> Albert: I'm sure she'll manage.
>
> Woman: I'd feel very relieved sir if I knew she was with you.
>
> Albert: Sorry, I'm not sure I understand you.
>
> Woman: I want to give you my daughter.

The woman's eyes travel behind Albert towards her daughter who is waiting a short distance away and she implores Albert to look at her. Albert turns to look and the camera pans to a young girl in a bright yellow dress. The girl's face is downcast and surrounding her, though seemingly oblivious and engrossed in conversation, are five men, one of whom is dressed in military fatigues, the others dressed in what appear to be security uniforms, perhaps representatives of the paramilitary army. On the wall is a large photo of the dictatorial Haitian President Jean Claude Duvalier. The location of the girl amongst this group of men, while they do not appear to be immediately threatening to her, augurs what might be the girl's fate if Albert fails to accept the woman's proposition. The woman continues her story:

> Woman: Unfortunately, being beautiful and poor in this country, she doesn't stand a chance … I always take her with me, so I can watch over her. But they won't think twice about killing me to grab her. I don't mind dying but I don't want them to have her. Please sir, take her with you.
>
> Albert (shaking his head): I can't do that Madame.
>
> Woman: Then may God be with you. Beware Sir, it's hard to tell the good masks from the bad. But everybody wears one.

The woman then walks away, and Albert watches her departure dispassionately and then turns back to finish writing on his chalkboard which he then holds up to welcome his new tourist arrival. It is not immediately clear what the woman means by her final cryptic statement and it is only when the film ends that we are able to deconstruct this message – but I will return to this later. What is important at this point is that the white tourist who Albert is collecting at the airport (Brenda) exists in stark juxtaposition with the Black woman that Albert has just encountered. This woman lives in fear for her teenage daughter (whose aspiration of becoming a nurse is unlikely to be realised). Through sheer desperation she is forced to 'prostitute' her teenaged daughter to an older man who is a stranger, on the basis only that he appears to be well dressed (affluent). Yet the arriving tourist, Brenda, is free from fear and is welcomed and whisked off to her hotel by Albert only encountering the poverty on the streets of the city through the windows of the vehicle, at a distance. Interestingly the name of the hotel where Brenda will reside for the duration of her holiday and where Albert is employed as a waiter cum driver is called 'Hotel Petit Anse' which in French means 'small cove'. This is a metaphor for the way in which the hotel acts as a haven or shelter for the Western tourists from the outside world, and specifically the harsh realities of the Haitian city of Port-au-Prince in which environs the hotel is located.

A technique that is used in the film is that of the monologue. The audience is drawn into the personal lives of the three North American female sex tourists (Brenda, Sue and Ellen) through the use of monologues and these women narrate their personal experiences and thoughts directly into the camera, thus establishing an intimacy with the audience. The only other person in the film who performs a monologue is Albert but what is interesting in his case is that his is delivered as a thought process rather than a speech act. Further he does not look directly into the camera when he delivers this monologue but is instead busy working in the hotel kitchen. Albert's thoughts provide a very powerful insight into his vituperative attitude towards white American tourists which contrasts with his outwardly welcoming performances. This is one instance which supports the statement made by the woman at the airport that we all wear masks. Indeed, Ellen also wears a mask – on the one hand she is contemptuous of the city of Port-au-Prince and its people and thus never leaves the hotel compound. On the other hand she is fascinated by, and consumed with love/lust, for the beach boy, Legba. In her monologue, Ellen recounts the following:

> I turned 55 last month. There's nothing in Boston for women over 40. Don't contradict me. I've checked out every bar in that goddamn, stuck up city. And there's nothing there that is even close to Legba. How was such a handsome boy born here? On this dung heap? I was always drawn to the South, but I neglected Port-au-Prince. To me it was a hick town. Nowhere city. An animal compound. Now for 6 years, I've come every year for the whole summer. I always stay in this hotel. It's quiet, clean, the beach is beautiful. The moment I get here, I feel at home.

Importantly, none of the Black women in the film deliver monologues, they never speak directly to the film's audience. We therefore have no access to their inner thoughts, their feelings or their experiences such that we can develop an intimate connection with them in similar fashion to how, through the monologues of the main white female characters the audience is made to empathise with their situations. As such the Black women are largely silenced in the film and their vulnerability is manifested in their erasure from the foreground.

The vulnerability of the Black female body is also represented by Legba's former girlfriend who has been coerced into sexual slavery by one of the Colonels in Duvalier's army, in exchange for expensive gifts. This young woman remains nameless in the film (another sign of the erasure of the Black female body) and in her conversation with Legba in the Colonel's expensive car with its blacked-out windows she tells the following story:

> Woman: One night, two months ago, I was at a wake at a friend's house. Colonel Beauvais was there, with his wife and his daughters. He looked so serious that I didn't think he would notice me. The next day Frank came to fetch me. I don't know how he found me. When these people want something they always get it. I had no choice.
>
> Legba: He raped you.
>
> Woman: No that's not how it happens. They give you jewels, smiles, gifts, roses. But you know that this man who's giving you gifts and smiles, just for kicks, may gun down any fool who crosses his path. So the roses and gifts are like machine gun against your neck.
>
> Legba: You seem to be doing fine
>
> Woman: No Legba, you're not the one to judge me. Don't' lecture me. Look at the people around us. Like my sister, Maryse, the live-in maid who has to sleep with her boss and his son!

Her story resonates with Legba's own situation as a beach boy who sells his sexual services to white women on the beach for money, nice clothes, and jewellery. In this way beach boys like Legba are feminised and thus their experiences are indistinguishable from the Black women in the film, demonstrating the fluidity of gender categorisations. One could argue that here Legba's situation is as complex as that of his ex-girlfriend – not only is he feminised through being controlled by the female tourists (like having a 'machine gun against his neck'), but his life is also in danger in the context of the ubiquitous political violence and instability of late 1970s Haiti.

The next time we encounter this woman, her dead body, like that of Legba's, has been dumped on the beach of the hotel where he had his sexual encounters with the female sex tourists. Both had been murdered elsewhere and taken to this very beach perhaps as a kind of warning to others who might dare to challenge the order of things. Ironically, in the conversation in the car, the woman had demanded that Legba promise her that they would meet again – and they had, not in life, but this time in death. Black women, and those like Legba who are feminised as such, are therefore disposable further highlighting their vulnerability.

Black female submission

The other dominant (re)presentation of Black women in the film is one of submission. The Black women who work at the resort are either waitresses or cooks and they are apparently managed by Albert who oversees the hotel restaurant. They are often portrayed in blurred, or distant shots in the background, or peripheral to the main characters in the film. Thus, their bodies are mostly indecipherable and have no material presence. The other Black women that we see in the film are in dance scenes at clubs (and only dancing with Black men), women who sell the fish brought in by their fishermen, women who sell their wares and vegetables/fruits in the markets in the city and two women who watch the young men and boys playing football in the streets. Some of these women speak a few lines in the film, but most of the women are generally silent. The other woman that we see is Legba's mother.

Only two of the Black women in the film have names – one is Denise, a waitress who is requested by Albert to take care of Brenda when she first arrives at the hotel and the other is Maryse, the sister of Legba's ex-girlfriend who is only referred to as part of the story that she recounts to him in the car. The audience never see Maryse. All these Black women are (re)presented as submissive, as accepting of their roles and positions in the society and in tourism. Most of them are working in servile roles and while they clearly know what is happening in the sex tourism industry and are also aware of the political situation in Haiti, they seem to accept both as normal and show no signs of resistance to either. The same cannot be said of the Black men portrayed in the film – Albert expresses his resistance to sex tourism in his monologue and in the way that he seeks to exclude Legba from eating in the restaurant; Legba shows resistance to attempts by the female sex tourists to control him. In the end he offers his last bastion of resistance -preferring to die than become a 'sex slave' for Ellen; a young boy Eddy (who frequently appears in the film and is apparently in 'training' to become a beach boy) shows his resistance when he tries to intervene in an incident with one of Duvalier's paramilitary troops who takes a drink without paying from a boy who is selling in the streets of the city. But none of the Black women show any signs of even covert resistance to the double exploitation of both the political regime in Haiti and the sex tourism industry.

There are a few instances in the film when the Black women demonstrate that they are conscious of white female sex tourism but accept it as normal. The first is when Legba is leaving Ellen's cottage after a night of sex and meets two of the waitresses as he passes the restaurant. The following is the conversation they have:

Legba: *Hi ladies*

Waitress 1: (kisses Legba on the cheek): *Had a nice night?*

Legba: *I'm wiped out*

Waitress 1: *Aahh*

Waitress 2 (Legba kisses her on the cheek): *Shall we trade places? You sweep the floor and I do the screwing.*

(Both waitresses laugh)

Legba: *Our lady guests won't agree*

Waitress 2: *Who were you with?*

Legba: *That's confidential*

(the women laugh again)

Towards the end of the film when Legba realises that he will soon be murdered by one of the Colonel's henchmen he goes to visit his mother. His mother looks tired and shows no sign of shock or surprise when he turns up after an absence of two weeks even though she says she was worried and had looked for him at the hospitals and in the prisons. His mother is aware of the danger that exists for her son as the following conversation demonstrates:

Mother: *I don't have much, but you can come back and live here. You should stay with me*

Legba: *Mum, cut it out*

Mother: *I'm afraid for you.*

Legba: *Mum!*

Mother; *You know I am afraid*

(Legba takes a wad of money out of his pocket and hands it to his mother)

Legba: *I came to give you this*

Mother: *Where did you get all that money?*

Legba; *I work*

His mother does not ask him what kind of work he does, she only gives him a long, telling look and then hugs him and thanks him for the money. Legba tells her he must go. She does not implore him to stay. Instead, she watches him from the door as he departs perhaps knowing that will be the last time that she sees her son.

Discussion and concluding remarks

Black feminist theory emerged due to the invisibility of Black women in both discourses of racism and of sexism. Black feminism argued for an intersectional approach that would recognise the way in which Black women suffered from interlocking oppressions – for Black feminists it was not about race or sexism but race ***and*** sexism (Crenshaw, 1989; Hill Collins, 1990). In the film, Heading South, white women are racialised as privileged while Black women are largely silenced and where they do appear in the film they are portrayed as scarcely having any agency over their own lives. I argue that this erasure of Black women as agentic subjects and the way they are (re)presented as disempowered has implications for the development of tourism that seeks to be sustainable. This is because of the importance of women (Ferguson & Alarcón, 2015) and of race (Jamerson, 2016) to tourism.

In Heading South Black women largely exist in the background. In the context of the political instability in Haiti Black women are (re)presented as vulnerable and as submissive vis a vis the white female sex tourist whose white privilege renders her immune to the perils of the dictatorial Haitian regime. Even the main Black male protagonist in the film is portrayed as feminised and thus rendered vulnerable in so far as he is economically dependent on the female sex tourists and can only escape his dependency through the extreme measure of death. These gendered and racialised tropes contrast starkly with the portrayals of the white female sex tourists who are free to move through time and space. At the end of the film, we do not know what happens to the Black women portrayed in the film apart from Legba's ex-girlfriend who is reunited with him in death. However, after Legba's death the three white women take different paths – Sue remains in the resort, and Ellen returns to her home in America. However, Brenda, almost immediately recovers from Legba's murder, and vows to continue her travels. Her final words are instructive as we see her on a boat headed to another Caribbean island, a slight smile on her face and her hair blowing in the wind:

> Brenda; Of course, I won't go back home. Besides I don't have a house anymore, or a husband. I want nothing to do with men from the North. I want to visit other islands in the Caribbean – Cuba, Guadeloupe, Barbados, Martinique, Trinidad, Bahamas. They have such lovely names. I want to know them all.

The Black women and Legba, who is feminised and racialised, cannot escape their material realities of poverty and danger. They have submitted to their condition. They cannot and will never ***know*** the other Caribbean islands.

The paper has several limitations which also constitute areas for further research. First, I have not focused on other socially constructed identifications such as sexuality, class, nationality and (dis)ability. Indeed, the film promoted heteronormativity and ableism and there were clear class differences between the white female sex tourists and the Black women (albeit in a postcolonial context race and class are often intertwined). In the context of CRT, examining how race intersects with a host of other identifications is important and further research could focus on how race intersects with other key identifications such as sexuality. What this discussion perhaps illustrates is the complexity of doing intersectional work and how in reality it is extremely difficult to separate identity categories as Trinh (1989) suggested.

I have focused here on the intersection between gender and race partly due to the ways in which I position myself, partly due to the issues that emerged from the film, and partly as these are the socially constructed identifications that are of most prominence in research that subscribes to CRT and the related logics of Black feminism. CRT is an important theoretical approach which seeks to highlight racial injustice in our societies and Black feminism (which is related to CRT) focuses on injustices that exist at the intersection between race and gender. Black women, as the film demonstrates, are often (re)presented in sustainable tourism and tourism more widely as an absent presence in so far as they are rarely portrayed as having agency. As issues of equity and justice are important for sustainable tourism then it is critical that we understand the extent to which Black women, who play a crucial role in tourism development in the Global South, are erased from the tourism landscape.

Second, I have not explored the film's reception by its audiences. Rather my focus has been on the film's (re)presentations and how this might be interpreted recognising that my interpretations have been influenced by my own positionality. There might be other interpretations that are equally plausible and that can be explored by future researchers. Third, and in a related point, as with any choice of methodology there are limitations. The use of storytelling through popular media such as film though a creative qualitative technique, is scarcely used in sustainable tourism research and is therefore subject to the usual criticisms associated with the use of qualitative methods some of which I have already enumerated. Another issue here is that there is insufficient evidence that supports a direct relationship between the discourse in films and praxis. However, it has been argued that media representations can influence perceptions of reality (Eschholz et al., 2002) and it would not be farfetched to suggest that the way in which Black women are (re)presented in such fictional films can serve to reinforce racial stereotypes that have material effects. Critical sustainable tourism researchers could use this creative technique to analyse other intersectional problematics.

Despite these limitations, this film has highlighted an issue that has been largely obscured in the literature on sustainable tourism – how intersectionality particularly that between race and gender provides an important lens through which we can understand issues of equity in sustainable tourism. That is, the focus of this film on female sex tourism has illustrated the power of whiteness and the way in which it relegates Black women to the margins and (re)presents them as vulnerable and disempowered in tourism. Such (re)presentation perhaps belies the reality of their experiences and role in the tourism sector in the postcolonial world. The white women in the film are (re)presented as having much more depth as characters in contrast to the Black women who are (re)presented as one-dimensional. Such limited and limiting (re)presentations has implications for the way in which tourism is developed and experienced in postcolonial societies which strive towards more sustainable and more equitable practices. This (re)presentation of Black women in the developing world is in stark contrast with the self-representations of Black women in the United States where there is talk of a new Black Travel Movement that is significantly powered by Black millennial women who express their travel experiences using digital and social media (see Gill, 2019).

I am cognisant of the problematics associated with establishing such binary oppositions between racialised terms such as black/white as I indicated earlier in the paper. My use of these concepts is political, mirroring that of Black British feminists (e.g., Mirza, 1997) and focused on highlighting the conditions of Black women in a postcolonial tourism context which is where I position myself. I also believe in the value of using the tenets of CRT to expose intersectional marginalisation (in this case between gender and race) with the goal of achieving social justice. In the title of this paper, I posed the question, referring to women in tourism – 'are we all in this together'? Hopefully, I have demonstrated that the answer to this question, at least for postcolonial Black women in tourism, is not positive. And until the answer to this question is affirmative then global moves towards equity in sustainable tourism will remain illusory.

Disclosure statement

No potential conflict of interest was reported by the authors.

References

Akama, J., Maingi, S., & Camargo, B. (2011). Wildlife conservation, safari tourism and the role of tourism certification in Kenya: A postcolonial critique. *Tourism Recreation Research*, *36*(3), 281–291. https://doi.org/10.1080/02508281.2011.11081673

Alarcón, D. M., & Cole, S. (2019). No sustainability for tourism without gender equality. *Journal of Sustainable Tourism*, *27*(7), 903–919.

Amoamo, M., & Thompson, A. (2010). (Re)imagining Maori tourism: Representation and cultural hybridity in postcolonial New Zealand. *Tourist Studies*, *10*(1), 35–55. https://doi.org/10.1177/1468797610390989

Anderson, E., & McCormack, M. (2010). Intersectionality, critical race theory, and American sporting oppression: Examining black and gay male athletes. *Journal of Homosexuality*, *57* (8), 949–967. https://doi.org/10.1080/00918369.2010.503502

Anthias, F., & Yuval-Davis, N. (1990). Contextualising feminism – gender, ethnic and class divisions. In T. Lovell (Ed.), *British feminist thought: A reader* (pp. 103–118). Blackwell.

Anzaldúa, G. (1990). Bridge, drawbridge, sandbar or island: lesbians of colour Hacienda Alianzas. In L. Albretch & R. M. Brewer (Eds), *Bridges of Power: Women's multicultural alliances*. Philadelphia, P.A.: New Society Publishers.

Arnfred, S. (2004). Introduction: Re-thinking sexualities in Africa. In S. Arnfred (Ed.), *Re-thinking sexualities in Africa* (pp. 7–29). Almqvist Wiksell Tryckeri AB.

Ateljevic, I., Morgan, N., & Pritchard, A. (Eds.) (2013). *The critical turn in tourism studies: Creating an academy of hope*. Routledge.

Aziz, R. (1997). Feminism and the challenge of racism: Deviance or difference. In H. S. Mirza (Ed.), *Black British Feminism: A reader* (pp. 70–77). Routledge.

Bandyopadhyay, R., & Nascimento, K. (2010). Where fantasy becomes reality: How tourism forces made Brazil a sexual playground. *Journal of Sustainable Tourism*, *18*(8), 933–949. https://doi.org/10.1080/09669582.2010.497220

Bell, D. (1995). Who's afraid of critical race theory? *University of Illinois Law Review*, *4*, 893–910.

Boluk, K., Cavaliere, C. T., & Duffy, L. N. (2019). A pedagogical framework for the development of the critical tourism citizen. *Journal of Sustainable Tourism, 27*(7), 865–881. https://doi.org/10.1080/09669582.2019.1615928

Bott, E. (2018). Among the piranhas: The troubling lifespan of ethnic tropes in "tribal" tourism to Vietnam. *Journal of Sustainable Tourism, 26*(8), 1291–1307. https://doi.org/10.1080/09669582.2018.1435669

Boukhris, L. (2017). The Black Paris project: The production and reception of a counter-hegemonic tourism narrative in postcolonial Paris. *Journal of Sustainable Tourism, 25*(5), 684–702. https://doi.org/10.1080/09669582.2017.1291651

Bright, C., Foster, K. N., Joyner, A., & Tanny, O. (2021). Heritage tourism, historic roadside markers and "just representation" in Tennessee, USA. *Journal of Sustainable Tourism, 29*(2-3), 428–447. https://doi.org/10.1080/09669582.2020.1768264

Browne, I., & Misra, J. (2003). The intersection of gender and race in the labour market. *Annual Review of Sociology, 29*(1), 487–513. https://doi.org/10.1146/annurev.soc.29.010202.100016

Burdsey, D. (2011). Applying a CRT lens to sport in the UK: The case of professional football. In K. Hylton., A. Pilkington., P. Warmington & S. Housee (Eds.), *Atlantic crossings: International dialogues on critical race theory* (pp. 39–60). Birmingham, UK: The University of Birmingham Higher Education Academic Network.

Carbado, D. W., & Roithmayr, D. (2014). Critical race theory meets social science. *Annual Review of Law and Social Science, 10*(1), 149–167. https://doi.org/10.1146/annurev-lawsocsci-110413-030928

Carr, A., Ruhanen, L., & Whitford, M. (2016). Indigenous peoples and tourism: The challenges and opportunities for sustainable tourism. *Journal of Sustainable Tourism, 24*(8-9), 1067–1079. https://doi.org/10.1080/09669582.2016.1206112

Carr, N. (2016). Sex in tourism: Reflections and potential future research directions. *Tourism Recreation Research, 41*(2), 188–198. https://doi.org/10.1080/02508281.2016.1168566

Chambers, D. (2021). Reflections on the relationship between gender and race in tourism. In P. Dieke, B. King & R. Sharpley (Eds). *Tourism in development: Reflective essays*. Wallingford: CAB International.

Chambers, D., & Buzinde, C. (2015). Tourism and decolonisation: Locating research and self. *Annals of Tourism Research, 51*, 1–16. https://doi.org/10.1016/j.annals.2014.12.002

Chambers, D., & Rakic, T. (2018). Critical considerations on gender and tourism. *Tourism Culture & Communication, 18*(1), 1–8. https://doi.org/10.3727/109830418X15180180585112

CineEuropa. (2006, 22 January). *Interview with Lauren Cantet by Fabien Lemercier (online)*. Retrieved October 10, 2020, from https://cineuropa.org/en/interview/61736/

Clift, S., & Carter, S. (2000). *Tourism and sex: Culture, commerce and coercion*. Cengage Learning.

Cohen, E. (2002). Authenticity, equity and sustainability in tourism. *Journal of Sustainable Tourism, 10*(4), 267–276. https://doi.org/10.1080/09669580208667167

Cole, S., & Ferguson, L. (2015). Towards a gendered political economy of water and tourism. *Tourism Geographies, 17*(4), 511–528. https://doi.org/10.1080/14616688.2015.1065509

Cole, S., & Morgan, N. (Eds.) (2010). *Tourism and inequality: Problems and prospects*. CABI.

Crenshaw, K. (1989). Demarginalizing the intersection of race and sex: A black feminist critique of antidiscrimination doctrine, feminist theory and antiracist politics. *University of Chicago Legal Forum, 1*(8), 139–167.

Crenshaw, K., Gotanda, N., Peller, G., & Thomas, K. (Eds.) (1995). *Critical race theory: The key writings that formed the movement*. Free Press.

Delgado, R. (2000). Storytelling for oppositionists and others: A plea for narrative. In R. Delgado & J. Stefancic (Eds.), *Critical race theory: The cutting edge* (pp. 60–70). Temple University Press.

Delgado, R., & Stefancic, J. (Eds.) (2000). *Critical race theory: The cutting edge* (2nd ed.). Temple University Press.

Devine, J., & Ojeda, D. (2017). Violence and dispossession in tourism development: A critical geographical approach. *Journal of Sustainable Tourism, 25*(5), 605–617. https://doi.org/10.1080/09669582.2017.1293401

Dillette, A. (2021). Roots tourism: A second wave of double consciousness for African Americans. *Journal of Sustainable Tourism, 29*(2-3), 412–427. https://doi.org/10.1080/09669582.2020.1727913

Dillette, A. K., Benjamin, S., & Carpenter, C. (2019). Tweeting the black travel experience: Social media counternarrative stories as innovative insight on #TravelingWhileBlack. *Journal of Travel Research, 58*(8), 1357–1372.

Duffy, L. N., Pinckney, H. P., Benjamin, S., & Mowatt, R. (2019). A critical discourse analysis of racial violence in South Carolina, USA: Implications for traveling while Black. *Current Issues in Tourism, 22*(19), 2430–2446. https://doi.org/10.1080/13683500.2018.1494143

Eschholz, S., Bufkin, J., & Long, J. (2002). Symbolic reality bites: Women and racial/ethnic minorities in modern film. *Sociological Spectrum, 22*(3), 299–334. https://doi.org/10.1080/02732170290062658

Ferguson, L., & Alarcón, D. M. (2015). Gender and sustainable tourism: Reflections on theory and practice. *Journal of Sustainable Tourism, 23* (3), 401–416. https://doi.org/10.1080/09669582.2014.957208

Figueroa-Domecq, C., Kimbu, A., de Jong, A., & Williams, A. (2020). Sustainability through the tourism entrepreneurship journey: A gender perspective. *Journal of Sustainable Tourism*. https://doi.org/10.1080/09669582.2020.1831001

Fraser, H. (2004). Doing narrative research: Analysing personal stories line by line. *Qualitative Social Work, 3*(2), 179–201. https://doi.org/10.1177/1473325004043383

Gao, J., & Kerstetter, D. (2016). Using an intersectionality perspective to uncover older Chinese female's perceived travel constraints and negotiation strategies. *Tourism Management, 57*, 128–138. https://doi.org/10.1016/j.tourman.2016.06.001

Gibson, H. (2001). Gender in tourism: Theoretical perspectives. In Y. Apostolopoulous, S. F. Sonmez, & D. J. Timothy (Eds.), *Women as producers and consumers of tourism in developing regions* (pp. 19–43). Praeger Publishers.

Gill, T. M. (2019). 'The world is ours, too': Millennial women and the new black travel movement. In D. Williis, E. Toscano, & K. B. Nelson (Eds.), *Women and migration: Responses in art and history. Global Institute for Advanced Studies*. Open Book Publishers.

Hall, C. M., & Tucker, H. (Eds.) (2004). *Tourism and postcolonialism, contested discourses, identities and representations*. Routledge.

Harris, C. (1995). Whiteness as property. In K. Crenshaw, N. Gotanda, G. Peller, & K. Thomas (Eds.), *Critical race theory: The key writings that formed the movement* (pp. 276–291). Free Press.

Henry, J. (2019). The unspeakable whiteness of volunteer tourism. *Annals of Tourism Research, 76*, 326–327. https://doi.org/10.1016/j.annals.2018.09.003

Herold, E., Garcia, R., & DeMoya, T. (2001). Female tourists and beach boys: Romance or sex tourism? *Annals of Tourism Research, 28*(4), 978–997. https://doi.org/10.1016/S0160-7383(01)00003-2

Higgins-Desbiolles, F. (2008). Justice tourism and alternative globalisation. *Journal of Sustainable Tourism, 16*(3), 345–364. https://doi.org/10.2167/jost749.0

Hill Collins, P. (1990). *Black feminist thought: Knowledge, consciousness, and the politics of empowerment*. Harper Collins.

Holden, S. (2006). Blinding lust in a sexual paradise lost. *New York Times*. 7th July 2006.

Hoppe, E. (2010). Tourism as racism: Fanon and the vestiges of colonialism. In E. Hoppe & T. Nicholls (Ed.), *Fanon and the decolonization of philosophy* (pp. 177–196). Lexington Books.

Hutchings, K., Moyle, C-l., Chai, A., Garofano, N., & Moore, S. (2020). Segregation of women in tourism employment in the APEC region. *Tourism Management Perspectives, 34*, 100655. https://doi.org/10.1016/j.tmp.2020.100655

Hylton, K. (2011). Too radical? Critical race theory and sport against racism in Ireland. In J. Long & K. Spracklen (Eds.), *Sport and challenges to racism. Global culture and sport* (pp. 229–246). Palgrave Macmillan.

Hylton, K. (2018). *Contesting 'race' and sport: Shaming the colour line*. Routledge.

Jamal, T., & Camargo, B. A. (2014). Sustainable tourism, justice and ethic of care: Toward the just destination. *Journal of Sustainable Tourism, 22*(1), 11–30. https://doi.org/10.1080/09669582.2013.786084

Jamerson, W. T. (2016). Race and critical tourism studies: An analytical literature review. *Sociology Compass, 10*(11), 1038–1045. https://doi.org/10.1111/soc4.12418

Jeffreys, S. (2003). Sex tourism: Do women do it too? *Leisure Studies, 22*(3), 223–238. https://doi.org/10.1080/026143603200075452

Khoo-Lattimore, C., Ling-Yang, E. C., & Sanggyeong Je, J. (2019). Assessing gender representation in knowledge production: A critical analysis of UNWTOS's planned events. *Journal of Sustainable Tourism, 27*(7), 920–938. https://doi.org/10.1080/09669582.2019.1566347

Kibicho, W. (2005). Tourism and the sex trade in Kenya's coastal region. *Journal of Sustainable Tourism, 13*(3), 256–280. https://doi.org/10.1080/01434630508668556

Ladson-Billings, G. (1998). Just what is critical race theory and what's it doing in a nice field like education? *International Journal of Qualitative Studies in Education, 11*(1), 7–24.

Ladson-Billings, G. (2009). "Who you calling nappy-headed?" A critical race theory look at the construction of black women. *Race, Ethnicity and Education, 12*(1), 87–99.

Lee, E. (2017). Performing colonisation: The manufacture of Black female bodies in tourism research. *Annals of Tourism Research, 66*, 95–104. https://doi.org/10.1016/j.annals.2017.06.001

Lyon, A., & Hunter-Jones, P. (2019). Critical research analysis and the questioning of dominant hegemonic discourses of sustainable tourism in the Waterberg Biosphere Reserve, South Africa. *Journal of Sustainable Tourism, 27*(7), 974–991. https://doi.org/10.1080/09669582.2018.1551896

Mellinger, W. (1994). Towards a critical analysis of tourism representations. *Annals of Tourism Research, 21*(4), 756–779. https://doi.org/10.1016/0160-7383(94)90082-5

Mirza, H. S. (Ed.). (1997). *Black British Feminism: A reader*. Routledge.

Mohanty, C. (1984). Under Western eyes: Feminist scholarship and colonial discourses. *Boundary 2, 12*(3), 333–358. https://doi.org/10.2307/302821

Mooney, S. (2018). Illuminating intersectionality for tourism researchers. *Annals of Tourism Research, 72*, 175–176. https://doi.org/10.1016/j.annals.2018.03.003

Mooney, S. (2020). Gender research in hospitality and tourism management: Time to change the guard. *International Journal of Contemporary Hospitality Management, 32*(5), 1861–1879. https://doi.org/10.1108/IJCHM-09-2019-0780

Mooney, S., Ryan, I., & Harris, C. (2017). The intersections of gender with age and ethnicity in hotel career: Still the same old privileges? *Gender, Work and Organisation, 24*(4), 360–375. https://doi.org/10.1111/gwao.12169

Morrison, T. (1992). *Playing in the dark: Whiteness and the literary imagination*. Harvard University Press.

Opperman, M. (1999). Sex tourism. *Annals of Tourism Research, 26*(2), 251–266.

Osagie, I., & Buzinde, C. (2011). Culture and postcolonial resistance: Antigua in Kincaid's a small place. *Annals of Tourism Research, 38*(1), 210–230. https://doi.org/10.1016/j.annals.2010.08.004

Parker, L., & Lynn, M. (2002). What's race got to do with it? Critical race theory's conflicts with and connections to qualitative research methodology and epistemology. *Qualitative Inquiry, 8*(1), 7–22.

Patil, V. (2011). Reproducing-resisting race and gender difference: Examining India's online tourism campaign from a transnational feminist perspective. *Signs, 37*(1), 85–210.

Pritchard, A. (2014). Gender and feminist perspectives in tourism research. In A. Lew & C. M. Hall (Eds.), *The Wiley Blackwell companion to tourism* (pp. 314–324). John Wiley and Sons.

Pritchard, A., & Morgan, N. J. (2000). Privileging the male gaze: Gendered tourism landscapes. *Annals of Tourism Research, 27*(4), 884–905. https://doi.org/10.1016/S0160-7383(99)00113-9

Pruitt, D., & LaFont, S. (1995). For love and money: Romance tourism in Jamaica. *Annals of Tourism Research, 22*(2), 422–440. https://doi.org/10.1016/0160-7383(94)00084-0

Ryan, C., & Hall, C. M. (2001). *Sex tourism: Marginal people and liminalities*. Routledge.

Sanchez Taylor, J. (2006). Female sex tourism: A contradiction in terms? *Feminist Review, 83*(1), 42–59. https://doi.org/10.1057/palgrave.fr.9400280

Sen, A. (2006). *Identity and violence: The illusion of destiny*. Penguin Books.

Small, S. (2013). Still back of the big house: Slave cabins and slavery in Southern heritage tourism. *Tourism Geographies, 15*(3), 405–442. https://doi.org/10.1080/14616688.2012.723042

Solorzano, D. G., & Yosso, T. J. (2001). Critical race and LatCrit theory and method: Counter storytelling. *International Journal of Qualitative Studies in Education, 14*(4), 471–495. https://doi.org/10.1080/09518390110063365

Suleri, S. (1992). Woman skin deep: Feminism and the postcolonial condition. *Critical Inquiry, 18*(4), 756– 769. https://doi.org/10.1086/448655

Syed, J. (2010). Reconstructing gender empowerment. *Women's Studies International Forum, 33*(3), 283–294. https://doi.org/10.1016/j.wsif.2010.03.002

Torabian, P., & Miller, M. (2017). Freedom of movement for all? Unpacking racialized travel experiences. *Current Issues in Tourism, 20* (9), 931–945. https://doi.org/10.1080/13683500.2016.1273882

Trinh, TM-h. (1989). *Woman, native, other: Writing postcoloniality and feminism*. Indiana University Press.

Urry, J. (1990). *The tourist gaze*. Sage.

Werry, M. (2011). *The tourist state: Performing leisure, liberalism and race in*. University of Minnesota Press.

Wijesinghe, S. (2020). Researching coloniality: A reflection on identity. *Annals of Tourism Research, 82*, 102901. https://doi.org/10.101016/j.annals.2020.102901

Yan, Y., Xu, J., & Zhou, Y. (2018). Residents' attitudes toward prostitution in Macau. *Journal of Sustainable Tourism, 26*(2), 205–220. https://doi.org/10.1080/09669582.2017.1338293

Yosso, T. J. (2002). Critical race media literacy: Challenging deficit discourse about Chicanas/os. *Journal of Popular Film and Television, 30*(1), 52–62. https://doi.org/10.1080/01956050209605559

Defacing: affect and situated knowledges within a rock climbing tourismscape

Michela J. Stinson and Bryan S. R. Grimwood

ABSTRACT

Rock climbing is frequently constructed as a tourism practice that exemplifies rational, solo, masculine quests to conquer the natural world. This paper troubles such gendered norms through an investigation of Southern Ontario's Niagara Escarpment (Canada) as a space of climbing tourism. Drawing on actor-network theory, our aim is to situate the drifts and dissolutions of affect through the narrative capacities of rock climbing. Specifically, we engage with the Escarpment as a rock climbing tourismscape to illuminate the unexpected and productive qualities of affect in rock climbing, a process we describe as *defacing*. Defacings reconfigure how climbing feels, shift perceptions of conservation and sustainability away from human interests, and ultimately alter climbers' relationships to natural spaces, prompting ways of knowing and being that trouble masculinist, rational conventions. In the context of welcoming creative solutions for promoting sustainable tourism, we illustrate how attending to the affective capacities of climbing can foster new, interesting, and vital possibilities.

Introduction

Climbing.
A sacred space (changeable).
A scared space (mobile).
It wrecks me,
completely throws me.
It's strange, because
in isolation:
none of the choreographies are strenuous, but…
the endless hustle,
the sustained stretch toward the top,
the run-out
the rush
and that exposed moment—
the *reality* of falling, (really)
qualitatively,
really—

drawing blood, and
just scraping at the edges of comfort.

When the call for papers for this special issue on gender and sustainable tourism landed in our inboxes, we were struck by the diversity of topics, themes, and approaches being invited as contributions. We identified within the call how gender-informed tourism research is about the way knowledge is produced and represented just as much as it is about what or who is studied (Figueroa-Domecq et al., 2015). While gender is frequently associated with structures and performances of identity, it is also is encoded within systems of understanding, action, and belief (Barad, 2007; Haraway, 1988; Pritchard & Morgan, 2000). Gender is not only about bodily demarcations, biologies, or identities: it also conditions ways of knowing, valuing, and experiencing the world (Haraway, 1988; Yudina et al., 2018). We approach gender as part of an embodied knowing and being which slants toward, welcomes, or orients us toward certain possibilities.

Taking up gender in relation to ways of knowing and being means, in some ways, opening up restrictive and prescriptive binaries: affect vs rationality, context vs abstraction, values-engaged vs objective or neutral (Haraway, 1988). Indeed, it is similarly prohibitive to rely on and reinstate the aforementioned binaries, as doing so infers a subjugated innocent position. Recognizing this, we draw on Haraway's (1988) use of *situated knowledges* to instead position gendered ways of knowing as both tenuously constructed and highly pervasive—simultaneously partial, paradoxical, and proximate. Situated knowledges embed a responsibility and an answerability into how both research and tourism are done, dethroning an illusory objective position in favour of a subjective, feminist locality that does not transcend the body (Jamal & Hollinshead, 2001). This "partial perspective"—an embodied feminism—"is not about fixed location in a reified body, female or otherwise, but about nodes in fields, inflections in orientations, and responsibility for difference in material-semiotic fields of meaning" (Haraway, 1988, p. 588). In this paper, we work through and with these gendered politics in the knowledge production of tourism.

Specifically, we examine practices of rock climbing on the Niagara Escarpment in Southern Ontario, Canada, a place of contested and constructed natures. Our research commits to advancing, engaging with, and enacting situated knowledges of rock climbing and—more broadly—sustainable tourism. In so doing, we seek to trouble the hegemonically-gendered ways through which rock climbing tourism tends to be conceived and practiced. As prevailing material and discursive practices of rock climbing are tethered to masculine quests to conquer nature, such pursuits are contingent upon and reproduce dichotomous categories of "nature" and "culture", and "emotion" and "rationality" among others. These gendered dichotomies become so entrenched within landscapes of climbing that they seem normal and common-sense; masculine ways of making the world are naturalized within climbing tourism spaces, much like they are in other tourism spaces (Pritchard & Morgan, 2000; Nettlefold & Stratford, 1999; Yudina et al., 2018). This paper is premised on the need to trouble such taken-for-granted understandings and practices of tourism, and how they recurrently circulate power to marginalize, oppress, and dispossess. A central aspect in undoing or re-ordering power relations—which feminist scholars have alerted us to for some time—is to illuminate voices, narratives, or ways of being that have been cast aside. While naming hegemonic power relations is one way to resist them, breathing life into alternatives is another, with a particularly productive agenda. Such intentions are informed by Foucault's (2007) suggestion that critique represents a moral imperative to not only question norms—to rattle "the acceptability of a system" (p. 61)—but also to articulate "a domain of possibility" (p. 66).

Through imagining these possibilities, conversations around power in climbing tourism emerge in tandem with concerns about sustainability, ethics, and conservation, most of which occur between climbers, coalition bodies, and land managers. Frequently, sustainable climbing practices focus on ensuring the ongoing availability of rock faces for the explicit purpose of ensuring people have unbridled access to climbing. These practices are predicated on asking "how much damage might we get away with?"—a decidedly utilitarian perspective. Not only are discussions of ethics and ownership within the climbing community weighted with binary assumptions that pit natural climbing spaces as separate from human, cultural practices of climbing, they also consistently devalue and undermine the power of emotion, embodiment, and the more-than-human in transforming how tourism spaces are constructed. Ignoring "how climbing feels"—affect—means that opportunities for considering alternative sustainabilities are partial and incomplete.

In this paper, we seek to detail how climbing tourism spaces are defaced by the persistent energetic exchange of affect, and how these defacings are prompted by the presence of nonhumans (Rossiter, 2007). We employ *defacing* as a concept that "organizes energies, fuses forces, and directs attention away from some things and towards others" (Rossiter, 2007, p. 299). Defacings are rubbings, abrasions, erosions—they are a persistent and changeable wearing-away through which "one can think a little differently about … the interrelations of humans and nature" (Rossiter, 2007, p. 298). In rock climbing, this destruction might take the form of rock modifications, the trampling of plants, or the changing musculature of a body in response to movement. Essentially, defacings are the ways in which climbers and climbing environments unavoidably change one another, materially and discursively (Rossiter, 2007).

Our aim is to present an active, narrative reworking of already-blurry culture/nature binaries, with the purpose of further disrupting climbing's cascade of related dualisms of masculine/feminine, mind/body, and knowing/being through *affective defacing*. In effect, we strive to illuminate something different in climbing tourism scholarship: a deep honesty and humility about how although we are continuously complicit in defacing—as ruination (Collard et al., 2015), as pollution (Evers, 2019), as contamination (Cunningham, 2018), as disturbance (Tsing, 2015)—there remains room to imagine otherwise. As Evers (2019) explores through the intermingling of surfers and pollution, ecological sensibilities are complex assemblages of more-than-humans, where even destructive agents like pollution have complex, often-positive effects. We take up Rossiter's (2007) assertion that while climbing necessarily involves defacings—of climber bodies, landscapes, and other actors—this is not inherently negative. Indeed, defacing is a relation that could sustain life, or perhaps welcome a surprising vitality (Shotwell, 2016). Recognizing that interrelations of humans and natures are ordered and practiced (Law, 2004), we use this paper to *practice something else* through narrating the sustainable possibilities that come from attending to affect as a type of defacing. We start by situating our research within relevant literatures and then scaffold our approach to actor-network theory, which generates productive narrative representations of climbing affect. By illuminating these narrative potentials, we pursue sustainable and abundant futures for all actors drawn up in the critically fraught worlds of tourism (Collard et al., 2015).

Scholarly backdrop

Climbing tourismscapes

Traditional, rigid definitions of tourism tend to disregard its diffuse, integrated appearance in the day-to-day lives of people and places. Scholars increasingly illuminate, however, that tourism is an emergence of endless relations (Grimwood et al., 2018; Rickly, 2016, 2017a; van der Duim et al., 2017). Following Franklin (2004), tourism can be conceptualized as an "ordering," that is, "as something that had to be made to happen, that belongs to a story of becoming … that once formed and unleashed on the world it [takes] on a life of its own" (p. 279). The concept of "tourismscape" (van der Duim, 2007) suggests a similar entanglement of local and translocal processes by which tourism places are made and remade. In this paper, we employ tourismscapes to attend to how tourism is ordered by many actors—humans, objects, landscapes, ideas—as a relation of time, space, and material (van der Duim, 2007).

Definitions of tourism involving climbing frequently fall under umbrellas of adventure and/or nature tourism (Vespestad et al., 2019). However, approaching climbing spaces as tourismscapes allow us to think through climbing tourism as proximate, mobile, and entangled with a changeable landscape (Ness, 2016). Tourism is not restricted by distance, locality, or the idea that it happens elsewhere; tourism is made and found wherever we choose to look (Larsen, 2008). If we engage tourismscapes as a way to position tourism as something that might emerge with the practice of climbing, we find a wealth of climbing literature that supports this position (Barratt, 2012; Ness, 2011, 2016; Rickly, 2017a). As Rossiter (2007) and Ness (2011) reveal, rock climbing is a relational practice that emerges with and through bodies and spaces: climbing is accomplished by connecting bodies, rocks, weather, gear, and feeling through a complicated series of movements. Climbing moves from the local to the expansive, allowing space for surprising, mundane, and unexpected happenings (slippages, encounters, falls) that might rework our worlds (Ness, 2016). If climbing practices produce tourismscapes, local material-discursive defacings have massive implications toward expansive sustainable possibilities, as the smallest changes to a local environment (like chopping limbs off a protected tree) might have detrimental effect to an expansive network of climbing (like climbing access being restricted).

Gendered natures of climbing

Rock climbing literature has presented gender in various ways, most notably in terms of how women experience climbing (Chisholm, 2008) and how climbing identities and spaces are inherently gendered (Appleby & Fisher, 2005). Similar discourses of gender, equality, and access are also discussed in tourism, both in terms of how tourism positions the body and how tourism landscapes themselves are gendered (Pritchard & Morgan, 2000; Veijola & Jokinen, 1994). This paper resonates more closely with arguments of climbing tourism spaces as hegemonically-gendered, though it shifts away from gendered demarcations of bodies or identities and instead wrestles with gendered ways of knowing and being in a climbing world.

Nettlefold and Stratford (1999) outline the two predominate ways in which climbing has historically been intertwined with nature. First, they discuss the position of the human as rational, dominant, and controlling of nature—a masculine fixture overpowering a feminine, subservient landscape (Nettlefold & Stratford, 1999). Without climbers, unclimbed spaces (feminine, chaotic wilds) remain *terra nullius*: blank, pure canvases set for human mastering (Rossiter, 2007). Second, Nettlefold and Stratford (1999) note the (masculine, Vitruvian) human's attempt to intermingle with nature through creativity, self-expression, and spiritual presence. Though this paints nature as less of a submissive actor, nonhuman natures are still rendered only as a means by which climbers achieve self-discovery, practice self-determination, and realize personal growth through experiences with "feminine nature-as-muse" (Nettlefold & Stratford, 1999). In both of these constructions, Nettlefold and Stratford (1999) reveal discourses of nature as the setting for humanity's dramas and achievements, and the continued reference to the contested dualisms of nature as female and human culture as male. This dualism is persistent in climbing literature, especially when extended to position climbing as a pursuit through which individuals might escape the automatic, rational effects of modernity (Heywood, 1994; Lewis, 2000). Climbing provides an immersion into natural, chaotic spaces where individuals are able to enact their logical, masculine, rational reasoning and skills (Kiewa, 2002), a privileging of the "male gaze" also seen in many tourism spaces (Pritchard & Morgan, 2000).

Clearly, climbing walks a precarious line as both an expression of freedom and an assertion of rationalization, a conquering of nature through numbering (i.e., grading) and "taming the wilds". For many climbers, responsibility to and for nature is assumed as a function of their interactions with—and perceived sensitivity towards—climbing spaces (Hutson & Montgomery, 2010; Rossiter, 2007), but often does not extend into concrete action. Climbers also frequently position themselves as more environmentally conscious and as engaging in more sustainable practices than non-climbers (Frauman & Rabinowitz, 2011), though these pronouncements are further demarcated through expressions of elitism and climbing purity (Heywood, 1994).

Despite some pushback, patterns of a subservient feminine nature continue to invite masculinized conceptualizations of climbing, a process that tends to reproduce gendered climbing tourismscapes (Bott, 2013; Nettlefold & Stratford, 1999). How climbing routes are named is a manifestation of these power relations—who and what is made visible through naming practices speaks to who is allowed to participate in certain spaces. Climbing guidebooks routinely neglect Indigenous land relations; actualize racism, sexism, and homophobia; perpetuate colonial narratives of conquering nature; and focus textual documentation on white, male, able bodies (Bott, 2013; Nettlefold & Stratford, 1999; Rickly, 2017a; Rossiter, 2007). Rocks are discovered, climbed, and named—thus they become claimed (Rossiter, 2007). Historically (and currently), the majority of first ascents are enacted by men, and men therefore more often than not prescribe route names, assign numeric difficulty (i.e., grade), and normalize what is documented in climbing (Nettlefold & Stratford, 1999). If women, non-male, non-binary, or contested (often feminine/feminized) bodies are considered in the textual histories of climbing, it is often as an afterthought, or a joke. This is clear when considering "girlfriend routes" (i.e., easy-graded routes designed specifically for the "inexperienced girlfriend" of an "accomplished" male climber) or controversial route names like Southern Ontario's "Female Belay Slave", "The Bitch Bent Over", or "Never Trust a Woman" (Alexandropoulos & Dwyer, 2016).[1]

It is unsurprising that other disputed dualisms—those of culture/nature, reason/emotion, or human/nonhuman—are prominent in discussions of climbing alongside that of the contested masculine/feminine (Lewis, 2000; Nettlefold & Stratford, 1999), though much recent work has been positioned to muddy them (Brighenti & Pavoni, 2018; Kiewa, 2002; Ness, 2011; Rickly, 2017a; Rossiter, 2007). As scholars like Rossiter (2007) show, traditional practices of nature conservation are challenged when the binary of culture/nature is troubled. Recent writings have repositioned the making of place as a human action to a process that is vital in its materiality, with an agency that is not personal but distributed in the world (Bennett, 2010; Lund & Jóhannesson, 2016). This work detaches nature's valuation solely from human usage (Smith et al., 2020), and alludes to how climbing routes themselves are neither built environments nor states of wilderness (Ness, 2016). This trend is also present in current climbing literatures which rework exhausted binaries—Ness (2011), for instance, speaks to the subjectivity of a climber being "situated in between the multiple human and nonhuman elements now integrally connected by the attempt [to climb the rock]" (p. 82). This distributed situatedness invites the inhuman animacy of rocks (Springgay & Truman, 2017)—a vital possession of "matter movement" inherent to all things that "unhinges the concept of affect from the human" (Springgay & Truman, 2017, p. 858). Disrupting the unassumed agency of rocks (and other nonhumans) allows rocks and landscapes to relate imaginative possibilities—they emerge as lively natures within which climbing becomes an affective field where a torrent of blurry bodies and unruly agencies intermingle to rework subject-object distinctions (Rossiter, 2007).

Affect: troubling gendered climbing tourismscapes

As a flickering of relational intensities, affect is notoriously hard to pin down. Seigworth and Gregg (2010) suggest that the simplest definition lies in potential "capacity to affect and be affected" (p. 2), a turn away from positivism and rationality in favour of momentary (yet lingering) forces of intense in-between-ness. Others explain affect as "not a personal feeling, nor … a characteristic; [affect] is the effectuation of a power of the pack that throws the self into upheaval and makes it reel" (Deleuze & Guattari, 1987, p. 240), or what Anderson (2014) considers the tension between the subjective and the objective. Indeed, "affect plays a vital if understated role in 'determining the relationship between our bodies, our environment, and others'" (Shouse, as cited in Spinney, 2015, p. 234). In the context of tourism, this also means positioning affect as a curious intensity never quite accounted for by destination, location, or promotional structure (Maksymowicz, 2019). For the purpose of this paper, we engage with affect as a forceful, partial, and provisional *becoming* (Latimer & Miele, 2013)—a relational potentiality of being with an energetic exchange that might prompt defacing.

Attention to affect in tourism is born from explorations of how tourism might be done differently (Buda et al., 2014; Tucker & Shelton, 2018). By grounding our scholarship in an attention to the generative capabilities of affect, we link these experiential becomings as alternative to previously held, dominant views of tourism as private, apart-from, or antagonistic to our public, daily lives (Tucker, 2009; Tucker & Shelton, 2018). In this way, "theorizing emotion and affect challenges assumptions that embodied feelings are a private matter, or that they simply begin within—and belong only to—individuals" (Buda et al., 2014, p. 103). Indeed, affects and emotions can be worked with to produce ethical and political effects, further arousing commitment to relational ontologies, contemplations of proximity tourism, and an attention to various bodies (Tucker & Shelton, 2018). Commitments to this affective turn are exemplified in discussions of the worldmaking capacities of shame (Tucker, 2009), explorations of destination attractivity (d'Hauteserre, 2015), the politics of affective proximity and hostile borders (Buda et al., 2014), and tourism erotics (Maksymowicz, 2019).

Affect runs through a diversity of current relational, material, and ethical arguments in tourism studies that draw attention away from what tourism is and towards what tourism can do (Grimwood et al., 2018). Informed by these turns, questions of conservation and sustainability have become wrought with a certain urgent recognition of diverse bodies, more-than-human positions, and alternative ways of knowing and being (Evers, 2019). Though certain studies show a marked shift in tourism conservation from promoting instrumental outcomes to those that favour a more values-based approach (Vespestad et al., 2019), Yudina and Grimwood (2016) note that most of these works remain decidedly anthropocentric. However, as work on sustainability and conservation in tourism has collided with the arrival of the Anthropocene (Moore, 2019; Smith et al., 2020), there is a need for further shifts toward situated, feminist-informed explorations of environmental conservation that incorporate recognition of the gendered ways in which nature is continually constructed (Yudina et al., 2018). If our intention is to consider how climbing feels in order to suggest new ways of thinking through climbing sustainabilities, attending to affect partially takes up this call.

The Niagara Escarpment tourismscape

The Niagara Escarpment is continuously made and narrated as a tourismscape—it is an expansive landscape that prompts the shifts and flows of tourism. Intensely dramatic, the blocky, steep walls of the Escarpment were urged into being by erosion of mineral deposits between three and five hundred million years old, resulting in cliffs of limestone, dolostone, shale, and sandstone (Bracken et al., 1991). The Escarpment is also home to some of North America's most endangered flora and fauna, including the Eastern White Cedar (Kelly & Larson, 1997). Along with its many nonhuman actors, the Escarpment is woven by and through certain industries, political classifications, and settlement histories. Designated as a part of the protected Ontario Greenbelt, the Escarpment hosts the Bruce Trail and contains multiple conservation areas, as well as quarries and aggregate yards. Land conflict remains a continuous contention in the Escarpment's ongoing history, with bodies like the Niagara Escarpment Planning and Development Act (NEPDA) and the Niagara Escarpment Commission (NEC) spearheading attempts to balance conservation, outdoor recreation, and industrial use (Hutson & Montgomery, 2010).

Due to its unique constitution and proximity to cities, the Escarpment also has a rich history as a destination for rock climbing, beginning as early as 1910 (Alexandropoulos & Dwyer, 2016; Oates & Bracken, 1997). These histories tell of competing versions of climbing that are more or less sympathetic to the intervention of nonhumans (i.e., the presence of permanently-fixed climbing gear, flora, and fauna), many of which employ rhetoric of access and conservation as a way to police which actors, materials, and ethics are acceptable in rock climbing. Generally, this discussion distills down to whether or not climbs should be "bolted"—affixed with permanent hardware—or remain "natural". The Ontario Alliance of Climbers (OAC) remains the primary body that suggests best practice for local (and visiting) climbers.[2] The OAC liaises with conservation areas, land managers, private citizens, and government bodies in order to secure access to outdoor areas for the purpose of climbing; hosts area clean up days; holds climbing teaching sessions; and promotes their Code of Ethics such that they position themselves as the authority on Escarpment climbing (OAC, 2019). Despite the OAC's self-positioning as an authority, climbing on the Escarpment remains mostly unregulated, and continues to be negotiated as route equippers develop new rock climbing areas and secure (or lose) access to others, depending on the ongoing behaviour of and rapport among these many stakeholders.

Ordering methodology

Actor-network theory

This research engaged actor-network theory (ANT) as its methodology. When employed as a methodology, ANT is notable in that it allows researchers to trace and explore complex relationships between humans and more-than-humans, a process sometimes referred to as the semiotics of materiality (Law, 1999). At its best and most useful, ANT engages critical and relational theories, ultimately "position[ing] the world as an outcome of a process of inquiry"—one that is relational, ontological, and generative (Ruming, 2009, p. 425). In this way, research is decidedly an act-of/active doing, as opposed to inquiry that presumes a researcher is detached and disengaged from the world (Ren, 2011). ANT places researchers in the messy middle of things, bringing together becomings of affect, surprising research outcomes, and the capacities of (human) embodied positions with respect to the emergent landscapes of both climbing and tourism (Müller & Schurr, 2016). By producing research through actor-network theory, we become attentive to and immersed in affective, embodied worldmaking (Stinson & Grimwood, 2019). This emphasis on the worldmaking capacity of methodological process is a personal and ethical priority—a process that Grimwood (2015) encapsulates as moral, transformative, political, and positively productive. Research methodologies make our worlds and also make other worlds possible (Law, 2004).

Engaging methods

While ANT prescribes no specific steps for choosing methods, it tends to orient researchers to ethnographic processes, including interviewing, site-visits, observational practices, and document analysis (Ren, 2011). Our choice of methods was informed by Beard et al. (2016). These authors recommend five considerations for pairing methods with ANT: that researchers rethink the spatiotemporality of the research field; that researchers "act-in-the-network" to produce reality; that researchers approach data sampling and selection as a process of "following actors"; that researchers "embrace materiality" with methods choices; and that researchers identify and trace key actors that constitute and construct an emerging network (Beard et al., 2016). Beard et al.'s (2016) suggestions direct us to practicing responsibility as researchers, ultimately demanding our constant attention to our generative practices of making worlds.

Figure 1. Bolt hanger with quickdraw

Similar to Ren's (2011) approach, we used a range of ethnographic data collection methods to translate the network of Niagara Escarpment climbing. Specifically, we assembled data through seven interviews and one focus group discussion with a total of twelve human participants; a review of multiple local climbing guidebooks; and numerous field notes, digital photographs, and video-recordings taken while immersed in the climbing context. These overt—perhaps tangible—data are held in contrast (but not in value) to the covert, relational, affective data that were captured through memory, tactile engaged witnessing (Bell et al., 2018), dreams, and imaginaries. Key to our inquiry, Bell et al. (2018) detail the process of engaged witnessing to consider the relationships between human and nonhuman actors, including how these relationships are interconnected and co-created. Engaged witnessing involves a dynamic, intentional openness to the possibility of being changed by an ongoing immersion in more-than-human entanglements, a deeply-situated version of participant observation (Bell et al., 2018). As a movement away from traditional participant observation, engaged witnessing welcomes substantial room to be moved by affects.

Locations, actors, and timelines

Data generation was performed across the Niagara Escarpment between October of 2018 and June of 2019. While the research field was initially supposed to include multiple climbing areas along the Niagara Escarpment, for reasons of weather interference, research anxiety (Stinson & Grimwood, 2019), and companion choice, the eventual location of data generation became Mount Nemo Conservation Area (just outside of Burlington, Ontario).

Early in our data-generation experience, we had multiple encounters with what would become our key actor—the bolt hanger. As we navigated the vast messes of possible data, the persistent presence of bolt hangers in conversation, text, and proximity meant that our attention was repeatedly drawn to their capacity to determine how climbing could feel—in Sayes' (2014) terms, bolt hangers gathered many other actors from numerous spatiotemporal locations and enacted an immediate ordering effect across our entire network.

Bolt hangers are small, folded circles of metal that are permanently affixed to rock walls in order for climbers to place a quickdraw device, and attach a rope (see Figure 1). This is done to ensure safety, so that falling climbers do not hit the ground, and are instead caught by this permanent protection. In Southern Ontario, modern bolt hangers are generally 3/8 inch thick high-corrosive resistant stainless steel, and are bored into limestone using a power drill, eventually affixed with expansion bolts. In hand, bolt hangers have a slight weight, rounded edges, and a transmissive, cool touch, like the back of a hand on a feverish forehead. As Alexandropolous and Dwyer (2016) note, "the mysterious metal object continues to attract attention" (cover). Loosely, bolt hangers determine the spatial boundaries of a climbing route, explaining visually to a climber some information about the intended path (Rickly, 2017a). Bolt hangers mark not only where certain climbs are, but also how climbing can feel and what climbing can be.

The intermittent presences and absences of bolt hangers is one way that climbing sustainability is ordered along the Niagara Escarpment, as bolt hangers contain and relationally enable specific narratives about the natures and cultures of climbing, and make climbing feel certain ways. While "successful practices of climbing" (i.e., completing the climb) ensure that physical use of the bolt hanger for its "intended purpose" (i.e., for catching a fall) is not performed, bolt hangers nevertheless embed considerable information—territorial, affective, and ethical maps of climbing—and subsequently enact pervasive affective orderings that deface and produce climbing tourismscapes.

Storying tourismscapes

Recognizing ANT as a generative practice, our research worked to trace and build stories that resonate with the vibrant and productive affects of climbing. Stories—as we've learned from tourism scholars working with narrative methodologies (Tucker & Shelton, 2018)—involve assembling and succumbing to affects: vast entanglements of atmosphere and feeling. Accordingly, the narrative representation we present below is an integration of creative, poetic possibilities. We engage what Law (2004) and Haraway (1991) consider blurring the dualism of reality and fiction, such that our practice of methods also re-works our worlds. Essentially, we affirm that "reality and fiction relate to one another. They are included in one another" (Law, 2004, p. 69).

This process of storying was deemed especially relevant for our research because of its congruency with the material-discursive practices of rock climbing. As Rickly (2017a) explains, the "discussion of routes climbed, in particular, includes recitation of portions or even the entire route, from the starting holds through the crux to the anchors, thereby illustrating the essential elements of narrative—setting, plot, characters, and so on" (p. 77). Narratives of climbing are not restricted to recitation, but are also embedded in the performative action of climbing processes, the drama of successfully completing routes, and the corporeal knowing of how climbing lingers in the body through musculature, abrasions, and injury (Rickly, 2017a). Much like the poetics inherent in the making of tourism (Lund & Jóhannesson, 2016), climbing writes a unique metaphorical-relational-material story of space. In this way, affective narratives and actor-networks are both decidedly productive and performative (Law, 1999), assembled with the narrative expressions of human (Tucker & Shelton, 2018), earthly (Lund & Jóhannesson, 2016), animal (Deleuze & Guattari, 1987), landscape (Nettlefold & Stratford, 1999) and object-based communiques (Bennett, 2010). Affective narratives also bring life to situated knowledges (Haraway, 1988)—ways of knowing and being that are deeply entangled in immersive reflexivity—that function outside of traditionally-gendered, rationalized ways of coding and categorizing.

Bolts, brightness, blood

Productive defacings

The following narrative presents a tracing of the Niagara Escarpment tourismscape through an exploration of the productive and destructive capacities of the bolt hanger as a material-discursive actor. Though d'Hauteserre (2015) cautions us that "the world expresses itself in numerous ways that cannot always be grasped with language," our personal, tentative worldings are consistently made poetic (p. 81). We remain convinced that narrative enacts one of many "forms of attunement to worldly compositions in which matter and thought are not opposed but emerge together as aspects and lines that throw together and are, one way or another, lived" (Stewart, 2014, p. 560). We draw mainly upon d'Hauteserre as our theoretical (and material/metaphorical) *belayer* for this working. In climbing, the belayer is responsible for ensuring the safety of the climber by managing the rope, essentially supporting the climb; as our theoretical belayer, we rely on d'Hauteserre's scholarship on affect as crucial assistance. In this climbing partnership we also have our narrator, who expresses a relational subjectivity as the unnamed, first-person storyteller of their embodied climbing experience. What follows is our translation of a purposive and grounded (re)assembling of the affective tourismscape of the Niagara Escarpment, as enabled though the action of climbing bolt hangers.

Ultimately, our networks are affective moments (Stewart, 2014): possibilities, interferences, and world-changing defacings. While climbing defacings are crudely identifiable as patches or swatches of damage (spray-paint, polish) they also have an active, vibrant, and ecological quality that works—they are liminal, they have verbage, they are ongoing (Bennett, 2010; Evers, 2019). By attending to the defacings of affects through the presence (or absence) of bolt hangers, we attend to the ways in which climbing tourism might become (or already may be) sustainable.

Assembling: Swan Song

It's late fall at Mount Nemo, and I'm wearing three jackets, my hood tied tight around my wind-burned face. Despite the terrible weather I am trying to outrun an even more disappointing climbing season; I am desperate to stretch these too-cold days into the reaches of November. The rock is damp and forbidding and I'm pulling quickdraws off of a length of cord as my belayer, Anne-Marie d'Hauteserre (2015), ushers in Spinoza, and Deleuze and Guattari to explain our path forward: "Affect here … is a line of force, a capacity to act, though to some extent it is derived from that original desire" (p. 79). She double-checks my safety knot (tied in an imperfect figure eight) and looks me in the eye. *Do you really want to climb this route? Is it what you desire?* I'm preparing to dance the *Swan Song*, a conniving, demanding moderate line of considerable length. I shiver and nod, feeling the cajoling of the wind, my palms still sweating, my body weighted under its action potential. d'Hauteserre (2015) seems pleased at my relational position, confident that we are within something diffused both socially and materially. I linger at the base of *Swan Song*, poised to climb, suspended with possibility, and I feel the raw, enriched, agentic hum of my body—I am an embodied transmitter, a capacitor, with my many senses open to the swift emergence and diffusion of affect prompted by the un/stable overlaps of my corporeal self in and among other bodies and bodily constituents (d'Hauteserre, 2015). I press my worn-away fingertips onto lichen-rich limestone and I flood the climbing space with possibility; I wear away and repopulate thousands of years of presence; I pass through my perspiration the anxiety of this tenuous new mattering (of me and the rock) and together d'Hauteserre (2015) and I and the many minglings of minerals assemble an encounter of surprising a/effectiveness—I almost let go of the wall.

As my belayer, d'Hauteserre is quick to remind me of my terrain. Her hand steady on the brake line (so as to not drop the belay) she coaches me: "Affect is beyond the senses that can be signified and is not consciously directed by actors upon others" (d'Hauteserre, 2015, p. 80). I prepare myself, steadily, and yet I remain completely unprepared—the climb itself (its bolts and bulges) is bound up and bursting, "intensely interrelated in the (affective) movements of the attempt" (Ness, 2011, p. 82). The rock around me is emergent, responsive, "an atmosphere [that] must be apprehended … [and] reworked as experience" (d'Hauteserre, 2015, p. 82). I unfasten a quickdraw from my harness and connect it with the first bolt hanger, suddenly remaking my body—my self—in the performance of the climb (Ness, 2011). As I step onto the polished rails of dusty Niagara Escarpment limestone, I both remember and re-matter *Swan Song*, drawn in by and drawing memory, experience, and influence into the affective defacing of climbing (d'Hauteserre, 2015). It is from here that other climbs I have engaged suddenly begin to assemble (and fall apart) once again.

Hiromi's Route—defacings of rationalization, tradition, and stability

As I approach the next bolt hanger of Swan Song, I feel my anxiety building—a fall from this position is unwise. The bolt is rusted and off-centre, and the hanger spins when I go to clip. (It moves within the rock, erratic, unhelpful.) My brain flashes scenes of the bolt pulling from the rock if I fall, but my body doesn't care. I clip and immediately relief courses through my limbs as the climb shifts into focus. The relief draws me into memory of another climb—one called *Hiromi's Route*—and suddenly I hear my friend Greg complaining. "[In] sport climbing we intuitively all just trust that bolts are solid. But you know less about that bolt than the gear you put in the wall yourself. Yet we all trust that bolted anchor way more, right?" Greg shakes his head, sarcastic: *"There are anchors in the walls. They've gotta be good. I don't need to check!"*

Bolt hangers deface conceptualizations of climbing as the unknown: they rationalize, modernize, and tame both rock and climber fear. Climbers, for the most part, have a limited ability to assess through visual inspection how "good" a bolt is (i.e., whether the bolt hanger has been affixed to the rock safely and effectively), but will frequently feel emotional relief after clipping bad bolts or rotten fixed gear. Clipping bolt hangers is often about psychological protection (Barratt, 2012)—despite physical assurance of safety, the bolt hanger also provides an immaterial, affective enactment of a material presence. *Relief.* My anxiety is consistently rendered in relation to bolt placement, presence, or perception: how run out lines are (i.e., how much space there is between bolts), or how good the bolting seems. While sport climbing, I do none of the protective work with my own body, something that would happen if I was climbing using impermanent gear. The ironic assertion that sport climbing's rationalization derails its emergence as salvation from everyday life (Heywood, 1994) is clear when I talk to Greg. As opposed to the defacing I experience by the action of the bolt hanger (a wearing away of my fear), Greg—because of his years of practice placing his own gear—experiences a detraction from the experience, a defacing that draws him out through fear and skepticism, instead of drawing him in through stasis and security.

Back in the imperfect present, d'Hauteserre feeds me rope and I clip again. *Thank God.* Fighting my way up between the second and third bolts on *Swan Song*, it is both easy to further remember Greg: hunched over his shoes, untying them after lowering off *Hiromi's Route*—a long, leisurely moderately-graded climb often used as a warm-up—shaking his head as he lambasts modern bolting practices. "Gone are the days of two, three, four metre run-outs," he complains, "And it's at a cost to who? To [route developers]. [They're] personally paying for this shit. … And it's giving people a false sense of security." My current section on *Swan Song* is only slightly run out, and despite my experience I still feel unprepared to complete it effectively, especially when my lack of height and meagre strength necessitate physical climbing sequences that take me even two inches above my bolt—the fear is palpable in my body, and it refracts off the rock, ricocheting through the area, rendering the entire climb shaky and unstable. (d'Hauteserrre remains patient and calm.)

Bolt hangers further wear away at the differences between indoor and outdoor spaces. Stylistically, bolting practices are changing to accommodate movement sequences more reproductive of indoor climbing spaces. Essentially, routes are being "over-bolted" (i.e., there are more bolts than necessary), which makes the climbing more accessible; *Hiromi's Route* is a prime example with its notorious 17 bolt hangers, when most routes at Mount Nemo have less than 10. Newer climbing areas and more recently bolted climbs are equipped with this gym-style over-bolting, equipping that mirrors the relative accomplishment of climbers and makes assumptions about their levels of anxiety, confidence, and experience climbing though generous, frequent placements (Alexandropoulos & Dwyer, 2016). This is in contrast to the initial style of sport bolting, wherein easier grades were more sparsely equipped—historically, easy routes were run-out, as they were for warming up, not for the pleasure of inexperienced climbers. Nick (Greg's belayer while we climbed *Hiromi's* that day) agrees. As an equipper and route developer who believes in using modern bolting when at all possible, he explains that not everything is controllable. He scoffs,

> What's wrong is the person who's climbing. They aren't able to assess [difficulty]. They're coming in with a perspective from climbing at the gym where bolts are every four feet and there's going to be perfect holds at every clipping stance and it's going to be easy. And it's like no, welcome to the outdoors.

Though this newer, generous bolting style is increasingly practiced and sanctioned by route developers, many "old guard" climbers (i.e., climbers with mostly traditional values, like Greg and Nick) still believe the process means that newer climbers are being robbed of the mental capacity to progress to harder climbing through learning to manage their stress while clipping and moving between bolts, completely changing the way that fear is integrated in relation to the bolt hanger. The increased accessibility prompted by modern bolting is also rarely seen as a good thing by the old guard, who cite hordes of new climbers as the reason for persistent environmental degradation, area overuse, and abhorrent Leave No Trace practices. But traditional climbing on the Escarpment—that is, un-rationalized climbing without using bolt hangers—is also not the bastion of authenticity it was once seen to be (Heywood, 1994).

Focusing on the possibilities embedded and facilitated by bolt hangers redirects attention away from conceptualizations of climbing that are over-focused on direct interaction with nature via the human body, as if the body is always assumed to be unmediated (Lewis, 2000). Bolting is seen a controlled, exceedingly safe method of consuming and rationalizing rock climbing by virtue of creating points of protection where there formerly were none (Heywood, 1994; Kiewa, 2002). By over-bolting, there is a direct defacing of both climber fear-control as well as Southern Ontario limestone—as climbing becomes more accessible routes become more highly trafficked, a recursive wearing-away at the feeling of climbing-and-climbing-spaces. Limestone material becomes resistant to touch (Rickly, 2017a)—the more a specific movement is required, the more difficult it becomes to enact said movement as the rock becomes slippery and polished, and the less physically salient it becomes despite the lingering symbolic mappings of guidebooks, photographs, and memories (Nettlefold & Stratford, 1999).

Bolt hangers project both ironic and complementary performances of climbing tourism as they re-render and deface the relational, emergent landscapes of climbing traditions, rationalizations, and securities (Ryan, 2002). As old-guard climbers lament modern bolting practices, they develop a curious affection for damage and polish, considering it both a restriction on newcomers and also a rite of passage, practices which Evers (2019) would consider a *living with* of destruction and defacing, hinting at a more-than-human mode of conservation via ruination. Accordingly, "spent" climbing spaces are subsequently both more and less "pure". Defacing itself becomes that which must be conquered; degradation becomes part of the subservient landscape, another natural facet of the cliff that must be overcome.

Crazy Doctor—defacings of ethics, nature, and purity

Mount Nemo is an intimidating climbing area with a dark atmosphere that dramatically displays the unique personality of Escarpment limestone (Oates & Bracken, 1997). Many of Nemo's routes are characterized by inconsistent bolting, complicated route-finding, "sandbagged" grading (i.e., grades that are said to be easier than they truly are), and unpredictable conditions; the overall feel of Mount Nemo remain imposing—a place of trial rippling with ego, pressure, and fear.

It *matters*. As d'Hauteserre (2015) so dutifully reminds me (my feet precariously perched on limestone nubbins), affect is not solely an emotional response, but something unpredictable, relational, and tactile—something beyond the representational scope of language. She feeds me rope as I move into space, and I have a sudden vision of Nick explaining to me how a climb's natural emergence is particularly altered by bolt hangers—not his feelings toward the climb, or even the sequence of movement (though those might change), but how the climb was done and undone for him in relation to the bolt hanger's presence. I asked him outright if he thought bolting changed the climb, and his response was enthusiastic: "Absolutely. It changes the aesthetic of the climb. It changes how the climb *feels*." He went on to relate these experiences to bodily positions and demands, feelings of immersion, and the tactile sensibility in the assembling of gear—how the route emerged differently through the collisions of these many factors (Barratt, 2012), or how his body, through bolt hangers (or self-placed, removable protection), co-constituted the climb (Barad, 2007).

And yet I think of Nick, moving towards the top of the cliff as he pulls his body beyond a bolt, his forearms taut with effort, his left foot stabbing blindly for stability. He is messy in his frustration, climbing *Crazy Doctor*, a notorious Mount Nemo route built from "chipping": some holds were manufactured with chisels and crowbars by the original developer. And yet this doesn't bother Nick… *the climb feels real*. Unmarred, still-limestone, the holds couldn't possibly be discerned as much of an intrusion as the bolt hangers. And yet… neither is innate.

In my memory, in the midst of many distractions, something happens: Nick stands (ever so slightly, and only for a second) on a bolt hanger. Realizing what he's done (despite not intending to do it), he steps off the climb and weights the rope, defeated: the pure, natural climb vanishes before his eyes the moment his misstep renders the climb unreal.

Bolt hangers invite affective defacings that challenge the purity of nature. We know—based on work by Rossiter (2007)—that defacings within rock climbing are unavoidable: this is especially clear with regard to bolting practices and access negotiations in Southern Ontario. As they persist in establishing how climbing tourism is ordered, material-discursive defacings "complicate predominant discourses about stewardship, sustainability, communion with nature, and wellbeing" among other things (Evers, 2019, p. 7). It is also clear that the assembled tourismscape of the Niagara Escarpment is continuously re/made as a climbing space that is artificially natural, seemingly considered to *be* without substantial human intervention—spaces of high climbing value are falsely identified as those that have yet to change through human disturbance (Cunningham, 2018). When climbing unbolted routes (like in traditional climbing), climbing movement becomes prescriptive by way of what is available through natural means—"If there are no natural fissures or points of weakness in the rock, then there is no protection. Nature dictates" (Lewis, 2000, p. 62). This is exceptionally notable as incorrect in spaces where climbing has only been made possible through quarrying, developer manufacturing, or chipping, though all climbing routes are "cleaned" (removed of plant life) or "modified" (stripped of loose, unsafe rock) in some way. We attend to the bolt hanger to reiterate Ryan's (2002) questions regarding the power of discourses that promote the continued separation of natures and cultures: "Are all 'unnatural' impacts negative and all 'natural' ones positive?" (p. 272). Bolt hangers, as a potential unnatural impact, often carry the burden of this line of thought. Perhaps, instead, bolt hangers further illuminate the already-present impurity of climbing natures by both reducing and increasing potential environmental impact, as is the case with many tourism and outdoor technologies (Ryan, 2002). How might this impact be measured? On whom (or what) might it be exacted?

Investing in sustainable practice rhetoric through re-working humanist ethics does little, save for re-inscribing notions of ownership, domination, and resource-extraction into neoliberal conservation processes. By attending to affects—how climbs *feel*—as a productive capacity in tourism sustainability, we may start to undertake new community building, destabilize dominant political structures, and resist commodification (Cunningham, 2018). To draw further on Cunningham (2018), "sets of affective relations are resistive to large-scale resource projects ... in that they are generative of alternative value templates and subjectivities that embrace common life, vitality, and livability" (p. 42). In his exploration of the destructive aftermath of volcanic eruption in Japan's highlands, Cunningham (2018) relies on Anna Tsing's concept of contamination to describe the ways in which residents, tourists, and landscape ecologies were affected, with an open-ended result of "crafting alternative worlds" (p. 50). By acknowledging and promoting living *with* these defacings, we confront and diminish the irony inherent in past conceptualizations of nature as discursively constructed by humans as entirely human-less (Ryan, 2002). If it does/not feel any different, might there be sustainable possibilities lingering in the destruction of chipping? What would happen if we allowed the unnatural presence of bolt hangers to lead us toward landscapes of further impurity, like restricting climbing to quarries, re-roughing smoothed edges, or opening-up inoperative bridges and urban structures to climbing?

(Re)assembling: Swan Song

As I approach the final length on *Swan Song* the difficult climbing tapers off and I find myself extended through movement—the process of shaking out my forearms from lactic-acid buildup, the rhythm of white-grey limestone, the texturing of my fingertips and palms rubbing against the rough/polish of limestone. I feel d'Hauteserre's satisfaction from thirty metres below, her connected confidence in our camaraderie. There is simultaneously a clarity and a vagueness gathered around the tenuous touch of shoe-rubber-on-rock—when I reach for my rope to clip the anchors I suddenly notice I have skipped the last bolt hanger. I am dis/connected through this omission—I evoke fear, confidence, and astonishment—and the climb has been rendered/rendered me something else, something not-quite-present before. Through missing the bolt hanger my making of *Swan Song* is forever changed—an affective dis/ordering through surrender—"neither pure cause or pure effect but part of the world in its open-ended becoming" (Barad, 2007, p. 150).

There's a good, small pocket in the roof that looks amazing but only feels good moving to the left and using it with your hand sideways. Above me there are two cracks that split like a tributary and they meander lazily in separate directions up the centre of the corner. Both have surprisingly good holds—in-cut and cupped, they make my hands into talons—but securing your position means hiking your left foot onto the side of the cornered roof and pushing with your leg, pushing like you are fighting (why does it feel this way?) bumping your right hand up the bad crack. It's strange and strong but overwhelmingly scary because of the coffin-sized ledges below and I have to figure out how to get more inside the corner and sort of half-wedge my way into it... I finally get my hands into position with my left jammed, like sliding my hand in a door jamb (enter the thought of losing my fingers) and I cruise out of the corner toward the fifth bolt hanger—glinting/glowing—and I lean into a deformed knob to place my quickdraw and clip, that crisp clasp, metallic, echoing off the cliff toward nothingness, the sound a relation referencing my belayer below. (What else is there?) I am still afraid. My shoes skate searching for stability on sloping ledges; the fear and cold have depleted my body and I blindly continue to creep up the crack weakness with the big good flakes toward the ledge above. I pull myself onto the ledge with a desperate scrabbling and try to step onto it without dragging my knee but I'm tired and my hands aren't working so I hurl my entire body over the ledge and into the strange, deep crack, an arm and a knee as I use pressure to push my shoulder into the groove and wedge my body into the fissure without using any of my fingers. There are no edges—the rock is smooth/soaked/solid. It feels absolutely awful, the body-jamming, and inside of the fissure I stand up and look across at the next section of rock shaking my hands, resting, waiting. My breath ragged, the cold chasing the fear into my chest, my guts a tangled knot. From the ledge you can stand and rest with no hands, still/stasis, looking out over the farmland that surrounds Mount Nemo. And I am feeling ecstatic, untouchable: to be up here, on this ledge, frustrated and unable to keep my fear leashed. In front of me there is a traverse: I will move sideways where there are numerous, endless choreographies as the prow creeps away from me toward a blunted vertical edge and clipping position ten feet left. The rock drops down, steep toward the ground and the feet are a series of micro-divots, tiny nubbins that I have to hit while I press the pads of my fingers into frozen, glassy edges, the bolt winking at me, teasing/taunting from across the face. (Trust.) I am paralyzed with possibility. I shake my hands and put them in my mouth for warmth (they taste of chalk, rust, shame) and I weight the first few edges—where is the bolt? I feel the tension rise into my forearms (taut and burning) and I move my hands slowly, infused with fear, rife with uncertainty as I creep sideways across the wall. At the vertical edge I'm shocked and lean my body onto the rock, gasping, and I reach for my knot at my harness's centre, pull rope frayed and chalk-thick toward the bolt, but I haven't yet placed the quickdraw and I blink: the bolt hanger blinds me, winking, knowing (a secret), and I fall.

Toward sustainable tourism affects

In this paper we have suggested a number of slippages, re-workings, and narrative blurrings prompted by affect, what we have deemed *defacings*. Like a small moment of security in a vertical world of luminous limestone, the anchor within all of these threads of thought is rock climbing: how climbing—and its constituent actors—might make space for new ways of thinking about tourism and sustainability. First, we have conceptualized gender in climbing as a way of knowing and being infused with affect. This works to detach notions of gender from strict bodily identifiers and moves discussions of gender in climbing toward prompting the rupture of binary categories of mind/body, knowing/being, and culture/nature. Second, we have opened the idea of sustainability to align not solely with the continuance of human access, but with the existence of possible just futures for more-than-humans as well. If we consider justice as a relation—*living* ecological knowledge (Morton, 2018)—then any considerations of sustainability that are not grounded in imaginative, multi-species futurism deeply miss the mark, as they serve only to re-inscribe neoliberal, anthropocentric values (Stinson et al., 2020). Finally, we have prompted understandings of tourism that are proximate, open-ended, and constantly assembled through a cacophony of actors. Orienting definitions of tourism toward emergent, worldmaking processes disrupts notions of tourism as that which happens elsewhere, and puts a concerted focus on how being there (wherever "there" is) might facilitate futures that were previously unthought (Smith et al., 2020).

Defaced tourismscapes continue to *work*—and contain *workings*—as they are political interventions invested with situated, translated theories of multiplicity, productivity, and mattering (Barad, 2007; Vikkelso, 2009). If we look to the bolt hanger as a beacon, we might learn to shift our own focus to how the affective capacities of climbing can foster ecologically vibrant, abundant possibilities for tourism (Evers, 2019). By attending to *how climbing feels* as opposed to *what climbing is*, we pull constructions of climbing away from conquering, rationalizing, and grading, and instead allow climbing to evoke a spectrum of embodied experiences that might move us. The reframing of rock climbing as an affective landscape of immersion and honest entanglement works to shift perceptions of rocks (and their environments) away from mere resources to be extracted, conquered, and overcome, and toward an acknowledgement that "[rocks] affect us materially, emotionally, and mentally ... and we cannot just abstract our-selves from these affects" (Smith et al., 2020, p. 9). Here, we insist that attending to this destruction—what Collard et al. (2015) call "reckoning with ruination" (p. 325)—must constantly involve questioning not how tourism itself might be made more sustainable (and assuming this possibility as fact), but rather questioning and illuminating the ontological relationship between tourism and sustainability. This is the difference between assuming sustainable tourism itself is an existent entity, as opposed to interrogating our definitional basis of sustainability and how defenders of traditional tourism industries might leverage such definitions as a veil, obscuring extractive process through greenwashing and neoliberal ecotourism agendas (Moore, 2019). We are inescapably impure—pretending we are not "shuts down precisely the field of possibility that might allow us to take better collective action against the destruction of the world in all its strange, delightful, impure frolic" (Shotwell, 2016, p. 8). Acknowledging our roles in the mess and the making of tourism worlds is not a free pass to further enact harm. (And recognizing entanglement is certainly not enough.) Indeed, our ethical responsibility lies in our *responses* to the systems in which we are entangled (Bennett, 2010; Collard et al., 2015). We can't control everything (nor will we ever know everything that is unfolding around us), but we can choose to situate differently (or at least be honest about our situatedness) (Stinson et al., 2020). Indeed, even by attending to climbing affects as a defacing of masculinist, rational ways of knowing and being, we may garner a renewed investment in understanding how tourism and sustainability might be related in a future that is currently contested, fraught, and precarious. Living with pollution, contamination, or defacing doesn't have to translate into fatalist attitudes that will bring about further ruination. Instead, it might prompt an orienting to how things *feel* through new modes of noticing, nudging, and narrating relations and their possibilities (Tsing, 2015). As Smith et al. (2020) warn us, there is no guarantee that this will save the world (or tourism, or climbing), nor is it clear that this saving is even possible (or up to us). But we still think it's worth a shot.

By invoking actor-network theory to illuminate and enact affective tourismscapes, we are placing ourselves in the midst of research relating to gender, tourism, and sustainability in a way that—we hope—resonates with honesty and humility. As researchers, we participate in bringing certain realities into being through our work, a much different process than the neutral or distant stance of observing, describing, or explaining the way things are. This partial and confounding process of situated knowledges is one through which we can be accountable: to our own contradictions, to the messiness of research, and to living with the pollution of bias itself (Haraway, 1988). Working toward sustainability is human work, in which we must recognize our situated humanity and its relations to power and potentiality (Tsing, 2015). Prompting affective defacings as one more *becoming* or *doing of* tourism, has the ability to alter "not merely knowledge of the world but the very *matter* of the world itself" (Showen & Mantie, 2019, p. 394). By allowing strange and surprising actors (like the climbing bolt hanger) to change the ways we move through, think about, and feel with the world, we illuminate alternative possibilities for engaging ethics, morality, and responsibility in sustainable climbing practices.

Notes

1. In summer 2020, in conjunction with Black Lives Matter protests across North America, multiple public guides have revisited racist, sexist, and generally offensive route names in climbing. Spurred by outrage directed at MountainProject—an online database of global climbing routes—activists have flagged more than 5,000 routes for potential re-naming, including some of those listed here.
2. With the partial exception of Lion's Head climbing area at the far north-western tip of the Escarpment, none of Ontario's climbing areas are proximate to the types of climber-focused campgrounds common of destination climbing areas (or indeed, to any substantial camping amenities at all). Consequently, there is a disconnect in Ontario around locality and area-ownership—despite proximity to urban centres, there are few "true locals" since the Escarpment does not attract the residential climbing communities of Kentucky's Red River Gorge, for example (Rickly, 2017b). In this sense, most Ontario climbers are tourists.

Disclosure statement

No potential conflict of interest was reported by the authors.

ORCID

Michela J. Stinson http://orcid.org/0000-0002-8449-4049
Bryan S. R. Grimwood http://orcid.org/0000-0003-1555-7541

References

Alexandropoulos, G., & Dwyer, J. (2016). *Ontario climbing: Vol 1 the southern escarpment*. If It Bleeds We Can Kill It Productions.

Anderson, B. (2014). Affective atmospheres. In B. Anderson (Ed.), *Encountering affect: Capacities, apparatuses, conditions* (pp. 137–162). Routledge.

Appleby, K. M., & Fisher, L. A. (2005). Female energy at the rock": A feminist exploration of female rock climbers. *Women in Sport and Physical Activity Journal*, *14*(2), 10–23. https://doi.org/10.1123/wspaj.14.2.10

Barad, K. (2007). *Meeting the universe halfway: Quantum physics and the entanglement of matter and meaning*. Duke University Press.

Barratt, P. (2012). 'My magic cam': A more-than-representational account of climbing assemblage. Area, *44*(1), 46–53. https://doi.org/10.1111/j.1475-4762.2011.01069.x

Beard, L., Scarles, C., & Tribe, J. (2016). Mess and method: Using ANT in tourism research. *Annals of Tourism Research*, *60*(1), 97–110. https://doi.org/10.1016/j.annals.2016.06.005

Bell, S. J., Instone, L., & Mee, K. J. (2018). Engaged witnessing: Researching with the more-than-human. Area, *50*(1), 136–144. https://doi.org/10.1111/area.12346

Bennett, J. (2010). *Vibrant matter*. Duke University Press.

Bott, E. (2013). New heights in climbing and tourism: Jordan's Wadi Rum. *Journal of Tourism and Cultural Change*, *11*(1–2), 21–34. https://doi.org/10.1080/14766825.2013.768253

Bracken, M., Barnes, J., & Oates, C. (1991). *The escarpment: A climber's guide*. Borealis Press.

Brighenti, A. M., & Pavoni, A. (2018). Climbing the city. Inhabiting verticality outside of comfort bubbles. *Journal of Urbanism: International Research on Placemaking and Urban Sustainability*, *11*(1), 63–80.

Buda, D. M., d'Hauteserre, A.-M., & Johnston, L. (2014). Feeling and tourism studies. *Annals of Tourism Research*, *46*(1), 102–114. https://doi.org/10.1016/j.annals.2014.03.005

Chisholm, D. (2008). Climbing like a girl: An exemplary adventure in feminist phenomenology. Hypatia, *23*(1), 9–40.

Collard, R., Dempsey, J., & Sundberg, J. (2015). A manifesto for abundant futures. *Annals of the Association of American Geographers*, *105*(2), 322–330. https://doi.org/10.1080/00045608.2014.973007

Cunningham, E. J. (2018). Nature interrupted: Affect and ecology in the wake of volcanic eruption in Japan. *Conservation and Society*, *16*(1), 41–51. https://doi.org/10.4103/cs.cs_16_50

d'Hauteserre, A.-M. (2015). Affect theory and the attractivity of destinations. *Annals of Tourism Research*, *55*(1), 77–89.

Deleuze, G., & Guattari, F. (1987). *A thousand plateaus: Capitalism and schizophrenia* (Trans. B. Massumi). University of Minnesota Press.

Evers, C. W. (2019). Polluted leisure. *Leisure Sciences*, *41*(5), 423–335.

Figueroa-Domecq, C., Pritchard, A., Segovia-Pérez, M., Morgan, N., & Villacé-Molinero, T. (2015). Tourism gender research: A critical accounting. *Annals of Tourism Research*, *52*(1), 87–103. https://doi.org/10.1016/j.annals.2015.02.001

Foucault, M. (2007). *The politics of truth*. Semiotext(e).

Franklin, A. (2004). Tourism as ordering: Towards a new ontology of tourism. *Tourist Studies*, *4*(3), 277–301.

Frauman, E., & Rabinowitz, E. (2011). A preliminary investigation of environmental and social practices among boulderers. *Journal of Outdoor Recreation, Education, and Leadership*, *3*(1), 12–25. https://doi.org/10.7768/1948-5123.1044

Grimwood, B. S. R. (2015). Advancing tourism's moral morphology: Relational metaphors for just and sustainable arctic tourism. *Tourist Studies, 15*(1), 3–26. https://doi.org/10.1177/1468797614550960
Grimwood, B. S. R., Caton, K., & Cooke, L. (2018). *New moral natures in tourism*. Routledge.
Haraway, D. (1988). Situated knowledges: The science question in feminism and the privilege of partial perspective. *Feminist Studies, 14*(3), 575–599.
Haraway, D. (1991). *Simians, cyborgs, and women: The reinvention of nature*. Routledge.
Heywood, I. (1994). Urgent dreams: Climbing, rationalization, and ambivalence. *Leisure Studies, 13*(3), 179–194.
Hutson, G., & Montgomery, D. (2010). Stakeholder views of place meanings along the Niagara Escarpment: An exploratory Q methodological inquiry. Leisure/Loisir, *34*(4), 421–442. https://doi.org/10.1080/14927713.2010.544121
Jamal, T., & Hollinshead, K. (2001). Tourism and the forbidden zone: The underserved power of qualitative inquiry. *Tourism Management, 22*(1), 63–82. https://doi.org/10.1016/S0261-5177(00)00020-0
Kelly, P. E., & Larson, D. W. (1997). Effects of rock climbing on presettlement Eastern White Cedar (Thuja occidentalis) on cliffs of the Niagara Escarpment, Canada. *Conservation Biology, 11*(5), 1125–1132. https://doi.org/10.1046/j.1523-1739.1997.96248.x
Kiewa, J. (2002). Traditional climbing: Metaphor of resistance or metanarrative of oppression? *Leisure Studies, 21*(2), 145–161. https://doi.org/10.1080/02614360210158605
Larsen, J. (2008). De-exoticizing tourist travel: Everyday life and sociality on the move. *Leisure Studies, 27*(1), 21–34.
Latimer, J., & Miele, M. (2013). Naturecultures? Science, affect, and the non-human. *Theory, Culture, & Society, 30*(7/8), 5–31.
Law, J. (1999). After ANT: Complexity, naming, and topology. *The Sociological Review, 47*(1_suppl), 1–14.
Law, J. (2004). *After method: Mess in social science research*. Routledge.
Lewis, N. (2000). The climbing body, nature, and the experience of modernity. *Body & Society, 6*(3-4), 58–80.
Lund, K. A., & Jóhannesson, G. T. (2016). Earthly substances and narrative encounters: Poetics of making a tourism destination. *Cultural Geographies, 23*(4), 653–669. https://doi.org/10.1177/1474474016638041
Maksymowicz, K. (2019). Pura vida: Affect, Puerto Viejo, and emergent tourism erotics. Téoros, *37*(2).
Moore, A. (2019). Selling Anthropocene spaces: Situated adventures in sustainable tourism. *Journal of Sustainable Tourism, 27*(4), 436–451.
Morton, T. (2018). *Being ecological*. Pelican Books.
Müller, M., & Schurr, C. (2016). Assemblage thinking and actor-network theory: Conjunctions, disjunctions, cross-fertilisations. *Transactions of the Institute of British Geographers, 41*(3), 217–229. https://doi.org/10.1111/tran.12117
Ness, S. A. (2011). Bouldering in Yosemite: Emergent signs of place and landscape. *American Anthropologist, 113*(1), 71–87. https://doi.org/10.1111/j.1548-1433.2010.01307.x
Ness, S. A. (2016). *Choreographies of landscape: Signs of performance in Yosemite National Park*. Berghahn Books.
Nettlefold, P. A., & Stratford, E. (1999). The production of climbing landscapes-as-texts. *Australian Geographical Studies, 37*(2), 130–141. https://doi.org/10.1111/1467-8470.00074
Oates, C., & Bracken, M. (1997). *A sport climber's guide to Ontario Limestone*. Borealis Press.
Ontario Alliance of Climbers. (2019). *Ontario alliance of climbers [website]*. https://www.ontarioallianceofclimbers.ca/
Pritchard, A., & Morgan, N. J. (2000). Privileging the male gaze: Gendered tourism landscapes. *Annals of Tourism Research, 27*(4), 884–905. https://doi.org/10.1016/S0160-7383(99)00113-9
Ren, C. (2011). Non-human agency, radical ontology, and tourism realities. *Annals of Tourism Research, 38*(3), 858–881. https://doi.org/10.1016/j.annals.2010.12.007
Rickly, J. M. (2016). Lifestyle mobilities: A politics of lifestyle rock climbing. Mobilities, *11*(2), 243–263. https://doi.org/10.1080/17450101.2014.977667
Rickly, J. M. (2017a). The (re)production of climbing space: Bodies, gestures, texts. *Cultural Geographies, 24*(1), 69–88. https://doi.org/10.1177/1474474016649399
Rickly, J. M. (2017b). I'm a Red River local": Rock climbing mobilities and community hospitalities. *Tourist Studies, 17*(1), 54–74. https://doi.org/10.1177/1468797616685648
Rossiter, P. (2007). On humans, nature, and other nonhumans. *Space and Culture, 10*(2), 292–305. https://doi.org/10.1177/1206331206298546
Ruming, K. (2009). Following the actors: Mobilizing an actor-network theory methodology in geography. *Australian Geographer, 40*(4), 451–469. https://doi.org/10.1080/00049180903312653
Ryan, S. (2002). Cyborgs in the woods. *Leisure Studies, 21*(3–4), 265–284. https://doi.org/10.1080/0261436022000030650
Sayes, E. (2014). Actor-network theory and methodology: Just what does it mean to say that nonhumans have agency? *Social Studies of Science, 44*(1), 134–149. https://doi.org/10.1177/0306312713511867
Seigworth, G. J., & Gregg, M. (2010). An inventory of shimmers. In M. Gregg & G. J. Seigworth (Eds.) *The affect theory reader* (pp. 1–25). Duke University Press.
Shotwell, A. (2016). *Against purity: Living ethically in compromised times*. University of Minnesota Press.
Showen, A., & Mantie, R. A. (2019). Playing in the posthuman band: Toward an aesthetics of intra-action in musical leisure. *Leisure Sciences, 41*(5), 385–401. https://doi.org/10.1080/01490400.2019.1627962
Smith, M., Speiran, S., & Graham, P. (2020). Megaliths, material engagement, and the atmospherics of neo-lithic ethics: presage for the end(s) of tourism. *Journal of Sustainable Tourism*, 1–16.

Spinney, J. (2015). Close encounters? Mobile methods, (post)phenomenology, and affect. *Cultural Geographies, 22*(2), 231–246.

Springgay, S., & Truman, S. E. (2017). *Stone Walks*: Inhuman animacies and queer archives of feeling. *Discourse: Studies in the Cultural Politics of Education, 38*(6), 851–863.

Stewart, K. (2014). Road registers. *Cultural Geographies, 21*(4), 549–563.

Stinson, M. J., & Grimwood, B. S. R. (2019). On actor-network theory and anxiety in tourism research. *Annals of Tourism Research, 77*(1), 141–143. https://doi.org/10.1016/j.annals.2018.12.003

Stinson, M. J., Grimwood, B. S. R., & Caton, K. (2020). Becoming common plantain: Metaphor, settler responsibility, and decolonizing tourism. *Journal of Sustainable Tourism*. 1–19.

Tsing, A. L. (2015). *The Mushroom at the End of the World: On the possibility of life in capitalist ruins*. Princeton University Press.

Tucker, H. (2009). Recognizing emotion and its postcolonial potentialities: Discomfort and shame in a tourism encounter in Turkey. *Tourism Geographies, 11*(4), 444–461.

Tucker, H., & Shelton, E. J. (2018). Tourism, mood, and affect: Narratives of loss and hope. *Annals of Tourism Research, 70*(1), 66–75. https://doi.org/10.1016/j.annals.2018.03.001

van der Duim, R. (2007). Tourismscapes: An Actor-Network perspective. *Annals of Tourism Research, 34*(4), 961–976. https://doi.org/10.1016/j.annals.2007.05.008

van der Duim, R., Ren, C., & Jóhannesson, G. T. (2017). A decade of interfering with tourism. *Annals of Tourism Research, 64*(0), 139–149. https://doi.org/10.1016/j.annals.2017.03.006

Veijola, S., & Jokinen, E. (1994). The body in tourism. *Theory, Culture, & Society, 11*(3), 125–151.

Vespestad, M. K., Lindberg, F., & Mossberg, L. (2019). Value in tourist experiences: How nature-based experiential styles influence value in climbing. *Tourist Studies, 19*(4), 453–474. https://doi.org/10.1177/1468797619837966

Vikkelso, S. (2009). Description as intervention: Engagement and resistance in Actor-Network analyses. *Science as Culture, 16*(3), 297–309.

Yudina, O., & Grimwood, B. S. R. (2016). Situating the wildlife spectacle: Ecofeminism, representation, and polar bear tourism. *Journal of Sustainable Tourism, 24*(5), 715–734. https://doi.org/10.1080/09669582.2015.1083996

Yudina, O., Grimwood, B. S. R., Berbary, L. A., & Mair, H. (2018). The gendered natures of polar bear tourism. *Tourism, Culture, & Communication, 18*(1), 55–56.

Gendering knowledge in tourism: gender (in)equality initiatives in the tourism academy

Katherine Dashper, Jane Turner and Yana Wengel

ABSTRACT

The tourism academy is a key site through which gender is produced, reproduced and, potentially, challenged. In this paper, we draw on Acker's (1990) concept of gendered organisations to present a case study of a tourism department preparing to apply for an international gender equality charter-accreditation, Athena SWAN. Ketso was used as a method to try to stimulate active involvement of all staff members and breakdown traditional hierarchies within the team, and to encourage honest discussion about gender and inequality in this context. This was only partially successful, however, and we discuss how explicit focus on gender (in)equality through this process both enabled discussion of usually ignored topics *and* revealed entrenched gender power dynamics and structural and institutional barriers to reform. The paper illustrates both the possibilities of gender equality initiatives like Athena SWAN to highlight many of the gendered practices of tourism academia and the limitations they hold for overcoming deep-rooted gender inequality.

Introduction

The tourism academy is highly gendered, yet gender as a topic of research and critical discussion is marginalised in tourism research (Figueroa-Domecq et al., 2015). Although tourism is a key sector through which gender is enacted – in terms of work, leisure experiences, symbolic messages etc. – gender remains a minority research interest in the academic field of tourism studies (Chambers & Rakić, 2018; Pritchard, 2018). It is risky for scholars to speak about gender in tourism, whether that be in terms of research or teaching (Jeffrey, 2017; Munar et al., 2017). However, gender is an integral aspect of all social institutions and interactions – including academia - and so should be central to research and critical reflection in any academic field. Morgan and Pritchard (2019) argue that male dominance of professorships and senior research positions in tourism and hospitality leads to "a situation that has significant implications for the kinds of knowledge we create" (p.39). In this paper, we respond to the calls of scholars including Chambers et al. (2017) to engage with gender theory to interrogate some of the practices and experiences of tourism academia.

The United Nation's Sustainable Development Goals (SDG) provide an aspirational roadmap to "promote prosperity while protecting the planet" (UN, n.d.). SDG 5 is about achieving gender equality and empowering women and girls. Alarcón and Cole (2019) argue that gender equality really underpins all 17 of the SDG, as "without gender equality, there can be no sustainability" (p.903). Academia produces and disseminates knowledge, and so has a broad ability to shape ideas and discourses which contribute to efforts towards achieving the SDG. Figueroa-Domecq et al. (2015) explain this as "the power to circumscribe; to slant; to reify; to elevate some issues and to deprecate others; to rule in certain ways of talking and to rule out and restrict others, normalising how we comprehend a certain field" (p.89). Limited critical reflection on gender and the tourism academy reinforces this power and entrenches gender inequality in ways of thinking about and understanding tourism. To contribute towards achieving the SDG it is imperative that tourism scholars reflect on gender inequality and gendered practices across contexts and institutional settings, including within the academy.

In this paper, we draw on a case study of an independent tourism department in a UK university preparing to apply for accreditation under the Athena SWAN Charter. This initiative recognises commitment and actions to advance the careers of women in higher education and research, and to working towards greater gender equality. Drawing on Acker's (1990) theory of gendered organisations, our analysis reveals some of the entrenched barriers to achieving gender equality in the tourism academy, even in the context of efforts aimed explicitly at addressing such inequality. During attempts to achieve the Bronze charter award (the first level of the three-tiered Athena SWAN initiative), we used Ketso workshops as a method to try and empower our colleagues to get involved in the process and to share their experiences and insights on the gendered aspects of tourism academia. Ultimately our attempts at achieving the Bronze award failed and gender (in)equality once again disappeared from focus in day-to-day discussions and strategic planning. The purpose of this paper is to reflect on how involvement in Athena SWAN enabled gender and inequality to be discussed in ways normally unimaginable in our academic lives, but also reflected the intractability of gender inequality in tourism and the wider academy. The paper thus contributes to critical discussions of tourism academia and the significance of gender in tourism organisational practices, many of which have negative consequences for knowledge creation, policy and actions that could contribute to the achievement of the SDG.

Theorising gender in the tourism academy

There has been growing interest in critically assessing different aspects of the tourism academy in recent years, as the dominance of certain groups and individuals can have profound effects on knowledge creation and the types of issues and different voices that are valued and prioritised. Tung and McKercher (2017) note that the rapidly changing environment of academia is characterised by poor job security, limited promotion prospects and increased pressure and expectations. This leads to what they call the "industrialisation of academic research" (p.323) and 'game playing' in terms of publishing and collaboration. In such an environment, senior academics are extremely powerful; regulating access to publication through journal editorships and dominating leadership positions in institutions and international networks. Munar et al. (2015) have demonstrated the dominance of men in these senior positions and the marginalisation of women as leaders, leading them to conclude that "gender matters in the tourism academy" (p.16).

Ek and Larson (2017) further illustrate the power of senior male figures within tourism. Their analysis of celebratory portraits of tourism scholars published in the journal Anatolia over four years found that such acclaim was reserved predominantly for male academics, with only 7% of published profiles being about women. What they call the 'Alpha males' of tourism are represented as pioneers of the field, guiding others and helping establish the canon of tourism knowledge. Tribe (2010) suggests that junior scholars feel a need to emulate the work of more senior figures in order to fit in and try and secure promotion. He argues that tourism remains an old boys' club, which is also very Anglo-Saxon dominated, and this stifles debate and limits change and innovation.

Successful academic careers are built around research profiles, and reputations gained predominantly through publication. Within the tourism field, a small number of journals are considered to be the most prestigious, and pressure to publish in these few outlets helps reinforce their dominance and the power of the gatekeepers of those journals. Nunkoo et al. (2020) research analysed gender differences in authorship, collaboration and research approach in articles published in the top three tourism journals over a 17-year period and found that male authors dominated outputs. Male authors were also more likely than female authors to adopt quantitative methods. Choice of research approach matters, as feminist scholars have long critiqued the gender-blind approach of supposedly objective quantitative methods that often mask the power relations inherent in research design and practices and which can lead to the silencing of whole groups of people (Madge et al., 2014). The dominant journals in the field of tourism clearly favour quantitative methods and positivist/post-positivist research designs and this has consequences for the topics studied and types of knowledge produced, which helps shape the field of tourism studies (Pritchard et al., 2011). Therefore, not only who publishes in the field's leading journals (majority male authors) but also what they research and how they produce and present their research matters and shapes the field in subtly masculine ways.

There are roughly equal numbers of male and female scholars in the tourism field, and female authorship has increased through time to nearly equal that of male authorship (Kirilenko & Stepchenkova, 2018). While it is undoubtedly positive that female authorship of tourism research is increasing, it remains problematic that publication in the leading tourism journals is dominated by male authors (Nunkoo et al., 2020). This is reinforced by citation counts which are often taken as a proxy for academic leadership and performance, forming an important aspect of hiring and promotion activities. Nunkoo et al. (2019) illustrate a gender gap in citation practices within tourism research whereby male authors are more likely to be cited than female authors, suggesting citation is often based less on quality and significance than on existing social norms and practices. Nunkoo et al. (2020, p. 1) conclude that "Research practices in tourism are inherently masculinized, posing challenges for the gender equality agenda".

There is thus widespread evidence of gender inequality in the tourism academy, as there is in academia more broadly (Pritchard et al., 2011). Academic culture is very masculine – competitive, individualistic, and requiring (more than) fulltime commitment to succeed (Van den Brink & Stobbe, 2009). The standard model of an academic career is presented as gender neutral but is in fact based on a masculine model. Bagilhole and Goode (2001) argue that a myth of individual merit disguises the importance of powerful male networks and conceals the support men often get from other men, as is evidenced in the citation practices and reification of senior male researchers reported in the field of tourism (Ek & Larson, 2017; Nunkoo et al., 2019). Collaboration is seen as key to academic success, but women often struggle to access influential collaborators and networks, which can negatively affect career progression (Hart, 2016; Zippel, 2019). Supposedly gender-neutral practices like recruitment are beset with gendered practices that can subtly sustain gender inequality (Van den Brink et al., 2010).

In order to try and redress gender inequality in academia different equality initiatives have been introduced. While appearing to provide a sensible strategy to increase gender representation across hierarchical levels, gender equality initiatives often fail and can be met with ambivalence and even resistance from staff. Van den Brink and Stubbe (2014) argue that men are more likely to be suspicious of gender equality programmes, seeing them as a threat to their own careers, whereas women and other members of marginalised groups may view the need for specific focused initiatives as criticism of their own abilities and careers. Gender equality programmes, such as mentoring and women's leadership schemes, can be problematic in this way as they single women out as a problem in need of special intervention to fix, distancing women from the supposedly gender-neutral figure of the ideal worker or leader. Women may thus respond negatively to such programmes and instead position themselves closer to the masculine model of leadership in order to reinforce their own credibility. In such ways, gender equality programmes can be understood as paradoxical in that they make gender visible in organisations and careers but at the same time fail to acknowledge, let alone address, the deeper gendered aspects of organisations that contribute to the persistence of gender inequality in careers and workplaces (Dashper, 2019).

Joan Acker's (1990) theory of gendered organisations provides a fruitful theoretical framework for understanding the persistence of gender inequality in academia and the failure of gender equality initiatives to redress this. Acker (1990) identified five processes that reproduce gender in organisations: the division of labour; cultural symbols; workplace interactions; individual identities; and, organisational logic. She argued that the practices, structures and policies that govern work in organisations are inherently gendered in ways that work to advantage men and disadvantage women. To say that an organisation is gendered means "that advantage and disadvantage, exploitation and control, action and emotion, meaning and identity are patterned through and in terms of a distinction between male and female, masculine and feminine" (Acker, 1990, p. 146). There is widespread empirical support for the theory of gendered organisations (see Britton & Logan, 2008), which has also been widely adopted in the study of academic careers (e.g. Hart, 2016; Zippel, 2019). It has not previously been applied to tourism academia, yet it provides a useful theoretical framework for exploring the gendered aspects of tourism academic careers and organisational practices, as we illustrate below.

Acker (1992) argues that gender is a foundational element of organisational structures and consequently of working lives and experiences, as it is "present in processes, practices, images and ideologies, and distributions of power" (p.567). This can be seen in the practices discussed above in relation to tourism academia, where senior positions and ideas about leaders in the field are built around masculine norms which reify metrics like publication counts and citations while failing to acknowledge the gendered underpinnings of those practices. Acker (2000) argues that gender equality programmes focus on perceived differences between men and women in terms of career-relevant characteristics, personalities and experiences but fail to recognise the systemic nature of gender inequality within organisations. Consequently, such initiatives are unlikely to produce the profound change needed for greater gender equality as they do not expose and attempt to tackle the gendered aspects of organisations and careers. Chambers et al. (2017) also caution against the dangers of some organisational practices, such as gender equality initiatives, for providing an appearance of tackling inequality while doing little to bring about real change. The gender equality initiative discussed below – Athena SWAN - is an attempt to tackle some of the gendered aspects of academic careers, but our experience resonates with the warnings offered by both Acker (2000) and Chambers et al. (2017) in that, despite good intentions, the initiative failed on an organisational level. In the next section, we explain the methods we adopted to research our experiences of attempting to engage with this particular gender equality initiative.

Case study and methods

The equality initiative: Athena SWAN

In 2005 the UK Equality Challenge Unit (now Advance HE) established the Athena SWAN (Scientific Women's Academic Network) Charter. The aim was to encourage and recognise a commitment to advancing the careers of women in science, technology, engineering, mathematics, and medicine (STEMM) in institutions of higher education (HE) and research. In 2015 the Charter expanded beyond the UK to Ireland and Australia and at the same time, broadened its recognition of the work that is undertaken to address gender equality in academia (Advance HE, 2020b). More recently, adaptations of the model have been seen in Canada and the United States

(Fernow, 2019; Xiao et al., 2020). Athena SWAN currently has over 160 members, holding over 800 awards between them (Advance HE, 2020a).

Originally an initiative to demonstrate commitment to gender equality and excellent working practices, over time Athena SWAN has become a strategic tool for HE institutions and funding donors seeking to address the gender gap. As such, research councils funding projects within HE in the UK now acknowledge the benefits of equality and diversity and take steps to ensure that funding beneficiaries and their institutions embed equality and diversity principles (Donald et al., 2011). In 2011 the National Institute for Health Research became the first funding body to stipulate that academic departments applying for funding must hold (at least) the Silver Award of the Athena SWAN Charter (Schmidt et al., 2019). Similar conditions are also being employed more widely within Europe, with 'CASPER', a HORIZON2020 project, also establishing a similar certification-award (European Commission, 2019).

Based on ten key principles, an application to the three-tiered Athena SWAN programme (Bronze, Silver Gold), acknowledges a commitment by an institution (or internal department) towards gender equality, representation and progression for all, through workplace policies, practice and culture (Advance HE, 2020a). The applications are peer reviewed by academics, industry experts, human resources and equality and diversity experts from Athena SWAN member institutions.

Munir et al. (2014) recognise that the effectiveness and impact of the Athena SWAN charter stems from the requirement to provide evidence and signpost efforts to improve gender balance, both at an institutional level and for individuals. Improvement of organisational structures and practices at several universities highlight greater awareness around gender equality. For individuals, the Charter is documented as having identified and exemplified role models and illustrated potential academic career pathways and opportunities for women.

An independent impact evaluation of Athena SWAN by Graves et al. (2019) found the Charter to be most effective when implemented by institutions in a holistic manner, with policies and practices developed to ensure that no member of staff or student is disadvantaged. Further, the Charter is widely accredited as a tool to facilitate open communication and scrutiny of workplace practices, but with recognition given to the need for resources and leadership support. To this end, the process and review of the Athena SWAN application in this case study is valuable in recognising the (in)equalities within a UK HE tourism department and the lessons that can be learnt and adopted by similar institutions seeking gender equality within tourism and the wider academy.

The department application in this study was for a Bronze level award, the first level of the initiative which demonstrates recognition of '... a solid foundation for eliminating gender bias and developing an inclusive culture that values all staff' (Advance HE, 2020a, n.p.) The Bronze level application is an extensive written application that focuses upon the department and the provision for supporting and advancing women's careers. A self-assessment team leads on a description of the department; an overview and account of the self-assessment process and provides a holistic overview of the department through quantitative (HR data) and qualitative data (policies, practices, systems and arrangements). The outcome is the development of a four-year plan to build-on current practice and aspirations for the next four years (Advance HE, 2020a).

The case study University is a member of the Athena SWAN Charter and holds an institution Bronze award. The tourism department initiated an independent application for its own Bronze level award in late 2017. At the time of the application, the department employed 63 members of staff (52% female) in academic and professional roles. The decision to apply for the Bronze award was in part strategic – to ensure we remained eligible to apply for grant funding – and in part based on the desire of some colleagues to reflect on and try and address issues of gender inequality in our working lives.

A self-selected departmental team, led by a senior colleague, was responsible for preparation of the Athena SWAN application. Specifically, the self-assessment team sought to collect

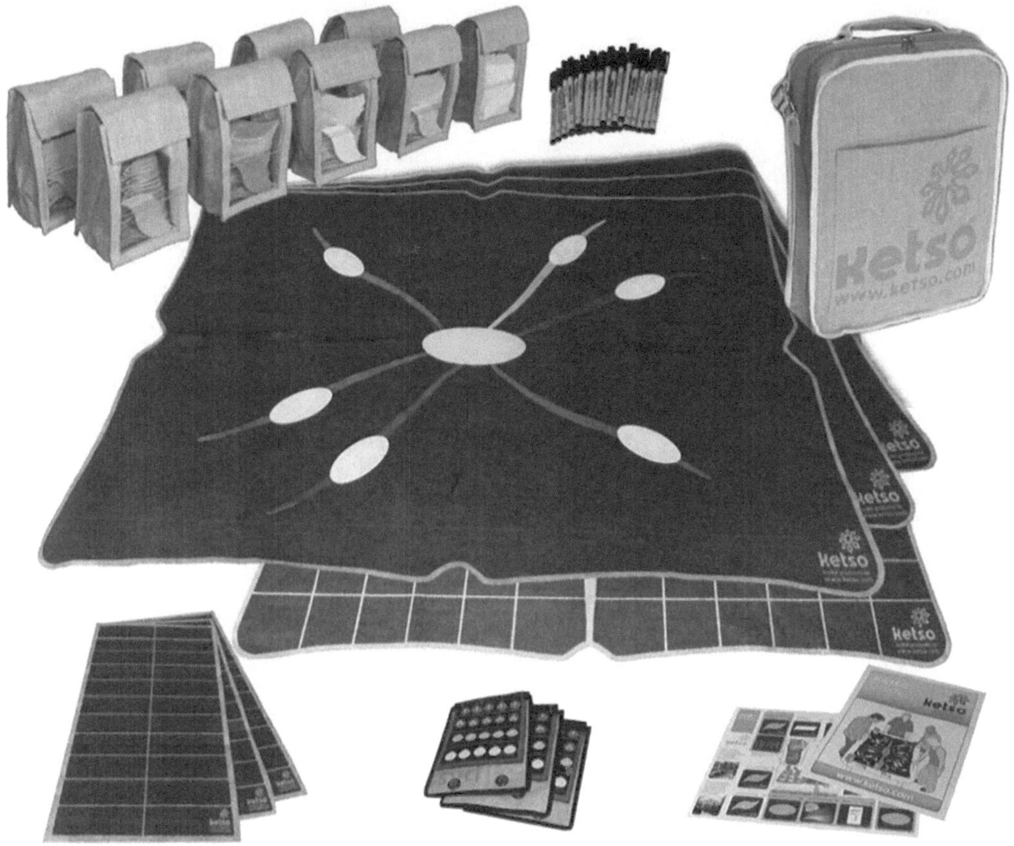

Figure 1. Standard Ketso Kit. Used with permission.

information around issues of representation, progression, and working environment. The group comprised of academic (14) and professional staff (5) of which 14 were female and 5 male. The authors of this paper were all active members of the self-assessment team. Staff were allocated deployment to work on the application, indicative of support for the initiative from senior management in the department.

To gather all the relevant information, the self-assessment team employed both qualitative and quantitative methods. Qualitative material was collected using an innovative technique called Ketso, whilst an online survey supplemented HR data in terms of quantitative material. Our focus in this paper is on the processes involved in preparing to apply for the Athena SWAN award and the use of Ketso workshops in particular.

Ketso workshops

The self-assessment team decided to collect qualitative encounters of academic staff, administrative staff and students in the department using the 'toolkit for creative engagement' Ketso (www.ketso.com). Two of the authors of this paper were charged with the task of conducting Ketso workshops. Ketso is a participatory action research tool underpinned by constructionist epistemology (Bates, 2016). The standard Ketso Kit, 'a workshop in a bag', consists of three felt mats, a grid mat, coloured plastic cards in the shape of leaves, plastic icons, felt stripes, as well as pens with water soluble ink and a guide (see Figure 1).

Figure 2. Ketso materials.

Ketso is a visual and interactive kit that encourages group interaction and stimulates discussion (Tippett et al., 2011; Zhang et al., 2016). It allows participants' individual voices to be heard and helps to elicit deeper meanings from the information provided by research participants (Tippett et al., 2007). As a qualitative toolkit, Ketso can be used for data collection and data analysis (Wengel et al., 2019).

Researchers using Ketso collect data and analyse it in themes during four key stages of a workshop (for the philosophy and process of Ketso see www.ketso.com). Each workshop is based on the analogy of the tree. The research question is a centerpiece of the workshop and represents the 'trunk'; participants' ideas are written on cards/'leaves' and are placed on felt strips/ 'branches' located on the felt.

Every Ketso workshop starts with warm-up questions to make sure participants understand the process. Participants get time to write their ideas on cards and then, one by one, they voice their ideas while placing 'leaves' on a mat. Guided by constructionist dynamics, participants discuss the proposed ideas and group similar ideas into themes. Hence, the leaves placed around a branch represent one theme. In this way the analysis and clustering of the answers in themes happens intuitively during the workshop.

The Ketso technique allows participants to engage in constructive dialogue where each voice can be heard as the workshop structure allows everyone to add to discussion, to see commonalities and notice differences. As Ketso attempts to provide a non-hierarchical and inclusive environment for discussion, we decided it was an ideal approach for the task of encouraging colleagues to talk about their views and experiences of equality and diversity, as we discuss further below. In this project, we conducted five Ketso workshops with academic and administrative staff and students. In total 29 participants (17 female and 19 male) participated. During the workshops, colleagues were asked to exchange ideas about gender, diversity and equality, prompted by the introductory question 'What does equality mean to you?' Each workshop lasted between 1-2 hours. The sessions were led by one of the authors, who is experienced in running Ketso workshops, with support from another of the writing team. We made written notes during the workshops. The 'data' that inform this paper consist of the Ketso cards and 'trees' constructed during workshops (see Figure 2), the written notes we took during the workshops and our subsequent reflections on the whole process of the (failed) Athena SWAN application. The data were manually coded and analysed for themes, patterns and trends, using the steps of Braun and Clarke (2006) thematic analysis. The writing team each analysed the data separately, before discussing and refining themes collaboratively, in order to ensure trustworthiness during thematic analysis (see Nowell et al., 2017; Shenton, 2004).

Ethical issues

Ethical approval was gained from the department's research ethics committee prior to data collection. Participation in Ketso workshops was voluntary, and we discussed issues of confidentiality with everyone involved and sought consent to develop this into a research paper. However, ethical issues are not resolved simply by asking participants to sign consent forms, and given this was a project that involved our colleagues and led to discussion of personal and sometimes emotional topics, we continually reflected on the ethics of the project and of writing this paper. Whilst research amongst academic peers does not necessarily raise ethical issues that are inherently different from any other form of research; it does arguably heighten sensitivity around issues such as confidentiality and anonymity in what Wiles et al. (2006, p284) identify as a 'research-wise' sample.

In terms of the wider Athena SWAN project and the Ketso workshops, we were aware of power issues inherent in asking colleagues to participate in a department-led and management-sanctioned initiative. We discussed the project openly with colleagues, explained the purpose of the workshops and why we felt it would be beneficial to write this paper and share insights on gendered academia in tourism. None of our colleagues expressed concern about this, but we are cognisant of the power we have as authors of this paper to represent the department and colleagues within it. Consequently, we have focused here more on the process of preparing to apply for Athena SWAN accreditation and avoided sharing stories or insights personal to individual colleagues. We also decided that our discussion of the project's findings presented below would not include data or reflections gained from Ketso workshop with students, who are certainly in a position of unequal power in relation to us as their lecturers.

Issues arose during the Ketso workshops that also caused us to reflect on some of the ethical dilemmas of researching with colleagues. The feeling of needing to give the 'right' answer amongst groups of peers of varying experience and status is something that needed to be given due consideration in the qualitative data collection process and something we worked to try and mitigate. 'Self-presentation' and the power imbalance that is often present in hierarchical relationships is well documented in terms of its influence in data collection (Coar & Sim, 2006) At the time of the research, all of the authors were employed within the department. Author A held a full-time academic post and had been a member of the department for approximately seven years. Author B was in a part-time academic position and had been within the team for approximately five years. Author C was a post-doc and had been within the department for just a few months. Consequently, although we had varied experience in terms of time and position, none of us were in a management role. Although we were actively involved in the self-assessment team, we were not leading on and responsible for delivery of the Athena SWAN application and were not involved in strategic discussions related to the process, or its reasons for failure. Our analysis and discussion below thus present a view from our position in this process, and it is possible that others would interpret the process differently.

There are ethical dilemmas in all research, and these may be heightened when research involves colleagues, friends or family members. However, as Knights and Clarke (2014) point out, academia as a setting should not be beyond the gaze of the academic researcher and proximity to and shared experiences with research participants can be beneficial in terms of critical insight.

Therefore, although we acknowledge the ethical dilemmas of researching with our colleagues and our place of work, we argue that this should not prevent us from critical reflection of those contexts (see also Dashper, 2019).

Findings

In the following sections, we discuss how the process of applying for Athena SWAN accreditation brought gender to the fore within the department and how the adoption of Ketso workshops enabled a variety of voices to be heard and experiences to be shared in ways not normally possible in routine department discussions and interactions. However, although there were some beneficial aspects of the process, ultimately the application floundered and was never submitted for accreditation. In the following two years, gender (in)equality has once again receded from focus in department discussions, and many of the issues raised through the Athena SWAN process remain unaddressed.

Making gender visible

Gender inequality is a pervasive feature of contemporary higher education, including within tourism departments. However, although – as discussed above - there is convincing evidence of such inequality in terms of hiring and promotion decisions, publication and citation practices and other factors generally associated with esteem and status, gender inequality is rarely spoken about in department meetings and in strategy and planning discussions. Applying for Athena SWAN accreditation required that gender become a focus for discussion and action, and this in itself was unusual and refreshing, prompting frank and open discussions that allowed colleagues to share stories, experiences and concerns.

The Ketso format proved beneficial in getting colleagues to open up and share stories. The session facilitators asked participants to write down and share things that the department currently does well in terms of equality. Participants were given a chance to think individually before writing on one of the Ketso leaves and sharing this with the rest of the group. The discussion then began around issues identified by colleagues, such as good communication across hierarchical levels and individual support in relation to caregiving responsibilities. Some participants struggled to think of positive things, but the Ketso format of turn-taking – where each person speaks in turn, sharing their thoughts and experiences – gave people time to think and to build on other people's comments.

Facilitators then asked participants to identify areas where things were done less well and needed improvement in terms of equality. A number of areas were identified, including inconsistency in deployment, challenges related to publication strategies, and perceived lack of transparency in things like awarding of travel and research grants. The discussion here was more free-flowing and participants commented on how it was refreshing to be able to discuss these issues that affect our daily working lives. After discussing these challenges and grouping them together around shared headings generated by workshop participants, the discussion turned to possible solutions. By this point, participants were very engaged and quick to respond, taking additional Ketso leaves to write down ideas in response to each other's comments and building up a network of responses. In this way, the Ketso workshop moved from Stage One: Identifying good practice; to Stage Two: Identifying areas for improvement; and finally Stage Three: Identifying solutions. This provided structure for discussions and ensured that the workshops did not become overly-focused on sharing grievances and looked instead to constructive actions that could be taken. Figure 2 illustrates the Ketso materials as used in one of the workshops.

Some participants were hesitant to participate and communicate at first, seemingly unwilling to share too much about their experiences, and it was here that the Ketso format proved useful in breaking down feelings of reluctance and maybe even anxiety about sharing personal insights and seeming critical of colleagues and the department. For example, in one workshop, one participant initially seemed reluctant to play an active role, waiting to see what other colleagues spoke about before committing his thoughts to the Ketso leaf. However, in line with the collaborative and democratic format of Ketso, it soon became his time to speak and, cautiously at first, he began to share his sense of imposter syndrome in an academic environment. As he spoke, he became more relaxed and opened up further, and other participants in the workshop shared similar anxiety and feeling like they are not enough in the competitive academic environment, with a female colleague sharing how her initial apprehension in relation to writing and publishing gradually reduced (although never disappeared) as she worked with others and built her confidence and skills. The discussion challenged the norm of masculine certainty that defines the academic role and opened up space for colleagues to share feelings of vulnerability in a way not normally possible in day-to-day academic life. It also showed the value of working collaboratively, leading to suggestions for increased mentoring opportunities within the department.

The pressures of contemporary academic life mean that academic staff rarely have time to connect outside of the day-to-day requirements of our roles. Burnout is an increasingly common problem for academic staff (Springer & Werner, 2020), associated with work intensification, casualisation and insecurity, and the demands of high numbers of diverse students (Taberner, 2018). An unexpected positive outcome of applying for Athena SWAN accreditation was that it provided space for us to talk with colleagues about issues and in a depth that there often does not seem time for. This helped draw attention to the positive aspects of team dynamics within the department and the supportive nature of the team, both in terms of individual colleagues and the broader culture of the department.

Highlighting gender power dynamics

An issue was raised around the dominance of a few individuals within team meetings and discussions. Several female colleagues raised concern about the ability of those 'who shout loudest' to get their view across at the expense of others, and the ways in which some male colleagues physically dominate space, through 'manspreading' and expressing dominance in terms of occupation of physical space (see Petter, 2017). Ketso helped address some of these issues, at least temporarily, but it did not eliminate all issues of dominance.

As each workshop member had an opportunity to speak in turn, based around the brief comment they wrote on the Ketso leaf, this helped ensure that everyone's voice was heard and no one could dominate discussions. This was largely successful but did highlight how, without the constraints of Ketso, it is easy for certain group members to overshadow others. This was illustrated in one workshop where a participant adopted a somewhat adversarial approach to discussions, disagreeing with other participants and keen to frame himself as marginalised through his maleness in relation to female colleagues and students. Speaking much more loudly than other participants and frequently interrupting others, it was easy to see how this kind of behaviour can dominate meetings and silence others, an issue discussed in other workshops. However, it was here that Ketso proved effective in minimising the disruption of this kind of behaviour as it was possible to restrict this participant's contributions by reminding him that if he was not placing a leaf down on the mat, then it was not his turn to speak. This was met with some resistance initially, but reverting to the requirements of the Ketso format helped minimise this by reminding participants that their views were all valued, but everyone had to wait their turn. That colleague went on to contribute some interesting insights and to participate more collaboratively in the workshop.

In a different workshop, another (male) participant disrupted the workshop process by attempting to take over the discussion and not follow the format and direction of the facilitators, who were women in more junior positions than this participant. This again shows how some people struggle to follow the structure of the activity and try to bring their personality and ideas to bear on other participants. This behaviour lead to tensions in 'constrained talk' as the communication was dictated by norms and rules of the Ketso methodology (Baxter & Montgomery, 1996) which should guide participants' 'intertextuality' (Bakhtin & Holquist, 1981). In such ways, the Ketso workshops helped both highlight and manage gender power dynamics in team settings.

Although there was broad support for the Athena SWAN initiative amongst many colleagues in the department, and there was benefit to coming together to talk and discuss issues that rarely get mentioned, there was also a sense that it was a little pointless. The gendered structures and practices of academia and broader society are deeply entrenched, and there was a feeling amongst some of our participants that initiatives like Athena SWAN would not bring about the kind of radical and far-reaching changes needed to overcome inequality. There was discussion around power issues and organisational cultures which value supposedly objective metrics like publication counts, impact factors and citation rates while failing to acknowledge the related gendered undertones, power issues and implications. This led to some colleagues feeling that while the Ketso workshops had been positive, even cathartic in many ways, it was unlikely that they would result in any concrete change. Issues like unrealistic performance expectations, competition and diversity of demands would require significant investment and action to tackle, and it was felt by many that this was unlikely to be forthcoming.

The (re)silencing of gender

The self-assessment team worked on the application for about 18 months, investing significant amounts of time and effort into the activity. Colleagues in the department participated in the staff survey and Ketso workshops and shared many personal stories and insights. Management within the department was supportive. Yet, it still failed. After review of the application-in-process the university's management took the decision not to continue. This was very disappointing for the self-assessment team and other colleagues in the department, yet it reflected the broader reach of the gendered organisation of the university and academia more widely. To be successful in the initiative required more than just the determination and efforts of the self-assessment team and depended on broader institutional support and impetus to support real change (Shen et al., 2009).

Although there was support for the process across hierarchical levels within the tourism department, this was not enough to ensure success in relation to the accreditation process or to address some of the complex issues it highlighted. The gendered structures of the organisation prioritise metrics and finances over less tangible factors, such as trying to achieve real equality. To be successful in this initiative would have required significant investment of resources – time, money and specialist expertise – that was not forthcoming. In an era of intense competition between universities, driven by league tables and performance metrics like NSS (National Student Survey), REF (Research Excellence Framework) and TEF (Teaching Excellence Framework) in the UK context, issues to do with gender inequality can slip down the hierarchy of priorities. In the next section, we discuss some of the implications of these kinds of experiences for gender (in)equality in the tourism academy.

Discussion

Our experiences of engaging in the process of preparing to apply for Athena SWAN accreditation brought about some positive benefits for the tourism department. Working collaboratively on the self-assessment team was an enjoyable and rewarding experience in many ways and led to an enhanced working environment and new research and mentorship relationships, one of the outcomes of which is this article. The Ketso workshops gave colleagues an opportunity to get together and share experiences and concerns in a low-risk environment, and many concrete suggestions for how we could improve our working lives within the department were put forward, such as formalised mentoring. However, the failure of the initiative was disappointing, and the subsequent retreat from prominence of gender issues within department meetings and in policy and action reinforced the sense that gender is seen as marginal to tourism academia, in day-to-day practice as well as in research (Munar et al., 2017).

The reasons that the initiative failed in this case were multiple and interconnected and included a lack of resources, issues with accessing required HR data and a sense amongst some colleagues that while Athena SWAN accreditation might be nice to have, and could even open up doors in terms of some funding opportunities, it was not a priority. Van den Brink and Benschop (2012) argue that gender equality practices are often ineffective against the multitude of gender inequality practices that characterise academia. While there was some support for the broad principles of Athena SWAN, and commitment from colleagues on the self-assessment team, ultimately, we were held back by the broader gender inequality practices of our environment that deprioritise gender as a key way to understand day-to-day working experiences. As Van den Brink and Benschop (2012, p. 89) argue

> This explains why it is often so difficult to undo inequality; simultaneous multi-faced gender inequality practices are ineffectively countered by gender equality practices because those lack teeth, especially in traditional masculine academic environments with 'thick', ponderous traditions and values.

Acker (2000) argues that gender equality initiatives often fail because they contradict many of the goals of the organisation. Those seeking change – the self-assessment team, in our case study – look to challenge current practices, such as workload models and hierarchical forms of communication. However, to bring about such change requires managerial support. As Acker (2000, p. 628) argues, "Gender equity of necessity redistributes power and rewards" and so may not actually be in the best interests of all, including those in positions of power who have the authority to implement change. Some men can be suspicious - even openly hostile – of gender equality initiatives which they think may disadvantage them (Van den Brink & Stobbe, 2014), and the examples above illustrate this in the responses to the Ketso workshops from some participants.

Applying for accreditation required sustained critical reflection on the practices, structures, culture and strategy of the department. Even if we had been successful in submitting our application for accreditation, and received a Bronze award, a huge amount of additional work would have been required to address some of the issues highlighted and work towards a truly equitable working environment. Other goals – such as improving publication counts or student attainment, for example – are easier to see and seem more immediate than gender equality (Acker, 2000). Consequently, despite good intentions, gender (In)equality slips out of focus and is replaced by more tangible priorities.

Athena SWAN was originally developed to address gender inequality in STEMM subjects in higher education, only expanding to cover other subjects in 2015. There is convincing evidence of gender inequality across different STEMM subjects, and there are numerous initiatives within and beyond HE to encourage more women into STEMM. Tourism is a very different domain. Although there is evidence of gender inequality in tourism academia, as discussed above, this inequality is in some ways harder to identify than in STEMM subjects. There are roughly equal numbers of men and women working in tourism academia, and publishing roughly equal numbers of papers (Kirilenko & Stepchenkova, 2018). This sounds very positive, in terms of gender equality. Yet numerical parity is not the same as equality of status, esteem and influence. Disparity in terms of publications in top-ranked journals, citations, conference keynote speakers and membership of leading subject organisations illustrates pervasive gender inequality in tourism academia beyond basic numerical representation (Munar et al., 2015; Nunkoo et al., 2019). However, this kind of inequality is less visible than numerical disparity, and is also much more difficult to address. An issue we faced in our efforts to apply for Athena SWAN accreditation was how to draw out some of the nuances of gender inequality in a department that contained roughly equal numbers of male and female staff. The Ketso workshops highlighted some of the subtle gender issues in this context, but the self-assessment team struggled to convince some colleagues that gender inequality was an issue in this department. In comparison to stark numerical inequality in other subjects, such as those in STEMM, in a field like tourism it can be hard to evidence many of the subtle and insidious aspects of bias and discrimination that shape academics' lives and contribute to ongoing inequality. Consequently, it can be difficult to convince those in power that gender inequality is an important issue in tourism and one that requires concerted effort to address.

Academia is a global professional environment and colleagues within this tourism department are judged in relation not just to their direct peers in this university but also those around the world. Many of the issues highlighted during this project, such as pressure to publish in a narrow set of male-dominated journals, or difficulties accessing funding and influential networks which are important for career progression, are global issues, beyond the reach of an individual academic, department or even institution. To try and address some of the pervasive gender inequality in tourism academia, collaborative initiatives will be needed on an international level. While Athena SWAN and other gender equality initiatives are important and can lead to positive change within individual departments and universities, they do not tackle the deep-rooted aspects of the gendered organisation of international higher education. Within tourism, as in all other subjects, academics are judged in relation to an ideal worker who is able and willing to put work before all else, is flexible and readily available to travel and work extremely long hours, and who manifests the masculine traits of leadership. Consequently, although seemingly gender-neutral, the ideal tourism academic is based on a masculine model that continues to marginalise women and minorities.

Our experiences of preparing to apply for Athena SWAN accreditation, as discussed in this paper, illustrate that tourism academics are subject to many of the same barriers and constraints as academics in more established fields. Change can happen on a small, local level, but the deeper issues that contribute to gender inequality remain unchallenged. Awards like Athena SWAN have a role to play in change but require commitment and input on a wider, possibly international, level. In the final section, we consider some of the implications of these issues for gender (in)equality and sustainability in tourism.

Conclusions

Figueroa-Domecq et al. (2015, p.89) describe the tourism academy as "exceptionally reluctant to engage in introspective gender-aware critique", and in this paper, we have responded to this by considering some of the positive and disappointing outcomes from our experiences of trying to apply for Athena SWAN accreditation within a tourism department in a UK university. Our discussion highlights some of the difficulties inherent in gender equality initiatives, especially in an academic context like tourism which is seen as female-dominated in terms of student numbers and relatively equal in terms of numerical gender representation of academic staff. However, as numerous other studies have illustrated, tourism is a gendered academic context and underpinned by pervasive inequalities that shape the lives of tourism academics. The paper makes three important contributions to understanding of tourism, gender and the academy.

First, guided by the concept of gendered organisations (Acker, 1990), our study highlights some of the limitations of gender equality initiatives for trying to overcome gender inequality in masculine organisational contexts like academia. Without significant senior management support and investment, we could not apply for Athena SWAN accreditation, despite the efforts of colleagues within the department. The qualitative research conducted to support the application also pointed to many subtle aspects of gender inequality that were both local to this tourism department (such as gender power relations between specific staff) and global to tourism academia (such as the power of male-dominated networks and publication practices). Real progress and change in terms of gender (in)equality will require both local and international efforts targeted at all aspects of academia, from research, to teaching, administration and factors associated with status and esteem. We thus encourage tourism academics at all levels – and particualry those in positions of influence, such as journal editors, heads of department and full professors – to proactively work to overcome entrenched inequality, through practices such as: mentoring; invitions to review, collaborate and give keynote addresses; openly discussing issues of inequality and discrimination, and possible ways to overcome this, on a regular basis; and, importantly, working to try and ensure tourism academia is a welcoming and supportive environment for all scholars.

Second, our study makes an important contribution to theoretical development. Our aborted attempt to apply for Athena SWAN accreditation exposes the paradoxical nature of such gender equality initiatives that both make gender visible, through explicit focus on gender as an important (in)equality issue, and at the same time reinforce many of the workings of the gendered organisation (see Dashper, 2019). Throughout the Athena SWAN process discussed above there was a necessary focus on some aspects of gender inquality, however, the failure to progress to application and accreditation, coupled with the way in which gender (in)equality disappeared again so quickly from focus within the department, illustrates the pervasiveness of the gendered organisation. Acker's (1990) identification of five key mechanisms through which gender is reproduced in organisations was formulated thirty years ago, but remains just as relevant today. Ketso workshops illustrated how colleagues experienced a gendered division of labour (for example, the self-assessment team for Athen SWAN consisted of 14 women but only 5 men, suggesting that gender equality is still seen as predominantly a women's issue); gendered cultural symbols (such as the dominance of the masculine model of the ideal academic); gendered workplace interactions (such as masculine dominance of space and discourse in meetings); gendered individual identities (in relation to issues ranging from childcare to student pastoral support) and, perhaps most significantly in this case, gendered organisational logic. Gender equality initiatives like Athena SWAN enable organisations to compartmentalise gender inequality and sidestep the need to acknowledge and then address the ways in which gender permeates all aspects of organisational life and practice. Our study thus contributes to the ongoing development of the theory of gendered organisations in contemporary neoliberal organisational contexts like tourism academia, and the paradoxical nature of initiatives that aim to tackle certain visible aspects of gender inequality whilst failing to address the more deep-rooted and systemic aspects of such inequality.

Third, in terms of methodology, our study illustrates the ways in which Ketso can be a useful tool for exploring issues of power and voice within groups and organisations. Ketso provided a safe space to share valuable insights into participants' lived experiences. As Bates (2016) summarises, Ketso is an ideal tool for collecting feedback, input, data, and information in an engaging, unique, and inclusive way. As a method, Ketso not only collects participants' ideas but also provides space and time to generate solutions and begin to address the highlighted issues (Wengel et al., 2019). The organisation and structure of the Ketso workshops are such that they seek to alleviate power dynamics and struggle that can often manifest in group discussions (Greer et al., 2017). Our study illustrates how the structure of Ketso can be effective for maintaining boundaries and limiting the ability of some group members to dominate discussion and silence others.Nevertheless, while Ketso claims to be an inclusive method allowing participants' voices to be heard, issues of power were still prevalent in the workshops. Attempts to dominate the workshop space could be seen as 'constrained talk' (Baxter, 2009). Hence, the 'intertextuality' of the discussion in the Ketso format helped to guide the meaning-making communication and deal with the dominating voices of some participants (Bakhtin & Holquist, 1981). Consequently, we suggest that Ketso offers a useful tool for research with groups where issues of power are apparent, and so can be an essential resource for tourism researchers.

Our discussion in this paper concentrates on the experiences of one tourism department in one UK university, yet the matters that arise from this case are revealing of wider issues. Gender inequality in tourism academia has numerous consequences. Academia is an important site for the production and reproduction of knowledge, and so can help shape broader discourses and practices within the wider tourism sector. Our students represent the future tourism workforce, as well as being current and future consumers of tourism products, and so the gendering of knowledge and tourism academia has an impact on their education and developing awareness of gender and inequality in the sector.

The gendered nature of tourism academia directs the kind of knowledge that is produced and valued; the voices that are heard; the debates that are deemed important and worthy of further consideration and investment. Pritchard (2018) suggests that tourism as a field has so far failed to engage with gender in a sustained and meaningful way. Gender analysis remains largely absent within tourism research and knowledge production and is rarely considered in relation to organisational practices and experiences (including those within academia), to the detriment of both the academic development of tourism studies and the practical advancement of the wider sector. This has implications for sustainability and the achievement of the SDG. Most obviously in terms of SDG 5, achieving gender equality requires focused attention to issues of gender and power across all sectors. Gender inequality will not be achieved within tourism without sustained effort and consideration; and lack of gender-aware research and debate means that, as in our experiences in academia, gender inequality easily slips out of notice in favour of other, often more tangible, priorities. However, gender equality is important for the achievement of all of the SDG (Alarcón & Cole, 2019), and underpins all efforts towards achieving more equitable societies, within and beyond tourism. Academia has a key role to play in shaping discourse, drawing attention to issues, providing evidence to both influence and support policy, and to help imagine and achieve a more gender-equal world. Our discussion in this paper illustrates that there remains much to be done to achieve gender equality within tourism academia and we join the call of other scholars, such as Chambers et al. (2017) among others, for more gender-aware research and practice to help transform the tourism academy.

Disclosure statement

No potential conflict of interest was reported by the authors.

ORCID

Katherine Dashper http://orcid.org/0000-0002-2415-2290
Jane Turner http://orcid.org/0000-0001-7753-747X
Yana Wengel http://orcid.org/0000-0002-8131-4137

References

Acker, J. (1990). Hierarchies, jobs, bodies: A theory of gendered organizations. *Gender & Society*, *4*(2), 139–158.

Acker, J. (1992). From sex roles to gendered institutions. *Contemporary Sociology*, *21*(5), 565–569. https://doi.org/10.2307/2075528

Acker, J. (2000). Gendered contradictions in organizational equity projects. Organization, *7*(4), 625–632. https://doi.org/10.1177/135050840074007

Advance HE. (2020a). *About Athena SWAN*. https://www.ecu.ac.uk/equality-charters/athena-swan/about-athena-swan/

Advance HE. (2020b). *Athena SWAN members*. https://www.ecu.ac.uk/equality-charters/athena-swan/athena-swan-members/

Alarcón, D. M., & Cole, S. (2019). No sustainability for tourism without gender equality. *Journal of Sustainable Tourism*, *27*(7), 903–919. https://doi.org/10.1080/09669582.2019.1588283

Bagilhole, B., & Goode, J. (2001). The contradiction of the myth of individual merit, and the reality of a patriarchal support system in academic careers: A feminist investigation. *European Journal of Women's Studies*, *8*(2), 161–180. https://doi.org/10.1177/135050680100800203

Bakhtin, M. M., & Holquist, M. (1981). *The dialogic imagination: Four essays*. University of Texas Press.

Bates, J. S. (2016). What's Ketso? A tool for researchers, educators, and practitioners. *Journal of Human Sciences and Extension*, *4*(2), 167.

Baxter, L. A. (2009). Relational dialectics. In S. W. Littlejohn & K. A. Foss (Eds.), *Encyclopaedia of communication theory* (pp. 837–840). SAGE.

Baxter, L. A., & Montgomery, B. M. (1996). *Relating: Dialogues and dialectics*. Guilford Press.

Braun, V., & Clarke, V. (2006). Using thematic analysis in psychology. *Qualitative Research in Psychology*, *3*(2), 77–101. [Database] https://doi.org/10.1191/1478088706qp063oa

Britton, D. M., & Logan, L. (2008). Gendered organizations: Progress and prospects. *Sociology Compass*, *2*(1), 107–121. https://doi.org/10.1111/j.1751-9020.2007.00071.x

Chambers, D., & Rakić, T. (2018). Critical considerations on gender and tourism: An introduction. Tourism Culture & Communication, *18*(1), 1–8. https://doi.org/10.3727/109830418X15180180585112

Chambers, D., Munar, A. M., Khoo-Lattimore, C., & Biran, A. (2017). Interrogating gender and the tourism academy through epistemological lens. Anatolia, *28*(4), 501–513. https://doi.org/10.1080/13032917.2017.1370775

Coar, L., & Sim, J. (2006). Interviewing one's peers: Methodological issues in a study of health professionals. *Scandinavian Journal of Primary Health Care*, *24*(4), 251–256. https://doi.org/10.1080/02813430601008479

Dashper, K., & Fletcher, T. (2019). Challenging the gendered rhetoric of success? The limitations of women-only mentoring for tackling gender inequality in the workplace. Gender, Work & Organization, *26*(4), 541–557.

Donald, A., Harvey, P. H., & McLean, A. R. (2011). Athena SWAN awards: Bridging the gender gap in UK science. Nature, *478*(7367), 36–36. https://doi.org/10.1038/478036b

Ek, R., & Larson, M. (2017). Imagining the Alpha male of the tourism tribe. Anatolia, *28*(4), 540–552. https://doi.org/10.1080/13032917.2017.1370778

European Commission. (2019). *Certification-award systems to promote gender equality in research*. https://cordis.europa.eu/project/id/872113

Fernow, J. (2019). *Creating a supportive and inclusive university culture*. https://starbios2.eu/2019/creating-a-supportive-and-inclusive-university-culture/

Figueroa-Domecq, C., Pritchard, A., Segovia-Pérez, M., Morgan, N., & Villacé-Molinero, T. (2015). Tourism gender research: A critical accounting. *Annals of Tourism Research*, *52*, 87–103. https://doi.org/10.1016/j.annals.2015.02.001

Graves, A., Rowell, A., & Hunsicker, E. (2019). *An impact evaluation of the Athena SWAN charter* [Online] Ortus Economic Research & Loughborough University. https://www.ecu.ac.uk/wp-content/uploads/2019/08/Athena-SWAN-Impact-Evaluation-2019.pdf

Greer, L. L., Van Bunderen, L., & Yu, S. (2017). The dysfunctions of power in teams: A review and emergent conflict perspective. *Research in Organizational Behavior*, *37*, 103–124. https://doi.org/10.1016/j.riob.2017.10.005

Hart, J. (2016). Dissecting a gendered organization: Implications for career trajectories for mid-career faculty women in STEM. *The Journal of Higher Education*, *87*(5), 605–634. https://doi.org/10.1353/jhe.2016.0024
Jeffrey, H. L. (2017). Gendering the tourism curriculum whilst becoming an academic. Anatolia, *28*(4), 530–539. https://doi.org/10.1080/13032917.2017.1370779
Kirilenko, A. P., & Stepchenkova, S. (2018). Tourism research from its inception to present day: Subject area, geography, and gender distributions. *PloS One*, *13*(11), e0206820. https://doi.org/10.1371/journal.pone.0206820
Knights, D., & Clarke, C. A. (2014). It's a bittersweet symphony, this life: Fragile academic selves and insecure identities at work. *Organization Studies*, *35*(3), 335–357. https://doi.org/10.1177/0170840613508396
Madge, C., Raghuram, P., & Skelton, T. (2014). Methods and methodologies in feminist geographies: Politics, practice and power. In *Women in geography study group, feminist geographies: Explorations in diversity and difference* (pp. 98–123). Abingdon, Oxon: Routledge.
Morgan, N., & Pritchard, A. (2019). Gender matters in hospitality. *International Journal of Hospitality Management*, *76*, 38–44. https://doi.org/10.1016/j.ijhm.2018.06.008
Munar, A. M., Budeanu, A., Caton, K., & Chambers, D. (2015). *The gender gap in the tourism academy: Statistics and indicators of gender equality. While Waiting for the Dawn*. University of Sunderland. Available at https://sure.sunderland.ac.uk/id/eprint/10369/1/FINAL%20GenderGapReport_WWFD.pdf [Accessed 12/10/20]
Munar, A. M., Khoo-Lattimore, C., Chambers, D., & Biran, A. (2017). The academia we have and the one we want: On the centrality of gender equality. Anatolia, *28*(4), 582–591. https://doi.org/10.1080/13032917.2017.1370786
Munir, F., Mason, C., McDermott, H., Morris, J., Bagilhole, B., & Nevil, M. (2014). *Advancing women's careers in science, technology, engineering, mathematics and medicine: Evaluating the effectiveness and impact of the Athena SWAN Charter*. Equality Challenge Unit.
Nowell, L. S., Norris, J. M., White, D. E., & Moules, N. J. (2017). Thematic analysis: Striving to meet the trustworthiness criteria. *International Journal of Qualitative Methods*, *16*(1), 160940691773384. https://doi.org/10.1177/1609406917733847
Nunkoo, R., Hall, C. M., Rughoobur-Seetah, S., & Teeroovengadum, V. (2019). Citation practices in tourism research: Toward a gender conscientious engagement. *Annals of Tourism Research*, *79*, 102755. https://doi.org/10.1016/j.annals.2019.102755
Nunkoo, R., Thelwall, M., Ladsawut, J., & Goolaup, S. (2020). Three decades of tourism scholarship: Gender, collaboration and research methods. *Tourism Management*, *78*, 104056. https://doi.org/10.1016/j.tourman.2019.104056
Petter, O. (2017, July 27). Revealed: The scientific explanation behind 'manspreading'. *The Independent*. https://www.independent.co.uk/life-style/manspreading-scientific-explanation-revealed-men-behaviour-public-transport-etiquette-a7862771.html
Pritchard, A. (2018). Predicting the next decade of tourism gender research. *Tourism Management Perspectives*, *25*, 144–146. https://doi.org/10.1016/j.tmp.2017.11.014
Pritchard, A., Morgan, N., & Ateljevic, I. (2011). Hopeful tourism: A new transformative perspective. *Annals of Tourism Research*, *38*(3), 941–963. https://doi.org/10.1016/j.annals.2011.01.004
Schmidt, E. K., Ovseiko, P. V., Henderson, L. R., & Kiparoglou, V. (2019). Understanding the Athena SWAN award scheme for gender equality as a complex social intervention in a complex system: Analysis of Silver award action plans in a comparative European perspective. Health Res Policy Sys, 18(19),ttps://doi.org/10.1186/s12961-020-0527-x
Shen, J., Chanda, A., D'netto, B., & Monga, M. (2009). Managing diversity through human resource management: An international perspective and conceptual framework. *The International Journal of Human Resource Management*, *20*(2), 235–251. https://doi.org/10.1080/09585190802670516
Shenton, A. K. (2004). Strategies for ensuring trustworthiness in qualitative research projects. *Education for Information*, *22*(2), 63–75. https://doi.org/10.3233/EFI-2004-22201
Springer, A., & Werner, I. (2020). Burnout among academics: An empirical study on the Universities of Poland. In M.H. Bilgin, H. Danis, E. Demir & U. Can (eds) *Eurasian business perspectives* (pp. 75–89). Springer.
Taberner, A. M. (2018). The marketisation of the English higher education sector and its impact on academic staff and the nature of their work. *International Journal of Organizational Analysis*, *26*(1), 129–152. https://doi.org/10.1108/IJOA-07-2017-1198
Tippett, J., Connelly, A., & How, F. (2011). You want me to do what? Teach a studio class to seventy students? *Journal for Education in the Built Environment*, *6*(2), 26–53. https://doi.org/10.11120/jebe.2011.06020026
Tippett, J., Handley, J. F., & Ravetz, J. (2007). Meeting the challenges of sustainable development—A conceptual appraisal of a new methodology for participatory ecological planning. *Progress in Planning*, *67*(1), 9–98. https://doi.org/10.1016/j.progress.2006.12.004
Tribe, J. (2010). Tribes, territories and networks in the tourism academy. *Annals of Tourism Research*, *37*(1), 7–33. https://doi.org/10.1016/j.annals.2009.05.001
Tung, V. W. S., & McKercher, B. (2017). Negotiating the rapidly changing research, publishing, and career landscape. *Tourism Management*, *60*, 322–331. https://doi.org/10.1016/j.tourman.2016.12.013
UN. (n.d.). Sustainable development goals. *United Nations*. https://sustainabledevelopment.un.org/?menu=1300

Van den Brink, M., & Benschop, Y. (2012). Slaying the seven-headed dragon: The quest for gender change in academia. *Gender, Work & Organization, 19*(1), 71–92.

Van den Brink, M., & Stobbe, L. (2009). Doing gender in academic education: The paradox of visibility. *Gender, Work & Organization, 16*(4), 451–470.

Van den Brink, M., & Stobbe, L. (2014). The support paradox: Overcoming dilemmas in gender equality programs. *Scandinavian Journal of Management, 30*(2), 163–174. https://doi.org/10.1016/j.scaman.2013.07.001

Van den Brink, M., Benschop, Y., & Jansen, W. (2010). Transparency in academic recruitment: A problematic tool for gender equality? *Organization Studies, 31*(11), 1459–1483. https://doi.org/10.1177/0170840610380812

Wengel, Y., McIntosh, A., & Cockburn-Wootten, C. (2019). Co-creating knowledge in tourism research using the Ketso method. *Tourism Recreation Research, 44*(3), 311–322. https://doi.org/https://doi.org/10.1080/02508281.2019.1575620 https://doi.org/10.1080/02508281.2019.1575620

Wiles, R., Charles, V., Crow, G., & Heath, S. (2006). Researching researchers: Lessons for research ethics. *Qualitative Research* , *6*(3), 283–299. https://doi.org/10.1177/1468794106065004

Xiao, Y., Pinkney, E., Kit Fong Au, T., & Siu Fai Yip, P. (2020). Athena SWAN and gender diversity: A UK-based retrospective cohort study. *BMJ Open, 10*(2), e032915. https://doi.org/10.1136/bmjopen-2019-032915

Zhang, K., Lynch, P., McIntosh, A., & Wengel, Y. (2016). Current and future potentialities of critical hospitality studies: Conference workshop report. *Hospitality & Society, 6*(1), 77–82.

Zippel, K. (2019). Gendered images of international research collaboration. *Gender, Work & Organization, 26*(12), 1794–1805.

Sexual harassment, psychological well-being, and job satisfaction of female tour guides: the effects of social and organizational support

Zaid Alrawadieh, Derya Demirdelen Alrawadieh, Hossein G. T. Olya, Gul Erkol Bayram and Onur Cuneyt Kahraman

ABSTRACT

Drawing on the Conservation of Resources Theory (COR), and a gender perspective, this study proposes and tests a conceptual model postulating relationships between sexual harassment, burnout, perceived social and organizational support, psychological well-being, and job satisfaction. A survey of Turkish female tour guides resulted in 221 valid questionnaires. The results reveal that female tour guides' sexual harassment experience has a negative impact on their job satisfaction and psychological well-being. Unlike perceived social support, perceived organizational support plays a significant and negative role in triggering sexual harassment. The findings also confirm the mediating effects of burnout on the relationship between sexual harassment and job satisfaction as well as the relationship between sexual harassment and psychological well-being. The study contributes to gender equality and sustainability research and offers several practical implications for stakeholders in the travel industry and public policy.

Introduction

As a vexing issue influencing both individuals and organizations with various social, organizational and legal implications, sexual harassment has received increasing attention in academia. While few would disagree that sexual harassment occurs in almost all workplaces (Kiely & Henbest, 2000), its existence and frequency are far more prevalent in the services industries in general, and tourism and hospitality in particular (Morgan & Pritchard, 2019; Poulston, 2008). Jobs in the tourism and hospitality industry have several distinct features which can potentially stimulate sexual harassment. These features include long working hours, fluctuating shifts, a highly gendered environment, and an explicit power imbalance (Gilbert et al., 1998; Ineson et al., 2013). A plethora of research has addressed sexual harassment in different settings within the industry, including hotels (Jung & Yoon, 2020; Nimri et al., 2020) restaurants (Slonaker et al., 2007; Szymanski & Mikorski, 2016) airlines (Gunnarsdottir et al., 2006), casinos (Stedham & Mitchell, 1998), and private clubs (Cannon et al., 1998).

Collectively, these studies agree on the negative impacts of sexual harassment on both individuals and firms. Yet, despite some research endeavors addressing this topic in the field of tourism and hospitality, there is a paucity of research focusing on the relationships between sexual harassment and psychological and organizational variables. More importantly, while previous research addresses the role of perceived social support (e.g. Cho, 2019) and organizational support (e.g. Hur et al., 2013) in reducing stressors in the workplace, there exists very limited empirical evidence on how perceived social support and organizational support can mitigate sexual harassment. Moreover, research delving into the experiences of tour guides is considerably scant. Notably, this specific job segment is largely under-presented both in tourism research and organizational behavior literature (Farkic, 2018).

The current investigation makes several original and incremental contributions. First, the study explores how sexual harassment experienced by female tour guides influence their psychological well-being and job satisfaction. By doing so, the study responds to calls for more research addressing sexual harassment in the field of tourism (Morgan & Pritchard, 2019; Pritchard, 2018) and calls for research focusing on the well-being of tourism employees (Sirgy, 2019). Second, the study draws on social support and organizational support theories and links them to the sexual harassment experience. To the authors' best knowledge, no previous study has addressed the role of social and organizational support in mitigating sexual harassment. Third, the study responds to recent calls to address sexual harassment as a barrier to gender equality and, therefore, sustainable tourism (Alarcón & Cole, 2019). Thus, the current study adds to an emerging stream of research integrating gender into the theory and practice of sustainable tourism (Ferguson & Alarcon, 2015). Fourth, the current study makes an incremental contribution by addressing the sexual harassment of tour guides. This is important given that jobs within tourism and hospitality may not be homogeneous. For instance, tour guides' interaction with customers is more intense and spans over a longer time compared to other jobs (Alrawadieh et al., 2020). Relatedly, the study draws on data from Turkey and thus echoes the call for more studies into gender equality within tourism in emerging economies (Je et al., 2020). Therefore, by focusing on the specific case of female tour guides, the present study makes a significant contextual contribution to the advancement of gender research in tourism and hospitality (Mooney, 2020). In sum, the study aims to provide different stakeholders in the travel industries, such as travel agencies and tour operators with empirical evidence-based insights into how to enhance tour guides' job satisfaction and psychological well-being by mitigating sexual harassment.

Conceptual background and hypotheses development

Sexual harassment, gender equality and sustainability

Despite the various conceptualizations around the concept, from a psychological point of view, sexual harassment is broadly defined as "unwanted sex-related behavior at work that is appraised by the recipient as offensive, exceeding her/his resources, or threatening her/his well-being" (Fitzgerald et al., 1997, p. 15). From a gender perspective, sexual harassment is viewed as a form of discrimination mainly against women (Larsen et al., 2019; Murrell et al., 1995). Therefore, sexual harassment is considered a barrier to gender equality. Despite efforts to combat sexual harassment, it is safe to say that it is still common in almost all countries and across all institutional environments and workplaces (Bell et al., 2002: Morgan & Pritchard, 2019). For instance, The U.S. Equal Employment Opportunity Commission reported that, in 2019 only, around 7,500 charges of sexual harassment were received and resolved, at a cost of over $68 million in monetary benefits (Equal Employment Opportunity Commission, 2019). A recent national survey in Australia indicated that one in three people had experienced sexual harassment at work in the last five years (Australian Human Rights Commission, 2018). Thus, combatting sexual harassment should be a key priority for societies aiming to achieve gender equality.

Gender inequality remains a global problem affecting women and girls in all countries (Koehler, 2016). Recognizing this challenge, the United Nation's 2030 Sustainable Development Agenda has included gender equality as one of the 17 Sustainable Development Goals (SDGs). These goals represent global challenges and barriers to sustainable development. Goal 5 expresses commitment to eliminate discrimination against women and girls, as well as all forms of violence against them in both public and private domains, including trafficking, sexual exploitation and other types of exploitation (United Nations, 2015). Thus, attention has been shifted toward gender equality as a means to achieve sustainable development. Recently, there has been a growing amount of research addressing the intersection between gender and sustainability (e.g. Koehler, 2016; Rosche, 2016). Drawing on a gender perspective, Alarcón and Cole (2019) explored the interconnections between the Sustainable Development Goals (SDGs) and tourism and concluded that achieving sustainability for tourism could not be possible without gender equality. They also identified sexual harassment that women, in particular, experience as one of the gaps that require more investigation in order to achieve gender equality and, therefore, sustainable tourism. Similarly, Ferguson and Alarcon (2015) raised the need for re-evaluating the concept of sustainable tourism by incorporating gender equality as a fundamental component.

These concerns warrant a need for a deeper understanding of the key barriers in the face of gender equality and sustainability in tourism. In this vein, the current investigation focuses on sexual harassment as one of the prominent problems for the tourism and hospitality workplace (Morgan & Pritchard, 2019; Ram, 2018). Research indicates that the nature of jobs within tourism and hospitality (e.g. sexualized role of employees, antisocial working hours) may significantly encourage the occurrence of sexual harassment incidents (Guerrier & Adib, 2000; Kensbock et al., 2015; Morgan & Pritchard, 2019; Poulston, 2008). However, despite its prevalence, sexual harassment goes largely unreported, making it difficult to accurately determine its extent and influence (Morgan & Pritchard, 2019). Regardless of these challenges, there have been recent calls for more research addressing sexual harassment. For instance, Pritchard (2018), and Morgan and Pritchard (2019) identified sexual harassment as one of the research areas that should be the focus of tourism gender scholars. The current investigation, therefore, addresses one of the most common forms of discrimination against women, thus contributing to the timely discussion around gender equality and sustainability (Alarcón & Cole, 2019).

The impact of social and organizational support on sexual harassment

Tourism and hospitality employees are acknowledged for performing highly emotional labor (Lee et al., 2016). To cope with this burden, they often rely on their social support networks as a means of relief and recovery (McGinley & Wei, 2018). Likewise, sexually harassed employees may use their social support networks, such as family members and friends, to reduce anxiety and stress (Cortina, 2004). Social support is broadly defined as "an exchange of resources between at least two individuals perceived by the provider or the recipient to be intended to enhance the well-being of the recipient" (Shumaker & Brownell, 1984, p. 13). According to Hobfoll's (1989) Conservation of Resources Theory (COR), individuals lose cognitive resources when dealing with a stressor (e.g. sexual harassment), and develop strategies to cope with it. In this vein, sexual harassment may constitute a resource loss whereas social support may constitute a form of resource gain. We assume that tour guides who perceive higher social support from their social networks are less likely to experience sexual harassment. Therefore, we propose the following hypothesis:

H1: Perceived social support decreases sexual harassment.

Like social support, the role of perceived organizational support in enhancing employees' well-being is acknowledged (Rhoades & Eisenberger, 2002; Stamper & Johlke, 2003). Organizational support is conceptualized as the extent to which the organization values the contributions of its employees and cares about their well-being (Eisenberger et al., 1986). From an organizational behavior perspective, perceived organizational support can potentially reduce stressors in the workplace (Kang et al., 2010; Stamper & Johlke, 2003) and enhance the well-being of the employees (Eisenberger et al., 1986; Rhoades & Eisenberger, 2002). Inherent characteristics of the service role in tourism, where pleasing the customer is the ultimate goal, provides the basis for sexual harassment (Yagil, 2008). Therefore, sexual harassment is often viewed as a workplace stressor (Gettman & Gelfand, 2007; Gutek & Koss, 1993; Willness et al., 2007) whereby the role of the organization in reducing its occurrence is crucial (Bell et al., 2002). Rabelo and Cortina (2014) found that employees who perceived low organizational support are more likely to experience sexual harassment. Overall, perceived organizational support may act as a preventive policy that attenuates sexual harassment (Stamper & Johlke, 2003). Thus, we assume that female tour guides who perceive their organizations as supportive are less likely to experience sexual harassment. Based on the preceding discussion, the following hypothesis is proposed:

H2: Perceived organizational support decreases sexual harassment.

Sexual harassment, burnout, job satisfaction, and psychological well-being

There is significant empirical evidence confirming the devastating impacts of sexual harassment on both organizational and personal levels. These impacts are largely interrelated in the sense that sexual harassment can potentially harm employees' well-being with spillover effects on their work-life and career development (Buchanan & Fitzgerald, 2008; Fitzgerald et al., 1988, 1997; Gettman & Gelfand, 2007). One of the prominent negative outcomes of sexual harassment is burnout (Hu et al., 2019; Yagil, 2008). As "a prolonged response to chronic emotional and interpersonal stressors on the job" (Maslach et al., 2001, p. 397), burnout is recognized as a significant organizational problem that gives rise to several unfavorable consequences such as job dissatisfaction, low commitment, and high turnover rates (Cheng & Yi, 2018; Lu & Gursoy, 2016). Considerable empirical evidence from different contexts confirms the relationship between sexual harassment and burnout (Hu et al., 2019; Savicki et al., 2003; Szymanski & Mikorski, 2016). In tourism and hospitality literature, burnout was also found as a potential consequence of the sexual harassment experience. For instance, Szymanski and Mikorski (2016) found that unwanted sexual advances can cause burnout in sexually objectifying restaurants. Similarly, Jung and Yoon (2020) found that sexual harassment as perceived by female hotel employees increased their burnout. Hence, we propose the following hypothesis:

H3: Sexual harassment increases tour guides' burnout.

Given the inherent characteristics of the service role, employees may often be vulnerable to misbehavior (Yagil, 2008), which leads to a hostile environment, thus influencing their attitudes towards their jobs (Ineson et al., 2013; Ram, 2018). In this vein, sexual harassment is widely accepted as one of the factors that negatively influence job satisfaction (Fitzgerald et al., 1997; Gettman & Gelfand, 2007; Willness et al., 2007; Yagil, 2008). Job satisfaction can be broadly defined as the overall favorable attitude of employees toward their job (Kong et al., 2018). There is also a large body of scholarly research both in the organizational behavior (e.g. Baruch-Feldman et al., 2002), as well as tourism and hospitality (e.g. Jung & Yoon, 2019; Kara et al., 2013; Koc & Bozkurt, 2017; Lu & Gursoy, 2016) confirming the negative relationship between burnout and job satisfaction. Therefore, we assume that female tour guides who experience sexual harassment and burnout are likely to have lower job satisfaction. Hence, we propose the following hypotheses:

H4: Sexual harassment reduces tour guides' job satisfaction.

H5: Burnout decreases tour guides' job satisfaction.

On a personal level, the psychological costs of sexual harassment experience on individuals are well documented (Gettman & Gelfand, 2007; Ineson et al., 2013). Specifically, the link between sexual harassment and psychological well-being has been confirmed in previous research (Fitzgerald et al., 1997; Munson et al., 2000; Rederstorff et al., 2007). For instance, in their study on hospitality employees in Greek Cyprus, Theocharous and Philaretou (2009) suggested that sexually harassed women are likely to experience decreased satisfaction in their lives, as well as an overall feeling of threat in their daily lives. Likewise, Cho (2002) noted that hotel female employees reported psychological problems such as tension, nervousness, and persistent anger and fear as the most prevalent consequences of sexual harassment experience. In the specific case of tour guides, there is very limited empirical evidence on the relationship between sexual harassment and psychological well-being. Broadly speaking, if sexual harassment is to be viewed as a job stressor, then it is inherently associated with a lower quality of life and psychological well-being (Min, 2014). Moreover, as discussed earlier, sexual harassment may be associated with higher levels of burnout. Along its negative effect on employees' organizational behavior (e.g. Humborstad et al., 2007; Lu & Gursoy, 2016), a plethora of empirical research confirms the negative impact of burnout on the psychological well-being among tourism and hospitality employees (e.g. Kara et al., 2013; Koc & Bozkurt, 2017; Pienaar & Willemse, 2008). As an organizational stressor, burnout does not only influence employees' current levels of psychological well-being but it also leads to a pessimistic attitude towards their well-being in the future (Koc & Bozkurt, 2017). In the specific case of tour guides, by the very inherent nature of their job, tour guides perform high emotional labor, and are often exposed to high levels of stress (Wong & Wang, 2009). In the current investigation, we assume that sexual harassment and burnout experienced by female tour guides are likely to harm their psychological well-being. Based on the aforementioned discussion, the following hypotheses are developed:

H6: Sexual harassment has a negative impact on psychological well-being.

H7: Burnout has a negative impact on psychological well-being.

Job satisfaction and psychological well-being are typically employed either as predictors (e.g. Wright & Cropanzano, 2000) or outcomes (e.g. Arenas et al., 2015) of other variables. The very association between these two constructs is, however, largely blurred (Kohan & O'connor, 2002). The unestablished causal nature of the relationship between job satisfaction and psychological well-being in terms of which one predicts the other contributes to this ambiguity (Bowling et al., 2010). According to Bowling et al. (2010) meta-analytic examination of this relationship and in line with the spillover hypothesis (Bowling et al., 2010; Judge & Watanabe, 1994), we assume a causal path from job satisfaction to psychological well-being and thus we hypothesize that professional well-being (i.e. job satisfaction) can act as an antecedent of well-being (i.e. psychological well-being). In other words, if tour guides are satisfied with their job, the likelihood of them being happy in their life is greater. With this realization, the following hypothesis is proposed:

H8: Job satisfaction improves psychological well-being.

Mediating effects of burnout

As argued earlier, numerous studies establish the direct effects of sexual harassment on job satisfaction (e.g. Fitzgerald et al., 1997; Willness et al., 2007) and psychological well-being (e.g. Sojo et al., 2016). Other studies examine the indirect effects of sexual harassment on proactive customer service performance (Li et al., 2016) and psychological distress (Jiang et al., 2015). However, none of these models assess the effects of burnout on the relationship between sexual harassment and job satisfaction and psychological well-being. While the mediating effects of

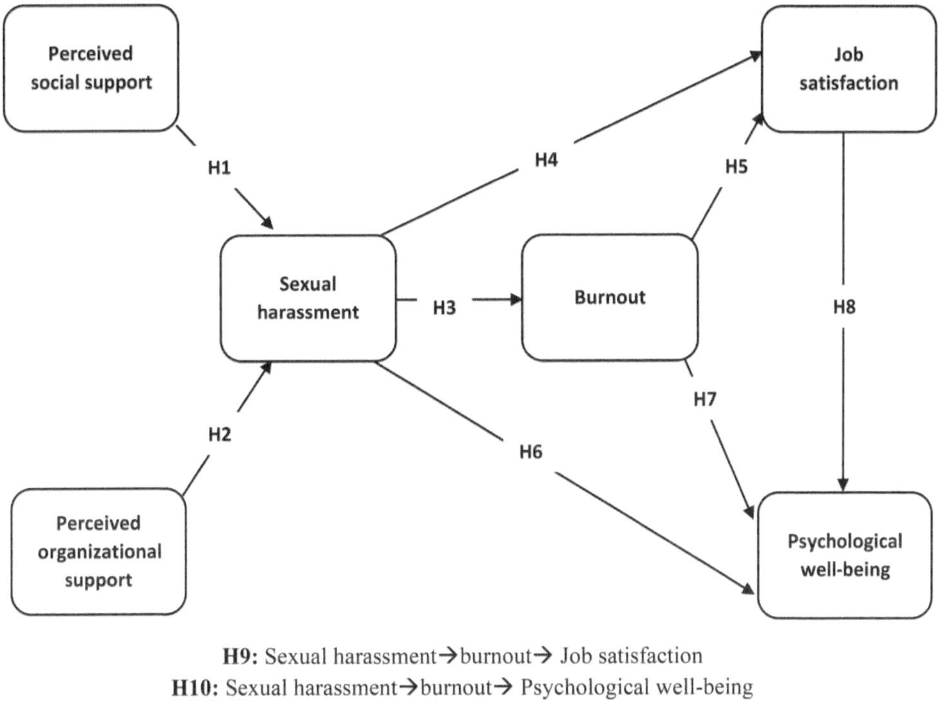

Figure 1. Conceptual model.

burnout were examined in some studies addressing workplace misbehavior such as incivility (Loh & Loi, 2018) and bullying (Mathisen et al., 2008), to the authors' best knowledge, no study has examined these effects in models involving sexual harassment. As depicted in Figure 1, burnout is purported to mediate the relationship between sexual harassment on the one hand, and job satisfaction and psychological well-being on the other hand. Based on the above discussions, the following hypotheses are formulated:

H9: Burnout mediates the relationship between sexual harassment and job satisfaction

H10: Burnout mediates the relationship between sexual harassment and psychological well-being.

Methods

Study context

The tour guiding profession in Turkey is regarded as both rewarding and challenging. Broadly speaking, the rate of remuneration is relatively high compared to the mid-range in the tourism sector, especially for tour guides mastering high-demanded languages such as Chinese. On the negative side, however, tour guides are often challenged by seasonality and market instability as well as by the demanding nature of the job (Alrawadieh et al., 2020), with female tour guides perceiving higher stress than males (Ayaz & Demir, 2019). To officially and professionally be a tour guide, candidates should either have an undergraduate degree in tour guiding or a certificate issued by the Union of Turkish Tourist Guides following the completion of an educational program organized by the Union. In terms of their licenses, tour guides can either be national or regional. The former are authorized to make tours all over the country whereas the latter are restricted to specific region(s) within Turkey. By November 2019, the total number of tour guides (both active and non-active) in Turkey was 11,024. Female tour guides constituted one-third

(3,647 guides). The number of active (practicing) tour guides were 8,358, of which less than 30 percent, or 2,461 were female. English is the primary language of tour guides followed by German and French (Union of Turkish Tourist Guides, 2019). Like in most countries, tour guides in Turkey are either salary-based or freelancers. Although this distinction was not considered in our study (which is a limitation that we acknowledge), freelance tour guides should not be viewed as self-employed in the sense that they may be working with a specific number of tour operators or travel agencies (Cetin and Yarcan, 2017). In this sense, perceived organizational support can be stemming either from a single organization (in the case of salary-based tour guides) or multiple organizations (in the case of freelancers).

With the absence of reliable data on sexual harassment incidents in tourism workplaces in general and tour guiding in particular, drawing an accurate portrait of the current circumstances in Turkey remains out of reach. Recent empirical works (e.g. Alrawadieh et al., 2020; Yıldırım & Özbek, 2019; Zengin et al., 2014) delving into female tour guides' experiences provide insufficient and even contradictive evidence making it difficult to draw definitive conclusions. From legislative and organizational perspectives, Turkish legislation does not seem to be clear on what should organizations do to prevent sexual harassment. Employers are encouraged but not legally required to take preventative action to deal with sexual harassment incidents (Süral & Kiliçoğlu, 2011). However, the Turkish labor law entitles the employee who experiences sexual harassment and who, despite reporting the incident to the employer, does not receive a fair solution, to resign with immediate effect and seek certain statutory payments (Turkish Labor Law, 2003). To the authors' best knowledge and following a desk-research into tour guides chambers and unions' published materials and web pages, it appears that there are no organizational policies or codes of conduct to prevent sexual harassment.

Measures

The constructs used in the current investigation were measured using multiple-item scales adopted from previous research. Sexual harassment was measured using a 21-item scale developed by Murry et al. (2001). The items were measured on a 5-point Likert-type scale, with the end poles labeled as "never" and "very often." Thus, higher scores indicated a greater level of sexual harassment. Sample items included "repeatedly told sexual jokes." Burnout was operationalized using 6 items adopted from Babakus et al. (2009). The construct was measured on a 5-point Likert-type scale, anchored from strongly disagree (1) to strongly agree (5). Psychological well-being was measured using a 5-item scale developed by Diener et al. (1985). The items were given on a 5-point Likert-type scale, with the end poles labeled as "strongly disagree (1)" and "strongly agree (5)." Job satisfaction was measured using a 5-item scale developed by Johlke and Duhan (2000). The original scale contains a 7-point Likert-type scale; but the items were adapted on a 5-point Likert-type scale, with the end poles labeled as "strongly disagree (1)" and "strongly agree (5)." Perceived social support was measured using a 12-item scale developed by Zimet et al. (1988). The items were given on a 5-point Likert-type scale, with the end poles labeled as "strongly disagree (1)" and "strongly agree (5)." Perceived organizational support was measured using a 6-item scale adapted from Shanock and Eisenberger (2006). The items were given on a 5-point Likert-type scale, with the end poles labeled as "strongly disagree (1)" and "strongly agree (5)." The survey also collected socio-demographic data such as gender, age, and education.

Sampling and data collection

The population of the current study is composed of all active female tour guides in Turkey (a total of 2461 as per November 2019). Prior to the data collection and to ensure the content validity of the measurement instrument, four female tour guides and three tourism and hospitality scholars were asked to examine the questionnaire. Based on their feedback, no major modifications were made. An online-based self-administered survey was used to collect the data. This

Table 1. Respondents' profile.

Age	N	%	Marital status	N	%
18–24 years old	27	12.2	Single	89	40.2
25–34 years old	71	32.1	Married	129	58.4
35–44 years old	60	27.1	Other	3	1.4
45–54 years old	47	21.3	Total	221	100.0
55–64 years old	14	6.3			
65 and above	2	1.0	Work experience as an official tour guide		
Total	221	100.0	Less than a year	24	10.9
			1–3 years	26	11.7
Education			4–6 years	36	16.2
High school graduate or less	3	1.3	7–9 years	41	18.6
College graduate- undergraduate	17	7.7	10–12 years	30	13.6
Postgraduate degree	149	67.4	13 years and over	64	29.0
Doctoral degree	43	19.5	Total	221	100.0
Other	9	4.1			
Total	221	100.0			

technique was deliberately used given the sensitivity of the topic and to reduce social desirability-bias (Tourangeau & Yan, 2007). Respondents were ensured complete confidentiality of the data they provided and anonymity of their identities. They were also given the option to provide an email address in case they wanted to receive a summary of the findings. Ethical approval for this study was obtained from the Istanbul University-Cerrahpaşa Review Board (2019/45). The data were collected over a period of two months from 18 October to 25 December 2019. By the cut-off date for data collection, 221 questionnaires were collected. To assess the sampling adequacy, the rule-of-thumb recommended by Cohen (1992) was adopted, whereby a sample size of 221 far exceeds the minimum required for a statistical power of 80% at the 5% level of significance.

Demographics of respondents are presented in Table 1. Respondents varied in terms of their age groups with around one-third of the sample aged between 25 and 34 years old. In terms of their educational level, the sample is highly educated with the majority of respondents (67.4%) holding either a postgraduate degree (67.4) or doctoral degree (19.5%). The respondents were mostly married (58.4%) or single (40.2%), with nearly one-third of them having work experience as an official tour guide for at least 13 years.

Data analysis

We tested the psychometric properties of the measurements including reliability and validity of the constructs and the measurement model. Cronbach's alpha (α) and composite reliability (CR) are used to check internal consistency of the constructs and values above .7 indicate reliability of the measures (Taheri et al., 2020). We performed exploratory factor analysis (EFA) to check scale composition of the scale items. Harman's single factor test is conducted to check the potential impact of common method bias (Olya & Al-Ansi, 2018). To test convergent and divergent validity of the constructs and fit validity of the measurement model, we performed confirmatory factor analysis (CFA) stage of CB-SEM (covariance based- structural equation modeling) using AMOS. After checking reliability and validity, we calculated descriptive statistics (means and standard deviations) and correlation analysis to investigate the relationship between the factors included in the research model. We conducted regression analysis to test the proposed hypotheses. We conducted Ordinarily Least Squares (OLS) analysis to test the proposed hypotheses (Olya, 2020). Basic assumptions of OLS (e.g. normality of data and associations between predictor and outcome variables) were evaluated (Olya & Han, 2020). OLS regressions used to test hypotheses 1–8 are outlined in Table A1, Appendix A. We used model 4 of PROCESS macro (Hayes & Rockwood, 2020) to test the mediation role of burnout on the association between sexual harassment, job satisfaction and psychological well-being (hypotheses 9–10).

Results

Measurement model testing

Before assessing the proposed model, reliability and validity of the measurements were evaluated. The values of Cronbach's alpha and composite reliability (CR) were above the recommended level of .7 which indicates reliability of all constructs. As shown in Table 2, no general factor (variance %>35 ~ 40) emerged which indicates common method bias is not a serious threat to the study measures. During the EFA which is conducted to check the scale composition, one item from job satisfaction, four items from perceived social support and ten items from sexual harassment were dropped due to low factor loading (λ<.4) or cross-loading. The CFA results confirm the result of the EFA as all items adequately and significantly loaded under

Table 2. Results of reliability and factor analyses.

Scale items	λ	%	α (CR)	SFL
Psychological well-being (AVE: .517; ASV:.039; MSV: .067)		8.474	831 (.839)	
In most ways my life is close to my ideal	.737			0.580***
The conditions of my life are excellent	.781			0.767***
I am satisfied with my life	.649			0.770***
So far I have gotten the important things I want in life	.703			0.789***
If I could live my life over I would change almost nothing	.842			0.661***
Job satisfaction (AVE: .652; ASV:.045; MSV: .080)		6.492	.874 (.876)	
I feel that my job is valuable	.641			.758***
In my job, I feel that I am doing something worthwhile	.781			.843***
I feel that my job is interesting	.668			.894***
I feel that my job is satisfying	.643			.723***
Burnout (AVE: .568; ASV:.031; MSV: .014)		1.389	.884 (.831)	
I feel emotionally drained from my work	.729			.824***
I feel used up at the end of the workday	.905			.766***
I feel burned out from my work	.886			.844***
I feel I treat some customers as if they are impersonal 'objects'	.711			.728***
I feel I have become uncaring toward people since I took this job	.638			.677***
I worry that this job is hardening me emotionally	.852			.665***
Sexual harassment (AVE: .515; ASV:.013; MSV: .014)		10.223	.884 (.907)	
Put down because my sex	.557			.640***
Treated different because of your sex	.656			.639***
Offensive sexist remarks	.723			.696***
Repeatedly told sexual jokes	.585			.757***
Sexual remarks, publicly or privately	.623			.708***
Unwelcome sex discussions	.703			.724***
Offensive sexual gestures	. 837			.608***
Display sexist materials	. 625			.580***
Remarks made about body/sex acts	.496			.678***
Asked for dates after I say no	.630			.553***
Stared at in a sexual way	.638			.569***
Perceived social support (AVE: .672; ASV:.071; MSV: .127)		13.393	.917 (.862)	
My family really tries to help me	.865			.855***
I get the emotional help and support I need from my family	.937			.857***
My friends really try to help me	.813			.799***
I can count on my friends when things go wrong	.896			.761***
I can talk about my problems with my family	.958			.829***
I have friends with whom I can share my joys and sorrows	.887			.802***
My family is willing to help me make decisions	.937			.820***
I can talk about my problems with my friends	.918			.778***
Perceived organizational support (AVE: .661; ASV:.040; MSV: .127)		17.289	.940 (.888)	
The organization values my contribution to its well-being	.833			.856***
The organization strongly considers my goals and values	.963			.952***
The organization really cares about my well-being	.963			.898***
The organization is willing to help me when I need a special favor	.867			.908***
The organization shows very little concern for me (R)	.789			.603***
The organization takes pride in my accomplishment at work	.815			.629***

Note: ***: $p < .001$. λ: factor lading, SFL: standardized factor loading: SFL, AVE: average variance extracted, ASV: average shared variance, MSV: maximum shared variance.

Table 3. Means, standard division and correlation matrix.

Variables	M	SD	1	2	3	4	5	6
1. Perceived social support	3.851	.920	1					
2. Perceived organizational support	3.458	.972	.392**	1				
3. Sexual harassment	1.655	.573	−.052	−.303**	1			
4. Burnout	1.952	.818	−.132*	−.308**	.279**	1		
5. Job satisfaction	3.357	.801	.234**	.398**	−.252**	−.374**	1	
6. Psychological well-being	3.813	.776	.273**	.382**	−.244**	−.214**	.525**	1

** Correlation is significant at the .01 level (2-tailed), * Correlation is significant at the .05 level (2-tailed).

related constructs (standardized factor loading> .4, p <.001). Construct validity including convergent and divergent validity are also assessed. The average variance extracted (AVE) is computed for each construct and all the values are above the recommended level of .5 (Table 2). In terms of divergent validity, average shared variance (ASV) and maximum shared variance (MSV) value for each construct is less than the correspondence AVE (Table 2). Empirical data fitted well with the proposed measurement model as fit statistics (X2/df: 2.564, PNFI: .678, PCFI: .756, RMSEA: .084) satisfied the commonly accepted levels (X2/df = 2–5, PNFI > .5, PCFI > .5, RMSEA: .05–.1).

Table 3 presents the means, standard deviations and correlation matrix of the study variables. Except sexual harassment and perceived social support, other study variables are significantly correlated. Specifically, perceived organizational support and sexual harassment are negatively related ($r = -.303$, $p < .01$). There is significant and positive relationship between sexual harassment and burnout ($r = .279$, $p < .01$). Job satisfaction and psychological well-being are negatively related to burnout and sexual harassment (Table 3).

Results of hypotheses testing

The results of the OLS regression analysis for testing hypotheses are illustrated in Figure 2. According to the results, sexual harassment is not significantly influenced by perceived social support ($\beta = .079$, $p = .260$), whereas it is significantly and negatively affected by perceived organizational support ($\beta = -.334$, $p < .001$). These results indicate that compared to social

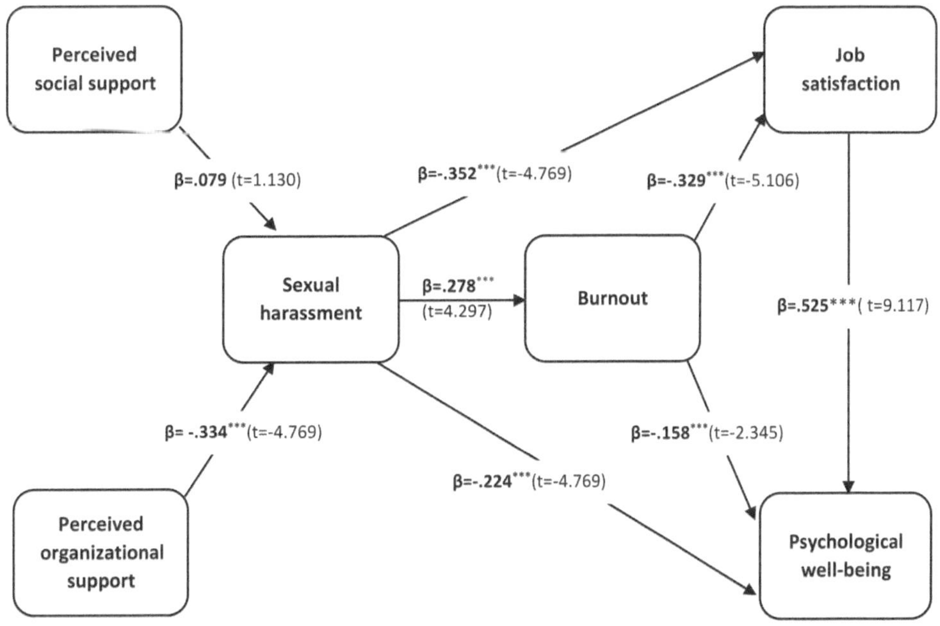

Figure 2. Results of regression analyses (***: $p < 0001$).

Table 4. Results of the mediation test of burnout in association with sexual harassment with job satisfaction and psychological well-being.

Mediator: Burnout		Outcome (Y): Job satisfaction		
Predictor (X): sexual harassment	Effect	s.e. (t)	LLCI	ULCI
Total effect	−.340***	.088 (−3.848)	−.515	−.166
Direct effect	−.216*	.087 (−2.477)	−.388	−.044
Indirect effect	−.124	. 037	−.200	−.051
Mediator: Burnout	Outcome (Y): Psychological well-being			
Predictor (X): sexual harassment	Effect	s.e. (t)	LLCI	ULCI
Total effect	−.340**	.091 (−3.720)	−.5210	−.160
Direct effect	−.278**	.094 (−2.955)	−.464	−.092
Indirect effect	−.061	.033	−.131	−.0004

Note: $*p < .05$, $**p < .01$, $*** < .0001$. s.e. = standard error. 5000 bootstrap samples.

support, organizational support significantly decreases sexual harassment, thus supporting hypothesis 2 but not hypothesis 1. Sexual harassment significantly and negatively effects job satisfaction ($\beta = -.352$, $p < .0001$) and psychological well-being ($\beta = -.224$, $p < .0001$) thus supporting hypothesis 4 and hypothesis 6. The results also support hypothesis 3, which predicted sexual harassment as having a significant and positive effect on burnout ($\beta = .278$, $p < .001$). As we predicted (hypothesis 5 and hypothesis 7), burnout has a significant and negative impact on job satisfaction ($\beta = -.329$, $p < .0001$) and psychological well-being ($\beta = -.158$, $p < 00.1$). Finally, job satisfaction has a significant and positive impact on psychological well-being ($\beta = .525$, $p < .001$), thus supporting hypothesis 8.

Results of mediation tests

Table 4 presents the results of mediation role of burnout on links of sexual harassment with job satisfaction and psychological well-being. Results of the bootstrapping method using Hayes's PROCESS macro shows that the indirect effect of sexual harassment on job satisfaction is significant (B = −.124, s.e. = .037, 95% Confidence interval = −.200, −.051). As shown in Table 4, the 95% CIs of the indirect effect does not involve zero. This result indicates that burnout mediates the effect of sexual harassment on job satisfaction, thus supporting hypothesis 9. The indirect effect of sexual harassment on psychological well-being is significant (B = −.061, s.e. = .033, 95% Confidence interval = −.131, −.0004). Similarly, burnout mediates the impact of sexual harassment on psychological well-being, thus supporting hypothesis 10.

Discussions, implications and conclusion

While sexual harassment and its relationship to various organizational and psychological variables have been the focus of a growing body of literature within tourism and hospitality scholarship (e.g. Cho, 2002; Jung & Yoon, 2019, 2020; Kensbock et al., 2015; Ram et al., 2016), much work is yet to be done to fully understand the nature and impacts of this phenomenon (Morgan & Pritchard, 2019; Pritchard, 2018). In particular, the relationships between sexual harassment, social and organizational support, job satisfaction, and psychological well-being remain understudied. Specifically, few studies have investigated sexual harassment and its impacts on tour guides (e.g. Alrawadieh et al., 2020; Cheung et al., 2018). Informed by the Conservation of Resources Theory (COR) and building on existing conceptualizations, the key objective of the current investigation was to test a conceptual model linking the sexual harassment experiences of tour guides, social support, organizational support, burnout, job satisfaction, and psychological well-being. The theoretical model was tested using data from Turkish female tour guides. Findings from the present study have significant theoretical and practical implications.

Theoretical implications

While the results show that perceived social support does not help in mitigating sexual harassment, they confirm that sexual harassment is significantly and negatively affected by perceived organizational support. As expected, the findings also reveal that sexual harassment is associated with burnout and has a negative impact on job satisfaction and psychological well-being. Overall, the results confirm that sexual harassment has a harmful impact on female tour guides' professional and psychological well-being, confirming a large body of research both in psychology and organizational behavior literature (Buchanan & Fitzgerald, 2008; Cho, 2002; Fitzgerald et al., 1988; Gettman & Gelfand, 2007; Sojo et al., 2016) as well as tourism and hospitality scholarship (Jung & Yoon, 2019, 2020; Ram, 2018; Theocharous & Philaretou, 2009).

Previous studies highlight the role of social and organizational support in reducing job-related stressors (Kang et al., 2010; McGinley & Wei, 2018), however, except for very few studies (Cortina, 2004; Livingston, 1982), little is known about the intersection between social and organizational support, and sexual harassment. The present study is perhaps the first to assess the relationships between perceived social support, perceived organizational support, and sexual harassment. In contrast to our prediction, social support does not influence sexual harassment, however, our findings confirm that organizational support decreases sexual harassment. In other words, the role of the informal social networks such as family members and friends is unlikely to contribute to reducing sexual harassment incidents whereas perceived formal support from travel agencies and tour operators is presumably effective in combatting sexual harassment. As discussed earlier, although there seem to be no organizational policies or codes of conduct to prevent sexual harassment in the specific context of tour guiding in Turkey, based on the findings of the present study, it may be plausible to assume that when an organization cares for its employees, they are likely to feel safer that they would receive support in case sexual harassment occurred, and if so they would be better able to protect themselves.

Research establishes a negative relationship between sexual harassment incidents and negative organizational outcomes such as job satisfaction and burnout (Fitzgerald et al., 1997; Gettman & Gelfand, 2007; Jung & Yoon, 2020; Szymanski & Mikorski, 2016; Willness et al., 2007). To our best knowledge, there has been no empirical study addressing these relationships in the specific case of tour guides. Our results support the notion that sexual harassment decreases job satisfaction and stimulates burnout, thus confirming the adverse impacts of sexual harassment in different settings. Confirming findings from previous research (Cho, 2002; Gunnarsdottir et al., 2006; Theocharous & Philaretou, 2009), our results indicate that female tour guides' sexual harassment experience has a negative impact on their psychological well-being. While this finding adds empirical evidence supporting a considerable amount of similar research endeavors, it contradicts the qualitative enquiry of Alrawadieh et al. (2020). According to their study, Alrawadieh et al. (2020) noted that female tour guides perceived sexual harassment incidents as being unthreatening thus perceiving minimal negative impact on their quality of life. It is safe to say that the sensitivity as well as the complexity of sexual harassment as a construct may explain the grey area between its normalization and the established discussion around its detrimental impacts. For instance, in their qualitative enquires, Folgerø and Fjeldstad (1995), and Alrawadieh et al. (2020) noted that informants initially denied having experienced sexual harassment while, eventually, reporting incidents that constituted objectively defined sexual harassment.

Overall, the present study makes two key contributions to existing theoretical assessments. First, our findings add to and extend previous theoretical considerations on the intersection between perceived support and sexual harassment experience (Cortina, 2004; Livingston, 1982). While these studies examined the types of support victims seek following a sexual harassment experience and the impact of such support on victims, our study goes beyond by investigating the role of social and organizational support in mitigating sexual harassment. Therefore, the present study contributes by theoretically proposing and empirically assessing the role of support received from organizations and social environment as a potential tool to proactively prevent sexual harassment.

Second, the current investigation views sexual harassment as a form of gender-based discrimination that harms female tour guides' professional and psychological well-being. Therefore, a significant contribution of the study stems from the fact that it adds to a research stream that addresses gender equality as a path to sustainability (Alarcón & Cole, 2019; Ferguson & Alarcon, 2015). In this vain, the study responds to calls in academia (Morgan & Pritchard, 2019; Pritchard, 2018) and joins similar efforts (Koehler, 2016; Rosche, 2016) to eliminate discrimination and violence against women in line with the 2030 Agenda for Sustainable Development (United Nations, 2015).

Managerial implications

Our findings confirm the harmful impacts of sexual harassment experiences on female tour guides' professional and psychological well-being. Specifically, sexual harassment is found to lead to increased burnout, decreased job satisfaction, and degraded psychological well-being. Stakeholders in the travel industry who are aware of the importance of tour guides' well-being as an essential asset for their businesses' success should, therefore, place more effort to combat sexual harassment. Thus, tour guides should be encouraged to report incidents of sexual harassment to their employers who, on their side, should intervene firmly and fairly. For instance, if the perpetrator is a hotel receptionist or a bus driver, the travel agency should report a complaint or switch to another service provider. If the perpetrator is a tour member, the travel agency should consider some options to ensure restoring their tour guides' dignity whilst sustaining a hospitable environment for the tour group. This may include talking to the perpetrator or the tour leader and providing an accompanying colleague while on the tour. It is also essential to anticipate the adverse effects of sexual harassment on tour guides by providing them with support in various forms (e.g. monetary or non- monetary compensation). Tour operators and travel agencies' proactive strategies can also be beneficial in reducing the likelihood of sexual harassment occurrence. These strategies may include sharing clear codes of ethics with outgoing tour operators (and eventually with tour members) as well as with local service providers.

Contrary to our predictions, perceived social support does not seem to mitigate sexual harassment. This finding indicates that workplace sexual harassment is primarily an organizational problem and thus support received from outside the workplace is unlikely to be effective. Indeed, our findings confirm the positive role of perceived organizational support in curbing sexual harassment. An obvious implication is that stakeholders within the travel industry such as travel agencies and tour operators should unequivocally communicate their support to their tour guides against deviant behaviors. Few would disagree that "sexual harassment is often disguised by the conspiracy of silence" (Gilbert et al., 1998, p. 49), especially when customers are the perpetrators. When it comes to serious problems violating individuals' dignity and threatening their well-being such as sexual harassment, "the customer is always right" mentality should not prevail. Therefore, travel agencies and tour operators should have zero tolerance towards sexual harassment despite its sources. This cannot be achieved without having clear policies and procedures that are made available for tour guides. These policies should vow commitment to providing a work environment free of sexual harassment.

Theoretically speaking, having internal policies to deal with sexual harassment is essential. On the practical side, however, tourism and hospitality businesses are often criticized for lacking such policies (Kensbock et al., 2015; Mkono, 2010). Sometimes, tour guides are not even aware whether their affiliated travel agencies have such policies or not (Cheung et al., 2018). Therefore, it might be necessary, and in the interest of both tour guides and businesses in the travel industries to enforce regulations and training programs to create a respectful work environment. For instance, tour guides chambers and unions (e.g. The Union of Turkish Tourist Guides' Chambers, in the case of Turkey) can introduce training programs into their affiliated travel agencies on the negative consequences of sexual harassment and how to communicate support for their tour guides who experience sexual harassment.

Limitations and areas of future research

The study has a few important limitations that can potentially encourage further empirical investigation into the area of sexual harassment. First, the data were collected solely from female tour guides based in Turkey. The findings are therefore confined to one gender, culture and professional context. Second, the study employed sexual harassment as a unidimensional construct. Considering the dimensions of sexual harassment (e.g. gender harassment and sexual assault) and their relationships with other organizational and psychological variables would provide a more accurate portrait of the effects of sexual harassment. Third, sexual harassment is a complex construct (Cheung et al., 2018), and so is the nature of the tour guiding profession (Ap & Wong, 2001), therefore, the current study may need to be validated and refined by considering specific types of tour guides (e.g. adventure tour guides) in other cultural contexts. Future research endeavors may extend the findings of the current study by examining whether the consequences of sexual harassment may vary based on the source of sexual harassment (e.g. customer, co-worker, and employer, other service providers) and/or the status power of the harasser. This is important given the nature of the tour guiding profession, which often involves high interaction not only with tour members but also with local service providers. Research may also look into how female tour guides cope with workplace sexual harassment incidents and as to whether their coping strategies may vary based on the frequency or severity of the sexual harassment. Finally, while the present study examines the impact of organizational and social support on sexual harassment, it remains unclear to which form of support do victims seek following a sexual harassment experience. Relatedly, while perceived organizational support is found to play a pivotal role in reducing incidents of sexual harassment (even with the absence of explicit policies in the area), it remains unclear why that is the case. This issue may also warrant further investigation.

Acknowledgement

This study was partially funded by Istanbul University-Cerrahpaşa Research Council (Project no: 35290).

Disclosure statement

No potential conflict of interest was reported by the author(s).

ORCID

Zaid Alrawadieh http://orcid.org/0000-0001-8355-9958
Derya Demirdelen Alrawadieh http://orcid.org/0000-0002-7554-2256
Hossein G. T. Olya http://orcid.org/0000-0002-0360-0744
Gul Erkol Bayram http://orcid.org/0000-0001-9764-2883
Onur Cuneyt Kahraman http://orcid.org/0000-0001-7773-8757

References

Alarcón, D. M., & Cole, S. (2019). No sustainability for tourism without gender equality. *Journal of Sustainable Tourism*, *27*(7), 903–919.

Alrawadieh, Z., Cetin, G., Dincer, M. Z., & Istanbullu Dincer, F. (2020). The impact of emotional dissonance on quality of work life and life satisfaction of tour guides. *The Service Industries Journal, 40*(1-2), 50–64. https://doi.org/10.1080/02642069.2019.1590554

Alrawadieh, Z., & Demirdelen Alrawadieh, D. (2020). Sexual harassment and wellbeing in tourism workplaces: The perspectives of female tour guides. In P. Vizcaino-Suárez, H. Jeffrey, & C. Eger (Eds.), *Tourism and gender-based violence – Challenging inequalities* (pp. 80–92). CABI.

Ap, J., & Wong, K. K. (2001). Case study on tour guiding: Professionalism, issues and problems. *Tourism Management, 22*(5), 551–563. https://doi.org/10.1016/S0261-5177(01)00013-9

Arenas, A., Giorgi, G., Montani, F., Mancuso, S., Perez, J. F., Mucci, N., & Arcangeli, G. (2015). Workplace bullying in a sample of Italian and Spanish employees and its relationship with job satisfaction, and psychological well-being. *Frontiers in Psychology, 6*, 1912. https://doi.org/10.3389/fpsyg.2015.01912

Australian Human Rights Commission. (2018). *Everyone's business: Fourth national survey on sexual harassment in Australian workplaces*. Australian Human Rights Commission. Retrieved July 6, 2020, from https://humanrights.gov.au/our-work/sex-discrimination/publications/everyones-business-fourth-national-survey-sexual

Ayaz, N., & Demir, C. (2019). Perceived work stress factors: A study on tour guides. *Journal of Tourism and Gastronomy Studies, 7*(1), 415–427. https://doi.org/10.21325/jotags.2019.370

Babakus, E., Yavas, U., & Ashill, N. J. (2009). The role of customer orientation as a moderator of the job demand–-burnout–performance relationship: A surface-level trait perspective. *Journal of Retailing, 85*(4), 480–492. https://doi.org/10.1016/j.jretai.2009.07.001

Baruch-Feldman, C., Brondolo, E., Ben-Dayan, D., & Schwartz, J. (2002). Sources of social support and burnout, job satisfaction, and productivity. *Journal of Occupational Health Psychology, 7*(1), 84–93. https://doi.org/10.1037//1076-8998.7.1.84

Bell, M. P., Quick, J. C., & Cycyota, C. S. (2002). Assessment and prevention of sexual harassment of employees: An applied guide to creating healthy organizations. *International Journal of Selection and Assessment, 10*(1&2), 160–167. https://doi.org/10.1111/1468-2389.00203

Bowling, N. A., Eschleman, K. J., & Wang, Q. (2010). A meta-analytic examination of the relationship between job satisfaction and subjective well-being. *Journal of Occupational and Organizational Psychology, 83*(4), 915–934. https://doi.org/10.1348/096317909X478557

Buchanan, N. T., & Fitzgerald, L. F. (2008). Effects of racial and sexual harassment on work and the psychological well-being of African American women. *Journal of Occupational Health Psychology, 13*(2), 137–151. https://doi.org/10.1037/1076-8998.13.2.137

Cannon, D. F., Ferreira, R. R., & Ross, L. E. (1998). An analysis of sexual harassment in private clubs. *Journal of Hospitality & Tourism Education, 10*(3), 63–71.

Cetin, G., & Yarcan, S. (2017). The professional relationship between tour guides and tour operators. *Scandinavian Journal of Hospitality and Tourism, 17*(4), 345–357.

Cheng, J. C., & Yi, O. (2018). Hotel employee job crafting, burnout, and satisfaction: The moderating role of perceived organizational support. *International Journal of Hospitality Management, 72*, 78–85. https://doi.org/10.1016/j.ijhm.2018.01.005

Cheung, C., Baum, T., & Hsueh, A. (2018). Workplace sexual harassment: Exploring the experience of tour leaders in an Asian context. *Current Issues in Tourism, 21*(13), 1468–1485. https://doi.org/10.1080/13683500.2017.1281235

Cho, M. (2002). An analysis of sexual harassment in Korean hotels from the perspective of female employees. *Journal of Human Resources in Hospitality & Tourism, 1*(3), 11–29.

Cho, S. (2019). Effects of social support and grateful disposition on employees' psychological well-being. *The Service Industries Journal, 39*(11-12), 799–819. https://doi.org/10.1080/02642069.2018.1444755

Cohen, J. (1992). A power primer. *Psychological Bulletin, 112*(1), 155–159. https://doi.org/10.1037/0033-2909.112.1.155

Cortina, L. M. (2004). Hispanic perspectives on sexual harassment and social support. *Personality & Social Psychology Bulletin, 30*(5), 570–584. https://doi.org/10.1177/0146167203262854

Diener, E. D., Emmons, R. A., Larsen, R. J., & Griffin, S. (1985). The satisfaction with life scale. *Journal of Personality Assessment, 49*(1), 71–75. https://doi.org/10.1207/s15327752jpa4901_13

Eisenberger, R., Huntington, R., Hutchison, S., & Sowa, D. (1986). Perceived organizational support. *Journal of Applied Psychology, 71*(3), 500–507. https://doi.org/10.1037/0021-9010.71.3.500

Equal Employment Opportunity Commission (2019). *Charges Alleging Sex-Based Harassment (Charges filed with EEOC) FY 2010 - FY 2019*. Retrieved March 6, 2020, from https://www.eeoc.gov/eeoc/statistics/enforcement/sexual_harassment_new.cfm

Farkic, J. (2018). Outdoor guiding as hospitality work. *Annals of Tourism Research, 73*, 197–199. https://doi.org/10.1016/j.annals.2018.04.004

Ferguson, L., & Alarcon, D. M. (2015). Gender and sustainable tourism: Reflections on theory and practice. *Journal of Sustainable Tourism, 23*(3), 401–416. https://doi.org/10.1080/09669582.2014.957208

Fitzgerald, L. F., Drasgow, F., Hulin, C. L., Gelfand, M. J., & Magley, V. J. (1997). Antecedents and consequences of sexual harassment in organizations: A test of an integrated model. *The Journal of Applied Psychology*, *82*(4), 578–589. https://doi.org/10.1037/0021-9010.82.4.578

Fitzgerald, L. F., Shullman, S. L., Bailey, N., Richards, M., Swecker, J., Gold, Y., Ormerod, M., & Weitzman, L. (1988). The incidence and dimensions of sexual harassment in academia and the workplace. *Journal of Vocational Behavior*, *32*(2), 152–175. https://doi.org/10.1016/0001-8791(88)90012-7

Fitzgerald, L. F., Swan, S., & Magley, V. J. (1997). But was it really sexual harassment? Legal, behavioral, and psychological definitions of workplace victimization of women. In W. O'Donohue (Ed.), *Sexual harassment: Theory, research, and treatment* (pp. 5–28). Allyn and Bacon.

Folgerø, I. S., & Fjeldstad, I. H. (1995). On duty—off guard: Cultural norms and sexual harassment in service organizations. *Organization Studies*, *16*(2), 299–313. https://doi.org/10.1177/017084069501600205

Gettman, H. J., & Gelfand, M. J. (2007). When the customer shouldn't be king: Antecedents and consequences of sexual harassment by clients and customers. *Journal of Applied Psychology*, *92*(3), 757–770. https://doi.org/10.1037/0021-9010.92.3.757

Gilbert, D., Guerrier, Y., & Guy, J. (1998). Sexual harassment issues in the hospitality industry. *International Journal of Contemporary Hospitality Management*, *10*(2), 48–53. https://doi.org/10.1108/09596119810207183

Guerrier, Y., & Adib, A. S. (2000). 'No, we don't provide that service': The harassment of hotel employees by customers. *Work, Employment and Society*, *14*(4), 689–705.

Gunnarsdottir, H. K., Sveinsdottir, H., Bernburg, J. G., Fridriksdottir, H., & Tomasson, K. (2006). Lifestyle, harassment at work and self-assessed health of female flight attendants, nurses and teachers. *Work (Reading, Mass.)*, *27*(2), 165–172.

Gutek, B. A., & Koss, M. P. (1993). Changed women and changed organizations: Consequences of and coping with sexual harassment. *Journal of Vocational Behavior*, *42*(1), 28–48. https://doi.org/10.1006/jvbe.1993.1003

Hayes, A. F., & Rockwood, N. J. (2020). Conditional process analysis: Concepts, computation, and advances in the modeling of the contingencies of mechanisms. *American Behavioral Scientist*, *64*(1), 19–54. https://doi.org/10.1177/0002764219859633

Hobfoll, S. E. (1989). Conservation of resources: A new attempt at conceptualizing stress. *The American Psychologist*, *44*(3), 513–524. https://doi.org/10.1037/0003-066X.44.3.513

Hu, Y.-Y., Ellis, R. J., Hewitt, D. B., Yang, A. D., Cheung, E. O., Moskowitz, J. T., Potts, J. R., Buyske, J., Hoyt, D. B., Nasca, T. J., & Bilimoria, K. Y. (2019). Discrimination, abuse, harassment, and burnout in surgical residency training. *New England Journal of Medicine*, *381*(18), 1741–1752. https://doi.org/10.1056/NEJMsa1903759

Humborstad, S., Humborstad, B., & Whitfield, R. (2007). Burnout and service employees' willingness to deliver quality service. *Journal of Human Resources in Hospitality & Tourism*, *7*(1), 45–64. https://doi.org/10.1300/J171v07n01_03

Hur, W. M., Moon, T. W., & Jun, J. K. (2013). The role of perceived organizational support on emotional labor in the airline industry. *International Journal of Contemporary Hospitality Management*, *25*(1), 105–123. https://doi.org/10.1108/09596111311290246

Ineson, E. M., Yap, M. H., & Whiting, G. (2013). Sexual discrimination and harassment in the hospitality industry. *International Journal of Hospitality Management*, *35*, 1–9. https://doi.org/10.1016/j.ijhm.2013.04.012

Je, J. S., Khoo, C., & Chiao Ling Yang, E. (2020). Gender issues in tourism organisations: Insights from a two-phased pragmatic systematic literature review. *Journal of Sustainable Tourism*, 1–24.

Jiang, K., Hong, Y., McKay, P. F., Avery, D. R., Wilson, D. C., & Volpone, S. D. (2015). Retaining employees through anti–sexual harassment practices: Exploring the mediating role of psychological distress and employee engagement. *Human Resource Management*, *54*(1), 1–21. https://doi.org/10.1002/hrm.21585

Johlke, M. C., & Duhan, D. F. (2000). Supervisor communication practices and service employee job outcomes. *Journal of Service Research*, *3*(2), 154–165. https://doi.org/10.1177/109467050032004

Judge, T. A., & Watanabe, S. (1994). Individual differences in the nature of the relationship between job and life satisfaction. *Journal of Occupational and Organizational Psychology*, *67*(2), 101–107. https://doi.org/10.1111/j.2044-8325.1994.tb00554.x

Jung, H. S., & Yoon, H. H. (2019). How does workplace harassment influence the employees' response in a deluxe hotel? *The Service Industries Journal*, *39*(11-12), 877–900. https://doi.org/10.1080/02642069.2018.1493103

Jung, H. S., & Yoon, H. H. (2020). Sexual harassment and customer-oriented boundary-spanning behaviors. *International Journal of Contemporary Hospitality Management*, *32*(1), 3–19. https://doi.org/10.1108/IJCHM-10-2018-0790

Kang, B., Twigg, N. W., & Hertzman, J. (2010). An examination of social support and social identity factors and their relationship to certified chefs' burnout. *International Journal of Hospitality Management*, *29*(1), 168–176. https://doi.org/10.1016/j.ijhm.2009.08.004

Kara, D., Uysal, M., Sirgy, M. J., & Lee, G. (2013). The effects of leadership style on employee well-being in hospitality. *International Journal of Hospitality Management*, *34*, 9–18. https://doi.org/10.1016/j.ijhm.2013.02.001

Kensbock, S., Bailey, J., Jennings, G., & Patiar, A. (2015). Sexual harassment of women working as room attendants within 5-star hotels. *Gender, Work & Organization*, *22*(1), 36–50.

Kiely, J., & Henbest, A. (2000). Sexual harassment at work: experiences from an oil refinery. *Women in Management Review, 15*(2), 65–80. https://doi.org/10.1108/09649420010319598

Koc, E., & Bozkurt, G. A. (2017). Hospitality employees' future expectations: Dissatisfaction, stress, and burnout. *International Journal of Hospitality & Tourism Administration, 18*(4), 459–473.

Koehler, G. (2016). Tapping the sustainable development goals for progressive gender equity and equality policy? *Gender & Development, 24*(1), 53–68.

Kohan, A., & O'connor, B. P. (2002). Police officer job satisfaction in relation to mood, well-being, and alcohol consumption. *The Journal of Psychology, 136*(3), 307–318. https://doi.org/10.1080/00223980209604158

Kong, H., Jiang, X., Chan, W., & Zhou, X. (2018). Job satisfaction research in the field of hospitality and tourism. *International Journal of Contemporary Hospitality Management, 30*(5), 2178–2194. https://doi.org/10.1108/IJCHM-09-2016-0525

Larsen, S. E., Nye, C. D., & Fitzgerald, L. F. (2019). Sexual harassment expanded: An examination of the relationships among sexual harassment, sex discrimination, and aggression in the workplace. *Military Psychology, 31*(1), 35–44. https://doi.org/10.1080/08995605.2018.1526526

Lee, J. J., Ok, C., & Hwang, J. (2016). An emotional labor perspective on the relationship between customer orientation and job satisfaction. *International Journal of Hospitality Management, 54*(3), 139–150. https://doi.org/10.1016/j.ijhm.2016.01.008

Li, Y., Chen, M., Lyu, Y., & Qiu, C. (2016). Sexual harassment and proactive customer service performance: The roles of job engagement and sensitivity to interpersonal mistreatment. *International Journal of Hospitality Management, 54*, 116–126. https://doi.org/10.1016/j.ijhm.2016.02.008

Livingston, J. A. (1982). Responses to sexual harassment on the job: Legal, organizational, and individual actions. *Journal of Social Issues, 38*(4), 5–22. https://doi.org/10.1111/j.1540-4560.1982.tb01907.x

Loh, J. M., & Loi, N. (2018). Tit for tat: Burnout as a mediator between workplace incivility and instigated workplace incivility. *Asia-Pacific Journal of Business Administration, 10*(1), 100–111. https://doi.org/10.1108/APJBA-11-2017-0132

Lu, A. C. C., & Gursoy, D. (2016). Impact of job burnout on satisfaction and turnover intention: Do generational differences matter? *Journal of Hospitality & Tourism Research, 40*(2), 210–235.

Maslach, C., Schaufeli, W. B., & Leiter, M. P. (2001). Job burnout. *Annual Review of Psychology, 52*(1), 397–422. https://doi.org/10.1146/annurev.psych.52.1.397

Mathisen, G. E., Einarsen, S., & Mykletun, R. (2008). The occurrences and correlates of bullying and harassment in the restaurant sector. *Scandinavian Journal of Psychology, 49*(1), 59–68. https://doi.org/10.1111/j.1467-9450.2007.00602.x

McGinley, S., & Wei, W. (2018). Emotional labor's impact on hoteliers outside the workplace. *International Journal of Contemporary Hospitality Management, 30*(9), 2965–2983. https://doi.org/10.1108/IJCHM-07-2017-0422

Min, J. (2014). The relationships between emotional intelligence, job stress, and quality of life among tour guides. *Asia Pacific Journal of Tourism Research, 19*(10), 1170–1190. https://doi.org/10.1080/10941665.2013.839459

Mkono, M. (2010). Zimbabwean hospitality students' experiences of sexual harassment in the hotel industry. *International Journal of Hospitality Management, 29*(4), 729–735. https://doi.org/10.1016/j.ijhm.2010.03.002

Mooney, S. K. (2020). Gender research in hospitality and tourism management: Time to change the guard. *International Journal of Contemporary Hospitality Management, 32*(5), 1861–1879. https://doi.org/10.1108/IJCHM-09-2019-0780

Morgan, N., & Pritchard, A. (2019). Gender matters in hospitality (invited paper for 'luminaries' special issue of International Journal of Hospitality Management). *International Journal of Hospitality Management, 76*, 38–44. https://doi.org/10.1016/j.ijhm.2018.06.008

Munson, L. J., Hulin, C., & Drasgow, F. (2000). Longitudinal analysis of dispositional influences and sexual harassment: Effects on job and psychological outcomes. *Personnel Psychology, 53*(1), 21–46. https://doi.org/10.1111/j.1744-6570.2000.tb00192.x

Murrell, A. J., Olson, J. E., & Frieze, I. H. (1995). Sexual harassment and gender discrimination: A longitudinal study of women managers. *Journal of Social Issues, 51*(1), 139–149. https://doi.org/10.1111/j.1540-4560.1995.tb01313.x

Murry, W. D., Sivasubramaniam, N., & Jacques, P. H. (2001). Supervisory support, social exchange relationships, and sexual harassment consequences: A test of competing models. *The Leadership Quarterly, 12*(1), 1–29. https://doi.org/10.1016/S1048-9843(01)00062-5

Nimri, R., Kensbock, S., Bailey, J., & Patiar, A. (2020). Management perceptions of sexual harassment of hotel room attendants. *Current Issues in Tourism*, 1–13. https://doi.org/10.1080/13683500.2020.1722619.

Olya, H. G. (2020). Towards advancing theory and methods on tourism development from residents' perspectives: Developing a framework on the pathway to impact. *Journal of Sustainable Tourism*, 1–21. https://doi.org/10.1080/09669582.2020.1843046

Olya, H. G., & Al-Ansi, A. (2018). Risk assessment of halal products and services: Implication for tourism industry. *Tourism Management, 65*, 279–291. https://doi.org/10.1016/j.tourman.2017.10.015

Olya, H. G., & Han, H. (2020). Antecedents of space traveler behavioral intention. *Journal of Travel Research, 59*(3), 528–544. https://doi.org/10.1177/0047287519841714

Pienaar, J., & Willemse, S. A. (2008). Burnout, engagement, coping and general health of service employees in the hospitality industry. *Tourism Management, 29*(6), 1053–1063. https://doi.org/10.1016/j.tourman.2008.01.006

Poulston, J. (2008). Metamorphosis in hospitality: A tradition of sexual harassment. *International Journal of Hospitality Management, 27*(2), 232–240. https://doi.org/10.1016/j.ijhm.2007.07.013

Pritchard, A. (2018). Predicting the next decade of tourism gender research. *Tourism Management Perspectives, 25*, 144–146. https://doi.org/10.1016/j.tmp.2017.11.014

Rabelo, V. C., & Cortina, L. M. (2014). Two sides of the same coin: Gender harassment and heterosexist harassment in LGBQ work lives. *Law and Human Behavior, 38*(4), 378–391. https://doi.org/10.1037/lhb0000087

Ram, Y. (2018). Hostility or hospitality? A review on violence, bullying and sexual harassment in the tourism and hospitality industry. *Current Issues in Tourism, 21*(7), 760–774. https://doi.org/10.1080/13683500.2015.1064364

Ram, Y., Tribe, J., & Biran, A. (2016). Sexual harassment: Overlooked and under-researched. *International Journal of Contemporary Hospitality Management, 28*(10), 2110–2131. https://doi.org/10.1108/IJCHM-05-2015-0240

Rederstorff, J. C., Buchanan, N. T., & Settles, I. H. (2007). The moderating roles of race and gender-role attitudes in the relationship between sexual harassment and psychological well-being. *Psychology of Women Quarterly, 31*(1), 50–61. https://doi.org/10.1111/j.1471-6402.2007.00330.x

Rhoades, L., & Eisenberger, R. (2002). Perceived organizational support: A review of the literature. *The Journal of Applied Psychology, 87*(4), 698–714. https://doi.org/10.1037/0021-9010.87.4.698

Rosche, D. (2016). Agenda 2030 and the sustainable development goals: Gender equality at last? An Oxfam perspective. *Gender & Development, 24*(1), 111–126.

Savicki, V., Cooley, E., & Gjesvold, J. (2003). Harassment as a predictor of job burnout in correctional officers. *Criminal Justice and Behavior, 30*(5), 602–619. https://doi.org/10.1177/0093854803254494

Shanock, L. R., & Eisenberger, R. (2006). When supervisors feel supported: Relationships with subordinates' perceived supervisor support, perceived organizational support, and performance. *Journal of Applied Psychology, 91*(3), 689–695. https://doi.org/10.1037/0021-9010.91.3.689

Shumaker, S. A., & Brownell, A. (1984). Toward a theory of social support: Closing conceptual gaps. *Journal of Social Issues, 40*(4), 11–36. https://doi.org/10.1111/j.1540-4560.1984.tb01105.x

Sirgy, J. M. (2019). Promoting quality-of-life and well-being research in hospitality and tourism. *Journal of Travel & Tourism Marketing, 36*(1), 1–13.

Slonaker, W. M., Wendt, A. C., & Baker, B. (2007). Employment discrimination in the restaurant industry. *Cornell Hotel and Restaurant Administration Quarterly, 48*(1), 46–58. https://doi.org/10.1177/0010880406297591

Sojo, V. E., Wood, R. E., & Genat, A. E. (2016). Harmful workplace experiences and women's occupational well-being: A meta analysis. *Psychology of Women Quarterly, 40*(1), 10–40. https://doi.org/10.1177/0361684315599346

Stamper, C. L., & Johlke, M. C. (2003). The impact of perceived organizational support on the relationship between boundary spanner role stress and work outcomes. *Journal of Management, 29*(4), 569–588. https://doi.org/10.1016/S0149-2063(03)00025-4

Stedham, Y., & Mitchell, M. C. (1998). Sexual harassment in casinos: Effects on employee attitudes and behaviors. *Journal of Gambling Studies, 14*(4), 381–400. https://doi.org/10.1023/a:1023025110307

Süral, N., & Kiliçoğlu, M. (2011). Prohibiting sexual harassment in the workplace in Turkey. *Middle Eastern Studies, 47*(4), 655–662. https://doi.org/10.1080/00263206.2011.590064

Szymanski, D. M., & Mikorski, R. (2016). Sexually objectifying restaurants and waitresses' burnout and intentions to leave: The roles of power and support. *Sex Roles, 75*(7-8), 328–338. https://doi.org/10.1007/s11199-016-0621-2

Taheri, B., Olya, H., Ali, F., & Gannon, M. J. (2020). Understanding the influence of airport servicescape on traveler dissatisfaction and misbehavior. *Journal of Travel Research, 59*(6), 1008–1028. https://doi.org/10.1177/0047287519877257

Theocharous, A., & Philaretou, A. G. (2009). Sexual harassment in the hospitality industry in the Republic of Cyprus: Theory and prevention. *Journal of Teaching in Travel & Tourism, 9*(3-4), 288–304.

Tourangeau, R., & Yan, T. (2007). Sensitive questions in surveys. *Psychological Bulletin, 133*(5), 859–883. https://doi.org/10.1037/0033-2909.133.5.859

Turkish Labor Law. (2003). Law No. 4857, Official Gazette, No. 25134, 10 June.

Union of Turkish Tourist Guides. (2019). *Rehber İstatistikleri*. Retrivied November 18, 2019, from http://tureb.org.tr/tr/RehberIstatistik/

United Nations. (2015). Transforming our world: The 2030 agenda for sustainable development. *General Assembley* 70 session, retrivied 6 March 2020 from https://www.un.org/ga/search/view_doc.asp?symbol=A/RES/70/1&Lang=E

Willness, C. R., Steel, P., & Lee, K. (2007). A meta-analysis of the antecedents and consequences of workplace sexual harassment. *Personnel Psychology, 60*(1), 127–162. https://doi.org/10.1111/j.1744-6570.2007.00067.x

Wong, J. Y., & Wang, C. H. (2009). Emotional labor of the tour leaders: An exploratory study. *Tourism Management, 30*(2), 249–259. https://doi.org/10.1016/j.tourman.2008.06.005

Wright, T. A., & Cropanzano, R. (2000). Psychological well-being and job satisfaction as predictors of job performance. *Journal of Occupational Health Psychology, 5*(1), 84–94. https://doi.org/10.1037//1076-8998.5.1.84

Yagil, D. (2008). When the customer is wrong: A review of research on aggression and sexual harassment in service encounters. *Aggression and Violent Behavior, 13*(2), 141–152. https://doi.org/10.1016/j.avb.2008.03.002

Yı ldı rı m, G. & Özbek, Ö (2019). Profesyonel turist rehberlerine yönelik cinsel taciz ile ilgili bir araştı rma. *Turist Rehberliği Dergisi, 2*(2), 72–87.

Zengin, B., Erkol, G., & Eker, N. (2014). *Rehberlik mesleğinde bir engel olarak cinsiyet ayrımcılığı: Bayan turist rehberleri üzerine bir araştı rma*. Proceedings of 15th National Tourism Conference. Nobel Akademik Yayı ncı lı k (pp. 1144–1159), 13-16 November, Ankara.

Zimet, G. D., Dahlem, N. W., Zimet, S. G., & Farley, G. K. (1988). The multidimensional scale of perceived social support. *Journal of Personality Assessment, 52*(1), 30–41. https://doi.org/10.1207/s15327752jpa5201_2

Appendix A

Table A1. OLS regressions correspondences with the hypotheses.

Hypothesis	Equation ($Y = \alpha + \beta X + \varepsilon$)	R^2	F
H1 & H2	Sexual harassment= 2.146 + .079 perceived social support −.334 Perceived organizational support + .085	.097	11.695***
H3	Burnout = 1.293 + .278 sexual harassment + .093	.078	18.468***
H4	Job satisfaction = 4.377 − .352 sexual harassment + .089	.063	14.811***
H5	Job satisfaction = 4.505 − .329 burnout + .059	.140	35.605***
H6	Psychological well-being = 3.921 − .244 sexual harassment + .092	.059	13.842***
H7	Psychological well-being = 3.767 − .158 burnout + .065	.046	10.521***
H8	Psychological well-being = 1.294 + .525 job satisfaction + .059	.275	83.115***

Note: Y: Dependent variable, α: regression intercept term, β: standardized regression coefficient, X: independent variable, ε : error term, ***: $p < .001$. All exogenous and endogenous variables are measured using survey (further details are provided in the measure section).

Gender issues in tourism organisations: insights from a two-phased pragmatic systematic literature review

Jess Sanggyeong Je, Catheryn Khoo and Elaine Chiao Ling Yang

ABSTRACT

Previous research has underlined gendered labour issues and provided recommendations for organisations to advance gender equality within the tourism and hospitality sector. Yet, little is known about how academic research has been translated into actual practice by tourism organisations. The study provides a pragmatic systematic review to compare tourism academic and grey literature from NGOs and large public tourism companies, with an aim of identifying gaps between academia and industry, and between rhetoric and practice. The analyses of 102 academic and 122 grey literature revealed that both types of literature lack focus on the cultural and individual dimensions of gendering process while explicitly focus on the organising system and developed economies. The review highlights insufficient details of company report and calls for developing concrete regulation or standard to measure and report gender equality practice in tourism. This review contributes to knowledge by charting the development of academic research in real-life situations to achieve equitable and sustainable workplace for both women and men in tourism.

Introduction

Tourism is a vital platform for women employment and entrepreneurial opportunities (Figueroa-Domecq et al., 2015). Yet, tourism employment depicts a mixed picture of women empowerment and exploitation (Hutchings et al., 2020). For one, women comprise of 54% of tourism workforce; however, they are under-represented in management roles and earn 14.7% less than men (UNWTO & UN Women, 2019). Women are over-represented in part-time, casual and seasonal jobs and likely to experience exploitation and poor working conditions (Santero-Sanchez et al., 2015). Accordingly, existing studies have recommended for tourism organisations such as women support systems or mentors (Remington & Kitterlin-Lynch, 2018), anti-discrimination and harassment policies (Quintana-García et al., 2018), diversity training (Madera, 2018), diversity council (Madera, 2013) and transparent remuneration process and report (Bakas et al., 2018). Nonetheless, no systematic approach has been made to investigate women employment issues in tourism and how the strategic recommendations from scholars have been enacted in real-life situations.

Past tourism studies suggested that gaps may exist between the rhetoric and practice for gender equality (Doherty, 2004; Khoo-Lattimore et al., 2019; Mkono, 2010). *Statement culture* and to a certain degree, *femwashing* have been noted in the tourism industry (Chambers et al., 2017; Khoo-Lattimore et al., 2019). Statement culture is where diversity and equality are mainly exercised through rhetoric or slogans instead of actual practice (Chambers et al., 2017). It also exemplifies femwashing where management practices continue to privilege the masculine as the authoritative norm in spite of their gender equality claim (Khoo-Lattimore et al., 2019). Notably, the UNWTO, which is the leading international tourism organisation, pledges the United Nations Sustainable Development Goal (UNSDG) 5 on promoting gender equality (UNWTO, 2019). However, our own recent study highlighted the lack of gender diversity among speakers at UNWTO events, which indicates an inconsistency in UNWTO's equality rhetoric and practice (Khoo-Lattimore et al., 2019). Likewise, the organisations' reports and audits on diversity have been criticised for lacking meaningful indicators to measure diversity performance or impacts (Grosser & Moon, 2005; Utting, 2007).

Informed by the research gaps identified above, this study has two main aims. The first aim is to examine peer-reviewed journal articles to comprehend how far tourism organisational studies (TOS) have explored gender issues and identify any under-researched area. The second aim is to investigate the extent to which the priorities given by scholars differ, if any, from practitioners'. By comparing academic and industry literature, this work contributes to existing knowledge by identifying research gaps not only within academic literature but also highlighting disparities between theory and reality, which might hinder the tourism industry from achieving equality. Guided by a pragmatic feminist perspective, this study contributes to existing knowledge by identifying gaps and opportunities to develop workable sustainability policies (Burritt & Schaltegger, 2010) for gender issues in tourism organisations. Although hospitality and tourism are not synonymous (Keiser, 1998), they are closely related and mutually dependent on each other (Baum, 2006). This paper used tourism as an umbrella term since tourism is a major force and directional mechanism in driving the hospitality industry (Ottenbacher et al., 2009).

Literature review

Gender in tourism research

Gender is embedded in the individual, interactional and institutional levels of societies (Risman, 2009). Thus, women and men experience differently as consumers and producers in the tourism and hospitality industry (Morgan & Pritchard, 2019). The norms and beliefs are (re)produced in every sphere of life, which in turn cement gendered societies (Wearing, 1998). In response, tourism studies have increasingly witnessed gender research on tourist behaviours, resident attitudes, managerial styles, entrepreneurship and sustainability (Figueroa-Domecq et al., 2015), largely focusing on similarities and differences between men and women (Calás & Smircich, 2006). Nevertheless, a recent systematic review of literature by Figueroa-Domecq et al. (2015) highlighted that studies on gender still remain marginal with little engagement of gender-aware frameworks. While Figueroa-Domecq et al. (2015) assessed the production and development of tourism gender research as a knowledge generating system, this review is specifically focused on the development of gender research in tourism organisations and its translation on the effective implementation of policies and strategies in organisations. Tourism gender research lacks the explicit support for qualitative and mixed methods, which are required to build inclusive views about gender issues with deeper understanding (Figueroa-Domecq et al., 2015). However, the methodological approach of TOS to discuss gendered employment issues is unknown to depict the holistic story of women workers. The empirical evidence on gender issues has been substantially gathered from developing regions such as Asia (Figueroa-Domecq et al., 2015). Still, women perform up to 80% of unpaid care work in Asia, which is a main barrier to develop their career in tourism (UNWTO & UN Women, 2019). As an extension of Figueroa-Domecq et al. (2015)'s analysis, this study examines what methodologies TOS have employed, in which location, to discuss gendered employment issues.

Three broad waves of feminist theories have shaped the understanding of gender issues in tourism: liberal feminism, standpoint feminism and post-structural feminism (Pritchard, 2014). This paper observes how these feminist theories are conceptualised within organisational policies and strategies for gender equality. Liberal feminism focuses on women's legal rights (Pritchard, 2014), which somewhat relates to the structural aspect of gender policies in tourism organisations. The structural policies include the visible and invisible objectives, rules and procedures that an organisation adopts to achieve gender equality (EIGE, 2019). Standpoint feminism acknowledges gendered power relations with different standpoints such as class (Marxist feminism) where women have been exploited in the capitalist patriarchy (Luxton, 2014). This relates to the provision of material conditions by organisations to emancipate women workers from subordination (Duffy et al., 2015), such as by flexible parental leave (Choi & Kim, 2012). Post-structural feminism identifies the cultural workings of gender power relations by deconstructing masculinised language or practice (Pritchard, 2014). This relates to cultural practice of organisations to impact attitudes and beliefs of people about gender equality (Wallace, 1998) through diversity training (Madera, 2018).

Tourism scholars also used the concept of *glass ceiling* to explain gender inequalities in organisations. The glass ceiling metaphor has been widely used to represent the invisible barriers that keep women and minorities from rising above certain level of positions (Jackson, 2001). However, glass ceiling is criticised for being too simplistic to explain discrimination such as horizontal segregation (Carvalho et al., 2019) where workers are distributed with certain occupations based on their gender. Other studies recognised gender as a product of power relations and criticised that organisations allow for masculinity to be implicitly reproduced as the norm in the management (Carvalho et al., 2019). The acknowledgement of oppression as a central feature of power politics is critical to propel tourism towards hopeful and sustainable workplaces (Pritchard & Morgan, 2013). Hopeful tourism is a transformative perspective and strives for the transformation of seeing and doing for sustainable tourism (Pritchard & Morgan, 2013). From the perspective of hopeful tourism, sustainable workplace requires the examination of multiple enquiries, and power relations entwined in the structure of gender and other classes for justice to build actionable approach (Pritchard & Morgan, 2013). Thus, this paper focusses on the analysis of gendering process by Acker (1990) in tourism organisations where there is a need to translate the gender awareness into practical solutions and systematic change for gender equality (Morgan & Pritchard, 2019).

Gendering processes in tourism workplaces

The hierarchy of organisations is believed to have *gendering* character where male power is (re)produced and women take more submissive roles than men (Carvalho et al., 2019). Acker (1990) has elucidated this proverbial existence of *gendering process*, which this study will utilise to capture how organisational processes and practices have reinforced gender inequality in tourism organisations. Gendering refers to the progression of socialisation according to the dominant gender norms (EIGE, 2019). It recognises gender differences as dynamic processes rather than static phenomena (Morgan & Knights, 1991). Gendering process not only constructs masculinity and femininity in a specific time and place, but also intersects with class, power and privilege (Creese, 1996). Organisations have a gendering character since gender is implicated in many forms and levels such as control, action, (dis)advantage, identity and emotions (Acker, 1990). Therefore, the identification of gendering process is important as it recognises how (in)equalities are created (in)visibly and how organisations play a role in gendering through practice (Carvalho et al., 2019). Acker (1990) suggested four aspects of gendering process: organising system, organisation culture, interactions on the job and gendered identities. The organising system is where the construction of gendered division occurs, for example, wage distribution, divisions of labour (horizontal segregation) and power distribution (vertical segregation) (Acker, 1990). The organisation culture refers to the construction of symbols and images that express, reinforce or occasionally oppose those divisions of gender (Acker, 1990), including implicit gender bias (Carvalho et al., 2019). Job interactions indicate all social patterns that enact dominance and submission, for example, gender differences during conversation in terms of interruptions, turn-taking and topic setting (Acker, 1990). Gendered identities refer to the gendered components of individual identities including consciousness of gender, choice of language-use and presentation of self as a gendered member (Acker, 1990). The identification of how gendering process is conceptualised in tourism research is necessary to delineate what practices and processes have reinforced inequalities in organisations.

Implementation gap for gender equality

The link between gendered employment and tourism was explicitly highlighted when UNWTO pledges UNSDG 5 gender equality through multiple reports (UNWTO & UN Women, 2011, 2019). In response, large tourism organisations have promoted gender equality as a corporate goal, published gender pay gap reports and included women empowerment in annual and sustainability reports (Accor Hotels, 2017; American Airlines Group, 2019; Marriott International, 2020). Considering the reports are referred to as "business cards" (Daub, 2007) which seek reputation and moral legitimacy, few studies examined how proactively tourism organisations investigate gendered employment issues; where they focus on; and how they justify their activities to various stakeholders who demand the information and statements on gender issues through reports. The gap between disclosure and performance might suggest that the impression of stakeholders are potentially managed through the selective and positive disclosure (Milne et al., 2009). Considering large public companies (PCs) do not see gender equality as a priority often (Grosser & Moon, 2008; PWC, 2017), it is critical to examine the relationship between the attitude of organisations and approaches to investigate gender issues. The recent reports highlight the widespread variations of inequality status according to geographical location (UNWTO & UN Women, 2019; World Economic Forum, 2020). While informality is a major trend across the regions including Europe, the essential tourism training is not available for African women and the individual awareness about gender equality needs to be increased in Asia and Pacific region (UNWTO & UN Women, 2019). Though the simple provision of equality initiatives does not guarantee the expected outcome due to the different state of law, labour market and participatory culture among male workers (Doherty, 2004), no systematic approach was conducted to examine how effectively tourism organisations respond to gender issues according to the geographical variation.

Informed by the identified research gaps and aims, this study has five research questions: (1) How far have TOS explored gender issues? (2) How are gendering processes conceptualised in tourism? (3) How are gender issues discussed in relation to geographical location? (4) What methods are employed to investigate gender issues in tourism? (5) How are strategic recommendations by TOS enacted in the tourism industry?

Methodology

Pragmatic systematic quantitative literature review

The study adopts a pragmatic systematic quantitative literature review (SQLR) as its methodology. The SQLR is instrumental in providing an exhaustive understanding of the subject of investigation (Pickering & Byrne, 2014). By mapping the boundaries of the existing literature, it

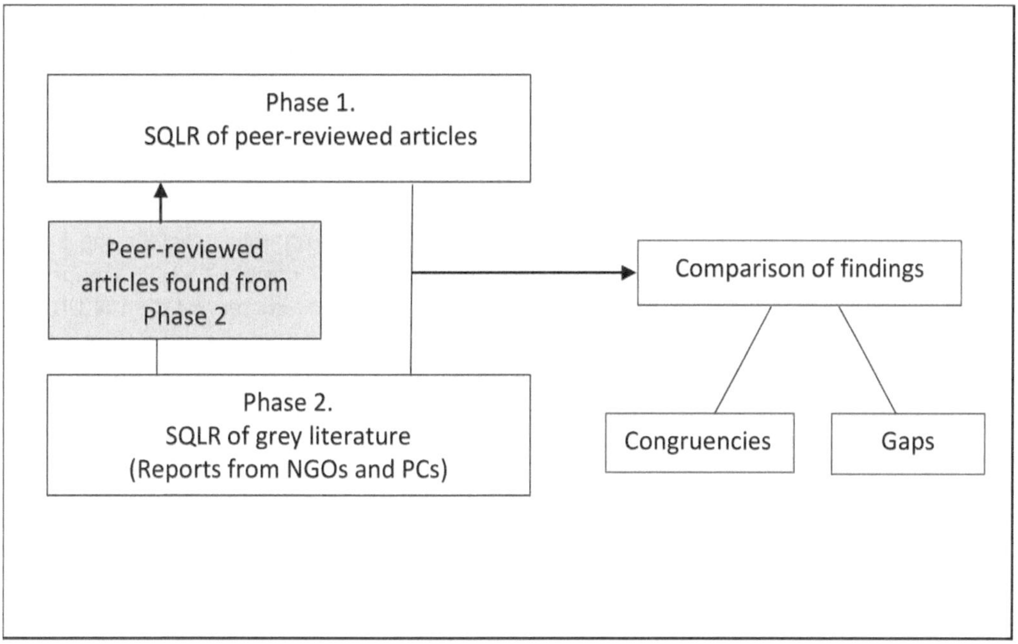

Figure 1. The process of pragmatic systematic literature review.

identifies where little research has been done and where new research is needed (Petticrew & Roberts, 2008); it is especially useful to examine the effectiveness of a policy, intervention or service (Petticrew & Roberts, 2008). A pragmatic approach, guided by the philosophy of action, involves understanding practical problems and contributes to develop solutions in organisational settings (Susman, 1983). Therefore, a pragmatic SQLR allows the identification of under-researched area as well as an understanding of practical problems or gaps to address issues of gender equalities. Figure 1 illustrates the review process of this multi-phase pragmatic SQLR developed by the authors. The review consisted of two phases: (1) peer-reviewed articles, (2) grey literature from PCs and NGOs in tourism. Both pragmatism and feminism privilege to eliminate oppressions, elaborate various perspectives and reach workable solutions (Rumens & Kelemen, 2010). This pragmatic feminist review can relate to hopeful tourism with the change of perspectives (Pritchard & Morgan, 2013) through combining different sources of data to offer a realistic approach to solve gender issues in tourism. A content analysis was conducted where particular items such as gendering process, geographical location, methods and strategic recommendations were coded, combined and abstracted into themes (Yang et al., 2017). Content analysis finds a theme based on the frequency of word occurrence, which enables both qualitative and quantitative analysis of data (Vaismoradi et al., 2013). The patterns of themes, which include overlapping themes between academic and grey literature were first quantified by the first author, and then cross-checked by all three authors to identify any gaps between academic and grey literature.

Phase 1 identified how prior TOS investigated gender issues in terms of themes, geographical locations, research methods, types of gendering process that (re)create inequality and recommendations for advancing gender equality. Phase 2 focused on how gender equality was articulated and implemented by NGOs and PCs to compare with the academic finding. It is a systematic review of "grey literature" that includes the unpublished research and report, such as by government, academia and business (Benzies et al., 2006). Many critics suggest excluding grey literature due to the heterogeneity quality (Conn et al., 2003). However, not all non-peer-reviewed journals are of poor quality and not all published journals are of high quality (Conn et al., 2003). Reviewing the reports is an important aid for advancing the equality agenda and improving human resource management (Kingsmill, 2001). Also, the reports from NGOs and PCs are valuable sources because the producers are the major actors of advancing equal opportunities for women (Grosser & Moon, 2005). NGOs play a critical role in creating gender equality regulation influencing companies (Grosser & Moon, 2005), engaging communities and national agendas (DeJaeghere & Wiger, 2013). PCs increasingly include gender equality agenda in corporate social responsibility (CSR) (Grosser, 2009). CSR refers to a company's responsiveness of financial, social and environmental outcome (Hughes & Scheyvens, 2018), which entails the impact assessment and reporting (Grosser, 2009). Many business leaders consider CSR as a priority in terms of its considerable publicity (Porter & Kramer, 2006) and in the implementation of UNSDGs (Schönherr et al., 2017). Therefore, reports from PCs and NGOs are useful to capture how gender equality is enacted at organisational level in tourism.

Phase 1: SQLR of peer-reviewed articles

To identify academic journals, the search terms {("hospitality" OR "tour*") AND ("organisation*" OR "corporate*" OR "company*") AND ("gender" OR "woman" OR "women" OR "female*" OR "divers*")} were utilised in Scopus database. Scopus was selected because of its comprehensive coverage of tourism and hospitality journals (Yang et al., 2017). The search was limited to titles, keywords and abstracts and the results were constrained to peer-reviewed articles and book chapters published in English regardless of years. The initial search of Scopus resulted in 464 peer-reviewed articles. However, a total of 487 articles were reviewed including additional 23 articles identified from Phase 2. Articles were excluded based on the relevance to the subject of investigation, which is gender issues in TOS. For example, sport management journals appeared in the results when "tour*" was referred to "tournament" while theatre and movie journals were captured when "tour*" was referred to "touring" theatre groups or movie stars; these journals were excluded. Articles focused on geographical, cultural, biological and environmental diversity in tourism were also excluded. Some articles mentioned "women" or "female*" only with reference to population ratio or economic activities without further analysis from the gender perspective. Others mentioned "gender" with reference to workforce diversity; however, they focused on other types of diversity such as age or disability instead of gender. As the aim of the review was to investigate gendered employment issues in TOS, gender studies on consumption or tourist behaviour were discarded. "organisation*", "company" or "corporate" were mentioned in these studies but only to highlight the research purpose, background or significance. After removing articles that are not available, 102 articles were identified as eligible for the synthesis. Figure 2 shows the Preferred Reporting Items for Systematic reviews and Meta-Analyses (PRISMA) flowchart adapted from Moher et al. (2009), which illustrates the flow of information through the different stages of SQLR with peer-reviewed TOS.

Phase 2: SQLR of grey literature

The principal feature of systematic review is rule-based searching when collecting materials (Adams et al., 2017). However, there is no specific guideline for systematic review of grey literature across disciplines (Godin et al., 2015; Mahood et al., 2014). It is challenging for the researchers to locate a wide range of data (e.g. industry reports, policy documents, etc.) in databases where the vast amount of information exist (Godin et al., 2015). Therefore, a grey literature search plan has been developed for this paper based on Godin et al. (2015), who aimed to improve the quality of large-scale review on guidelines of school-based breakfast programs in Canada. The grey literature was located through grey literature databases, Google search engine and targeted websites to ensure a robust, transparent and replicable search process (Godin et al., 2015). Figure 3 is the documenting of each search process, which includes the types of

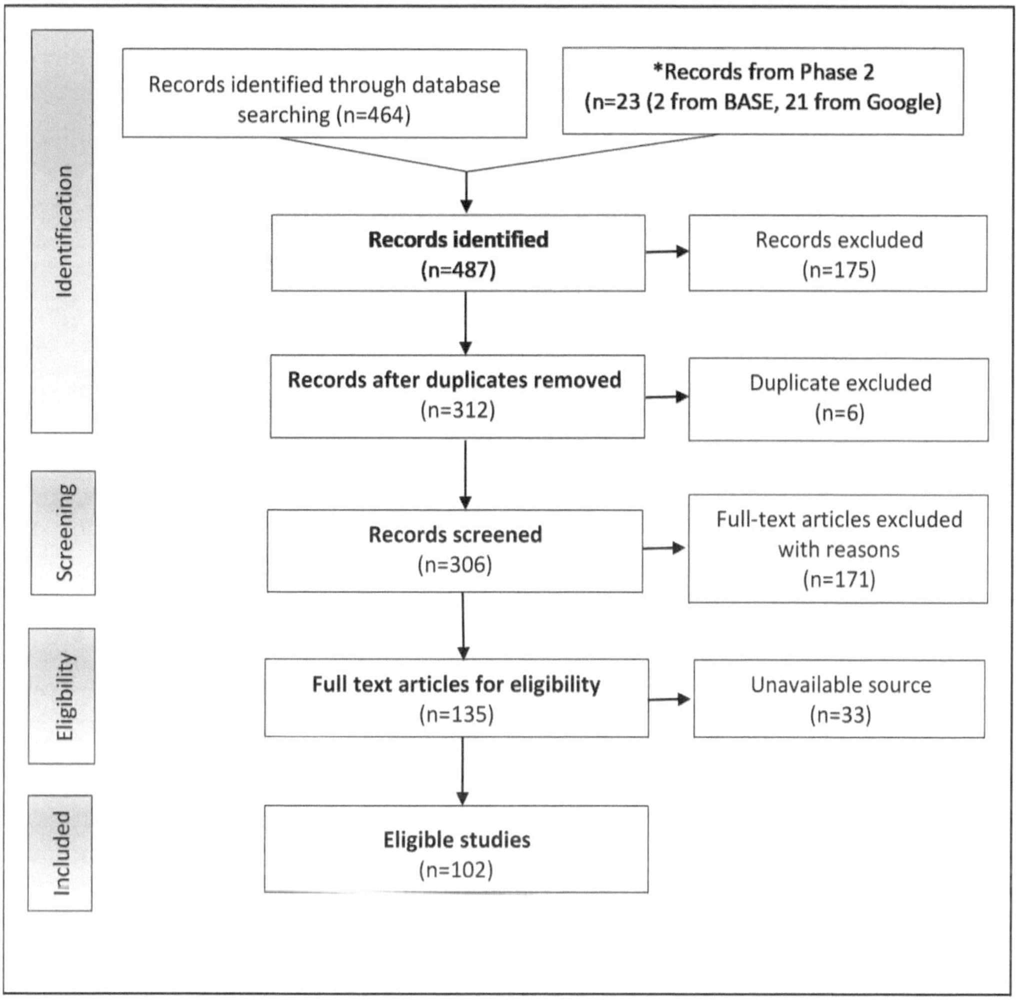

Figure 2. PRISMA flowchart of SQLR with TOS for gender issues.
Note: *Please see the section 2.1.2.1

documents (NGOs or PCs) and the number of inclusion and exclusion with reasons to ensures the compliance with systematic review reporting (Godin et al., 2015).

Databases. The search terms "gender" AND "tourism" were utilised in grey literature databases, including Eldis, Open Grey and Bielefeld Academic Search Engine (BASE) (London South Bank University, 2019). The results from each database were exported into an Excel spreadsheet and the title of all searches was reviewed to remove duplicates. The databases only located NGOs' documents. Eldis generated 12 documents nevertheless, 10 out of 12 documents were about sex trafficking and only two documents were relevant. Open Grey identified seven documents, but none were relevant. The outcome from BASE yielded 63 documents, which was the most extensive result among the grey literature databases. Three out of 63 articles were removed for duplication and 44 articles were also removed as they focused on tourism consumption. Additionally, two documents were moved to Phase 1 as they were working papers to share ideas between scholars but have subsequently been published as peer-reviewed articles. One more document in Turkish was discarded due to the language limitation of the researchers. A total of 15 documents were collected from grey literature databases.

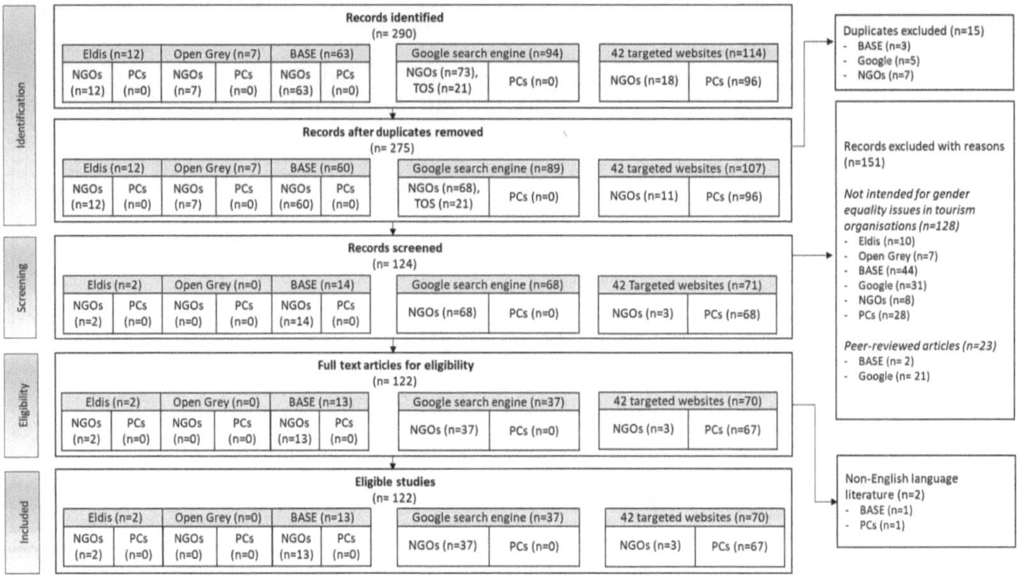

Figure 3. A flowchart of SQLR with grey literature.

Google search engine. The search of grey literature was complemented by Google search engine. The search terms "gender" AND "tourism" AND "report" OR "policy" were used and resulted in 70,800,000 results. Considering the volume of results, the researchers followed a Google search process suggested by Furgal et al. (2010). To capture the most relevant sources, Furgal et al. (2010) recommend including the first 200 hits for grey literature review. After 200, each consecutive source should be considered for relevance until 25 unrelated hits are found consecutively (Furgal et al., 2010). In this study, most of the first 200 hits were news articles, blogs and the government documents for tourism development planning or statistical report of international and domestic travellers. A total of 94 potentially relevant articles were found of which 73 were published by NGOs and 21 were peer-reviewed articles. Five out of 73 documents overlapped with the result from BASE and were removed. Thirty-one out of 68 documents were removed for not relating to the research topic and 21 journal articles were moved to Phase 1 at the screening stage. Google search resulted in 37 NGO publications for synthesis.

Targeted websites. *NGOs in tourism.* The third strategy was identifying and browsing targeted websites of NGOs and PCs to collect relevant documents. A list of 10 tourism-related international NGOs was identified from the first and second strategies (see Appendix A). The NGOs' websites were manually searched with the key terms "tourism" AND "gender" OR "women" OR "female" resulting in 18 potentially relevant documents. Seven out of 18 documents were excluded because they overlapped with the Google search results while another eight out of 11 were discarded for not relating to the review topic. Only three documents were included from targeted NGOs' websites.

Public companies in tourism. Forbes' *Global 2000 List – The World's Largest Public Companies* was used to shortlist the renowned publicly owned companies in tourism. According to UNWTO and ILO (2014, p. 19), the "tourism characteristic industry" is wide-ranging that includes 12 subsectors ranging from railway transport to food and beverage services. Some of the service products such as food and beverage involve substantial sales purchased by non-visitors (UNWTO & ILO, 2014). On the contrary, the demand of airlines and hotels significantly comes from the visitors compared to food and beverage or transport services (Keith & Ioannides, 1998). Therefore, in this study, the websites of 10 hotels and 22 airlines (see Appendix B) listed in Forbes' Global 2000 List were reviewed as it is not possible to obtain data from all 12 subsectors in tourism.

The 32 PCs' websites were manually searched using key terms "gender" or "equality" and "report" to locate documents on gender equality policy or empowerment strategy for women. A total of 96 documents were identified as potentially relevant, including codes of conduct (24), gender pay gap reports (12), annual reports (29), CSR reports (24), sexual harassment prevention and retaliation policy (2), women empowerment initiative reports (2), and diversity and inclusion policy reports (3). A code of conduct is an essential ethics guideline for organisations and also works as an external announcement of corporate value (ECI, 2019) such as gender equality policy for fair employee recruitment and management. Gender pay gap reports show PCs' acknowledgement and commitments to achieve gender equality through the disclosure of wage distribution. Annual reports were browsed since they may disclose information of the gender representation of board members and employees. Each document was screened and 28 out of 96 documents were excluded for not having information or assessment about gender equality: nine codes of conduct, 14 annual reports, five CSR reports. One more annual report in French was discarded due to the language limitation of the researchers. The search in targeted PCs' websites resulted in 67 documents. The grey literature search strategies located a total of 122 documents: 55 NGOs and 67 PCs documents.

Findings

Gender issues in TOS

The first aim of this review was to analyse gender issues in TOS to identify the extent of research and any under-researched area. Table 1 summarises the research topics and issues covered in 102 articles with different gender foci. The three categories of gender foci in Table 1 were adapted from Yang et al.'s (2017) systematic literature review of gender and risk in tourism. The reviewed literature consisted of 54 women-focused, 33 gender-focused and 15 gender-as-variable studies.

Five research themes were identified through the content analysis of TOS with different gender foci. While gender is embedded in all themes, gender discrimination as a theme encompasses studies that bring the issue of gender to the forefront. Women-focused studies dominated the majority of TOS and exclusively emphasised discriminatory experiences in tourism organisations. The studies predominantly investigated job segregation (González-Serrano et al., 2018), implicit gender bias and rule (Carvalho et al., 2019) and gender pay gap (Bakas et al., 2018), which negatively impacted women's career in tourism. Gender-focused studies investigated employee perception and behaviour such as job satisfaction, commitment (Petrović et al., 2014) and CSR performance outcome (Rosa & Bernini, 2018) from a gender perspective rather than focusing on women per se. Gender-as-variable studies used gender as a variable without further analysis from a gender perspective (Yang et al., 2017). These studies mainly focused on employee's job satisfaction (Appiah, 2019; Jung & Yoon, 2015). According to the cursory literature review and research aims, both academic and grey literature were analysed to see to what extent gender issues were investigated: how gendering process was investigated; from where; with what approach; and whether strategic recommendations given by scholars differ from practitioners.

Comparative analyses of TOS, NGOs and PCs

The second aim of the review was to compare TOS with grey literature to ascertain the alignment between academic research and practitioners' foci. The findings are presented in three categories: TOS ($n = 102$), NGOs ($n = 55$) and PCs ($n = 67$). NGO publications were distinguished

Table 1. Research themes and sub-topics by different gender foci.

	Women-focused studies		Gender-focused studies		Gender-as-variable studies	
Research themes and sub-topics	No.	%	No.	%	No.	%
Human resource management	11	8%	4	3%	3	2%
Recruitment	8	6%	1	1%	1	1%
Retention	2	1%				
Strategic human resource management	1	1%				
Turnover			2	1%		
Training and development			1	1%	2	1%
Employee perception and behaviour	16	11%	20	14%	8	6%
Job satisfaction and commitment	8	6%	12	9%	5	4%
Performance outcome	4	3%	4	3%	2	1%
Emotional management	2	1%				
Future employee expectation	1	1%	1	1%		
Career advancement	1	1%				
Organisational citizenship			2	1%	1	1%
Employee well-being			1	1%		
Other managerial practice	8	6%	3	2%	2	1%
Diversity management	4	3%	2	1%	1	1%
Marketing	2	1%			1	1%
Leadership style	1	1%	1	1%		
Total quality management	1	1%				
Gender discrimination	35	25%	0	0%	0	0%
Vertical segregation	8	6%				
Implicit bias and rule	8	6%				
Gender pay gap	5	4%				
Horizontal segregation	5	4%				
Sexual harassment	3	2%				
Discriminative culture	3	2%				
Work–life balance	2	1%				
Gendered identity	1	1%				
Business ethics	10	7%	10	7%	4	3%
Sustainable development	5	4%	4	3%	2	1%
Corporate governance and CSR	5	4%	6	4%	2	1%
Theories	6	4%	0	0%	0	0%
*Total	86	61%	37	26%	17	12%

Note: *The total number exceeds the number of studies since a study may explore multiple concepts.

from those published by PCs. While the former focused on gender-specific issues, the latter generally included gender equality a part of CSR projects. Differentiating NGOs and PCs is important as it allows a deeper understanding of how gender issues were explored and differed from academic research respectively. All three literature types are compared across gendering process, geographical location, methodological approach and strategic recommendations for promoting gender equality.

Gendering processes

Table 3 summarises findings from 22 TOS, 24 NGOs and 29 PC publications where gendering process was explicitly discussed. Both academic and grey literature were analysed according to Acker's (1990) gendering process in organisations, which consists of four dimensions: organising system, organisation culture, interactions on the job and gendered identities. Gender pay gap (Accor Hotels, 2017; Alba Sud, 2018; González-Serrano et al., 2018), power distribution (Bartis, 2018; Oriental Land Co., 2019; People 1st, 2017) and horizontal job segregation (American Airlines Group, 2019; Baum, 2013; Perkov et al., 2016) were commonly researched by both academic and grey literature. Compared to organising system that can be observable through descriptive data, little attention was paid to organisation culture, job interactions or identities, which are somewhat more invisible. To be specific, both types of literature emphasised the importance of supportive culture for women (Carvalho et al., 2019; Davis et al., 2016; UN

Table 2. Gendering processes discussed in TOS, NGOs and PCs.

	TOS		NGOs		PCs	
Gendering process that continually recreates gender inequalities in organisations	No.	%	No.	%	No.	%
Organising system	27	19%	34	24%	26	19%
Distribution of decision-making and supervisory power	8	6%	6	4%	10	7%
Gender pay gap	5	4%	5	4%	12	9%
Division of men and women's job	5	4%	4	3%	4	3%
Demand of availability	4	3%	9	6%		
Explicit and implicit rules at work	3	2%	3	2%		
Part time vs. full time	2	1%	7	5%		
Organisation culture	7	5%	4	3%	0	0%
Unexamined beliefs about gender differences and ideal worker	3	2%	3	2%		
Invisible inequalities	4	3%	1	1%		
Interactions on the job	5	4%	6	4%	1	1%
Interaction between colleagues and those at different levels of power	2	1%	1	1%		
Harassment during interaction with colleagues	3	2%	5	4%	1	1%
Gendered identities	3	2%	0	0%	0	0%
Individual gendered identities constructed and brought into the workplace	3	2%		0%		
*Total	57	41%	54	39%	28	20%

Note: *The total number of references exceeds the total number of documents because one document may discuss more than one gendering process.

Women, 2018), however, less efforts were made in investigating the gendering process in organisational culture. PCs repeatedly highlighted their zero tolerance for any form of harassment and discrimination and yet, only one company disclosed how many legal complaints of sexual harassment were received and managed in the reports (Marriott International, 2018).

Geographical locations

Geographical locations were analysed as gender issues and policy responses vary across countries (Dustin, 2006). 77% (79/102) TOS clearly identified the study contexts while 31% (18/55) NGOs and 39% (26/67) PCs reported the specific contexts. The remaining documents were excluded from this analysis because they were cross-border studies with either multiple or unspecified locations. Overall, the geographical locations covered by TOS were wide-ranging compared to NGOs and PCs. Both academic and grey literature predominantly focused on European contexts. The UK and the USA were the only two countries that were covered by all three. Spain was the most frequently researched by TOS while the UK was frequently studied by NGOs and PCs. Meanwhile, the focus of NGOs was different from that of the private sectors – the former concentrated on under-developing economies while the latter, on developed ones. Even in Asia and Pacific, PCs concentrated on certain countries, namely Japan and Australia, which are economically advanced regions (UN, 2019). PCs hardly examined gender issues in the Middle East and Africa. This could be related to the lack of requirement to report gender issues by the governments. For example, the UK requires large firms with more than 250 employees to report gender pay gaps since 2017 (GOV.UK, 2019). The obligatory report presents hourly wage pay gaps, proportions of women according to quartile pay bans and bonus gaps (GOV.UK, 2019). The Workplace Gender Equality Agency in Australia also requires non-public sector employers with 100 or more to submit a questionnaire concerning gender equality each year.

Methodological approaches

Table 4 summarises the research methods and approaches found in TOS, NGOs and PCs. The identification of methodological approach is important to have preliminary idea of how tourism academics and organisations understand gendered employment issues. While academic research used a wider range of methods in researching gender issues, quantitative approaches remained dominant. TOS often employed surveys to identify employees' behaviour, job satisfaction, CSR

Table 3. Geographical location of literature found in TOS, NGOs and PCs.

	TOS		NGOs		PCs	
Geographical locations	No.	%	No.	%	No.	%
Europe	29	24%	3	2%	17	14%
Spain	7	6%				
Greece	3	2%				
Italy	2	2%				
UK	2	2%	2	2%	17	14%
Croatia	2	2%				
Portugal	2	2%				
Romania	2	2%				
Scotland	2	2%				
Dominican Republic	1	1%				
Finland	1	1%				
Ireland	1	1%				
Lithuania	1	1%				
Morocco	1	1%				
Norway	1	1%				
Serbia	1	1%				
Sweden			1	1%		
Asia	19	15%	5	4%	2	2%
Taiwan	4	3%				
China	3	2%			1	1%
Hong Kong	2	2%				
South Korea	2	2%				
Vietnam	1	1%				
India	1	1%	2	2%		
Iran	1	1%				
Macao	1	1%				
Malaysia	1	1%				
Sri Lanka	1	1%				
Thailand	1	1%				
Nepal	1	1%	1	1%		
Bangladesh			1	1%		
Pakistan			1	1%		
Japan					1	1%
North America	12	10%	1	1%	1	1%
USA	12	10%	1	1%	1	1%
Canada					3	2%
Latin America	1	1%	3	2%	3	2%
Barbados	1	1%				
Mexico			2	2%		
Honduras			1	1%		
Chile					3	2%
Middle East	5	4%	1	1%	0	0%
Egypt	2	2%				
Turkey	2	2%				
Cyprus	1	1%				
Saudi Arabia			1	1%		
Africa	6	5%	5	4%	0	0%
South Africa	3	2%	1	1%		
Zimbabwe	2	2%				
Ethiopia	1	1%				
Tanzania			1	1%		
Uganda			1	1%		
Kenya			1	1%		
Zambia			1	1%		
South Pacific	7	6%	0	0%	3	2%
Australia	5	4%			3	2%
Fiji	1	1%				
New Zealand	1	1%				
Total	79	64%	18	15%	26	21%

Table 4. Research methods found in TOS, NGOs and tourism company reports.

	TOS		NGOs		PCs[a]	
Research methods	No.	%	No.	%	No.	%
Research with primary data	85	42%	28	14%	0	0%
Quantitative methods	46	23%	2	1%	0	0%
Survey	43	21%	1	0%		
Experimental design	2	1%				
Multi-quantitative methods	1	0%	1	0%		
Qualitative methods	21	10%	12	6%	0	0%
Interview	9	4%	4	2%		
Case study	4	2%	4	2%		
Multi-qualitative methods	3	1%	1	0%		
Focus group	1	0%		0%		
Documentary analysis	1	0%	2	1%		
Ethnography	1	0%				
Observation	1	0%				
Vignette	1	0%				
Participatory action research			1	0%		
Mixed methods	18	9%	14	7%	0	0%
Research with non-primary data	17	8%	27	13%	45	22%
Descriptive statistics	8	4%	10	5%	12	6%
Conceptual papers	3	1%	9	4%		
Review	6	3%	7	3%		0%
Annual report			1	0%	14	7%
CSR report					19	9%
Total	102	50%	55	27%	45	22%

Note: [a]45 out of 67 PC documents reported methodological approaches.

performance, and the relationship between diversity climate and organisation's performance. NGOs often used interviews and mixed methods utilising a combination of surveys, impact assessments, descriptive statistics, interviews and document analyses. PCs stated that some opinions in the reports were collected from inside and outside stakeholders through interviews, workshops and surveys. While PCs claimed to be committed to gender equality, they rarely conduct primary research (at least not explicitly reported so) as evidenced by the publication of mainly non-primary data. For instance, the gender pay gap reports were published with data such as employee's wage, board membership or job position, which were already available as corporate governance data. The lack of primary research for PCs is of concern since they can lead to systematic change and improve the quality of reports to convince stakeholders who are concerned about gender issues.

Strategic recommendations for gender equality

Table 5 shows policies or strategies for promoting gender equality found in TOS, NGOs and PCs. The policies or strategies were categorised into structural, material and cultural aspects. The diversity report requires the disclosure of minimal amount of information such as targets or performance data including unaccountable impacts (Adams & Harte, 1999). Therefore, the analyses focus on whether the documents include the minimal amount of information with statistical or narrative accounts.

Overall, most recommendations from academia have been mentioned by NGOs and PCs. Not every academic literature proposed policies or strategies rather, they focused on the analysis of gendering problem or gender differences in organisational behaviour. Both academia and industry focused on structural policies to advance gender equality. The leadership development programme for women was commonly emphasised by TOS, NGOs and PCs to empower women employees. Nonetheless, TOS rarely explained what kind of skills women need to advance their

Table 5. Policies and strategies found in TOS, NGOs and PCs regarding gender equality.

	TOS		NGOs		PCs	
Policies and strategies	No.	%	No.	%	No.	%
Structural	64	12%	125	23%	222	40%
Provide leadership development programmes for women	8	1%	7	1%	19	3%
Promote women to middle management and executive positions	6	1%	7	1%	15	3%
Encourage full-time women workers	5	1%	1	0%	3	1%
Involve employees in formulating diversity initiatives	4	1%			18	3%
Establish partnership with government and NGOs to promote gender equality	4	1%	6	1%	9	2%
Ensure diversity in recruitment	3	1%	1	0%	9	2%
Ensure diversity of suppliers including women-owned business	3	1%			3	1%
Report gender pay gap	2	0%	8	1%	12	2%
Ensure board member diversity	2	0%			10	2%
Provide networking opportunity for women	1	0%	6	1%	7	1%
Ensure employees to be aware of policy against sexual harassment and discrimination	1	0%	5	1%	5	1%
Hire gender and diversity specialist to remit the gap	1	0%	1	0%	5	1%
Flexible working hour	1	0%	4	1%	4	1%
Minimum wage legislation for strong protection of women than equal pay according to law	1	0%	2	0%	1	0%
Address women's work safety issues and policy guideline			4	1%	1	0%
Legalise gender equity			2	0%	15	3%
Abide by the law of maximum working hours			1	0%	1	0%
Abide by the law of no discrimination and equal opportunity					18	3%
Set up conferences or events for women					8	1%
Offer mentorship for women employees					5	1%
Material	11	2%	35	6%	27	5%
Provide funding opportunity for women to establish their own shop and service	6	1%	8	1%	3	1%
Provide flexible parental and maternity leave	2	0%	7	1%	6	1%
Provide equal wages for equal work and experience	2	0%	10	2%	4	1%
Provide family care such as childcare service reimbursement	1	0%	9	2%	13	2%
Provide academic sponsorship for women employees			1	0%	1	0%
Cultural	14	3%	30	5%	23	4%
Provide diversity training	9	2%	8	1%	7	1%
Establish partnership with educational institution	4	1%	9	2%	8	1%
Create supportive organisational culture	1	0%			2	0%
Run campaigns to increase awareness about gender equality			5	1%	4	1%
Stop using gender-stereotyped image in marketing			5	1%		
Appoint equality and diversity advocates including men			3	1%	2	0%
*Total	89	16%	190	34%	272	49%

Note: *The total number of references exceeds the total number of documents because one document may discuss more than one strategy.

careers. PCs sometimes described the types of programme they provided for example, foreign language, marketing, customer service and business management. However, few companies described who and how many received those training and the training results.

Likewise, promoting women to executive positions was highlighted as a strategy for gender equality in tourism. Nonetheless, only 6 out of 32 companies shared their current and the target number of women at executive roles. Among PCs who reported gender pay gaps, very few companies set specific targets for reducing the gap. PCs also showed their efforts in setting up events for women to network in the tourism industry. Yet, they did not describe the gender representation of speakers. The absence of a critical framework to organise events is of concern because it may raise implicit gender bias on the notion of meritocracy of White male experts (Khoo-Lattimore et al., 2019).

Diversity council was highlighted by TOS (Madera, 2013) and frequently promoted by PCs to resolve inequality issues while engaging employees in decision-making process (Marriott International, 2020). Diversity council serves as a resource or taskforce to mandate diversity initiatives and provide supports and guidance where necessary (Sippola & Smale, 2007). Nonetheless, PCs did not explain how the members were selected, and where and how they support equality initiative for the organisation. Flexible working hour was often mentioned as part of family-friendly policy for both women and men. Neither academia nor industry described what "flexibility" is and whether all employees from every department can be supported given the long working-hour culture in the tourism industry (Zopiatis & Constanti, 2006).

PCs utilised material resources to create family-friendly workplaces, but they also did not describe who and how many employees received those benefits. Few companies mentioned how many employees return to work after taking maternity leave, for how long they take, or whether the companies provide the returners care. Diversity training was specifically highlighted by TOS and NGOs to tackle implicit gender bias, however, the receivers and results of the training were not found in PC documents. Establishing partnerships with external organisations such as universities was frequently mentioned to reduce horizontal segregation. Nevertheless, few PCs reported on the number of women who received the benefits from the partnership and/or attained their career development goals.

Discussion and implications

This study was set out to systematically review gender issues in TOS and to compare academic research and industry literature with an aim to investigate if the priorities given by scholars differ from practitioners. The analysis of research themes shows that the majority of TOS are women-focused studies nevertheless, only a handful of studies and reports have explicitly addressed the gendering process while the importance of gender diversity at the board of directors or top management team was often emphasised to generate the desirable organisational outcome. The comparative analyses show both congruencies and incongruencies between academic and grey literature. The overall geographical focus of scholarly research was aligned with the industry but, both evidenced a lack of focus on developing economies. Both types of literature focused on the organising system of gendering process. Yet, other dimensions of gendering process, such as organisational culture, interaction between colleagues and individual identities were paid much less attention. The incongruency was observed in methodological approach to explore gender issues. PCs did not seem to conduct any primary data collection except utilising the available corporate governance data such as wage and the job position of women. Most gender equality strategies proposed by TOS were mentioned in NGOs and PCs. Nevertheless, recommendations made by TOS seem to be few and lack precise advice for tourism organisations to action on. The study also revealed insufficient standards or details about the implementation and reporting of gender equality strategies among tourism organisations in comparison to the claimed commitment.

Lack of investigation of the gendering process

The identification of gendering process enables organisations to achieve gender equality since gender underlies their daily practices (Carvalho et al., 2019). Peer reviewed articles on gender inequalities have not examined the full dimension of gendering process in tourism (Carvalho et al., 2019). Both research and industry focused on organising systems to address gender inequality issues while less attention was paid to organisational culture, interaction with colleagues and individual identities. This is of concern because a gap is likely to be found in the knowledge and perception of people regarding diversity, which leads to an "implementation gap" between the policy and practice (Conley et al., 2007, p. 605). In other words, the organisational change in terms of equality and diversity policy does not necessarily guarantee the supportive working culture that embraces the differences (Conley et al., 2007).

This review suggests more research about organisational culture that is intrinsically linked with the success of gender equality strategy (Moser & Moser, 2005) is needed. Organisational culture is a system of shared symbols and meanings in which organisations are "symbolically constituted and reproduced through interaction" (Hood & Kober, 1994, p. 161). Organisational culture can be an effective tool to examine the effectiveness of policy since it is conceived as the employees' point of view regarding practice (Van den Berg & Wilderom, 2004). Therefore, we need to better understand how management behaviours are interpreted by individual employees, especially from a feminist lens. The informal aspect of organisational culture can be just as important as formal policies for women to buy-in and take-up gender equality strategies offered by their organisations (Holt & Thaulow, 1996). A feminist analysis of organisational culture is a necessary process for advancing theories of power and inequality because organisational culture, as it stands currently, has been the creation and maintenance of patriarchal ideologies (Green & Cassell, 1996). The examination of organisational culture may unveil other invisible, unexplored gendering processes (Green & Cassell, 1996; Moser & Moser, 2005). Qualitative methods such as in-depth group or individual interviews can be useful for researchers to discover the unique characteristics of the culture and devise new direction in organisations (Van den Berg & Wilderom, 2004). In terms of PCs, the investigation of inequalities is especially crucial to mainstream gender equality (EIGE, 2019). The deeper understanding of gendering process would help organisations make inequalities into visible power relations, which open up a space for systematic change (Smircich et al., 2014). The primary research will improve the quality of company reporting (Thompson & Zakaria, 2004) and support their existing data such as gender pay gap capturing women's perceptions and experiences in current organisations (Carvalho et al., 2019).

Little attention in developing economies

In terms of the geographical location of investigation, both academic and grey literature focused on developed economies such as the UK and the USA. Europe is considered a ground-breaker in terms of gender equality (UNWTO & UN Women, 2019). TOS and PCs indicated that Europe appear to understand the importance of integrating gender equality into the tourism and hospitality industry (UNWTO & UN Women, 2019). A recent report suggests that policy-making actions for gender equality have focused on developing countries such as Nicaragua through international cooperation with the European Commission (UNWTO & UN Women, 2019). Nevertheless, this review emphasises that much more research and work are needed in emerging economies such as Latin America and Africa, where women often have limited access to social security, education and quality tourism training compared to other regions (UNWTO & UN Women, 2019). In Asia and the Pacific regions, women often work in a poor environment and they may suffer harassment and violence in the tourism industry (UNWTO & UN Women, 2019). Pressure is growing to acknowledge these issues and take action where there is most needed.

A research partnership can increase the focus of academia and industry in developing economies. A partnership can produce high-quality research at lower cost, with the great influence on the policy and execution (Costello & Zumla 2000). Although PCs denoted a partnership with other stakeholders such as universities, the partnership was mainly articulated through the provision of scholarship or leadership programmes for women with limited collaborative research that audit and improve gender equality performance. This is an underutilised area given that universities and public research centres play key roles in creating knowledge for organisational innovation (Sobaih & Jones, 2015). Similarly, NGOs seek deeper changes in the social and economic relations providing critical awareness and knowledge (Stromquist, 2006). Especially for developing countries where more attention is needed by PCs, the collaborative research with universities and NGOs can be developed into sponsored research ventures which can provide job opportunities for tourism graduates including young women while expanding companies' knowledge as well (Sobaih & Jones, 2015).

Absence of concrete compliance for gender equality

PCs showed their attempts to advance gender equality through various types of strategies. However, they have not yet "mainstreamed" gender equality. Gender mainstreaming is a transforming process that moves beyond individual rights and has positive institutional actions to address the issues of disadvantaged group (Grosser & Moon, 2005). It is more than a simple adoption of new policies (Walby, 2005); it is the systematic integration of gender equality into processes and policies (Rees, 2005). Adams and Harte (1999) suggested that successful reporting of diversity and equality practice should include information, such as the achievement of policies, targets and monitoring systems. Although PCs identify gender equality as one of their corporate goals, most of their reports are missing such information in narrative and statistical accounts. Their lack of accounts regarding gender equality can be detected when they only disclosed gender pay gap in regions where there is political or legal pressure such as in the UK. The substandard reports may indicate some extent of femwashing practice (Khoo-Lattimore et al., 2019).

The notion of femwashing could be rooted from the lack of precise advice from scholarly research and the absence of coercive regulation to report gender issues by governments or leading inter-governmental organisation (Grosser & Moon, 2008). Although TOS encompassed structural, material and cultural aspects of gender equality and inclusion strategies, they lack detailed information of how to conduct those policies. The comprehensive regulation to effectively implement and measure gender equality goal is missing; and existing analysis tool or reporting standard are not specifically designed for the tourism industry. UN provides organisations with the Women's Empowerment Principles Gender Gap Analysis Tool (WEPs Tool). It is designed to help companies to assess gender equality performance, identify strengths and gaps and guide future actions with concrete goals and targets (UN Global Compact, 2019). Still, the tool is a voluntary and learning platform and does not serve as a reporting mechanism or certification for organisations to promote gender equality goal (UN Global Compact, 2019). The Global Reporting Initiative (GRI) is one of the well-known sustainability reporting frameworks that companies use to publish CSR reports (Daizy, 2014). The set of GRI standards only require the organisation to include the quantitative information of diversity categories such as gender and age group of employees. If power or distributional issues are at the centre of institutional analysis, the compliance would become an important variable (Mahoney & Thelen, 2009). Institutional change can occur through the creation of rules or the extension of existing rules (Waylen, 2014). A critical and concrete reporting requirement is necessary to measure gender issues because "what gets measured gets managed" (Grosser & Moon, 2008, p. 181).

Conclusion

Past studies explored women's issues and provided various strategic recommendations to empower women in tourism organisations; yet, little is known to what extent academic research was translated into actual practice. This is the first study to comprehensively map gender issues in TOS and compare academic and industry literature to identify gaps between theory and practice and actionable approach to create hopeful tourism workplaces. This pragmatic SQLR has its value in developing the holistic understanding of gendered employment issues through multiple tourism knowledge and offering the attainable approach such as research partnership to achieve sustainable workplace in tourism. The pragmatic SQLR, which compared academic and grey literature sought the relevance of tourism research to practice gender equality policies and strategies in real-life situations. The rigour of findings is achieved through the cross-checking by authors and the audit trail of academic and grey literature analysis through PRISMA flowcharts. Through the comparison of 102 papers from academia and 122 documents from the industry, the review revealed a gap in research, and between the rhetoric and practice to promote gender equality. The analysis of TOS highlighted the lack of research on gendering process, attention in developing economies as well as the precise recommendation for organisations to implement, measure and report gender equality practices. Although the strategic recommendations from academia were mostly covered by grey literature, the study acknowledges that the simple provision of formal policies was not enough to challenge the status quo. Considering the deep-rooted gendering character in organisations (Acker, 1990), more research about organisational culture is necessary to render the power relations of organisation visible. Since the implementation of gender equality strategies was not always clearly articulated in PC publications, more research on effective practice and the development of measurement is necessary to remove the notion of femwashing. A research partnership between the academia and private sectors is proposed to better understand the gender issues in tourism and to develop meaningful indicator with narrative and statistical accounts. This collaboration will facilitate the mainstreaming of gender equality in the tourism industry.

The current review assumed the structure of gender as a binary relation regardless of other identities, personalities or sexual preferences (Budgeon, 2014). As gender issues do not stand alone but often intersect with other forms of inequality such as racism, future research and practice are encouraged to employ an intersectional lens to facilitate a more thorough and robust exploration of diversity issues (Mooney, 2018). Furthermore, the review excluded government documents as the aim was to examine the extent of equality practice at organisational level. Accordingly, the investigation of gendering process was highlighted in organisations, which excludes the macro-level, structural causes of gender inequality. Future research can include the understanding of how individuals of organisations are affected by and in turn affect the structure of society (Dunn et al., 1993). The retrieval of tourism literature could be limited since keywords such as hospitality or women; and organisation as spelt in the USA were not specifically included in the search. Although Google search is disputed because the relevance of data is disseminated by a complex system of algorithms and software rather than human gatekeepers, the authors have tried to capture the most relevant data, including the first 200 hits and considering the relevance of each source until 25 unrelated hits are found, as suggested by Furgal et al. (2010). What is missing from the data is an analysis of other factors that may contribute to gender issues in respective countries regardless of their economic advancements. This micro-level analysis could be important for country and state-level implementation. The study examined large PCs to gain an insight into how gender equality policies were enacted because they are influential in setting national/international standards, policies, and regulations, and impact small business (Martin, 2003). Future research is encouraged to include small to medium size companies where a high percentage of women pursuing career opportunities as entrepreneurs in tourism (UNWTO & UN Women, 2019). This review is exclusively based on documents published in English, which calls for a global knowledge platform where researchers can access and exchange relevant documents for gender equality regardless of language limitation. Meanwhile, future research is encouraged to review related documents in other languages to achieve sustainable workplaces that promote gender equality in overall economies. Despite these limitations, this is the first systematic comparative review to gain an insight into the translation of gender equality research in tourism organisations. This review highlights that gaps between the gender equality rhetoric and practice and sheds light on the underlying structural and cultural problems. Future directions for research and practice were provided to achieve an equitable, inclusive, sustainable and hopeful tourism workplace.

Disclosure statement

No potential conflict of interest was reported by the authors.

ORCID

Catheryn Khoo http://orcid.org/0000-0003-2858-870X
Elaine Chiao Ling Yang http://orcid.org/0000-0001-5245-1688

References

Accor Hotels. (2017). 2017 Gender pay gap report. http://www.accorhotels.group/-/media/Corporate/Legal-terms

Acker, J. (1990). Hierarchies, jobs, bodies: A theory of gendered organizations. *Gender & Society*, *4*(2), 139–158. https://doi.org/10.1177/089124390004002002

Adams, C. A., & Harte, G. (1999). Toward corporate accountability for equal opportunities performance (ACCA Occasional Research Paper No. 26). Association of Chartered Certified Accountants (ACCA).

Adams, R. J., Smart, P., & Huff, A. S. (2017). Shades of grey: Guidelines for working with the grey literature in systematic reviews for management and organizational studies. *International Journal of Management Reviews*, *19*(4), 432–454. https://doi.org/10.1111/ijmr.12102

Alba Sud. (2018). Gender dimensions in tourism work. http://www.albasud.org

American Airlines Group. (2019). *2018 Gender pay gap report*. www.americanairlines.co.uk/customer-service/about-us

Appiah, J. K. (2019). Community-based corporate social responsibility activities and employee job satisfaction in the US hotel industry: An explanatory study. *Journal of Hospitality and Tourism Management*, *38*, 140–148. https://doi.org/10.1016/j.jhtm.2019.01.002

Bakas, F. E., Costa, C., Breda, Z., & Durão, M. (2018). A critical approach to the gender wage gap in tourism labor. *Tourism Culture & Communication*, *18*(1), 35–49. https://doi.org/10.3727/109830418X15180180585167

Bartis, H. H. (2018). A socio-demographic analysis of the board of directors of selected destination marketing organisation (DMOs). *African Journal of Hospitality, Tourism and Leisure*, *7*(2), 3.

Baum, T. (2006). *Human resource management for tourism, hospitality and leisure: An international perspective* (1st ed.). Thomson Learning.

Baum, T. (2013). International perspectives on women and work in hotels, catering and tourism. https://digitalcommons.ilr.cornell.edu/cgi/viewcontent.cgi?article=1264&context=intl

Benzies, K. M., Premji, S., Hayden, K. A., & Serrett, K. (2006). State-of-the-evidence reviews: Advantages and challenges of including grey literature. *Worldviews on Evidence-Based Nursing*, *3*(2), 55–61. https://doi.org/10.1111/j.1741-6787.2006.00051.x

Budgeon, S. (2014). The dynamics of gender hegemony: Femininities, masculinities and social change. Sociology, *48*(2), 317–334. https://doi.org/10.1177/0038038513490358

Burritt, R. L., & Schaltegger, S. (2010). Sustainability accounting and reporting: Fad or trend? *Accounting, Auditing & Accountability Journal*, *23*(7), 829–846. https://doi.org/10.1108/09513571011080144

Calás, M. B., & Smircich, L. (2006). From the 'woman's point of view' ten years later: Towards a feminist organization studies. In S. Clegg, C. Hardy, & T. Lawrence (Eds.), *The Sage handbook of organization studies* (pp. 284–346). Sage.

Carvalho, I., Costa, C., Lykke, N., & Torres, A. (2019). Beyond the glass ceiling: Gendering tourism management. *Annals of Tourism Research*, *75*, 79–91. https://doi.org/10.1016/j.annals.2018.12.022

Chambers, D., Munar, A. M., Khoo-Lattimore, C., & Biran, A. (2017). Interrogating gender and the tourism academy through epistemological lens. Anatolia, *28*(4), 501–513. https://doi.org/10.1080/13032917.2017.1370775

Choi, H. J., & Kim, Y. T. (2012). Work-family conflict, work-family facilitation, and job outcomes in the Korean hotel industry. *International Journal of Contemporary Hospitality Management*, *24*(7), 1011–1028. https://doi.org/10.1108/09596111211258892

Conley, H., Colgan, F., Creegan, C., McKearney, A., & Wright, T. (2007). Equality and diversity policies and practices at work: Lesbian, gay and bisexual workers. *Equal Opportunities International*, *26*(6), 590–609. https://doi.org/10.1108/02610150710777060

Conn, V. S., Valentine, J. C., Cooper, H. M., & Rantz, M. J. (2003). Grey literature in meta-analyses. *Nursing Research*, *52*(4), 256–261. https://doi.org/10.1097/00006199-200307000-00008

Costello, A., & Zumla, A. (2000). Moving to research partnerships in developing countries. *BMJ*, *321*(7264), 827–829. https://doi.org/10.1136/bmj.321.7264.827

Creese, G. (1996). Gendering collective bargaining: From men's rights to women's issues. *Canadian Review of Sociology/Revue canadienne de sociologie*, *33*(4), 437–456. https://doi.org/10.1111/j.1755-618X.1996.tb00956.x

Daizy, N. D. (2014). Sustainability reporting framework: Comparative analysis of global reporting initiatives and dow jones sustainability index. *International Journal of Science, Environment and Technology*, *3*(1), 55–66.

Daub, C.-H. (2007). Assessing the quality of sustainability reporting: An alternative methodological approach. *Journal of Cleaner Production*, *15*(1), 75–85. https://doi.org/10.1016/j.jclepro.2005.08.013

Davis, P. J., Frolova, Y., & Callahan, W. (2016). Workplace diversity management in Australia: What do managers think and what are organisations doing? *Equality, Diversity and Inclusion: An International Journal*, *35*(2), 81–98. https://doi.org/10.1108/EDI-03-2015-0020

DeJaeghere, J., & Wiger, N. P. (2013). Gender discourses in an NGO education project: Openings for transformation toward gender equality in Bangladesh. *International Journal of Educational Development*, *33*(6), 557–565. https://doi.org/10.1016/j.ijedudev.2013.02.002

Doherty, L. (2004). Work-life balance initiatives: Implications for women. *Employee Relations*, *26*(4), 433–452. https://doi.org/10.1108/01425450410544524

Duffy, L. N., Kline, C. S., Mowatt, R. A., & Chancellor, H. C. (2015). Women in tourism: Shifting gender ideology in the DR. *Annals of Tourism Research*, *52*, 72–86. https://doi.org/10.1016/j.annals.2015.02.017

Dunn, D., Almquist, E. M., & Chafetz, J. S. (1993). Macrostructural perspectives on gender inequality. In P. England (Ed.), *Theory on gender/feminism on theory* (1st ed.). Hawthorne.

Dustin, M. (2006). Gender equality, cultural diversity: European comparisons and lessons. http://sro.sussex.ac.uk/id/eprint/64051

ECI. (2019). *Code construction and content*. https://www.ethics.org/resources/free-toolkit/code-of-conduct/

EIGE. (2019). *Institutional transformation*. https://eige.europa.eu/gender-mainstreaming/toolkits/gender-institutional-transformation

Figueroa-Domecq, C., Pritchard, A., Segovia-Pérez, M., Morgan, N., & Villacé-Molinero, T. (2015). Tourism gender research: A critical accounting. *Annals of Tourism Research*, *52*, 87–103. https://doi.org/10.1016/j.annals.2015.02.001

Furgal, C. M., Garvin, T. D., & Jardine, C. G. (2010). Trends in the study of Aboriginal health risks in Canada. *International Journal of Circumpolar Health*, *69*(4), 322–332. https://doi.org/10.3402/ijch.v69i4.17672

Godin, K., Stapleton, J., Kirkpatrick, S. I., Hanning, R. M., & Leatherdale, S. T. (2015). Applying systematic review search methods to the grey literature: A case study examining guidelines for school-based breakfast programs in Canada. *Systematic Reviews*, *4*(1), 138. https://doi.org/10.1186/s13643-015-0125-0

González-Serrano, L., Villacé-Molinero, T., Talón-Ballestero, P., & Fuente-Cabrero, C. D. L. (2018). Women and the glass ceiling in the community of Madrid hotel industry. *International Journal of Human Resources Development and Management*, *18*(1/2), 91–111. https://doi.org/10.1504/ijhrdm.2018.10013649 https://doi.org/10.1504/IJHRDM.2018.092289

GOV.UK. (2019). *Gender pay gap service*. https://gender-pay-gap.service.gov.uk/Employer/WDQu77k5/2017

Green, E., & Cassell, C. (1996). Women managers, gendered cultural processes and organizational change. *Gender, Work & Organization*, *3*(3), 168–178. https://doi.org/10.1111/j.1468-0432.1996.tb00057.x

Grosser, K. (2009). Corporate social responsibility and gender equality: Women as stakeholders and the European Union sustainability strategy. *Business Ethics: A European Review*, *18*(3), 290–307. https://doi.org/10.1111/j.1467-8608.2009.01564.x

Grosser, K., & Moon, J. (2005). Gender mainstreaming and corporate social responsibility: Reporting workplace issues. *Journal of Business Ethics*, *62*(4), 327–340. https://doi.org/10.1007/s10551-005-5334-3

Grosser, K., & Moon, J. (2008). Developments in company reporting on workplace gender equality? *Accounting Forum*, *32*(3), 179–198. https://doi.org/10.1016/j.accfor.2008.01.004

Holt, H., & Thaulow, I. (1996). Formal and informal flexibility in the workplace. In S. Lewis & J. Lewis (Eds.), *The work-family challenge: Rethinking employment* (1st ed., pp. 79–92). Sage.

Hood, J. N., & Kober, C. S. (1994). Patterns of differential assimilation and acculturation for women in business organisations. *Human Relations*, *47*(2), 159–181. https://doi.org/10.1177/001872679404700202

Hughes, E., & Scheyvens, R. (2018). Development alternatives in the Pacific: How tourism corporates can work more effectively with local communities. *Tourism Planning & Development*, *15*(5), 516–534. https://doi.org/10.1080/21568316.2018.1478881

Hutchings, K., Moyle, C.-L., Chai, A., Garofano, N., & Moore, S. (2020). Segregation of women in tourism employment in the APEC region. *Tourism Management Perspectives*, *34*, 100655. https://doi.org/10.1016/j.tmp.2020.100655

Jackson, J. C. (2001). Women middle managers' perception of the glass ceiling. Women in Management Review, *16*(1), 30–41.

Jung, H. S., & Yoon, H. H. (2015). The impact of employees' positive psychological capital on job satisfaction and organizational citizenship behaviors in the hotel. *International Journal of Contemporary Hospitality Management*, *27*(6), 1135–1156. https://doi.org/10.1108/IJCHM.2014.0019

Keiser, J. D. (1998). Hospitality and tourism: A rhetorical analysis and conceptual framework for identifying industry meanings. *Journal of Hospitality & Tourism Research*, *22*(2), 115–128. https://doi.org/10.1177/109634809802200201

Keith, G. D., & Ioannides, D. (1998). *The economic geography of the tourist industry: A supply-side analysis* (1st ed.). Routledge.

Khoo-Lattimore, C., Yang, E. C. L., & Je, J. S. (2019). Assessing gender representation in knowledge production: A critical analysis of UNWTO's planned events. *Journal of Sustainable Tourism*, *27*(7), 920–938. https://doi.org/10.1080/09669582.2019.1566347

Kingsmill, D. (2001). *A review of women's employment and pay*. University College London. http://dera.ioe.ac.uk/id/eprint/10105

London South Bank University. (2019). *Tourism and hospitality management: Theses, dissertations and grey literature*. https://libguides.lsbu.ac.uk/tourism/theses

Luxton, M. (2014). Marxist feminism and anticapitalism: Reclaiming our history, reanimating our politics. *Studies in Political Economy*, *94*(1), 137–160. https://doi.org/10.1080/19187033.2014.11674957

Madera, J. M. (2013). Best practices in diversity management in customer service organizations: An investigation of top companies cited by Diversity Inc. Cornell *Hospitality Quarterly*, *54*(2), 124–135. https://doi.org/10.1177/1938965513475526

Madera, J. M. (2018). What's in it for me? Perspective taking as an intervention for improving attitudes toward diversity management. *Cornell Hospitality Quarterly*, *59*(2), 100–111. https://doi.org/10.1177/1938965517730319

Mahoney, J., & Thelen, K. (2009). *Explaining institutional change: Ambiguity, agency, and power* (1st ed.). Cambridge University Press.

Mahood, Q., Van Eerd, D., & Irvin, E. (2014). Searching for grey literature for systematic reviews: Challenges and benefits. *Research Synthesis Methods*, *5*(3), 221–234. https://doi.org/10.1002/jrsm.1106

Marriott International. (2018). *Harassment prevention*. www.marriott.com/marriott/aboutmarriott

Marriott International. (2020). *Empower through opportunity*. https://serve360.marriott.com/empower/

Martin, P. Y. (2003). "Said and done" versus "saying and doing" gendering practices, practicing gender at work. *Gender & Society*, *17*(3), 342–366. https://doi.org/10.1177/0891243203017003002

Milne, M. J., Tregidga, H., & Walton, S. (2009). Words not actions! The ideological role of sustainable development reporting. *Accounting, Auditing & Accountability Journal*, *22*(8), 1211–1257. https://doi.org/10.1108/09513570910999292

Mkono, M. (2010). An analysis of Zimbabwean hotel managers' perspectives on workforce diversity. *Tourism and Hospitality Research*, *10*(4), 301–310. https://doi.org/10.1057/thr.2010.18

Moher, D., Liberati, A., Tetzlaff, J., & Altman, D. G., & PRISMA Group. (2009). Preferred reporting items for systematic reviews and meta-analyses: The PRISMA statement. *Annals of Internal Medicine*, *151*(4), 264–269. https://doi.org/10.7326/0003-4819-151-4-200908180-00135

Mooney, S. (2018). Illuminating intersectionality for tourism researchers. *Annals of Tourism Research*, *72*(C), 175–176. https://doi.org/10.1016/j.annals.2018.03.003

Morgan, G., & Knights, D. (1991). Gendering jobs: Corporate strategy, managerial control and the dynamics of job segregation. *Work, Employment and Society*, *5*(2), 181–200. https://doi.org/10.1177/0950017091005002003

Morgan, N., & Pritchard, A. (2019). Gender Matters in Hospitality (invited paper for 'luminaries' special issue of International Journal of Hospitality Management). *International Journal of Hospitality Management*, *76*, 38–44. https://doi.org/10.1016/j.ijhm.2018.06.008

Moser, C., & Moser, A. (2005). Gender mainstreaming since Beijing: A review of success and limitations in international institutions. *Gender & Development*, *13*(2), 11–22. https://doi.org/10.1080/13552070512331332283

Oriental Land Co. (2019). 2018 *Annual report*. www.olc.co.jp/library/annual

Ottenbacher, M., Harrington, R., & Parsa, H. (2009). Defining the hospitality discipline: A discussion of pedagogical and research implications. *Journal of Hospitality & Tourism Research*, *33*(3), 263–283.

People 1st. (2017). Women working in the hospitality & tourism sector. www.people1st.co.uk/insight-opinion

Perkov, D., Primorac, D., & Perkov, M. (2016). Position of female managers in Croatian tourism. *International Journal of Economic Perspectives, 10*(1), 62–70.

Petrović, M. D., Jovanović, T., Marković, J. J., Armenski, T., & Marković, V. (2014). Why should gender differences in hospitality really matter? A study of personnel's service orientation and job satisfaction in hotels. *Economic Research-Ekonomska Istraživanja, 27*(1), 799–817. https://doi.org/10.1080/1331677X.2014.975516

Petticrew, M., & Roberts, H. (2008). *Systematic reviews in the social science: A practical guide.* (1st ed., Vol. 43). Choice Reviews Online.

Pickering, C., & Byrne, J. (2014). The benefits of publishing systematic quantitative literature reviews for PhD candidates and other early-career researchers. *Higher Education Research & Development, 33*(3), 534–548. https://doi.org/10.1080/07294360.2013.841651

Porter, M. E., & Kramer, M. R. (2006). The link between competitive advantage and corporate social responsibility. *Harvard Business Review, 84*(12), 78–92.

Pritchard, A. (2014). Gender and feminist perspectives in tourism research. In A. A. Lew, C. M. Hall, & A. M. Williams (Eds.), *The Wiley Blackwell companion to tourism* (1st ed., pp. 314–324). John Wiley & Sons, Ltd.

Pritchard, A., & Morgan, N. (2013). Hopeful tourism: A transformational perspective. In Y. Reisinger (Ed.), *Transformational tourism: Tourist perspectives* (1st ed., pp. 3–14). CAB International.

PWC. (2017). *Women in hospitality, travel and leisure 2020.* https://www.voced.edu.au/content/ngv%3A80645

Quintana-García, C., Marchante-Lara, M., & Benavides-Chicón, C. G. (2018). Social responsibility and total quality in the hospitality industry: Does gender matter? *Journal of Sustainable Tourism, 26*(5), 722–739. https://doi.org/10.1080/09669582.2017.1401631

Rees, T. (2005). Reflections on the uneven development of gender mainstreaming in Europe. *International Feminist Journal of Politics, 7*(4), 555–574. https://doi.org/10.1080/14616740500284532

Remington, J., & Kitterlin-Lynch, M. (2018). Still pounding on the glass ceiling: A study of female leaders in hospitality, travel, and tourism management. *Journal of Human Resources in Hospitality & Tourism, 17*(1), 22–37. https://doi.org/10.1080/15332845.2017.1328259

Risman, B. J. (2009). From doing to undoing: Gender as we know it. *Gender & Society, 23*(1), 81–84. https://doi.org/10.1177/0891243208326874

Rosa, F. L., & Bernini, F. (2018). Corporate governance and performance of Italian gambling SMEs during recession. *International Journal of Contemporary Hospitality Management, 30*(3), 1939–1958. https://doi.org/10.1108/IJCHM-03-2017-0135

Rumens, N., & Kelemen, M. (2010). American pragmatism and feminism: Fresh opportunities for sociological inquiry. *Contemporary Pragmatism, 7*(1), 129–148. https://doi.org/10.1163/18758185-90000159

Santero-Sanchez, R., Segovia-Pérez, M., Castro-Nuñez, B., Figueroa-Domecq, C., & Talón-Ballestero, P. (2015). Gender differences in the hospitality industry: A job quality index. *Tourism Management, 51*, 234–246. https://doi.org/10.1016/j.tourman.2015.05.025

Schönherr, N., Findler, F., & Martinuzzi, A. (2017). Exploring the interface of CSR and the sustainable development goals. *Transnational Corporations, 24*(3), 33–47. https://doi.org/10.18356/cfb5b8b6-en

Sippola, A., & Smale, A. (2007). The global integration of diversity management: A longitudinal case study. *The International Journal of Human Resource Management, 18*(11), 1895–1916. https://doi.org/10.1080/09585190701638101

Smircich, L., Holvino, E., & Calás, M. B. (2014). *From theorising gender in organisations to theorising gendering organisations: Produced, producing and reproducing* (1st ed.). Oxford University Press.

Sobaih, A. E., & Jones, E. (2015). Bridging the hospitality and tourism university–industry research gap in developing countries: The case of Egypt. *Tourism and Hospitality Research, 15*(3), 161–177. https://doi.org/10.1177/1467358415578188

Stromquist, N. P. (2006). Gender, education and the possibility of transformative knowledge. Compare: A Journal of Comparative and International Education, *36*(2), 145–161. https://doi.org/10.1080/03057920600741131

Susman, G. I. (1983). Action research: A sociotechnical systems perspective. In G. Morgan (Ed.), *Beyond method: Strategies for social research* (pp. 95–113). Sage.

Thompson, P., & Zakaria, Z. (2004). Corporate social responsibility reporting in Malaysia: Progress and prospects. *Journal of Corporate Citizenship, 2004*(13), 125–136. https://doi.org/10.9774/GLEAF.4700.2004.sp.000014

UN. (2019). World economic situation prospects. https://www.un.org/development/desa/dpad/wp-content/uploads/sites/45/WESP2019_BOOK-ANNEX-en.pdf

UN Global Compact. (2019). *Empower women in the workplace, marketplace and community.* https://www.unglobalcompact.org/what-is-gc/our-work/social/gender-equality

UN Women. (2018). Emerging solutions for gender equality. https://www.heforshe.org/sites/default/files/2018-10/HeForShe%20Emerging%20Solutions%20Report%202018%20-%20Full%20Report.pdf

UNWTO. (2019). *Tourism in the 2030 agenda.* https://www.unwto.org/tourism-in-2030-agenda

UNWTO & ILO. (2014). Measuring employment in the tourism *industries - Guide* with best practices. https://www.ilo.org/global/industries-and-sectors/hotels-catering-tourism/WCMS_329309/lang–en/index.htm

UNWTO & UN Women. (2011). *Global report on women in tourism 2010.* https://www.e-unwto.org/doi/pdf/10.18111/9789284420384

UNWTO & UN Women. (2019). Global report on women in tourism - Second edition. https://www.e-unwto.org/doi/pdf/10.18111/9789284420384

Utting, P. (2007). CSR and equality. *Third World Quarterly, 28*(4), 697–712. https://doi.org/10.1080/01436590701336572

Vaismoradi, M., Turunen, H., & Bondas, T. (2013). Content analysis and thematic analysis: Implications for conducting a qualitative descriptive study. *Nursing & Health Sciences, 15*(3), 398–405. https://doi.org/10.1111/nhs.12048

Van den Berg, P. T., & Wilderom, C. P. (2004). Defining, measuring, and comparing organisational cultures. *Applied Psychology, 53*(4), 570–582. https://doi.org/10.1111/j.1464-0597.2004.00189.x

Walby, S. (2005). Gender mainstreaming: Productive tensions in theory and practice. *Social Politics: International Studies in Gender, State and Society, 12*(3), 321–343. https://doi.org/10.1093/sp/jxi018

Wallace, T. (1998). Institutionalizing gender in UK NGOs. *Development in Practice, 8*(2), 159–172. https://doi.org/10.1080/09614529853783

Waylen, G. (2014). Informal institutions, institutional change, and gender equality. *Political Research Quarterly, 67*(1), 212–223. https://doi.org/10.1177/1065912913510360

Wearing, B. (1998). *Leisure and feminist theory.* Sage.

World Economic Forum. (2020). *Global gender gap report.* http://www3.weforum.org/docs/WEF_GGGR_2020.pdf

Yang, E. C. L., Khoo-Lattimore, C., & Arcodia, C. (2017). A systematic literature review of risk and gender research in tourism. *Tourism Management, 58*, 89–100. https://doi.org/10.1016/j.tourman.2016.10.011

Zopiatis, A., & Constanti, P. (2006). Mission impossible? Motivating hospitality managers in Cyprus. *Tourismos: An International Multidisciplinary Journal of Tourism, 2*(1), 31–46.

Appendix A. The list of international NGOs in tourism.

1.	World Tourism Organisation (UNWTO)
2.	International Tourism Organisation (ITO)
3.	World Travel & Tourism Council (WTTC)
4.	Global Sustainable Tourism Council (GSTC)
5.	World Trade Organisation (WTO)
6.	Caribbean Tourism Organisation (CTO)
7.	World Bank Group
8.	UN Women
9.	Asia-Pacific Economic Cooperation (APEC)
10.	International Labour Organisation (ILO)

Appendix B. The list of world's largest public hotel and airline companies by Forbes (2019).

Name of company	Industry	Country	#Rank
Carnival	Hotels & Motels	United States	#312
Marriott International	Hotels & Motels	United States	#411
Royal Caribbean Cruises	Hotels & Motels	United States	#573
Shenzhen Overseas Chinese Town Co., Ltd	Hotels & Motels	China	#808
Hilton Worldwide	Hotels & Motels	United States	#857
Accor Hotels	Hotels & Motels	France	#1041
Norwegian Cruise Line Holdings	Hotels & Motels	United States	#1075
Oriental Land Co., Ltd	Hotels & Motels	Japan	#1174
Great Eagle Holdings	Hotels & Motels	Hong Kong	#1707
Hyatt Hotels	Hotels & Motels	United States	#1787
Delta Air Lines	Airlines	United States	#203
United Continental Holdings	Airlines	United States	#306
American Airlines Group	Airlines	United States	#373
Southwest Airlines	Airlines	United States	#398
Deutsche Lufthansa	Airlines	Germany	#413
International Airlines	Airlines	United Kingdom	#428
All Nippon Airways	Airlines	Japan	#702
China Southern Airlines	Airlines	China	#713
China Eastern Airlines	Airlines	China	#768
Japan Airlines	Airlines	Japan	#776
Ryanair Holdings	Airlines	Ireland	#841
Air France-KLM	Airlines	France	#950
Turkish Airlines	Airlines	Turkey	#1034
Singapore Airlines	Airlines	Singapore	#1078
Qantas Airways	Airlines	Australia	#1282
Cathay Pacific Airways	Airlines	Hong Kong	#1285
Hainan Airlines	Airlines	China	#1393
Korean Air	Airlines	South Korea	#1446
Air Canada	Airlines	Canada	#1578
Latam Airlines	Airlines	Chile	#1622
Alaska Air Group	Airlines	United States	#1672
EasyJet	Airlines	United Kingdom	#1912

Adapted from "The World's Largest Public Companies" by Forbes, 2019. https://www.forbes.com/global2000/list/#industry:Apparel_2FFootwear%20Retail.

Part III

Challenging Gender Norms

Part III

Challenges to Standard Theories

The sustainability of gender norms: women over 30 and their physical appearance on holiday

Jennie Small

ABSTRACT

The present study examines women's experiences of their physical appearance on holiday questioning whether gendered norms, focussed on the "ïdeal" body, are sustainable if women are to experience fully the health and wellness benefits promised in a holiday. While the research focus on body image has most often been adolescents and young women, this study explores the experiences of midlife and older women. Employing the feminist research method, memory-work, women individually wrote memories of recent experiences of physical appearance while on holiday and discussed them in collectives. The findings suggest that while some of the older women appear to have resisted societal directives for the body by focussing on what the body could do, rather than what the body looked like, for most, the dominant image of the body was oppressive with memories of appearance related to body weight/shape and clothing. The findings suggest that women, as they age, are confronted by gender, age and health discourses which are difficult to resist. Whether or not women can challenge core societal beliefs about appearance has implications for whether they can experience the freedom and lack of constraints popularly associated with a holiday.

Introduction

Traditionally, sustainability has been understood as "development that meets the needs of the present without compromising the ability of future generations to meet their own needs" (World Commission on Environment & Development, 1987, p. 43). Within tourism studies and practice, the sustainability focus has been the economic, environmental, and social dimensions of travel routes and destinations. Within this mainstream understanding of tourism sustainability, a consideration of gender has been absent despite the extensive body of work in gender and tourism over the past few decades. Feminists have argued that for tourism development to be sustainable, gender equality must be addressed (Alarcón & Cole, 2019; Ferguson & Alarcón, 2015). Their focus has been gender equality and empowerment of women who work/reside in the host community. As those concerned with gender and tourism acknowledge, 'all parts of the tourism experience are influenced by our collective understanding of the social construction of gender' (Kinnaird & Hall, 1996, p. 100). In other words, tourism is a gendered experience for all: operators, employees, residents as well as tourists. In this paper, I continue the argument for gender analysis and equality as fundamental components in sustainable tourism but disrupt the focus on women as residents/workers to examine women as tourists. Here, I question the sustainability of the gendered social norms which govern the *tourist* experience. Through a study of women's experience of their bodies/physical appearance on holiday, I consider whether Western norms which articulate (the ideal) physical appearance for women are sustainable for women's well-being today and in the future.

Women's bodies have generated an increasing amount of scholarly attention in recent decades (Grogan, 2017) due to Western culture's fixation on a thin-body/sexual-body ideal for women and the impact that the ensuing normative discontent can have on their quality of life. In particular, the focus has been on adolescent and young bodies neglecting those of older women (Kilpela et al., 2015). Here I address this gap by focussing on women tourists over the age of 30. Taking a social constructionist approach, the paper extends the traditional psychological research on body image with its focus on the individual – "a person's perceptions, thoughts, and feelings about his or her body" (Grogan, 2017, p. 4) – to consider women's bodies as a social practice. It is thus concerned with "embodiment": "individuals' interactions with their bodies and through their bodies with the world around them" (Davis, 1997, p. 9). Acknowledging the centrality of this social body to the tourist experience, the paper extends the work of critical tourism studies (Pritchard et al., 2007).

It is through social interaction and discourse that gender, including the "look" of gender, is leant; women learn what is expected to be a woman and act in expected ways. As Butler (1999) would argue, gender is performative, "the repeated stylization of the body, a set of repeated acts within a highly rigid regulatory frame that congeal over time to produce the appearance of substance, of a natural sort of being" (Butler, 1999, p. 44). These norms become embodied, internalised, reifying what it means to be a woman. As she explains, "performativity must be understood not as a singular or deliberate 'act,' but, rather, as the reiterative and citational practice by which discourse produces the effects that it names" (Butler, 2011, p. xii). In the case of this paper, the focus is the gender performativity of appearance. Butler's (1999) notion of gender performativity is allied with the notion of habitus (Bourdieu, 1980) by which an individual acquires and internalises social practices (such as those to do with body image). Bourdieu's notion of the body as a form of (physical) capital with symbolic value and social rewards (such as having the ideal body), and fields (the space where capital is distributed) also assist in our understanding of how women come to view their bodies (Bourdieu, 1978, 1986). We need to situate women's bodies in a cultural and historical context to understand women's relationship to their bodies and their physical appearance. In the tourist environment, women come from different cultures and engage with women (and men) from other cultures whose social norms of physical appearance may differ (for example, see Brown & Osman, 2017). Whether women adhere to, or can resist, these norms is also culturally based.

The system of surveillance is a means by which gender can be understood, constructed and performed "thus highlighting the performativity of gender" (Meyers, 2017, p. 1) but, as Meyers says, surveillance, too, is performative. Women become the bearers of the discourse, socially reproducing the norms. Women not only gaze on others but also on themselves. As Foucault (1980) explains, dominant societal discourses render our bodies "docile" and "normalized", through bodily discipline and social and self-surveillance, whereby each individual exercises surveillance "over, and against himself" (Foucault, 1980, p. 155). Waskul and Vannini (2006) refer to "the looking glass body" to describe this reflexive nature of women's embodiment. Objectification Theory (Fredrickson & Roberts, 1997) explains women's "objectified body consciousness": "body surveillance" (seeing themselves as others see them), "internalisation of cultural standards" (in which the woman comes to see these standards as her choice) and "beliefs about appearance control" (in which the woman feels she is responsible for the way her body appears) (McKinley & Hyde, 1996). McKinley (2004) explains the consequence of internalisation:

> internalization predisposes women to connect achievement of these standards with their sense of self-worth. Because the standards are narrow and difficult to achieve, women experience a sense of empowerment as they approximate them but more often shame when they do not. (2004, p. 57).

The media with its focus on youthfulness and the ideal body is often cited as contributing to young women's problematic relationship with their physical appearance. At the same time, the underrepresentation of older women's bodies (Tortajada et al., 2018; Van Bauwel, 2018) may protect older women from unrealistic norms. Nonetheless, the media is clear as to which bodies have symbolic value. While Jermyn (2016) reports instances of a growing movement in the media to include older women, she notes that the women reflect post-feminism's "makeover" culture and are not representative of older women. Van Bauwel (2018)) also notes that, when older women are represented, it is often unfavourably and in relation to advertisements for anti-ageing products. In tourism studies, scholars interested in embodiment have examined the representation of tourist bodies in the tourism media (Small et al., 2008). Here too, a significant amount of tourism advertising is centred on representations of the ideal woman - attractive, white, able-bodied, slim, heterosexual, and *young*. Women's lifestyle magazines similarly promote this homogenous body as the acceptable, ideal holiday beach-body (Jordan, 2007; Small, 2017). In magazines where images of women over the age of 40 in swimsuits do appear, the body is *de*sexualised – the focus is on control of the ageing body – concealing, rather than revealing – reflecting Western society's views on the ageing female (Small, 2017). Which body is welcome (or not) in the holiday space becomes obvious to the viewer. Coleman (2009) reminds us that bodies and images are not separate entities acting on each other in a linear fashion but, rather, are entwined; "bodies become known, understood and lived through their relations with images" (Coleman, 2009, p. 3).

However, while there have been representational studies, there has been little research on how the tourist *experiences/feels about* their physical appearance on holiday and whether the holiday is a time and space in which women can escape the thin-body ideal. One study which addressed this issue (Small, 2016) found that the holiday was not a site of resistance for young women. The present paper extends the above study to examine how midlife and older women experience their physical appearance on holiday and the implications for their wellbeing. With an ageing society and an active older leisure and travel population, there is a need to explore the embodied experience of this previously neglected group of holidaymakers.

Older women and their bodies

Citing Woodward, Jermyn (2016) argues that even second wave feminism and post-feminism have ignored older women, perpetuating the marginalisation of women in the later stages of the life-course. Of those studies which have focussed on the body image of Western, older women, the conclusion is: women of all ages have internalised the "thin ideal" (Ferraro et al., 2008). In a review article of the literature on body image in adult women, Kilpela et al. (2015) confirm the findings of earlier studies that "the prevalence of body dissatisfaction remains consistently high in women from young adulthood through mid- and late-life" (2015, p. 146). The rates of dissatisfaction have also been found to increase markedly since the 1970s (Tiggemann, 2004). Ferraro et al. (2008) confirmed the findings of other studies that older women experience more body dissatisfaction than older men. According to Kilpela et al. (2015) "body image in adult women may differ from, and possibly be more complex, than that of younger women" (2015, p. 145). There are a number of reasons for such complexity as women's bodies age (Pruis & Janowsky, 2010). These may include increased body mass, a redistribution of body fat to the torso, decreased muscle mass, changes in skin colouration, elasticity and firmness and changes to hair colour, texture and thickness. From earlier in life, women's bodies may be altered from the experience of childbirth. Nash (2015) confirms the findings of other studies (Kilpela et al.,

2015; Woekel & Ebbeck, 2013) that postpartum women experience general dissatisfaction with the appearance of their bodies and struggle to cope with their bodily changes. Today, there is increasing pressure on pregnant women to bounce back to their previous body shape and weight soon after the birth of their babies. This is reinforced by media images of celebrities who are back in their bikinis soon after the baby's birth, and the marketing of "mommy makeovers" (Kilpela et al., 2015). As Nash states, images of "yummy mummies" have become "highly visible, 'public' obsessions evident across many representations and discourses of contemporary Western motherhood" (2015, p. 33). The findings related to body image and menopause do not appear as consistent. Although body changes that occur at this time conflict with the Western standard of beauty, a review of studies concluded that, due to a variety of factors that affect menopause, "menopause neither invariably nor uniquely engenders body dissatisfaction in women" (Kilpela et al., 2015, p. 154).

Other changes which may affect the body include medical procedures (such as mastectomy) and medications. In addition, life stressors and demands of employment and family may have reduced opportunities for self-care. As the body changes, so it is likely to be moving away from the young thin-ideal. As Pruis and Janowsky (2010, p. 226) state, "Collectively, these factors suggest that aging primes older women for negative body image". Hurd Clarke et al. (2009) report the ways women aged 71 to 93 chose clothing to mask or compensate for "bodily transgressions" that had occurred as a consequence of ageing. While older women may be more vulnerable to body shape and size as their bodies change, Ferraro et al. (2008) claim one may not know whether body image issues are associated with youthfulness and beauty or concern for health. Both discourses are circulating in Western society and are intertwined. As Carter (2016, p. 201) explains, "the medicalization of aging has led to the reframing of youthfulness and anti-aging as tied to good health". By implication, ageing is constructed as negative, a time of decline and loss of both beauty and health.

While studies confirm that body dissatisfaction is prevalent throughout women's lives, it appears that as women age, the importance placed on physical appearance may lessen. Tiggemann (2004) concluded that with age, women move further from the ideal but, at the same time, body appearance becomes less important to their identity. As she explains: "these two processes .... counterbalance one another to produce a stable level of body dissatisfaction" (2004, p. 35). Grippo and Hill (2008) similarly found that, despite older women reporting levels of self-objectification and habitual body monitoring similar to younger women, body monitoring was less central to body dissatisfaction for older women than midlife women. As women age, they may use cognitive control strategies to increase body acceptance (Peat et al., 2008). They may feel less pressure to conform to the media's ideal image (Pruis & Janowsky, 2010). Gosselink et al. (2008) concluded that women aged 73 and older were relieved to eschew appearance standards but, at the same time, expressed ongoing concerns for their appearance. Studying women aged 18–75, Tiggemann and McCourt (2013) found that appreciation for one's body increased with age. They suggest that perhaps around the age of 50 (age of menopause) women can (or are forced to) come to terms with their body's departure from the societal ideal. Certainly, Montemurro and Gillen (2013) found that women in their 50 s and 60 s, while lamenting that ageing took them further away from looking attractive to others and being sexually desirable (a thin body), were most likely to resist the normative body by accepting their bodies and appreciating what their bodies had done for them through life. A lack of sexually appealing older role models had the benefit of lifting the pressure to conform to the ideal body. The authors surmised: "It seems looking sexually appealing becomes less important to women as they age, yet, looking attractive remains a concern for women of any age" (Montemurro & Gillen, 2013, p. 20). Nonetheless, there may be individual differences, with women high on objectified body consciousness more anxious about ageing (McKinley & Lyon, 2008).

Women's physical appearance in leisure and tourism

Leisure is a valued aspect of people's lives at all ages and often seen as a counter site, or heterotopia (Foucault, 1984), where constraints are reduced and exercise of choice and transgression are possible. Yet, while there is evidence of women's empowerment in leisure activities through resisting bodily norms, women's negative feelings about their appearance can constrain their participation or reduce their enjoyment (James, 2000; Liechty et al., 2006). Studying the role of body image in the leisure of women aged 60–69, Liechty and Yarnal (2010) found that while body image was more closely related to perceptions of health and physical ability than to appearance, for many of the women "appearance concerns led to reduced participation in activities requiring a bathing suit" (Liechty & Yarnal, 2010, p. 443).

The holiday, as a form of leisure undertaken away from home and the everyday, offers a particular promise of escape, relaxed rules, and anonymity. That it is often promoted as a romantic getaway, a sexualised space (Jordan & Aitchison, 2008), is relevant in understanding women's bodies in that space. The limited research on women's lived *experience* of their physical appearance confirms the importance of appearance in the holiday. Berdychevsky et al. (2015) found that, even while the holiday allowed women to invert their sexual roles, "sexual confidence was contingent upon … [women's] self-perception as sexy/feminine/attractive, while their bodies had to be in the best shape for holidays". Small's (2007, 2016) studies of Australian women highlight the strength of gendered appearance norms in women's holiday experiences. In her study of young women, Small (2016) concluded that women felt positive about their bodies when they conformed to the body ideal and negative when they failed to meet the standards. Central to their experience were feedback/comments and perceived surveillance by others, and social comparison with other women's bodies. Surveillance of women's bodies has also been highlighted in studies of solo women travellers (such as Jordan & Aitchison, 2008). It is clear that women travelling on their own are very aware of others gazing at their bodies. Women's experiences need to be viewed within their cultural context as Abramovici (2007) demonstrates in her detailed account of the tanning process of Italian women. To understand the process, one needs to appreciate "the importance of beauty in Italian society, the patriarchal and masculinist ways of Italian society, and the sensuality and passion with which life is lived in this country" (Abramovici, 2007, p. 123).

This paper is premised on the view that the meaning of the body and its appearance is socially constructed, thus is culturally and historically specific. Women perform gender internalising the repetition of appearance practices of their culture. The paper also maintains that if a holiday can contribute to one's quality of life and wellbeing (Dolnicar et al., 2012; Gilbert & Abdullah, 2004), then women are entitled to these benefits. The present research extends the research on body image beyond that of young women to hear the voices of midlife and older women. It explores the women's memories of both negative and positive experiences of physical appearance on holiday. In doing so, it follows the relatively recent shift of focus in body image research away from body dissatisfaction, to consider *positive* body image (Grogan, 2017; Halliwell, 2015).

Methodology

Following the theoretical underpinning of the study, the social constructionist, feminist research method, memory-work (Haug, 1987), was employed to investigate women's socially shared meanings of their physical appearance in the holiday environment.

Study participants

Participants were four groups of women aged 30–49 (total of 18) and three groups of women aged 50–64 (total of 15). In each group there was at least one woman known to the researcher who, in turn, invited others to form a group. In some groups, all members were known to each other, in others, some were strangers. In either situation, memory-work groups have been found to work well together (Crawford et al., 1992). The participants were Australian, white, middle-class, able-bodied and heterosexual women. Group homogeneity was cited by Crawford et al. (1992) as their preferred option for their memory-work studies. Similar subjectivities allow for a more thorough understanding of the collective especially when the group size is small.

Memory-work

The underlying theory of memory-work is that "subjectively significant events, events which are remembered, and the way they are subsequently constructed, play an important part in the construction of self" (Crawford et al., 1992, p. 37). With the emphasis on reflection in the construction of self, memories become the data. They are studied in their own right and not judged against the 'real/true' past event; memories are the reality. The focus of the present study was memories of physical appearance on holiday. Since the method considers that we are experts in our own experiences, the barriers between the subject and the object of research are broken down. Memory-work stresses *collective* theorising. With this inter-subjective focus, memory-work is differentiated from other research methods such as in-depth interview (Small, 2004). Memory-work is also a feminist method. It is hoped that through sharing experiences and interpretations with women similar to oneself, women can come to a new understanding of their experiences, how they have been constructed and can possibly be resisted.

Memory-work phases

There are three phases in memory-work. In *Phase 1* (before convening as a group) each participant writes a memory. (In this case the memories were of a positive experience and a negative experience in the last two years relating to *my physical appearance on holiday.*) The topic acts as a trigger to generate memories and later discussion. Participants are asked to *write 1–2 pages* on each of the two memories. The memories are written according to a set of rules (Haug, 1987) which includes writing in the third person using a pseudonym. In *Phase 2,* the participants convene (in this study, in groups of 4–6), *read and discuss the written memories* drawing out their shared social understandings, theorising as a collective. In this study, to encourage frankness and avoid embarrassment from what could be considered a sensitive topic, the participants read and discussed the written memories of anonymous members of another same-aged participant group. The method suggests that groups identify: similarities/commonalities and dissimilarities in the memories; clichés and contradictions; and absences – what is not being written (but which might be expected). In addition, the women were asked: to consider whether they identified with the memories they were reading; to explain those memories, and to contemplate the importance of physical appearance to the success of a holiday for women of their age. The discussion lasted up to one and three-quarter hours and took place at the researcher's university or the home of one of the group members. After introducing the group to the process, they were left on their own to discuss and theorise by themselves. I, the researcher, returned at the end to elicit a summary of the discussion to ensure my understanding. The discussion was audio-recorded and later transcribed. In one of the 50–64 age groups, I was a participant. In the final phase, *Phase 3,* I collated the material, drawing out the themes across the groups. (Where quotations are used in the following section of the paper, those from the collective discussion are italicised to distinguish them from the individual written memories.)

Data analysis

The research questions guided thematic analysis of the written memories and collective understanding of the memories. The data was first analysed by the collective itself in Phase 2 as they looked for similarities in the written memories and in their discussion. Secondly, the group summarised their discussion in my presence to ensure my comprehension of their identified themes, thus, increasing the research credibility (Nowell et al., 2017). I then continued the thematic analysis immersing myself in the data and organising the data from each group under the identified themes. I confirmed themes within and across groups following the checklist of criteria for good thematic analysis as proposed by Braun and Clarke (2006). The trustworthiness of the data was strengthened by these layers of analysis and the employment of a recognised reputable research method.

Findings

The written memories and discussion covered holidays that were domestic and international, from touring to beach holidays. Overwhelmingly, the holidays were taken in the company of friends or family. From the written memories and the collective discussion, it was clear that both groups of women understood societal prescriptions for women's appearance – how they *should* look – toned and fit, slim, tanned, wearing the right clothes, and youthful.

Resistance to the body ideal

While understanding gendered norms which directed their appearance, there were some, especially in the older cohort, who appeared to resist these ideals and focus on their fitness rather than appearance. While fitness can also equate to "looking good", these memories of fitness were concerned, not so much with *looking* fit but rather, with what the body could *do*.

> I wasn't concerned about my physical appearance, whether my hair was wind-blown, face flushed, or in the appropriate clothes. My focus was on the sensation of walking, strolling at my own pace, looking in shops on the way … My attention was outward, seeing what I could observe, rather than on my own appearance. However, I can remember being pleased that I had on the right shoes and comfortable clothing, not too tight, so that I could enjoy the walk. I felt a sense of achievement when I arrived. I can remember thinking, 'Not too bad for over 60', and being grateful that I was in such good health and spirits. [Age 50–64]

> *… we go on big cycling holidays, walking … all of that's far more important now than what you look like … how good it makes you feel.* [Age 50–64]

In both age groups, there were some women who while recognising that they did not meet expected standards of appearance, claimed they "didn't care".

> There was nothing that she could do until nightfall to resurrect the situation [wet hair] … It was all quite liberating, actually. Especially, once she persuaded herself that nobody actually cared. [Age 50–64]

> I felt very uncomfortable in my grey pants which accentuated my big thighs and bottom... My husband laughed jokingly and keeping telling me that he didn't notice any difference, in fact that I looked just as beautiful as always... I felt that people were looking at me, although to tell you the truth I didn't really care. [Age 30–49]

The women were inspired when they saw women in other countries oblivious to how they looked.

> *… [in Europe at the beach] there were these really old ladies sitting there … one didn't even have a bikini but she'd taken off her top and she was there in her bra, just soaking up the rays …. And I just thought 'Wow, I want to be like that'!* [Age 30–49]

Both groups of women felt that they were less focussed on their physical appearance than they had been at younger ages when they were single, the object of the sexual gaze, and establishing who they were.

F1: So once you've found your partner and established your world, had your kids, and ... you've created something for yourself... you're not so buffeted by the winds of ...

F2: Yeah. I agree with that.

F1: ... you actually can stand firm and say 'Well, actually ...'

F2: Yeah. 'This is me'.

F1: ... 'this is who I am', and ...

F3: 'And it's far more than what I look like.' [Age 30–49]

While many felt more confident by 30-49, they acknowledged that today they had the burden of Facebook for posting holiday photos. They also felt there was more pressure for their generation to look good than in the past. Despite examples of resistance, the extent to which the women could subvert appearance norms was questionable.

Adherence to the body ideal

It was clear that approximating the ideal look remained important for most of the 30-49 age group and many of the 50–64 age group. The women wanted to "fit in", look like others in their holiday environment. They felt positive about their appearance when they met (or came close to) the ideal body. Although the two groups of women were at different life stages, at the core were similar key concerns – body size and weight.

> She went to a beach resort destination for her honeymoon and loved prancing around in her bikini. She felt confident and attractive. She felt others look at her admiringly at times... She was staying in the penthouse suite with private rooftop spa, she would often walk around naked in front of her husband, never considering there was anything to slightly embarrass her. Her body was not perfect by any means but it was athletic, toned and fit. She was tanned and healthy and slim and knew she looked good. She loved to go out at night and she felt sexy and wore clothes to enhance her figure... [Age 30–49]

> She had started to diet in the spring of the previous year and was working out at the gym. She managed to lose 8 kilos and felt really good about her weight loss, and she liked her new body. She felt comfortable in her summer gear, particularly in her swimming costume... She enjoyed her meal-times too as she didn't have to worry about what she ate and didn't feel like a greedy guts if she had the occasional ice cream, or fish and chips. [Age 50–64]

If the women lacked the ideal body, they could still experience their appearance positively provided they could *hide* their "flaws".

> She wore the kaftan at night with leggings and it did the job of covering up tuck shop arms and tummy but also not making her look too enormous. [Aged 30–49]

Inherent in the memories of *negative* experiences was the perceived failure of women to meet the prescribed weight norms.

> .... She feels old and sometimes ugly. Most of it has to do with the weight, some of it is the wrinkles.... She constantly worries about whether she will ever be able to look half decent in swimmers again. Her husband is kind but of course he notices that she is not the same girl in the swimmers that he met. She doesn't want any photos taken of her without being clothed. Definitely not in the swimmers. Carrying extra weight changes her confidence on everything. [Age 30–49]

Life stage events could exacerbate these feelings. In the 30–49 age group, the tiring nature of child-rearing negated any possibility of feeling attractive or sexy. The women compared their pre-baby to post-baby bodies. They spoke of scars from C-sections and stretch marks which prevented them from topless sunbaking. While some claimed that after a while they didn't care, there was still distress, in varying degrees.

> It was the first holiday post baby and of course post baby body. Her body had really let her down.... It is disgusting to look at. She can't stand to look in the mirror but does so as self-torture. This is horrible. And

recrimination for not getting back into shape seven months after the baby was born… She can't wear a bikini anymore… She is happy to stay around the resort rather than going out. She does not feel attractive or young anymore due to her figure and the three million lines that have appeared on her forehead. Jesus, it's all gone to shit. Her husband is kind but of course he notices that she is not the same girl in the swimmers that he met. [Age 30–49]

The women complained of the discourse exhorting them post-partum to return speedily to pre-pregnancy weight (as achieved by celebrities) and argued for a self-compassionate approach for mothers. Beyond the age of pregnancy and motherhood, similar memories of weight gain were evident among the older cohort.

She knew she had run out of time to lose the weight that she had put on during the past year. She was panicking even… She felt that she had let herself down by not doing something about her weight gain sooner… She felt awful that she would be embarrassing her family once they saw how fat she looked in her swimming costume and summer gear… On arrival, she didn't even want to go near the beach. Any excuse like, 'it's too hot now', 'I'll go later', were uttered frequently… Mealtimes were not as pleasant as they used to be, as she was conscious of not eating too much, in case everyone was thinking, she shouldn't be eating that, she'll get even fatter… She couldn't wait to get home so she could start her diet in time for next year. [Age 50–64]

It was not only the women's weight and shape that featured in the positive and negative memories but also their clothing. Clothing could cover or expose, be appropriate or inappropriate, be fashionable or not. It could define one as a seasoned traveller or an inexperienced traveller.

…it felt good to be wearing the latest gear on the [ski] slopes. Her husband even commented that she looked like a professional in her new pants and jacket… She was the most relaxed and happy she had been in a long time. Even her daughter commented that she looked good. [Age 30–49]

Not fitting in was a negative experience.

[She] was so underdressed compared to everyone else. Just jeans, boots and a simple top on. Everyone else was in their tiny summer dresses and they all looked so cute as if they'd just stepped out of a London night club. 'God, just look at me. I'm so plain and daggy', [she] thought to herself. [Age 30–49]

Clothing could influence their identity as a traveller.

She had packed too few choices to wear. In Chengdu when meeting up with the other travellers for the first time, she felt her heart sink when she noticed some of the others were seasoned travellers who had the proper light-weight, perfectly matching, travel wardrobe. They looked so composed and clever. She felt like a dag. [Age 50–64]

While *shopping* for clothes featured strongly in the memories, these were not positive experiences despite many wanting the pleasure of purchasing clothes when on holiday. Shopping in Asia was difficult and discouraging because Australian women were all much larger in size to Asian women. Treatment by sales staff could be intimidating and demoralising; the women took these experiences personally.

[In the Philippines]: … a shop assistant approached, a beautiful Filipino girl… she uttered 'too big, you too fat, try men's department down stairs'… Absolutely crushed of spirit and enjoyment the girls walked home, not speaking, feeling fat, being stared at and feeling out of place. [Age 30–49]

[In Paris] The actual shopping experience became more and more humiliating as she tried on coat after beautiful coat, hunching shoulders, pulling in waist, yanking up bra… all to no avail. She began to feel as though everyone in the store knew what was happening. In fact, as she was taking a coat from a rack, a French woman pulled it from her, as if to say 'this will not fit you'… To alleviate a sense of humility and of being 'huge', [she] pretended to herself she was shopping for a non-existent daughter. [Age 50–64]

Especially for the older group, clothes could be challenging as body shape and size changed with age. Women reported that clothes were not made for bodies of their age, clothes were either too revealing or matronly They wanted to "feel nice" on holiday but if they were "overweight", clothes could be uncomfortable.

As demonstrated above, the perceived gaze and comments of others could dramatically affect how the women understood their physical appearance. Feedback also came in the form of the holiday photos posted on social media where their appearance was "captured forever" for all to see and comment upon. Viewing the photos could change what had previously been considered a positive holiday into a negative experience. Even a chance event like catching her reflection in a mirror quickly changed one woman's self-image.

> She… saw herself in the mirror and was aghast at what she saw… she seemed to have aged overnight. She saw wrinkles that she hadn't seen before and realised this was the view that everyone else must have of her – old! [Age 50–64]

Preparing the body

Prior to the holiday, the women, to various degrees, engaged in bodily preparations – diet/exercise, haircut, hair removal, fake tan, manicure, and purchase of clothes. The women recognised that for some there was quite a bit of work and anxiety around this but there was a general understanding that a degree of preparation was required. Some explained that being prepared allowed them to relax on holiday.

> *Pack 15 different outfits just in case... I wanted it to be a good holiday. I didn't want to be worried about… not being ready for whatever situation may have come up.* [Age 30–49]

> [The holiday] is 'time-out' for physical appearance as long as I've banked the fitness beforehand, and as a result feel that I'm looking my best on the whole. [Age 30–49]

Preparation could add to the excitement of the upcoming holiday.

> *Get my hair cut and that sort of stuff… that nice build up I think is the thing that is quite a pleasant sensation… It sort of starts to put you in the frame of being somewhere else.* [Age 50–64]

The holiday context

The women agreed that how they felt about their appearance was related to the type of holiday and the people in that context who might judge them on their physical appearance. Being with familiar people such as family, a supportive partner or close friends was more relaxing and comfortable than with others. Conscious of their ageing, changing body, meeting up with old friends whom they hadn't seen for a long time could induce anxiety. Many of the negative experiences were associated with a beach holiday where the body was on display. The beach was signified in Australian culture as "a place of sun-bronzed… goddesses and gods" where, due to a warm climate for much of the year, women had little "downtime". Nonetheless they considered there were differences *between* beaches. They differentiated Australian beach culture from European where women were less concerned about body shape. Family style beaches were more relaxing than beaches "to be seen at" such as those in Brazil where one woman reported:

> Every day going to the beach was a mission and she had to worry with the brand of her bikini… [if] the bikini was matching the bag and the sandals. Going to the beach was worse than going to a party! [Age 30–49]

Glamorous destinations, (New York and some European countries) could prove challenging. Women in both groups discussed the cultural cringe that Australian women can feel in such destinations – perceiving themselves as the "country bumpkin"/"dowdy country cousin". These Western women could feel uncomfortable about their bodies in Islamic destinations.

> … there was not one moment when in the back of her mind she was not filled with a consciousness of the physical presence of her body and hair... Men did not look her in the eye, but they looked at her body and some made difficult comments in the street. [Age 30–49]

As she danced, the shawl moved, uncovering part of her arms, and she was aware that this was unlike her Muslim friends, whose arms, legs, and head were covered… [She] didn't feel so good about her body. [Age 50–64]

Importance of physical appearance on holiday

Attention to appearance was inescapable as women pre-holiday had to make choices when packing. It was also suggested that maybe there was more time on a holiday to think about their appearance. The salience of appearance to their enjoyment of the overall holiday was expressed by some.

A holiday is something special. You want to make an effort to make yourself look special [Age 30–49]

I remember the holiday as how I looked at that time… [Age 30–49]

Certainly, the women of both age groups, perceived gender differences: women were more self- absorbed, felt more pressure and experienced more angst about appearance on holidays than did men.

Gee, it would be so much easier to be a bloke. Flat comfortable shoes, shorts and a clean t-shirt and voila, you are ready for holidays. [Age 30–49]

F1: And men on holidays, …

F1: … showing their beer guts and they're growing their beard …

F4: And they don't brush their hair.

F2: Generally [Age 30–49]

Impact on wellbeing

It was clear that appearance norms were constraining for most – where they went, what they wore and what they did. To escape surveillance, a number avoided swimming. Some of the older women reported rarely wearing a swimsuit or not owning one. They were often annoyed at themselves when succumbing to social standards, for example, "covering up in an Islamic country (where it "felt like a cop-out to her feminist independence"). Some of the women were clearly conflicted. Memories were often a contradiction: distress at not measuring up to the ideal and then the assertion that they "didn't care".

She didn't feel good about herself. She wished she didn't let this get to her, but it did. [Age 30–49]

Their attitudes to their appearance could potentially impact on others.

… she never went into the water. She wanted to play with the kids and splash, they would love it but something stopped her. Maybe she felt over dressed to get in the water, maybe… she didn't feel playful. It annoyed her. (Age 30–49)

Self- talk could help women manage their feelings.

Despite assurances from her husband that she looked lovely, she did not feel lovely… it affected her whole mood and was threatening to spoil the entire evening for both of them… a stern, internal discussion with herself, 'Get over it, who cares what you look like, as long as he likes what you look like… '[Age 30–49]

Discussion

Women's memories of their physical appearance on holidays are about embodiment and internalisation of norms of what it means to be a woman (Bourdieu, 1980; Butler, 1999), how a woman should look. As was found with younger women (Small, 2016), most positive memories reflected approximation to the ideal-body discourse: slim, toned, tanned, and appropriately dressed (or a camouflaging of the negative body) while negative memories were the inverse. With a few exceptions, the findings in the present study are consistent with earlier research that dissatisfaction with the way one looks is stable across the lifespan (Tiggemann, 2004; Tiggemann & Lacey, 2009). As Jermyn (2016, p. 586) found, "all but the very oldest women are increasingly compelled to adhere to the same strictures of public scrutiny, self-surveillance, and the incessant reworking of the self". The social norms for appearance did not deviate for age or life-stage events such as pregnancy and birth and yet the written memories and discussion reflected a consciousness of women's changing, ageing, "deteriorating" body. As women aged, their physical capital and symbolic value of the body declined. Bourdieu's theory of social reproduction which stresses the importance of the body in the formation of the body in social inequalities (Shilling, 2003) is useful in understanding the memories, as women, to a greater or lesser extent, deviated from the ideal body associated with youthfulness. There were rewards associated with having physical capital (Bourdieu, 1978) and costs for many when it was lacking.

While physical appearance can encompass any aspect of the look of the body, it was body size/weight and clothes which were the focus of the memories. Tiggemann and Andrew (2012) stress an interrelationship between attitudes to clothes and body. In a study of women's memories of holidays across their life-course, clothes featured strongly (Small, 2007). In the present study, clothing was important as a means of "fitting in"; the women wanted to look like those around them in the holiday context. The women also highlighted the role of clothing, in revealing or concealing the body. Those who have studied older women (Hurd Clarke et al., 2009; Lö vgren, 2016) have highlighted the way clothing has been used to mask the stigma of the ageing body. Consistent with Small's (2016) findings with younger women, unless women were satisfied with their weight, the item most problematic on holiday was the (non-concealing) swimsuit. As discussed elsewhere (Frederick et al., 2006) the wearing of a swimsuit can be a stressful experience. Shopping for clothes could also be a problematic practice (as Tiggemann and Lacey, 2009, previously found) exemplified here in unsuccessful shopping ventures where clothes were found too small or sales staff were "snooty". The fragility of women's identity was noticeable in many of the memories and discussion. There were many examples of women (in the 30–49 age group) imagining the gaze of their male partner and seeking reassurance that their physical appearance was 'ok'. Across both groups, some random event such as, seeing one's reflection in a mirror or viewing the holiday photos, could quickly alter, what had been until then, a positive perception of self and holiday.

One could argue that rather than holidays being an escape, an opportunity to resist the body politics of the dominant discourse, the holiday is a social field (Bourdieu in Shilling, 2003) bestowing even greater value on the bodily form. As was said, holidays are a *special time* in which women want to look good. The fantasy of "the perfect holiday" also incorporates an image of "the perfect body". In many ways, holidays are constructed as sexualised space, especially beach holidays where bodily exposure in a public place is licensed. Certainly, many women engaged in bodily practices in preparation for the holiday. Physical attributes were "worked on, improved, and leveraged" (Malacrida & Low, 2008, p. 306) to accumulate capital (Bourdieu, 1978). Holidays are also a time for photographs with images of one's body "captured forever". Yet, not all holidays were the same; context was important. The culture of the destination and its attitudes to a woman's body were significant in the experiences of many of these Western women. Even within a destination there could be subtle differences between spaces (such as beaches) in the physical capital required. The salience of physical appearance was also related to type of holiday, activities undertaken and the people in that space. How relaxed the women were with their appearance depended on their relationship with their companions; being with familiar people such as family was more comfortable than with others.

For a number of women, especially as they got older, *avoidance* of certain types of holiday, destinations, or activities was a strategy to escape negative embodied experiences. As others have found, older women, due to appearance concerns, will often avoid activities requiring a bathing suit (Liechty & Yarnal, 2010). There were few reported memories of what might be considered "true" transgression. As Liechty and Yarnal (2010) had similarly found, many of the women accepted the constraint, rather than actively negotiating through it. Rather than adjusting attitudes/rejecting societal ideals (such as, feeling free to visit any beach in a swimsuit whatever size, shape or look of the body), they changed behaviour to accommodate these ideals (such as, avoiding certain beaches, covering up on the beach, or not swimming at all).

While embedded beliefs and attitudes about bodily appearance are extremely difficult to dislodge, Foucault (1980) argues that individuals have the capacity to resist the dominant body discourse. According to Butler (1999), the norms that govern gendered performance can be destabilised. As she sees it, the body is "a continual and incessant *materializing* of possibilities" (Butler, 1990, p. 272). Bourdieu would argue that despite the difficulty of changing the deeply-embedded and long-established nature of habitus, the body is "never fully finished"; the symbolic value attached to body appearances can change. As Shilling (2003, p. 121) explains, "The value attached to particular bodies changes over time; as fields within societies change, so may the forms of physical capital they reward". There may be generational (Shilling, 2003) and age changes. As seen in this study, there were (some) older women redefining physical appearance away from looks to bodily strength, fitness and empowerment suggesting, as Tiggemann (2004) found, that although they might have been dissatisfied with their bodies, physical appearance had become less important with a consequent decrease in self-objectification and habitual monitoring. They accepted their ageing bodies; as one said, *"you can make peace with your body."* However, there were other older women (and midlife women) who, while knowing they should accept their bodies, found it difficult, suggesting that there are important distinctions between intellectual and emotional aspects of body image (Grippo & Hill, 2008). As Grippo and Hill (2008) found, feminist consciousness might not be enough to resist these entrenched discourses. Speaking of their own participants aged 40–87, they claimed:

> In essence, many of these women have been raised in a sexist society and this objectification has become such a part of their daily lives that even though they know intellectually that they should not let it effect their views on their bodies, emotionally it is hard to escape (Grippo and Hill (2008, p. 180).

Within neo-liberal capitalism there is immense pressure on the individual to take responsibility for their bodies. Aligned with this ideology, post-feminism views women in the 21st century as having agency, power, self-determination and choice; they can enjoy the body work associated with the ideal body as a fun, pampering or deserved experience (Gill, 2007). The onus is on the self-regulating individual rather than societal structures; if the individual puts their mind to it, anything can be achieved. We also see this ideology in the health discourse which extols weight loss (the slim-looking body is the healthy body) and in the consumer culture that presses women to purchase body products and cosmetic surgery, especially as they age. That women have internalised neo-liberal thinking of individualism and self-responsibility for their bodies is reflected in the present study where women acknowledged pride in achievement of the prescribed body and shame and guilt – "beating themselves up" when they hadn't. While age may permit resistance, this is not always easy as older women are also confronted with discourses of ageing, especially the "successful ageing" discourse which encourages women to do all they can to maintain a youthful appearance. Especially amongst the older cohort, staying active, autonomous, and responsible, as in leisure travel, fits with the doctrine of seniors "ageing well". Many have criticised the narrow, binary discourses of ageing ("successful" versus "decline"). Sandberg (2013) proposes an alternative conceptualisation, "affirmative old age" which "aims to acknowledge the material specificities of the ageing body and is an attempt to theorise the ageing body in terms of difference but without understanding it as a body marked by decline, lack or negation" (p. 12). Although the discourses of ageing are associated with later life, they could be seen emerging in the narratives of younger women as they experienced bodily changes that come with age. While some may debate whether women's experiences of the body are better explained through "oppression" or "agency", the negative (and some of the positive) experiences reported in this study appear oppressive and thus fit more closely with second wave feminism than with post-feminism which prefers to see women's actions as self-determined. Examples of resistance provide hope for the prospect of change but, from a radical, not post-feminist perspective (Colebrook, 2010).

The present study contributes to the broader field of body image research by examining women's physical appearance within a life setting – the holiday. Hopefully, it extends our understanding of midlife and older tourists, the latter who are traditionally marginalised in their "embodiments, emotions, identities and narratives" (Sedgley et al., 2011, p. 432). More specifically, it contributes to a discussion on gender and tourism sustainability by questioning the sustainability of Western society's gendered body norms for women's wellbeing on holiday. The message from this study is that we need to be aware of the intersection of gender and age when considering holiday experiences and the potential physical and psychological consequences of the discourses surrounding women's body and body work. The feminist aim of memory-work is that, through sharing experiences, women become stronger and better able to understand the construction of self. While acknowledging the power of women's entrenched beliefs about the body, it is hoped that hearing others' experiences, especially those who had redefined physical capital, might assist women to dislodge the culturally engrained constraints of "beauty" to enable them to experience, fully, the freedom of a holiday.

The paucity of literature on the *subjective* experience of ageing (Cameron et al., 2019) is reason to pursue the experience of older women in holiday settings. The older one gets, the less represented they are in tourism research (Sedgley et al., 2011). Future research should be extended to those over the age of 65 for whom appearance remains an issue (Gosselink et al., 2008; Hurd Clarke et al., 2009) and who comprise a significant number of leisure travellers. While not intentionally selected by body size, none of the women in the study would be considered in the obese category. With the increasing number of people in the population moving further from the "ideal" it would be useful to understand their experience of physical appearance in the holiday environment. While some differences were found between the younger and older cohorts in this study, it is not clear whether the differences can be explained by age or generation, thus the need for longitudinal studies. Finally, to broaden our understanding of women's holiday experience of body and appearance, research could be extended to subjectivities beyond the white, able-bodied, heterosexual, middle-class, Australian women of this study.

Conclusion

Holidays are constructed as a time of escape and freedom from life at home providing health and wellness benefits which contribute to our quality of life. Yet the findings from this study suggest that the gender and ageing discourses which prescribe a narrow physical appearance for women are not sustainable for the well-being of women today or in the future. While the focus of the study – physical appearance – might be thought of as only one aspect of the holiday experience, it was clear in the written memories and discussion that participants' feelings about their appearance could impact on how they viewed the holiday as-a-whole. Beliefs and feelings about their physical appearance were related to their confidence and comfort, their ability to relax, the type of holiday and activities chosen/avoided, the clothes worn, and potentially, their relationships. Clearly the female body is a problematic subject that continues to demand feminist attention and is not passé as post-feminism assumes. As argued by Lazar (2007, p. 160) post-feminism "muddies questions of power and ideology in contemporary gender relations and stalls critical social awareness". While society's body directives go far beyond tourism and holidays, the tourism industry can go some way to persuade women of an alternative embodiment by promoting positive, realistic images of women's bodies which reflect health and ability – what the body can do – as opposed to the narrow "body ideal". As women age and move further from youthfulness, this might lessen their perceived need to engage in body work in order to maintain social value and retain visibility.

Disclosure statement

No potential conflict of interest was reported by the authors.

ORCID

Jennie Small http://orcid.org/0000-0002-6200-5398

References

Abramovici, M. (2007). The sensual embodiment of Italian women. In A. Pritchard, I. Ateljevic, N. Morgan & C. Harris (Eds.), *Tourism and gender: Embodiment, sensuality and experience* (pp. 107–125). CABI Publishing.

Alarcón, D., & Cole, S. (2019). No sustainability for tourism without gender equality. *Journal of Sustainable Tourism, 27*(7), 903–919. https://doi.org/10.1080/09669582.2019.1588283

Berdychevsky, L., Gibson, H., & Poria, Y. (2015). Inversions of sexual roles in women's tourist experiences: Mind, body, and language in sexual behaviour. *Leisure Studies, 34*(5), 513–528. https://doi.org/10.1080/02614367.2014.938770

Bourdieu, P. (1978). Sport and social class. *Social Science Information, 17*(6), 819–840. https://doi.org/10.1177/053901847801700603

Bourdieu, P. (1980). *The logic of practice*. Polity.

Bourdieu, P. (1986). The forms of capital. In J. Richardson (Ed.), *Handbook of theory and research for the sociology of capital* (pp. 241–258). Greenwood Press.

Braun, V., & Clarke, V. (2006). Using thematic analysis in psychology. *Qualitative Research in Psychology, 3*(2), 77–101. https://doi.org/10.1191/1478088706qp063oa

Brown, L., & Osman, H. (2017). The female tourist experience in Egypt as an Islamic destination. *Annals of Tourism Research, 63*, 12–22. https://doi.org/10.1016/j.annals.2016.12.005

Butler, J. (1990). Performative acts and gender constitution: An essay in phenomenology and feminist theory. In S. Case (Ed.), *Performing feminisms: Feminist critical theory and theatre* (pp. 270–283). John Hopkins University Press.

Butler, J. (1999). *Gender trouble: Feminism and the subversion of identity*. Routledge.

Butler, J. (2011). *Bodies that matter: On the discursive limits of "sex."* Routledge.

Cameron, E., Ward, P., Mandville-Anstey, S., & Coombs, A. (2019). The female aging body: A systematic review of female perspectives on aging, health, and body image. *Journal of Women & Aging, 31*(1), 3–17. https://doi.org/10.1080/08952841.2018.1449586

Carter, C. (2016). Still sucked into the body image thing: The impact of anti-aging and health discourses on women's gendered identities. *Journal of Gender Studies, 25*(2), 200–214. https://doi.org/10.1080/09589236.2014.927354

Colebrook, C. (2010). Toxic feminism: Hope and hopelessness after feminism. *Journal for Cultural Research, 14*(4), 323–335. https://doi.org/10.1080/14797581003765291

Coleman, R. (2009). *The becoming of bodies: Girls, images, experiences*. Manchester University Press.

Crawford, J., Kippax, S., Onyx, J., Gault, U., & Benton, P. (1992). *Emotion and gender: Constructing meaning from memory*. Sage.

Davis, K. (1997). Embody-ing theory: Beyond modernist and postmodernist readings of the body. In K. Davis (Ed.), *Embodied practices: Feminist perspectives on the body* (pp. 1–23). Sage.

Dolnicar, S., Yanamandram, V., & Cliff, K. (2012). The contribution of vacations to quality of life. *Annals of Tourism Research, 39*(1), 59–83. https://doi.org/10.1016/j.annals.2011.04.015

Ferguson, L., & Alarcón, D. (2015). Gender and sustainable tourism: Reflections on theory and practice. *Journal of Sustainable Tourism, 23*(3), 401–416. https://doi.org/10.1080/09669582.2014.957208

Ferraro, F., Muehlenkamp, J., Paintner, A., Wasson, K., Hager, T., & Hoverson, F. (2008). Aging, body image, and body shape. *The Journal of General Psychology, 135* (4), 379–392. https://doi.org/10.3200/GENP.135.4.379-392
Foucault, M. (1980). The eye of power. In C. Gordon (Ed.), *Power/knowledge: Selected interviews and other writings 1972-1977 Michel Foucault.* Trans. C. Gordon, L. Marshall, J. Mepham, & K. Soper, (pp.146–165). Pantheon Books.
Foucault, M. (1984). Of other spaces, heterotopias. *Architecture, Mouvement, Continuité, 5*, 46–49.
Frederick, D., Peplau, L., & Lever, J. (2006). The swimsuit issue: Correlates of body image in a sample of 52,677 heterosexual adults. *Body Image, 3*(4), 413–419. https://doi.org/10.1016/j.bodyim.2006.08.002
Fredrickson, B., & Roberts, T. (1997). Objectification theory: Toward understanding women's lived experiences and mental health risks. *Psychology of Women Quarterly, 21*(2), 173–206. https://doi.org/10.1111/j.1471-6402.1997.tb00108.x
Gilbert, D., & Abdullah, J. (2004). Holidaytaking and the sense of well-being. *Annals of Tourism Research, 31*(1), 103–121. https://doi.org/10.1016/j.annals.2003.06.001
Gill, R. (2007). Postfeminist media culture: Elements of sensibility. *European Journal of Cultural Studies, 10*(2), 147–166. https://doi.org/10.1177/1367549407075898
Gosselink, C., Cox, D., McClure, S., & De Jong, M. (2008). Ravishing or ravaged: Women's relationships with women in the context of aging and Western beauty culture. *International Journal of Aging & Human Development, 66*(4), 307–327. https://doi.org/10.2190/AG.66.4.c
Grogan, S. (2017). *Body image: Understanding body dissatisfaction in men, women and children.* (3rd ed.). Routledge.
Grippo, K., & Hill, M. (2008). Self-objectification, habitual body monitoring, and body dissatisfaction in older European American women: Exploring age and feminism as moderators. *Body Image, 5*(2), 173–182. https://doi.org/10.1016/j.bodyim.2007.11.003
Halliwell, E. (2015). Future directions for positive body image research. *Body Image, 14*, 177–189. https://doi.org/10.1016/j.bodyim.2015.03.003
Haug, F. (1987). *Female sexualization: A collective work of memory.* Trans. E. Carter. Verso.
Hurd Clarke, L., Griffin, M., & Maliha, K. (2009). Bat wings, bunions, and turkey wattles: Body transgressions and older women's strategic clothing choices. *Ageing and Society, 29*(5), 709–726. https://doi.org/10.1017/S0144686X08008283
James, K. (2000). You can *feel* them looking at you": The experiences of adolescent girls at swimming pools. *Journal of Leisure Research, 32*(2), 262–280. https://doi.org/10.1080/00222216.2000.11949917
Jermyn, D. (2016). Pretty past it? Interrogating the post-feminist makeover of ageing, style, and fashion. *Feminist Media Studies, 16*(4), 573–589. https://doi.org/10.1080/14680777.2016.1193371
Jordan, F. (2007). Life's a beach and then we diet: Discourses of tourism and the 'beach body' in UK women's lifestyle magazines. In A. Pritchard, I. Ateljevic, N. Morgan & C. Harris (Eds.), *Tourism and gender: Embodiment, sensuality and experience* (pp. 92–106). CABI Publishing.
Jordan, F., & Aitchison, C. (2008). Tourism and the sexualisation of the gaze: Solo female tourists' experiences of gendered power, surveillance and embodiment. *Leisure Studies, 27*(3), 329–349. https://doi.org/10.1080/02614360802125080
Kilpela, L., Becker, C., Wesley, N., & Stewart, T. (2015). Body image in adult women: Moving beyond the younger years. *Advances in Eating Disorders: Theory, Research and Practice, 3* (2), 144–164. https://doi.org/10.1080/21662630.2015.1012728
Kinnaird, V., & Hall, D. (1996). Understanding tourism processes: A gender-aware framework. *Tourism Management, 17*(2), 95–102.
Lazar, M. (2007). Feminist critical discourse analysis: Articulating a feminist discourse praxis. *Critical Discourse Studies, 4*(2), 141–164. https://doi.org/10.1080/17405900701464816
Liechty, T., Freeman, P., & Zabriskie, R. (2006). Body image and beliefs about appearance: Constraints on the leisure of college-age and middle-age women. *Leisure Sciences, 28*(4), 311–330. https://doi.org/10.1080/01490400600745845
Liechty, T., & Yarnal, C. (2010). The role of body image in older women's leisure. *Journal of Leisure Research, 42*(3), 443–467.
Lövgren, K. (2016). Comfortable and leisurely: Old women on style and dress. *Journal of Women & Aging, 28*(5), 372–385. https://doi.org/10.1080/08952841.2015.1018029
Malacrida, C., & Low, J. (2008). Consumer bodies. In C. Malacrida & J. Low (Eds.), *Sociology of the body: A reader* (pp. 305–307). Oxford University Press.
McKinley, N. (2004). Feminist perspectives and objectified body consciousness. In T. Cash & T. Pruzinsky (Eds.), *Body image: A handbook of theory, research, and clinical practice* (pp. 55–62). The Guilford Press.
McKinley, N., & Hyde, J. (1996). The objectified body conscious scale. *Psychology of Women Quarterly, 20*(2), 181–215. https://doi.org/10.1111/j.1471-6402.1996.tb00467.x
McKinley, N., & Lyon, L. (2008). Menopausal attitudes, objectified body consciousness, aging anxiety, and body esteem: European American women's body experiences in midlife. *Body Image, 5*(4), 375–380. https://doi.org/10.1016/j.bodyim.2008.07.001

Montemurro, B., & Gillen, M. (2013). Wrinkles and sagging flesh: Exploring transformations in women's sexual body image. *Journal of Women & Aging, 25*(1), 3–23. https://doi.org/10.1080/08952841.2012.720179

Meyers, K. (2017). *I Figured You Were Probably Watching Us": Ex Machina and the Performativity of Lateral Surveillance* [M.A. Thesis]. College of William and Mary.

Nash, M. (2015). Shapes of motherhood: Exploring postnatal body image through photographs. *Journal of Gender Studies, 24*(1), 18–37. https://doi.org/10.1080/09589236.2013.797340

Nowell, L., Norris, J., White, D., & Moules, N. (2017). Thematic analysis: Striving to meet the trustworthiness criteria. *International Journal of Qualitative Methods, 16*(1), 160940691773384–160940691773313. https://doi.org/10.1177/1609406917733847

Peat, C., Peyerl, N., & Muehlenkamp, J. (2008). Body image and eating disorders in older adults: A review. *The Journal of General Psychology, 135*(4), 343–358. https://doi.org/10.3200/GENP.135.4.343-358

Pritchard, A., Morgan, N., Ateljevic, I. & Harris, C. (Eds.). (2007). *Tourism and gender: Embodiment, sensuality and experience*. CABI Publishing.

Pruis, T., & Janowsky, J. (2010). Assessment of body image in younger and older women. *The Journal of General Psychology, 137*(3), 225–238. https://doi.org/10.1080/00221309.2010.484446

Sandberg, L. (2013). Affirmative old age – The ageing body and feminist theories on difference. *International Journal of Ageing and Later Life, 8*(1), 11–40. https://doi.org/10.3384/ijal.1652-8670.12197

Sedgley, D., Pritchard, A., & Morgan, N. (2011). Tourism and ageing: A transformative research agenda. *Annals of Tourism Research, 38*(2), 422–436. https://doi.org/10.1016/j.annals.2010.09.002

Shilling, C. (2003). *The body and social theory.* (2nd ed.). Sage.

Small, J. (2004). Memory-work. In J. Phillimore & L. Goodson (Eds.), *Qualitative research in tourism: Ontologies, epistemologies and methodologies* (pp. 255–272). Routledge.

Small, J. (2007). The emergence of the body in the holiday accounts of women and girls. In A. Pritchard, N. Morgan, I. Ateljevic & C. Harris (Eds.), *Tourism and gender: Embodiment, sensuality and experience* (pp. 73–91). CABI Publishing.

Small, J. (2016). Holiday bodies: Young women and their appearance. *Annals of Tourism Research, 58*, 18–32. https://doi.org/10.1016/j.annals.2016.01.008

Small, J. (2017). Women's beach body in Australian women's magazines. *Annals of Tourism Research, 63*, 23–33. https://doi.org/10.1016/j.annals.2016.12.006

Small, J., Harris, C., & Wilson, E. (2008). A critical discourse analysis of in-flight magazine advertisements: The "social sorting" of airline travellers? *Journal of Tourism and Cultural Change, 6*(1), 17–38. https://doi.org/10.1080/14766820802140422

Tiggemann, M. (2004). Body image across the adult life span: Stability and change. *Body Image, 1*(1), 29–41. https://doi.org/10.1016/S1740-1445(03)00002-0

Tiggemann, M., & Andrew, R. (2012). Clothing choices, weight, and trait self-objectification. *Body Image, 9*(3), 409–412. https://doi.org/10.1016/j.bodyim.2012.02.003

Tiggemann, M., & Lacey, C. (2009). Shopping for clothes: Body satisfaction, appearance investment, and functions of clothing among female shoppers. *Body Image, 6*(4), 285–291. https://doi.org/10.1016/j.bodyim.2009.07.002

Tiggemann, M., & McCourt, A. (2013). Body appreciation in adult women: Relationships with age and body satisfaction. *Body Image, 10*(4), 624–627. https://doi.org/10.1016/j.bodyim.2013.07.003

Tortajada, I., Dhaenens, F., & Willem, C. (2018). Gendered ageing bodies in popular media culture. *Feminist Media Studies, 18*(1), 1–6. https://doi.org/10.1080/14680777.2018.1410313

Van Bauwel, S. (2018). Invisible golden girls? Post-feminist discourses and female ageing bodies in contemporary television fiction. *Feminist Media Studies, 18*(1), 21–33. https://doi.org/10.1080/14680777.2018.1409969

Waskul, D., & Vannini, P. (2006). Introduction: The body in symbolic interaction. In D. Waskul & P. Vannini (Eds.), *Body/embodiment: Symbolic interaction and the sociology of the body* (pp. 1–18). Ashgate.

Woekel, E., & Ebbeck, V. (2013). Transitional bodies: A qualitative investigation of postpartum body self-compassion. *Qualitative Research in Sport, Exercise and Health, 5*(2), 245–266. https://doi.org/10.1080/2159676X.2013.766813

World Commission on Environment and Development. (1987). *Our common future*. Oxford University Press.

Cultivating success: personal, family and societal attributes affecting women in agritourism

Ann E. Savage, Carla Barbieri and Susan Jakes

ABSTRACT

The public's burgeoning interest in authentic place-based experiences and local foods, as well as farmers' interest in increasing their incomes continue driving the development of agritourism. Although women often initiate agritourism, scant information identifies the factors contributing to or hindering their success. Therefore, this study combines feminist and systems approaches to identify the factors affecting women achievements in their functions as farmers and entrepreneurs and in various aspects of their lives. With such an aim, 216 women farmers in North Carolina (USA) were surveyed in 2017. Descriptive and inferential statistics of data collected indicate that women felt moderately successful in their farmer and entrepreneur roles, while self-fulfillment and business continuance appeared as the most important dimensions of their success. Significant models indicate that attributes at the personal, farm household, and society levels predict women's perceived functional success and the importance of different life aspects to their success. Study results move the literature of women in agritourism beyond entrepreneurial motivations by providing a thorough understanding of how gendered nuances in agriculture affect farm women's success in a holistic manner. This study also contributes to the sustainability of the agritourism practice by identifying attributes that increases chances of success among women farmers.

Introduction

Family farms are seeking to advance their household incomes by reallocating farm resources (land, labor, or capital) to develop new on-farm enterprises that can allow them to respond to emerging market opportunities and capture customers more directly (Barbieri et al., 2008; Meert et al., 2005). An on-farm enterprise that continues increasing prominence in the United States of America (USA) involves the offering of education and recreation to visitors (e.g., school tours, corn mazes, wineries), usually referred as agritourism (Gil Arroyo et al., 2013). The increase of agritourism operations emerges from supply and demand forces occurring in parallel. From the supply side, farmers are venturing into agritourism mainly due to economic motivations (McGehee & Kim, 2004; Nickerson et al., 2001; Ollenburg & Buckley, 2007). On the demand side, the public's increased interest in (re)connecting with local food systems, especially among urban dwellers, is encouraging travel to working farms (Kline et al., 2016; Schilling et al., 2012).

Ontological and epistemological advances in the scholarship of agritourism have occurred along with its sophistication in the practice (Barbieri, 2020). Such paralleled progress has consolidated agritourism as a type of agricultural enterprise occurring on working agricultural settings, rather than rural landscapes (Gil Arroyo et al., 2013). This agricultural dependence has stimulated the development of a wide range of tourism offerings throughout the world where localized uniqueness emerges from the farm's resources and the surrounding cultural and natural landscapes (Barbieri & Streifeneder, 2019; Gao et al., 2014). Agritourism research is predominantly framed within utilitarian lenses which stress the different economic, socio-cultural and environmental benefits that agritourism delivers to farmers, visitors, and the greater society (Barbieri, Stevenson, & Knollenberg, 2019). Altogether, these benefits support the greater sustainability of agritourism as compared to other agricultural enterprises (Barbieri, 2013). Agritourism increases farm profits, creates jobs, preserves farmlands, conserves natural and cultural heritage, boosts local economies, and stimulates the consumption of local and sustainably-produced foods, to a name a few (Brune et al., 2020; Kline et al., 2016; LaPan & Barbieri, 2014; Schilling et al., 2012; Veeck et al., 2006). Still more investigation on issues related to agritourism development and management is needed to increase its sustainability (Yang, 2012; Phelan & Sharpley, 2012).

Agritourism sustainability can be enhanced by further examining and then supporting the success of women operators. More often, women drive agritourism development not only as the initiators, but also as the ongoing managers who constantly seek innovations to maintain visitors' interests (McGehee et al., 2007). This may be due to their historic role in re-purposing farm resources to contribute to farm survival (Anthopoulou, 2010; Gasson & Winter, 1992; Wright & Annes, 2016). Despite women's importance in agritourism, evidence indicates that their efforts are not fully rewarded. A study conducted across the USA showed that women in agritourism make less gross income than their male counterparts (Barbieri & Mshenga, 2008). Although not directly investigated, potential reasons for the gap in agritourism economic returns between genders could be women's limited access to agricultural and financial networks (Che et al., 2005), and their expected domestic roles as caregivers (Anthopoulou, 2010; Gasson & Winter, 1992). The hegemonic masculinity still prevailing in some rural areas and within agricultural policies could also explain this economic gap as disproportionate access to resources advances men's opportunities for agricultural engagement and entrepreneurial success (Bock, 2015; Halim, 2016; Little & Jones, 2000). As a social construct, hegemonic masculinity (re)produces beliefs that "tend to legitimatize patriarchy as the apparent 'natural' order of things" (Vavrus, 2002, p. 357). Main patriarchal practices prevailing in agriculture and rural entrepreneurship include prejudices toward changing gender roles within the family (e.g., caregiving vs. breadwinning) and the farm (e.g., physical vs. soft tasks), gender-based inequalities to access resources (e.g., local networks, subsidies), and perceived limited skills to undertake agricultural and entrepreneurial tasks (Anthopoulou, 2010; Bock, 2004; Carter, 2017; Halim, 2016).

The economic underperformance of women in agritourism could also be attributed to gendered differences in entrepreneurial motivation. Women in agritourism prioritize providing employment for family members and giving back to the community to a greater extent than their male counterparts (McGehee et al., 2007). Those priorities also determine women's notion of success as having a more comprehensive approach beyond economic indicators (Halim et al., 2020). Research in business success has long focused on measuring the profitability and growth of businesses, where men are commonly found to be more successful than women. This financial success undervalues other factors, such as one's presence in the community and mentoring of future agricultural operators (Aldrich, 1989). However, the similar survival rates between women and men owned firms suggest that profit and growth may not be the most accurate indicators of success among women (Kalleberg & Leicht, 1991). This has prompted a robust research line to understand gender differences on the motivations driving business development, with key findings of women prioritizing personal growth and social contributions at the expense of financial goals (Buttner & Moore, 1997; Carter et al., 2003; Greene, Hart, Gatewood, Brush, & Carter et al.,

2003). Yet, the literature has called for a more thorough investigation of the extent to which women perceive their success, particularly in the context of patriarchal industries, and how their success transforms patriarchal societies (Halim et al., 2020; Snyder-Hall, 2010).

While the literature on women in agriculture and agritourism in the USA has expanded in recent years, more research-based education on how to support women as agritourism operators is needed to increase their chances of success and to achieve gender equity in agriculture. Given the existence of gender comparison studies pointing out existing gaps, a feminist approach is suitable to identify factors enabling or hindering the success of agricultural women in view of patriarchal systems dictating norms restraining women's behaviors and beliefs (Ahl, 2006; Snyder-Hall, 2010). Although information on challenges related to gender roles and the hegemonic masculinity of entrepreneurial agriculture is available in the literature (Anthopoulou, 2010; Bock, 2004, 2015; Carter, 2017; Little & Jones, 2000), limited information exists on the context of agritourism related to women's success (Halim et al., 2020). In filling such a knowledge gap, it is critical to recognize the interplay of the many factors that may affect women's performance (Eger et al., 2018; Mooney, 2016). Thus, this feminist quantitative investigation will move forward current knowledge in agritourism by identifying the attributes of the farmer (values), farm household (family dynamics, farm business), and society (trends, challenges) that exert an influence on women's success. Doing so, will provide pertinent information to support women operators in fortifying their agritourism enterprises sustainably.

Literature review

Constructing women's success: functional and life-aspects considerations

Success is a complex construct that can be evaluated in multiple ways. One way is focusing on the extent to which individuals perceive achievement in a defined role (e.g., caregiver, farmer) they hold, which is known as *Functional Success*. However, a set of factors, such as personal values, family attributes or cultural norms can shape the extent of perceived functional success (Eccles, 1987, 1994). Specifically, the dissonance between fulfilling roles traditionally considered masculine and their feminine identity can affect the extent of women's functional success (Horner, 1972). This is the case of women in agritourism, whose functional success as farmers and entrepreneurs confronts the historically dominant male representations in the practice of agriculture and business and requires re-defining these roles in the current agricultural context (Sachs et al., 2016; Wright & Annes, 2016). Thus, the literature has called for a more thorough investigation of the extent to which women perceive their functional success, particularly in the context of patriarchal industries, and how their success transforms patriarchal societies (Snyder-Hall, 2010).

Likewise, masculine approaches centered on profits and growth have dominated the assessment of business success (Aldrich, 1989; Kalleberg & Leicht, 1991). Yet, evidence suggests that women have a more holistic vision of business success that also accounts for non-monetized *life-aspects* such as personal growth and social contributions (Carter et al., 2003; Walker & Brown, 2004). Although most agritourism studies recognize that farmers pursue both internal (e.g., family connections) and external (e.g., market opportunities) goals (Nickerson et al., 2001; Ollenburg & Buckley, 2007; Tew & Barbieri, 2012), very few studies have accounted for gender differentiation in this pursuit. As such, the literature calls for further scrutiny beyond gendered motivations (McGehee et al., 2007) and the degree to which internal factors in terms of personal fulfillment (e.g., pursuing happiness) and social contribution (e.g., broad community impact) effect women's sense of holistic success (Halim et al., 2020).

Success among women in agriculture

In the context of agriculture, success incorporates perceptions of farmer's social contributions and personal achievements. Social contributions relate to the farmer's sense of responsibility toward the community, including the public (Walker & Brown, 2004). While social contribution has little relevance in determining the success of women in traditional businesses (Buttner & Moore, 1997), it is very important for women in agricultural businesses because it gives them the opportunity to expand their personal connections, educate customers, and promote cultural traditions and heritage (Anthopoulou, 2010; Gasson, 1973). Social contribution is also very important to the success of women in agritourism. Women tend to emphasize the opportunity that agritourism brings to educate visitors on issues related to food and farming (McGehee et al., 2007; Wright & Annes, 2016) and to mentor young employees (Halim et al., 2020).

Perceptions of personal achievements as an indicator of women's success in agriculture includes several components. First, it means breaking the hegemonic gendered ideology of both entrepreneurship and agriculture (Brandth & Haugen, 2011; Stead, 2017) by seeking recognition for women's on-farm contributions from their male counterparts and from the wider community (Driga et al., 2009; Wright & Annes, 2016). In turn, such recognition is necessary for women to immerse themselves in community networks that enable agritourism success (Che et al., 2005). Personal achievement also comes with the self-fulfillment of one's deepest desires (Buttner & Moore, 1997; Greene et al., 2003). The latter could be described as an overall feeling of satisfaction and enjoyment from work (Halim, 2016; Markantoni & van Hoven, 2012), or more specifically in terms of work-life balance (Walker & Brown, 2004) or of professional growth (Buttner & Moore, 1997; Halim, 2016).

Influencers of success of women in agritourism: a systemic approach

The use of systemic approaches to examine agricultural issues is essential given the interrelatedness among the farmer, farm household and community (Dogliotti et al., 2014), and to account for the mix of internal and external factors exercising pressure on the farm business and farmer's lifestyle (Ikerd, 1993). As such, systemic approaches have been valuable to ensure that each aspect of agriculture (e.g., labor, land) is assessed when farms shift with entrepreneurial diversification (Giampietro, 1997). Building on agricultural systemic approaches, Barbieri (2017) developed the Agritourism System's Approach that seeks to capture the interconnections among the entrepreneurial farmer (as the system's nucleus), within expanding concentric circles (*layers*) representing the farm household (including both the family dynamics and business). Thus, holistic evaluations of agritourism should take into account aspects from each layer as they can come together to either facilitate or limit the success of agritourism ventures (Barbieri, 2017).

At the farmer (nucleus) level it is important to take into account the symbolic values and self-identity of farmers. These values and identities have formed throughout the historical evolution of agriculture in the USA (Burton, 2004) and are changing due to major industry (e.g., high-mechanization) and political (e.g., selected subsidies) forces (Burton & Wilson, 2006). Although research efforts have been devoted to gain a deeper understating of the evolving farmer's identities (e.g., producer, steward) and roles (Ferrell, 2012), limited information is available on women's identities on the interaction of their numerous farm roles with these identities. Women's identity on the farm are oftentimes tied to her role in the farm household as mother, wife or nurturer, making her family caregiver role as primary and farmer as secondary (Anthopoulou, 2010; Gasson & Winter, 1992). In agritourism, farmer identities may also shift as women take more prevalent responsibilities in business decision-making (Bock, 2004; Brasier et al., 2014).

Given the interdependence between business profit growth and household utility maximization within family farming, the *farm household* construct encompasses aspects of both, the farm business and the farm family (Benjamin, 1994). As such, factors at the farm household layer affecting women's success may include those associated with family dynamics as well as those related to the business assets (Barbieri, 2017). Agritourism operations become more complicated because of the need to reallocate family and business resources across different enterprise lines (Barbieri et al., 2008; Schilling et al., 2014). The administrative burdens that agritourism bears to the farm household represent increased time dedicated to record keeping, which is often placed on the woman in the family (Whatmore, 1991). In addition, the work seasonality of agritourism can increase the challenge of finding reliable labor to work on the farm (Halim, 2016; Kline & Milburn, 2010). Suess-Reyes and Fuetsch (2016) also suggested family dynamics related to intergenerational authority and adaptability within the farm business should be further investigated.

Societal traits (e.g., gendered systems, social trends) can hinder or facilitate the success of women in agritourism. Some farming communities uphold practices that perpetuate both, the patriarchal nature of agriculture and gendered expectations in the farm family (Little & Jones, 2000; Wright & Annes, 2014). Patriarchal agricultural systems tend to ostracize women operators implementing managerial changes (Carter, 2017), which obstructs their ability to access pooled resources (e.g., knowledge) from community networks (Wang, 2010). Ostracism is especially impactful for women in agritourism because access to local networks and resources are critical for agritourism success (Che et al., 2005; Li & Barbieri, 2019). Gendered expectations can challenge women in agriculture as they are often responsible for the domestic chores of the family (caregiving role) in addition to their involvement in farm tasks (Bock, 2004; Dogliotti et al., 2014; Gasson & Winter, 1992; Whatmore, 1991). Gendered expectations particularly affect women in agritourism because the additional workload to accommodate farm visitors adds to women's existing family and agricultural responsibilities (Anthopoulou, 2010; Halim, 2016).

Several societal trends currently taking place in agriculture are influencing the success of women in agritourism. Although women have always participated in many farming tasks, their increased presence as primary operators is promoting their strategies for progressing out of their hidden place and aiding their move toward more agency, especially in local foods production and sustainable agriculture (Ball, 2014; Hoppe & Korb, 2013). Yet, the impact of women's leadership in sustainable agriculture and community based food system organizations has not reached local, state or national governments as these continue developing and enforcing masculine policies that inherently challenge women's success (Alston, 2003; Bock, 2015). This context of women's increased agricultural prominence within masculine dominant systems calls for the (re)evaluation of women's situation through feminist lenses to uncover how agricultural trends and gendered challenges are affecting their abilities to attain this success.

Data and methods

Study purpose and epistemological framework

Based on the extant literature, this study was designed to contribute to the understanding of the success of women in agritourism addressing three specific objectives. Given that agritourism is considered a form of entrepreneurial farming (Barbieri et al., 2008; Tew & Barbieri, 2012), the first objective is to evaluate women's perceived success in their roles as a farmer and as an entrepreneur (*functional success*). Recognizing that success is a complex construct built upon the fulfillment of various individual pursuits (Carter et al., 2003; Sachs et al., 2016; Walker & Brown, 2004), the second objective is to evaluate the importance of different life aspects to the contribution of women's success (*life-dimensional success*). Given that assessments concerning agritourism (success, failures) should be framed within a system approach comprising elements at the personal, farm household and society levels (Barbieri, 2017), the last objective is to identify farmer values, farm family dynamics and business characteristics, and societal traits associated with women's success (Figure 1). In this study, references to 'women' purposely reflects the

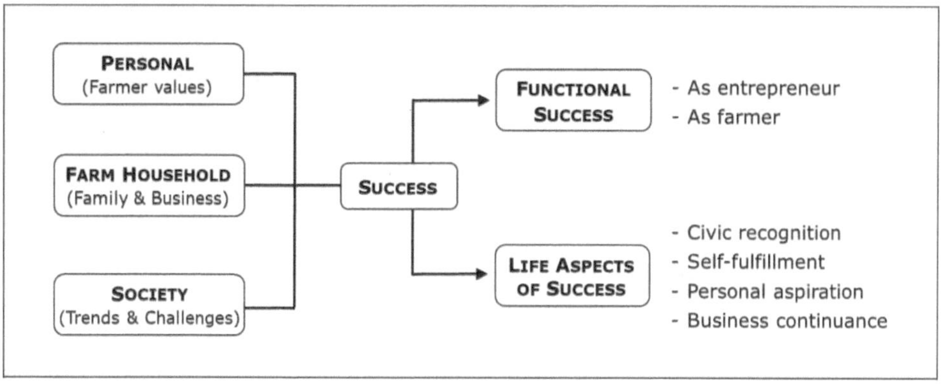

Figure 1. Study research model.

socially-constructed notion of gender beyond virtue of biological sex, denoting how roles for women and men are enacted and perceived according to social norms (Mooney, 2020).

This study, designed as descriptive and relational in nature, was framed within a feminist approach that allows challenging social norms that neglect or negate women's experiences beyond the mere description of women's condition or between-gender differences (Aitchison, 2000). Adopting a feminist approach is particularly important for this study because the extant entrepreneurial literature tend to make between-gender comparisons superficially, overlooking cultural standards of masculine power and ideologies (Stead, 2017). Having a more inquisitive perspective is especially critical in research involving agricultural contexts where women have long been relegated to the private sphere (Alston, 2003). More specifically, this study adopts a radical or critical feminist paradigm as the authors posit patriarchy as the social structure that have created gendered social roles (Mooney, 2020) and caused women's subordination and oppression in the private and public spheres within their lives (Aitchison, 2000, Parry et al., 2013).

By positioning this study within a critical feminist lens, the authors explicitly seek the political change of improving women's condition in society (Aitchison, 2000) through the identification of the factors constraining and enabling the success of women in agritourism. In brief, adopting a critical feminist approach is especially important in this study to highlight and examine women's roles in entrepreneurial and agricultural contexts that are often overlooked and misunderstood (Brandth, 2002; Midgley, 2006; Walker & Brown, 2004), and to account for the continued existence of agricultural and social patriarchal systems (Halim, 2016; Snyder-Hall, 2010). This critical feminist approach of the agritourism context that incorporates socially-constructed gendered norms (challenges and opportunities) at the personal, farm household, and society levels responds to the call to advance the investigation of gender by using multi-level (micro, meso, macro) approaches (Mooney, 2020).

Sampling and survey instrument

The theoretical study population was women working on agritourism farms in North Carolina (NC), which actual number and characteristics are unknown. Given that a list of such population is not readily available, study participants were identified through snowball sampling. This non-probability sampling technique starts with identifying an initial group of people fitting the study criteria (i.e., being a woman, a farmer, and involved in agritourism), who along with their participation are asked to share the survey instrument with more possible participants or suggest contacts to the research team (Babbie, 2013). The study's initial list was constructed following a systematic internet search for NC agritourism farms using key words (e.g., corn mazes, u-pick) and from specialized listings (e.g., NC County Visitor Bureaus). The initial sample frame included 243 farms, which included the contact information of farmers without any gender reference. Then, study participants were asked to refer the survey to other women farmers they may know either by forwarding the survey link directly to them or by providing their contact information to the research team. To expand the sample, the research team asked selected agriculture and agritourism organizations (e.g., Blue Ridge Women in Agriculture, NC Agritourism Networking Association) and agents of the NC Cooperative Extension to promote the survey and encourage participation among their constituents.

Informed by the literature, a survey instrument was developed to collect information on participants' perception of success through two types of indicators. General assessments of functional success (i.e., extent of success tied to an individual's role as a farmer and as an entrepreneur) were queried through five-point unipolar scales to capture the cumulative presence of success (thus no need of neutral point) from "not at all successful" (1) to "extremely successful" (5). The level of importance of 19 life aspects contributing to women's success (Buttner & Moore, 1997; Halim, 2016; Walker & Brown, 2004) was queried through a series of five-point Likert scales in a bipolar mode which opposing anchors (1 = very unimportant; 5 = very important) and neutral option (3 = neither) are suitable to capture participants' attitudes. These indicators represent aspects of women's personal achievement (e.g., similar recognition as male farmers, pursuing happiness) and social contribution (e.g., mentoring young employees; setting an example for other women farmers).

The survey also collected information on a suite of personal, farm household and societal characteristics that may affect the success of women in agritourism to account for the intertwined relationships across farmers, farm family and business, and community occurring in agricultural enterprises, including agritourism (Barbieri, 2017; Dogliotti et al., 2014; Suess-Reyes & Fuetsch, 2016). At the personal level, respondents were queried on the cumulative relevance of a mix of farmer values to capture women farmers' identities (Burton, 2004; Burton & Wilson, 2006; Brasier et al., 2014). Specifically, value assessments of nine items denoting conservationist, productivist, and civic-minded identities were gathered (McGuire et al., 2015) through a series of five-point unipolar Likert scales (1 = not at all valuable; 5 = extremely valuable). Farm household information collected included indicators of the family dynamics (e.g., number of generations in farming), farm economic standing (e.g., farm gross sales), and extent of agritourism involvement (e.g., seasonality) that previous studies identified as determinants of success (Che et al., 2005; Li & Barbieri, 2019; Schilling et al., 2014; Tew & Barbieri, 2012; Wang, 2010). The survey also queried the extent to which societal traits either constrain or facilitate the success of women farmers in agritourism (Ball, 2014; Bock, 2004; Brasier et al., 2014; Carter, 2017; Hoppe & Korb, 2013; Little & Jones, 2000) through four-points unipolar scales to capture their cumulative effect in reality (1 = not at all; 4 = very much). Constraints included gendered traits related to women's caregiver role (6 items) and the patriarchal agricultural system (8 items) while facilitators were related to public awareness (4 items) and the changing social fabric (3 items) that were compiled from the literature (Anthopoulou, 2010; Halim, 2016; Halim et al., 2020; Wright & Annes, 2014; Whatmore, 1991). Socio-demographic attributes (e.g., age, level of formal education) were also queried.

Data collection and statistical analysis

Data were collected using printed and electronic questionnaires in parallel to account for participants' response preferences, streamline time efficiency for data input, and reduce expenses (Dillman et al., 2009). Both formats had identical content and word choices; yet, the online version included logic-branching patterns to skip sections that were not applicable to respondents. Prospective participants were invited to participate using online and mailed communications upon availability of appropriate contact information. An incentive of a chance to win one of five \$50 gift cards was used to encourage participation. Data were collected in early 2017 and spanned three months. A total of three e-mail reminders, a postcard and a second round of surveys were sent to non-respondents. A total of 180 valid responses were obtained, from which 140 respondents were from the initial sampling list (59.3% response rate). Levene's Test for Equality Variance conducted between mailed ($n = 75$) and online ($n = 65$) respondents of key variables supported merging both datasets. Similar tests conducted between early and late responses provided assurance of non-response bias (Armstrong & Overton, 1977).

Given the snowball sampling technique used, several respondents did not fit the population criteria (e.g., women farmers not involved in agritourism, male agritourism operators). Thus, statistical analysis included only 116 cases of women farmers involved in agritourism. First, descriptive statistics were conducted to profile respondents' socio-demographic characteristics, personal values, perceived success, and factors enabling or constraining their perceived success. Then, exploratory factor analysis with varimax rotation was used to reduce the life-aspects contributing to women's success, agritourism challenges and agriculture trends to fewer dimensions to understand the dimensional structure of the constructs as well as facilitate further analysis. Eigenvalues over one and loadings over 0.5 were used as thresholds in the factor analysis; a pairwise method was used to handle missing values due to the limited number of respondents (Field, 2013). Cronbach's alphas were computed to confirm the reliability of the farmers' values dimensions (conservationist, productivist, civic minded), challenges related to gender (caregiver role, patriarchal agricultural system) and societal trends facilitating agritourism success (public awareness, social fabric). Aggregated means of each of the later dimensions (values, challenges, facilitators) were computed for further analysis.

Finally, a series of multiple linear regressions were used to identify the extent to which personal values, farm family attributes and societal traits (independent variables) predicted functional success and importance of life-aspects in women's success (dependent variables). Specific independent variables included the aggregated means of the personal values dimensions (productivist, conservationist, civic-minded), three descriptors of the farm family (likelihood of passing the farm on, extent of agritourism diversification, agritourism seasonality), and the aggregated means of four societal traits comprising perceived facilitators (public awareness, changing social fabric) and gendered challenges (caregiver role, patriarchal agricultural system).

Results

Respondents profile

The typical respondent was a middle-aged ($M = 49$ years old) and highly educated woman (Table 1). A relatively large proportion of respondents (35.9%) were young women (less than 45 years old) while most (55.7%) were middle aged (between 45 and 64 years old); only 8.4% were of retirement age (65 years old). Over two-thirds (74.8%) of respondents reported at least a four-year college degree and 32.2% held a post-graduate degree. The vast majority (90.1%) of respondents lived with their spouse or significant other. A relative small proportion (8.1%) lived with at least one child six years old or younger; one quarter (25.2%) reported living at home with at least one child of at least 16 years old. The largest proportion of responding women work on the farm either full-time (42.3%) or part-time (17.7%); they also reported, although to a lesser extent, off-farm employment either full-time (12.0%) or part-time (10.3%). Although most respondents (52.6%) were first generation farmers, half (50.8%) reported that they would likely or very likely to pass the farm on to the next generation. Most respondents (52.8%) indicated making less than \$50,000 in gross farm sales in 2016; a relatively small proportion (16.3%) reported farm gross income of at least \$250,000.

Table 1. Respondent's socio-demographic and farm family characteristics.

Farmer & Farm Family Indicators	Number	Percent
Farmer's Age ($n = 106$)		
Less than 35 years old	15	14.2%
35–44 years old	23	21.7%
45–54 years old	29	27.4%
55–64 years old	30	28.3%
65 years old or older	9	8.4%
Mean (in years)		***(49.0)***
Farmer's Level of Education ($n = 115$)		
High school graduate or less	0	0.0%
Some college	16	13.9%
Technical degree (2-year degree)	13	11.3%
Four-year college degree	49	42.6%
Post graduate studies	37	32.2%
Household Composition ($n = 111$)[a]		
Spouse or significant other	100	90.1%
Children 6 years or younger	9	8.1%
Children 7–15 years old	26	23.4%
Children 16–20 years old	16	14.4%
Children 21 years or older	12	10.8%
Type of Employment ($n = 116$)[a]		
Full-time farming	74	42.3%
Part-time farming	31	17.7%
Full-time off-farm job	21	12.0%
Part-time off-farm job	18	10.3%
Homemaker	16	9.1%
Retired	11	6.3%
Other	4	2.3%
Generations on the Farm ($n = 116$)		
First generation	61	52.6%
2 generations	9	7.8%
3 generations	17	14.7%
4 generations or more	26	22.4%
Do not know	3	2.5%
Likelihood of Farm Succession ($n = 116$)		
Very likely	39	33.6%
Likely	20	17.2%
Undecided	17	14.7%
Unlikely	15	12.9%
Very unlikely	17	14.7%
Do not know	8	6.9%
Farm Gross Income in 2016 ($n = 110$)		
Less than $1,000	3	2.7%
$1,000–9,999	20	18.2%
$10,000–49,999	35	31.9%
$50,000–99,999	20	18.2%
$1,00,000–249,999	14	12.7%
$250,000–499,999	11	10.0%
$500,000 or more	7	6.3%

[a]Adds to more than 100% because participants could check more than one response.

The survey captured a mélange of agritourism operations regarding maturity, seasonality, number of visitors and economic relevance (Table 2). In terms of years involved in this industry, 42.1% were emerging agritourism farms with less than five years of receiving visitors while 31.6% have been doing so for at least ten years ($M = 9.0$ years). Most (52.2%) have limited agritourism offerings with three or less different types of activities. As for seasonality, 20.0% offer agritourism for less than four months per year while 34.5% do so year round ($M = 7.9$ months per year). Agritourism variability was likewise diverse in terms of number of visitors; while 26.3% hosted less than 150 visitors on their farm in 2016, 40.0% indicated hosting at least 1,000 visitors ($M = 3{,}550$ visitors). Consistent with the number of visitors reported, 33.1% reported that all or most of their farm revenues come from their agritourism activities while only 5.2% indicated not

Table 2. Agritourism profile of participating farmers.

Agritourism Indicators	Total	Percent
Years Offering Agritourism ($n = 114$)		
Less than 2 years	12	10.5%
2–4 years	36	31.6%
5–9 years	30	26.3%
10–19 years	22	19.3%
20 years or more	14	12.3%
Mean (in years)		***(9.0)***
Number of Agritourism Activities Offered ($n = 113$)		
1–3 activities	59	52.2%
4–6 activities	44	38.9%
7–10 activities	10	8.9%
Mean (in activities)		***(3.7)***
Agritourism Seasonality - Months per Year ($n = 110$)		
Less than 4 months	22	20.0%
4–6 months	19	17.3%
7–11 months	31	28.2%
12 months (year round)	38	34.5%
Mean (in months)		***(7.9)***
Number of Visitors in 2016 ($n = 95$)		
Under 150 visitors	25	26.3%
150–499 visitors	23	24.2%
500–999 visitors	9	9.5%
1,000–4,999 visitors	20	21.1%
5,000 or more visitors	18	18.9%
Mean (in number of visitors)		***(3,550)***
Proportion of Farm Revenues from Agritourism ($n = 114$)		
All	11	9.6%
Most	27	23.5%
Some	47	40.8%
Very little	24	20.9%
None	6	5.2%

receiving revenues from agritourism. Respondents also reported offering a variety of farm recreation and hospitality, agricultural education, and outdoor recreation activities (Table 3). Overall, the most commonly offered activities were educational activities, such as classes, workshops, and school tours (81.4%), followed by festivals or events (53.9%) and farm based recreational activities (45.2%).

In terms of farming values, the most valued practices among participants were being active in the community ($M = 4.48$) and minimizing soil erosion ($M = 4.25$) which represent indicators of different value dimensions (Table 4). Reliability tests showed a moderate to strong internal consistency of the overall farmers' values scale ($\alpha = 0.722$) and of each of the three dimensions: civic minded ($\alpha = 0.\ 774$), conservationist ($\alpha = 0.776$) and productivist ($\alpha = 0.618$). Taken by dimensions, results indicate that responding women highly regard civic-minded ($M = 4.13$, $SD = 0.797$) and conservationist ($M = 4.12$, $SD = 0.899$) values. Specifically, the majority indicated that being active in their community (62.8%), maintaining organic matter (55.8%) and minimizing soil erosion (52.1%) were extremely valuable to them as farmers. The productivist dimension ($M = 3.26$, $SD = 0.860$) was the least regarded; yet most respondents indicated that obtaining the highest yields per acre was very or extremely valuable (57.0%).

Women's success

Women felt at least moderately successful in their roles (*functions*) as farmers and entrepreneurs (Table 5). About half of respondents indicated feeling at least very successful as a farmer (42.3%; $M = 3.42$) and entrepreneur (45.7%; $M = 3.41$). Respondents considered different aspects of their own lives contributing to their sense of overall success (Table 6). More specifically, remaining

Table 3. Agritourism activities offered by participating women farmers.

Agritourism Activities ($n = 115$)	Number	Percent[a]
Farm Recreation and Hospitality		
Festivals or events	62	53.9%
Farm-based recreational activities	52	45.2%
Meals	33	28.7%
Overnight stays	18	15.7%
Non-farm recreational activity	9	7.8%
Agricultural Education		
Educational activities	92	81.4%
Farm hands-on activities	51	44.3%
Observation of agricultural processes	26	22.6%
Outdoor Recreation		
Nature observation activities	36	31.3%
Physically active activities	15	13.0%
Wildlife extractive activities	11	9.6%
Summer camps	6	5.3%
Other		
Other activities	6	5.3%

[a]Adds to more than 100% because participants could check more than one response.

Table 4. Farming values among responding women.

Values by Dimensions ($n = 115$)[a]	Not at all valuable	Slightly valuable	Moderately valuable	Very valuable	Extremely valuable	Mean[b]	*S.D.*
Civic-minded **(α = 0.774)**							
Be active in your community	0.0%	3.5%	8.0%	25.7%	62.8%	4.48	0.792
Be a leader in your community	3.4%	6.3%	15.2%	31.3%	43.8%	4.05	1.081
Participate in farm-related organizations	0.9%	8.7%	22.6%	36.5%	31.3%	3.89	0.980
Composite Mean						***4.13***	***0.797***
Conservationist **(α = 0.776)**							
Maintain organic matter	5.3%	3.5%	15.9%	19.5%	55.8%	4.17	1.149
Minimize soil erosion	1.8%	1.8%	18.6%	25.7%	52.1%	4.25	0.940
Minimize nutrient runoff	2.8%	11.9%	16.5%	26.6%	42.2%	3.94	1.149
Composite Mean						***4.12***	***0.899***
Productivist **(α = 0.618)**							
Have the highest yields per acre	8.8%	4.4%	29.8%	31.6%	25.4%	3.61	1.172
Use the latest technology	4.5%	19.3%	32.1%	23.9%	20.2%	3.36	1.143
Have the most up-to-date equipment	9.8%	33.0%	36.7%	12.5%	8.0%	2.76	1.059
Composite Mean						***3.26***	***0.860***

[a]Overall reliability $\alpha = 0.722$.
[b]Measured on a 5-point unipolar scale, ranging from 'Not at all valuable' (1) to 'Extremely valuable' (5).

mentally creative ($M = 4.75$), pursuing happiness ($M = 4.73$) and building long lasting relationships with customers ($M = 4.73$) were the most important life aspects to their sense of success. Conversely, receiving similar recognition as male farmers ($M = 3.78$), being an "agritourism" expert ($M = 3.78$) and receiving community recognition ($M = 3.75$) were the least important aspects contributing to their sense of success, yet still important. Factor analysis of the life aspects contributing to women's sense of success resulted in a significant model of four factors (total variance = 52.8%; $\alpha = 0.835$; KMO = 0.791; $p < 0.001$). Being an expert in some aspect of farming, remaining mentally creative, having flexibility in work hours, and passing the farm to the next generation within the family were removed from further analysis because did not load in any factor or showed multiple loadings.

The first factor was named *Civic Recognition* because it captured women's involvement in different communities, such as the farming community and women farmers, as well as their broader community (variance = 28.3%; eigenvalue = 5.383; $M = 4.11$). The second factor, *Self-fulfillment* described the overall feeling of satisfaction and enjoyment that women value in different aspects of their lives, such as happiness and having good work-life balance (variance = 10.1%;

Table 5. Indicators of functional success among responding women.

Functional Success ($n = 117$)	Not at all successful	Slightly successful	Moderately successful	Very successful	Extremely successful	Mean[a]	*S.D.*
As a farmer	0.9%	10.3%	46.6%	30.2%	12.1%	3.42	0.866
As an entrepreneur	0.9%	15.5%	37.9%	32.8%	12.9%	3.41	0.933

[a]Measured on a 5-point unipolar scale, ranging from 'Not at all successful' (1) to 'Extremely successful' (5).

Table 6. Rotated factor matrix of the importance of life aspects contributing to women's success.

Success by Factors ($n = 117$)[a]	Mean[b]	Factor Loadings	Explained Variance	Eigenvalue
Civic Recognition	***4.11***		***28.3%***	***5.383***
Giving back to community	4.50	0.559		
Being part of the local farming community	4.34	0.683		
Setting an example for other women farmers	4.30	0.718		
Gaining respect from other farmers	4.01	0.800		
Receiving similar recognition as male farmers	3.78	0.679		
Being recognized by my community	3.75	0.792		
Self-fulfillment	***4.62***		***10.1%***	***1.925***
Pursuing happiness	4.73	0.752		
Having a good work-life balance	4.60	0.757		
Family supporting my farming career	4.52	0.658		
Personal Aspirations	***4.34***		***8.3%***	***1.576***
Educating the public about farming	4.64	0.751		
Having an active lifestyle	4.59	0.658		
Being an "agritourism" expert	3.78	0.588		
Business Continuance	***4.41***		***6.0%***	***1.147***
Building long-lasting relationships with customers	4.73	0.703		
Earning a good income from the farm	4.34	0.532		
Mentoring young employees	4.15	0.535		

[a]Total variance explained: 52.8%; KMO = 0.791; overall $\alpha = 0.835$.
[b]Measured on a 5-point Likert bipolar scale, ranging from 'Very unimportant' (1) to 'Very important' (5).

eigenvalue = 1.925; $M = 4.62$). The *Personal Aspirations* factor included elements related to individual's ambition such as being physically active and becoming an agritourism expert (variance = 8.3%; eigenvalue = 1.576; $M = 4.34$). The final factor, *Business Continuance*, comprised three items that contribute to the viability of the business in terms of customers, employees and farm income (variance = 6.0%; eigenvalue = 1.147; $M = 4.41$).

Responding women perceived that there are societal patterns hindering their success as farmers resulting from their expected *Caregiver Role* ($\alpha = 0.868$, $M = 3.22$) and the *Patriarchal Agricultural System* ($\alpha = 0.767$, $M = 2.98$), although the latter to a lesser extent (Table 7). In order, the more pressing challenges related to their ability to balance farm and household tasks ($M = 3.54$) and off-farm and on-farm work ($M = 3.45$), expectations as caregivers ($M = 3.36$), and receiving cooperation from their spouse/partner ($M = 3.34$). Women felt least challenged by in-group support in terms of knowledge sharing from parents ($M = 2.66$) and the number of women famers ($M = 2.39$). On the contrary, respondents perceived that current societal trends, in terms of increased *Public Awareness* ($\alpha = 0.697$, $M = 3.66$) and the evolving *Social Fabric* ($\alpha = 0.816$, $M = 3.56$), are facilitating the success of their agritourism ventures (Table 8). The most impactful trends were the demand for local farm products ($M = 3.77$), public interest in local agriculture ($M = 3.73$), and access to social media ($M = 3.72$).

Farmer, family and society attributes associated with women's success

Simultaneous multiple linear regressions regressing farmer values, farm household characteristics and societal traits on women's functional success resulted in two significant models (Table 9) pertaining to their perceived success as a woman farmer ($R^2 = .197$, $p = .048$) and as a woman entrepreneur ($R^2 = .637$, $p < .001$). When controlling for other variables, positive associations

Table 7. Societal challenges affecting the success of women farmers.

Societal Challenges ($n = 115$)[a]	Not at all	Very little	Some	Very much	*Mean*[b]	*S.D.*
***Caregiver Role* (α = 0.868)**					***3.22***	
Balancing farm & household tasks	1.8%	10.0%	20.9%	67.3%	3.54	0.750
Expectations as a caregiver	7.2%	7.2%	27.9%	57.7%	3.36	0.902
Cooperation from spouse/partner	10.9%	7.3%	19.1%	62.7%	3.34	1.016
Demand of child care	12.8%	9.2%	25.7%	52.3%	3.17	1.053
Falling short on caring for the family	7.3%	13.6%	30.9%	48.2%	3.20	0.937
Falling short on others' expectations	16.7%	18.5%	32.4%	32.4%	2.81	1.072
***Patriarchal Agricultural System* (α = 0.767)**					***2.98***	
Managing off-farm & on-farm work	3.6%	11.8%	20.9%	63.6%	3.45	0.841
Physical demand of farm-work	1.8%	10.8%	43.2%	44.1%	3.30	0.734
Access to grants	9.0%	18.0%	25.2%	47.7%	3.12	1.007
Ability to inherit farmland	14.0%	11.2%	29.0%	45.8%	3.07	1.066
Availability of farmers' networks	6.3%	21.6%	34.2%	37.8%	3.04	0.924
Lack of respect towards farmers	19.6%	13.1%	29.9%	37.4%	2.85	1.131
Knowledge sharing from parents	21.9%	21.0%	26.7%	30.5%	2.66	1.134
Number of farmers of the same gender	30.8%	23.4%	21.5%	24.3%	2.39	1.164

[a]Overall reliability of gendered challenges was $\alpha = 0.879$.
[b]Measured on a 4-point unipolar scale ranging from 'Not at all' (1) to 'Very much' (4).

Table 8. Societal trends facilitating agritourism success.

Societal Trends ($n = 115$)[a]	Not at all	Very little	Some	Very much	Mean[b]	*S.D.*
***Public Awareness* (α = 0.697)**					***3.66***	
Demand of local farm products	0.0%	1.7%	20.0%	78.3%	3.77	0.465
Public interest in local agriculture	0.0%	1.8%	23.0%	75.2%	3.73	0.482
Access to social media	0.0%	1.8%	24.6%	73.7%	3.72	0.489
Farmers seeking direct markets	0.9%	4.4%	39.5%	55.3%	3.49	0.627
***Social Fabric* (α = 0.816)**					***3.56***	
Women's involvement in farming	0.0%	7.0%	29.8%	63.2%	3.56	0.625
Entrepreneurial mindset of young farmers	0.0%	4.4%	36.0%	59.6%	3.55	0.581
Women's leadership in agritourism	0.9%	6.1%	34.2%	58.8%	3.51	0.655

[a]Overall reliability was $\alpha = 0.792$.
[b]Measured on a 4-point unipolar scale ranging from 'Not at all' (1) to 'Very much' (4).

were found for conservationist values with women's success as farmers (β = .219, p = .057) and civic-minded (β = .440, p < .001), and productivist values (β = .402, p < .001) as entrepreneurs. The existing patriarchal agricultural system was perceived as negatively impacting the success of women as farmers (β = −.216, p = .090) and entrepreneurs (β = −.340, p = .002). On the contrary, trends on public awareness of agriculture were perceived as an enabler of women's success as entrepreneurs (β = .177, p < .082).

Significant models (Table 10) were also obtained on the extent to which farmer values, farm household characteristics and societal traits were perceived to affect the life aspects of women's success in terms of civic recognition (R^2 = .461, p < .001), personal aspirations (R^2 = .252, p = .006), self-fulfillment (R^2 = .197, p = .045), and business continuance (R^2 = .297, p < .001). When controlled for other variables, the Civic-minded and Productivist values were positively associated with the importance of Civic Recognition (β = .308, p = .001; β = .242, p = .007, respectively) and Business Continuance (β = .248, p = .018; β = .222, p = .030, respectively) in women's sense of success. In addition, the stronger the Productivist value the stronger the sense of fulfillment on Personal Aspirations (β = .269, p = .011). The extent of agritourism diversification exerted a negative impact on Civic Recognition (β = −.229, p = .021) while a positive one on Personal Aspirations (β = .272, p = .020). Regarding societal trends, positive associations were found between the changing social fabric and the sense of women's success in terms of Civic Recognition (β = .278, p = .006) and Self-fulfillment (β = .276, p = .023) and the patriarchal agricultural system and Personal Aspirations (β = .242, p = .048).

Table 9. Farmer, family and societal attributes associated with women's functional success.

	Functional Success (standardized β and significance)	
Independent Variables	As a Farmer	As an Entrepreneur
Farmer Values		
Conservationist	0.219*	−0.138
Civic-minded	0.158	0.440***
Productivist	0.127	0.402***
Farm Family Attributes		
Likelihood of passing the farm on	0.153	0.127
Extent of agritourism diversification	−0.052	0.058
Months per year of agritourism activities	−0.049	0.124
Societal Traits		
Caregiver role	0.056	0.066
Patriarchal agricultural system	−0.216*	−0.340**
Public awareness	0.170	0.177*
Changing social fabric	−0.092	−0.016
Model Statistics		
R	0.444	0.637
R^2	0.197	0.406
df	10	10
F	1.967	5.468
p-value	0.048	<0.001

*$p < 0.100$.
**$p < 0.050$.
***$p < 0.001$.

Table 10. Farmer, family and societal attributes associated with women's life aspects of success.

	Life Aspects of Success (standardized β and significance)			
Independent Variables	Civic Recognition	Personal Aspirations	Self- Fulfillment	Business Continuance
Farmer Values				
Conservationist	0.079	−0.030	0.184	0.099
Civic-minded	0.308*	0.151	0.019	0.248*
Productivist	0.242*	0.269*	0.057	0.222*
Farm Family Attributes				
Likelihood of passing the farm on	0.125	0.092	0.107	0.156
Extent of agritourism diversification	−0.229*	0.272*	−0.057	−0.088
Seasonality of activities	0.098	−0.026	−0.037	−0.009
Societal Traits				
Public awareness	0.076	0.065	0.020	−0.010
Changing social fabric	0.278*	0.011	0.276*	0.139
Caregiver role	0.001	−0.130	0.089	0.129
Patriarchal agricultural system	0.125	0.242*	−0.013	0.073
Model Statistics				
R	0.679	0.502	0.444	0.545
R^2	0.461	0.252	0.197	0.297
df	10	10	10	10
F	6.924	2.731	1.991	3.422
p-value	< 0.001	0.006	0.045	0.001

*$p < 0.050$.

Discussion and implications

The young age and high education level of participants were consistent with the overall young contingent of high-skilled farmers entering agritourism (Tew & Barbieri, 2012), and the substantial number of young women entering farming (Ball, 2014; Hoppe & Korb, 2013; Sachs et al., 2016) and entrepreneurship in rural areas (Gupta & York, 2008). Such demographic composition along with the high presence of first generation full-time farmers is a major change in the narrative of women captured in the literature (Ahl, 2006; Ball, 2014; Brandth, 2002; Whatmore, 1991). Yet, women's modest perceptions of functional success found in this study stress the need to continue efforts to eliminate remnants of agricultural patriarchal norms in which women farmers are portrayed as the business assistants (Brasier et al., 2014; Gasson, 1973; Sachs et al., 2016). This is especially important when women's functions as farmers and entrepreneurs need to be juxtaposed within a complex system of personal, community and business aspirations that define women's sense of success as this study showed.

Participants' inclination for civic and conservationist values found in this study supports women's preferences for using sustainable practices (Ball, 2014; Ferrell, 2012; Sachs, 2016) and their strong consideration for their community and the wider public (Johnson et al., 2016; Wright & Annes, 2016). These results reaffirm the existence of gendered agricultural paradigms in which women prioritize smaller scale production to ensure production quality and customers' satisfaction (Anthopoulou, 2010; Halim, 2016). At the same time, these results expand women's agricultural values beyond idealism as participants also value taking actions to enhancing the wellbeing of their community, the farming environment, and their business (Sachs et al., 2016). Women's emphasis on business continuance found in this study concurs with their historic dedication to provide for the farm family by finding ways to supplement incomes and make ends meet (Bock, 2004; Sachs et al., 2016). Recognizing respondents' aspirations for community welfare, conservation and agrarian productivism is important as they were found to predict women's success in their roles as farmers and entrepreneurs, as well as contribute to their personal, community and business aspirations.

This study's results confirmed that social values in agricultural settings uphold patriarchal ideals. Caregiving expectations are still a challenge for women farmers, even to a greater extent than the patriarchal agricultural system, supporting the additional burden women experience because farm tasks add to their household responsibilities (Ball, 2014). Yet, women's caregiver role did not appear to influence their perceived success, which contributes to the evidence that farm women are shifting identities in farm household dynamics from the private (farm family caregiver) into the public (farm business operator) realm of agriculture (Brasier et al., 2014). It may also indicate that although women recognize these challenges, they are not internalizing the impact on their success because they are used to negotiating these challenges within their personal or family realms. As such, the dynamics of societal structures in women's success need to be further explored to elucidate ways to disrupt the masculine hegemony of farming and entrepreneurship in the construction of femininity in rural contexts (Bock, 2015; Brasier et al., 2014; Little & Jones, 2000; Sachs et al., 2016).

Respondents' high levels of diversified agricultural production were expected as this variety entices visitors' participation (Barbieri et al., 2008; Tew & Barbieri, 2012) and allows farm women to move beyond production for family sustenance (Alston, 2003; Ferrell, 2012; Hoppe & Korb, 2013). This is in line with the positive association found with women's personal aspirations. Yet, the negative association with women's perception of civic recognition could be related to the resistance of agricultural communities to accept agricultural changes, such as entrepreneurial diversification led by women (Carter, 2017). Notably, participants recognized that changing societal trends are conducive to their entrepreneurial success and contribute to their community recognition and their self-fulfillment. Having more accepting community fellows (e.g., more open to agriculture entrepreneurship) is critical for the sustainability of agritourism as the lack of community acceptance is a major challenge hindering the economic viability of women in agritourism (Halim, 2016).

Study implications

This study' findings contribute to the scholarship and practice of women in agritourism. Scholarly, this study advanced the existing knowledge on agritourism performance beyond farmers' motivations and goals (Tew & Barbieri, 2012; McGehee et al., 2007) and overall entrepreneurial performance (Walker & Brown, 2004). Adopting a critical feminism perspective allowed juxtaposing factors operating at the micro (farmer), meso (family), and macro (society) levels (Mooney, 2020) to dissecting how women farmers perceive success based on their distinct yet interrelated roles as farmers and entrepreneurs. Specifically, this study identified that farmer values, farm household attributes, and societal traits determine women's success in their farmer and entrepreneurial roles as well as their different life-dimensions constructing their holistic perceptions of success. Worth noting, this study's incorporation of farmer values to examine women's success moves forward the post-productivist research agenda (Burton, 2004; McGuire et al., 2015) finding that civic-related values are of particular importance to women farmers' perceived success. This finding also adds to the existent evidence of agritourism as a sustainable tourism venture (Barbieri, 2013) taken that women farmers in this study seek to expand the benefits of agritourism beyond their individual farms and beyond economic indicators.

Adopting a feminist lens from a critical standpoint also allowed positing gendered norms associated with patriarchal systems prevailing in agriculture and overall society as underlying factors (Halim, 2016; Snyder-Hall, 2010) that need to be identified, and thus change, to improve women's condition in society (Aitchison, 2000; Mooney, 2020; Parry et al., 2013). The omnipresent effect of the patriarchal system has been consistently reported in the agriculture (Sachs et al., 2016), entrepreneurial (Buttner & Moore, 1997; Greene et al., 2003) and tourism (Ferguson & Alarcon, 2015; Figueroa-Domecq et al., 2015) literature. This study extended the existing knowledge of the limitations women in agriculture face beyond economics, most commonly gravitating around accessing grants and land, by identifying women's struggles related to the social perceptions of what constitutes "agriculture" and women's place in such social structure. This information is critical to (re)design outreach efforts disseminating the many strategies women have used to overcome gendered challenges and succeed in the field, and to increase awareness of the still existing gendered biases. It is also advisable to encourage young women farmers to undertake leadership roles within their local and state agricultural or business associations and in policymaking positions to increase their community recognition. Educational programs tailored to women farmers can help to increase gender equality and women's leadership in agriculture when designed around problem-solving discussions and network building.

Study limitations and insights for future research

Caution is advised when generalizing study results and its implications beyond NC in view of a few study limitations and delimitations. First, the absence of a directory of women in agritourism in NC prevented determining the size and access of the study population, thus the proportion surveyed. The research team placed special effort to recruit women farmers across all the state from different statewide and regional organizations (e.g., NC Cooperative Extension, Blue Ridge Women in Agriculture) on top of an intensive web-search. Yet, a closer examination of the respondents indicates a high presence of young, highly educated, and first generation farmers that may have affected the results related to the caregiver role and the patriarchal agricultural system. These results call for further investigation regardless if such a high presence is related to an over-representation in the study sample or a reflection of the evolving social fabric of agricultural entrepreneurs (Ball, 2014; Gupta & York, 2008; Hoppe & Korb, 2013; Sachs et al., 2016).

Specifically, future research could examine the extent to which young woman farmers with college education are better equipped to voice and challenge patriarchal norms in their communities. Likewise, further investigation could also be useful to identify advantages that new entrants to agriculture, especially women, have in innovating agriculture entrepreneurship (e.g., agritourism) as the extant literature mainly focuses on barriers (e.g., limited access to local networks). Finally, it would be of utmost importance to examine the extent to which different family life cycle stages (e.g., childless, pre-school children at home) intervene in women's perceptions and attainment of success, which this study did not capture. Filling those knowledge gaps can elucidate on ways to support women's success in agritourism, which ultimately is conducive to community development in a sustainable manner.

Secondly, it is also important to bear in mind that the examination of women's challenges and opportunities in this study was delimited to the innermost layers–farmer, farm household, society–of the agritourism systems approach. Moving forward, future studies should also include the outermost layers defined by the economic, socio-cultural, and environmental tenets of sustainability that may exert additional pressure on farmers. For example, the strengthening of social justice grassroots movements currently happening across the USA may empower women farmers to bend patriarchal norms and more actively seek a public presence in entrepreneurship. This potential impact could be re-examined in contrast with the impact of different feminism waves (e.g., women's vote, rights on family planning) in the USA. Finally, extrapolating results beyond NC should be done with caution due the contextual nature of agriculture and agritourism in particular, as both activities depend on physical (e.g., climate, soil) and cultural (e.g., consumer preferences) attributes. For example, this study could be replicated across regions with different geopolitical schemes (e.g., social vs. capital macro-schemes) as the prevailing tacit norms (e.g., caregiving gender roles) and explicit regulations (e.g., level of child-care subsidies) can influence women's success.

Conclusion

This study applied feminist (Aitchison, 2000; Alston, 2003) and systems (Barbieri, 2017) approaches to gain a holistic understanding of the success of women in agritourism taking into consideration their farmer and entrepreneurial roles as well as different aspects of their lives.

Altogether, study results contribute to the current literature of women in agriculture, their roles on the farm and in the household, challenges within the current patriarchal agricultural system, and opportunities with the current societal trends. The application of feminist approaches in agricultural-related studies are critical to compensate for a sector where masculine purviews and practices prevail (Brandth, 2002). Considered as a form of entrepreneurial farming (Barbieri et al., 2008), the literature on agritourism framed within feminism is extremely limited. As such, this feminist-framed study was suitable to provide insights to strive for gender equity by informing agritourism-related policy to foster more prosperous and equitable rural development.

Furthermore, the application of the agritourism systems approach (Barbieri, 2017) to women farmers contributed to existing agricultural holistic frameworks that examine the farm in its entirety and delineate how interdependent elements affect the individual famer as well as the overall farm health and survival (e.g., Dogliotti et al., 2014; Ikerd, 1993). By applying this systemic approach within a feminist lens, this study further advanced existing knowledge of the patriarchal systems affecting women beyond negative impacts (Bock, 2015). Furthermore, adopting a critical standpoint allowed identifying opportunities that women can capitalize on to nullify the effects of such dominant systems. Thus, both approaches enabled the identification of the gendered norms and nuances that span the private and public lives of women in agritourism and revealed how these infiltrate their different farmer and entrepreneurial roles. Specifically, results elucidate the dynamics women navigate to contribute to their different life aspects of their own success, their farm family and business, and to the farming community and wider public. In doing so, this study expands the empirical data on the subordination of women in agritourism and agribusiness, which is useful to reduce between gender disparities that prevent moving forward the quest of sustainability.

Disclosure statement

No potential conflict of interest was reported by the authors.

Funding

This investigation was funded by the College of Natural Resources - College of Agriculture and Life Sciences Cross College Enrichment Grants, North Carolina State University.

ORCID

Carla Barbieri http://orcid.org/0000-0001-8079-2316

References

Ahl, H. (2006). Why research on women entrepreneurs needs new directions. *Entrepreneurship Theory and Practice, 30*(5), 595–621. https://doi.org/10.1111/j.1540-6520.2006.00138.x

Aitchison, C. (2000). Postructural feminist theories of representing others: A response to the 'crisis' in leisure studies' discourse. *Leisure Studies, 19*(3), 127–144. https://doi.org/10.1080/026143600500230440

Aldrich, H. (1989). Networking among women entrepreneurs. *Women-Owned Businesses, 103*, 132.

Alston, M. (2003). Women in agriculture: The 'New Entrepreneurs. *Australian Feminist Studies, 18*(41), 163– 152. https://doi.org/10.1080/08164640301726

Anthopoulou, T. (2010). Rural women in local agrofood production: Between entrepreneurial initiatives and family strategies. A case study in Greece. *Journal of Rural Studies, 26*(4), 394–403. https://doi.org/10.1016/j.jrurstud.2010.03.004

Armstrong, J. S., & Overton, T. S. (1977). Estimating nonresponse bias in mail surveys. *Journal of Marketing Research, 14*(3), 396–402. https://doi.org/10.1177/002224377701400320

Babbie, E. R. (2013). *The practice of social research*. Cengage Learning Asia Pte Ltd.

Ball, J. A. (2014). She works hard for the money: Women in Kansas agriculture. *Agriculture and Human Values, 31*(4), 593–605. https://doi.org/10.1007/s10460-014-9504-8

Barbieri, C. (2013). Assessing the sustainability of agritourism in the US: A comparison between agritourism and other farm entrepreneurial ventures. *Journal of Sustainable Tourism, 21*(2), 252–270. https://doi.org/10.1080/09669582.2012.685174

Barbieri, C. (2017). Agritourism and sustainable rural development: The agritourism system's approach. In: Solha, K., Elesbão, I., & de Souza, M. (Eds.), *O Turismo Rural Comunitário como Estratégia de Deselvolvimiento* (p. 240). Universidade Federal do Rio Grande do Sul.

Barbieri, C. (2020). Agritourism research: A perspective article. *Tourism Review – Jubilee Special Edition, 75*(1), 149–152.

Barbieri, C., Mahoney, E., & Butler, L. (2008). Understanding the nature and extent of farm and ranch diversification in North America. *Rural Sociology, 73*(2), 205–229. https://doi.org/10.1526/003601108784514543

Barbieri, C., & Mshenga, P. M. (2008). The role of the firm and owner characteristics on the performance of agritourism farms. *Sociologia Ruralis, 48*(2), 166–183. https://doi.org/10.1111/j.1467-9523.2008.00450.x

Barbieri, C., Stevenson, K. T., & Knollenberg, W. (2019). Broadening the utilitarian epistemology of agritourism research through children and families. *Current Issues in Tourism, 22*(19), 2333–2336.

Barbieri, C., & Streifeneder, T. (2019). Agritourism advances around the globe: A commentary from the editors. *Open Agriculture, 4*(1), 712–714. https://doi.org/10.1515/opag-2019-0068

Benjamin, C. (1994). The growing importance of diversification activities for French farm households. *Journal of Rural Studies, 10*(4), 331–342. https://doi.org/10.1016/0743-0167(94)90043-4

Bock, B. B. (2004). Fitting in and multi-tasking: Dutch farm women's strategies in rural entrepreneurship. *Sociologia Ruralis*, *44*(3), 245–260. https://doi.org/10.1111/j.1467-9523.2004.00274.x

Bock, B. B. (2015). Gender mainstreaming and rural development policy; the trivialisation of rural gender issues. *Gender, Place and Culture*, *22*(5), 731–745. https://doi.org/10.1080/0966369X.2013.879105

Brandth, B. (2002). On the relationship between feminism and farm women. *Agriculture and Human Values*, *19*(2), 107–117. [Mismatch] https://doi.org/10.1023/A:1016011527245

Brandth, B., & Haugen, M. S. (2011). Farm diversification into tourism - Implications for social identity? *Journal of Rural Studies*, *27*(1), 35–44. https://doi.org/10.1016/j.jrurstud.2010.09.002

Brasier, K. J., Sachs, C. E., Kiernan, N. E., Trauger, A., & Barbercheck, M. E. (2014). Capturing the multiple and shifting identities of farm women in the Northeastern United States. *Rural Sociology*, *79*(3), 283–309. https://doi.org/10.1111/ruso.12040

Brune, S., Knollenberg, W., Stevenson, K., Barbieri, C., & Schroeder-Moreno, M. (2020). The influence of agritourism experiences on consumer behavior towards local food. *Journal of Travel Research*. https://doi.org/10.1177/0047287520938

Burton, R. J. F. (2004). Seeing through the 'good farmer's' eyes: Towards developing an understanding of the social symbolic value of 'productivist' behaviour. *Sociologia Ruralis*, *44*(2), 195–215. https://doi.org/10.1111/j.1467-9523.2004.00270.x

Burton, R. J. F., & Wilson, G. A. (2006). Injecting social psychology theory into conceptualisations of agricultural agency: Towards a post-productivist farmer self-identity? *Journal of Rural Studies*, *22*(1), 95–115. https://doi.org/10.1016/j.jrurstud.2005.07.004

Buttner, H. E., & Moore, D. P. (1997). Women's organizational exodus to entrepreneurship: Self-reported motivations and correlates with success. *Journal of Small Business Management*, *35*(1), 34–46.

Carter, A. (2017). Placeholders and changemakers: Women farmland owners navigating gendered expectations. *Rural Sociology*, *82*(3), 499–523. https://doi.org/10.1111/ruso.12131

Carter, N. C., Gartner, W. B., Shaver, K. G., & Gatewood, E. J. (2003). The career reasons of nascent entrepreneurs. *Journal of Business Venturing*, *18*(1), 13–39. https://doi.org/10.1016/S0883-9026(02)00078-2

Che, D., Veeck, A., & Veeck, G. (2005). Sustaining production and strengthening the agritourism product: Linkages among Michigan agritourism destinations. *Agriculture and Human Values*, *22*(2), 225–234. https://doi.org/10.1007/s10460-004-8282-0

Dillman, D. A., Smyth, J. D., & Christian, L. M. (2009). *Internet, mail, and mixed-mode surveys - The tailored design method*. John Wiley & Sons, Inc.

Dogliotti, S., García, M. C., Peluffo, S., Dieste, J. P., Pedemonte, A. J., Bacigalupe, G. F., Scarlato, M., Alliaume, F., Alvarez, J., Chiappe, M., & Rossing, W. A. H. (2014). Co-innovation of family farm systems: A systems approach to sustainable agriculture. *Agricultural Systems*, *126*, 76–86. https://doi.org/10.1016/j.agsy.2013.02.009

Driga, O., Lafuente, E., & Vaillant, Y. (2009). Reasons for the relatively lower entrepreneurial activity levels of rural women in Spain. *Sociologia Ruralis*, *49*(1), 70–96. https://doi.org/10.1111/j.1467-9523.2008.00475.x

Eccles, J. S. (1987). Gender roles and women's achievement-related decisions. *Psychology of Women Quarterly*, *11*(2), 135–172. https://doi.org/10.1111/j.1471-6402.1987.tb00781.x

Eccles, J. S. (1994). Understanding women's educational and occupational choices. *Psychology of Women Quarterly*, *18*(4), 585–609. https://doi.org/10.1111/j.1471-6402.1994.tb01049.x

Eger, C., Miller, G., & Scarles, C. (2018). Gender and capacity building: A multi-layered study of empowerment. *World Development*, *106*, 207–219. https://doi.org/10.1016/j.worlddev.2018.01.024

Ferguson, L., & Alarcon, D. M. (2015). Gender and sustainable tourism: Reflections on theory and practice. *Journal of Sustainable Tourism*, *23*(3), 401–416. https://doi.org/10.1080/09669582.2014.957208

Ferrell, A. K. (2012). Doing masculinity: Gendered challenges to replacing burley tobacco in central Kentucky. *Agriculture and Human Values*, *29*(2), 137–149. https://doi.org/10.1007/s10460-011-9330-1

Field, A. (2013). *Discovering statistics using IBM SPSS statistics*. Sage.

Figueroa-Domecq, C., Pritchard, A., Segovia-Pérez, M., Morgan, N., & Villacé-Molinero, T. (2015). Tourism gender research: A critical accounting. *Annals of Tourism Research*, *52*, 87–103. https://doi.org/10.1016/j.annals.2015.02.001

Gao, J., Barbieri, C., & Valdivia, C. (2014). Agricultural landscape preferences: Implications for agritourism development. *Journal of Travel Research*, *53*(3), 366–379. https://doi.org/10.1177/0047287513496471

Gasson, R. (1973). Goals and values of farmers. *Journal of Agricultural Economics*, *24*(3), 521–542. https://doi.org/10.1111/j.1477-9552.1973.tb00952.x

Gasson, R., & Winter, M. (1992). Gender relations and farm household pluriactivity. *Journal of Rural Studies*, *8*(4), 387–397. https://doi.org/10.1016/0743-0167(92)90052-8

Giampietro, M. (1997). Socioeconomic pressure, demographic pressure, environmental loading and technological changes in agriculture. *Agriculture, Ecosystems and Environment*, *65*(3), 201–229. https://doi.org/10.1016/S0167-8809(97)00050-9

Gil Arroyo, C., Barbieri, C., & Rozier Rich, S. (2013). Defining agritourism: A comparative study of stakeholders' perceptions in Missouri and North Carolina. *Tourism Management*, *37*, 39–47. https://doi.org/10.1016/j.tourman.2012.12.007

Greene, P. G., Hart, M. M., Gatewood, E. J., Brush, C. G., & Carter, N. M. (2003). Women entrepreneurs: Moving front and center: An overview of research and theory. *Coleman White Paper Series*, *3*, 1–47.

Gupta, V. K., & York, A. S. (2008). The effects of geography and age on women's attitudes towards entrepreneurship: Evidence from the state of Nebraska. *The International Journal of Entrepreneurship and Innovation*, *9*(4), 251–262. https://doi.org/10.5367/000000008786208777

Halim, M. F., Barbieri, C., Morais, D. B., Jakes, S., & Seekamp, E. (2020). Beyond economic earnings: The holistic meaning of success for women in agritourism. *Sustainability*, *12*(12), 4907.

Halim, M. F, Barbieri, C., Morais, D. B., Jakes, S., & Seekamp, E. (2020). Beyond economic earnings: The holistic meaning of success for women in agritourism. *MDPI Sustainability – Special Issue on Gender and Rural Development: Sustainable Livelihoods in a Neoliberal Context*, *12*(12), 4907; DOI: 10.3390/su12124907.

Halim, M. F. (2016). *Ushering NC women in agritourism towards success: Challenges and opportunities* [Unpublished master's thesis]. North Carolina State University.

Hoppe, R. A., Korb, P. (2013). *Characteristics of women farm operators and their farms*. US Department of Agriculture, Economic Research Service. Retrieved February 9, 2017, from https://www.ers.usda.gov/webdocs/publications/43749/37013_eib111_summary.pdf?v=41390.

Horner, M. (1972). Toward an understanding of achievement-related conflicts in women. *Journal of Social Issues*, *28*(2), 157–175. https://doi.org/10.1111/j.1540-4560.1972.tb00023.x

Ikerd, J. E. (1993). The need for a systems approach to sustainable agriculture. *Agriculture, Ecosystems & Environment*, *46*(46), 147–160.

Johnson, L. B., Schnakenberg, G., & Perdue, N. (2016). *Toward a relational agro-food system: The case of the Blue Ridge Women in Agriculture High Country Farm Tour* [Doctoral dissertation]. Michigan State University.

Kalleberg, A. L., & Leicht, K. T. (1991). Gender and organizational performance: Determinants of small business survival and success. *The Academy of Management Journal*, *34*(1), 136–161.

Kline, C., Barbieri, C., & LaPan, C. (2016). The influence of agritourism on niche meats loyalty and purchasing. *Journal of Travel Research*, *55*(5), 643–658. https://doi.org/10.1177/0047287514563336

Kline, C., & Milburn, L. (2010). Ten categories of entrepreneurial climate to encourage rural tourism development. *Annals of Leisure Research*, *13*(1–2), 320–348. https://doi.org/10.1080/11745398.2010.9686850

LaPan, C., & Barbieri, C. (2014). The role of agritourism in heritage preservation. *Current Issues in Tourism*, *17*(8), 666–673. https://doi.org/10.1080/13683500.2013.849667

Li, J., & Barbieri, C. (2019). Incentives to join associations: The case of agritourism. *Tourism Insights*, *9*(1), 5.

Little, J., & Jones, O. (2000). Masculinity, gender, and rural policy. *Rural Sociology*, *65*(4), 621–639. https://doi.org/10.1111/j.1549-0831.2000.tb00047.x

Markantoni, M., & van Hoven, B. (2012). Bringing "invisible" side activities to light. A case study of rural female entrepreneurs in the Veenkoloniën, the Netherlands. *Journal of Rural Studies*, *28*(4), 507–516. https://doi.org/10.1016/j.jrurstud.2012.05.006

McGehee, N. G., & Kim, K. (2004). Motivation for agri-tourism entrepreneurship. *Journal of Travel Research*, *43*(2), 161–170. https://doi.org/10.1177/0047287504268245

McGehee, N. G., Kim, K., & Jennings, G. R. (2007). Gender and motivation for agri-tourism entrepreneurship. *Tourism Management*, *28*(1), 280–289. https://doi.org/10.1016/j.tourman.2005.12.022

McGuire, J. M., Morton, L. W., Arbuckle, J. G., & Cast, A. D. (2015). Farmer identities and responses to the social–biophysical environment. *Journal of Rural Studies*, *39*, 145–155. https://doi.org/10.1016/j.jrurstud.2015.03.011

Meert, H., Van Huylenbroeck, G., Vernimmen, T., Bourgeois, M., & van Hecke, E. (2005). *Farm Household Survival Strategies and Diversification on Marginal Farms*, Journal of Rural Studies, *21*(1), 81–97.

Midgley, J. (2006). Gendered economies: Transferring private gender roles into the public realm through rural community development. *Journal of Rural Studies*, *22*(2), 217–231. https://doi.org/10.1016/j.jrurstud.2005.08.014

Mooney, S. (2016). Nimble' intersectionality in employment research: A way to resolve methodological dilemmas. *Work, Employment and Society*, *30*(4), 708–718. https://doi.org/10.1177/0950017015620768

Mooney, S. (2020). Gender research in hospitality and tourism management: Time to change the guard. *International Journal of Contemporary Hospitality Management*, *32*(5), 1861–1879. https://doi.org/10.1108/IJCHM-09-2019-0780

Nickerson, N. P., Black, R. J., & McCool, S. F. (2001). Agritourism: Motivations behind farm/ranch business diversification. *Journal of Travel Research*, *40*(1), 19–26. https://doi.org/10.1177/004728750104000104

Ollenburg, C., & Buckley, R. (2007). Stated economic and social motivations of farm tourism operators. *Journal of Travel Research*, *45*(4), 444–452. https://doi.org/10.1177/0047287507299574

Parry, D. C., Johnson, C. W., & Stewart, W. (2013). Leisure research for social justice: A response to Henderson. *Leisure Sciences*, *35*(1), 81–87. https://doi.org/10.1080/01490400.2013.739906

Phelan, C., & Sharpley, R. (2012). Exploring entrepreneurial skills and competencies in farm tourism. *Local Economy: The Journal of the Local Economy Policy Unit*, *27*(2), 103–118. https://doi.org/10.1177/0269094211429654

Sachs, C., Barbercheck, M., Braiser, K., Kiernan, N. E., & Terman, A. R. (2016). *The rise of women farmers and sustainable agriculture*. University of Iowa Press.

Schilling, B. J., Attavanich, W., & Jin, Y. (2014). Does agritourism enhance farm profitability? *Journal of Agricultural and Resource Economics*, *39*(1), 69–87.

Schilling, B. J., Sullivan, K., & Komar, S. (2012). Examining the economic benefits of agritourism: The case of New Jersey. *Journal of Agriculture, Food Systems, and Community Development*, *3*(1), 199–214.

Snyder-Hall, R. C. (2010). Third-wave feminism and the defense of "Choice. *Perspectives on Politics*, *8*(1), 255–261. https://doi.org/10.1017/S1537592709992842

Stead, V. (2017). Belonging and women entrepreneurs: Women's navigation of gendered assumptions in entrepreneurial practice. *International Small Business Journal*, *35*(1), 61–77. https://doi.org/10.1177/0266242615594413

Suess-Reyes, J., & Fuetsch, E. (2016). The future of family farming: A literature review on innovative, sustainable and succession-oriented strategies. *Journal of Rural Studies*, *47*, 117–140. https://doi.org/10.1016/j.jrurstud.2016.07.008

Tew, C., & Barbieri, C. (2012). The perceived benefits of agritourism: The provider's perspective. *Tourism Management*, *33*(1), 215–224. https://doi.org/10.1016/j.tourman.2011.02.005

Vavrus, M. D. (2002). Domesticating patriarchy: Hegemonic masculinity and television's "Mr. Mom. *Critical Studies in Media Communication*, *19*(3), 352–375.

Veeck, G., Che, D. & Veeck, A. (2006). America's changing farmscape: A study of agricultural tourism in Michigan. *The Professional Geographer*, *58*(3), 235–248. DOI: 10.1111/j.1467-9272.2006.00565.x

Walker, E., & Brown, A. (2004). What success factors are important to small business owners? *International Small Business Journal: Researching Entrepreneurship*, *22*(6), 577–594. https://doi.org/10.1177/0266242604047411

Wang, C. (2010). Daughter exclusion in family business succession: A review of the literature. *Journal of Family and Economic Issues*, *31*(4), 475–484. https://doi.org/10.1007/s10834-010-9230-3

Whatmore, S. (1991). Life cycle or patriarchy? Gender divisions in family farming. *Journal of Rural Studies*, *7*(1–2), 71–76. https://doi.org/10.1016/0743-0167(91)90043-R

Wright, W., & Annes, A. (2014). Farm women and agritourism: Representing a new rurality. *Sociologia Ruralis*, *54*(4), 477–499.

Wright, W., & Annes, A. (2016). Farm women and the empowerment potential in value-added Agriculture. *Rural Sociology*, *81*(4), 545–571. https://doi.org/10.1111/ruso.12105

Yang, L. (2012). Impacts and Challenges in Agritourism Development in Yunnan. Tourism Planning & Development, *9*(4), 369–381. https://doi.org/10.1080/21568316.2012.726257

The contribution of all-women tours to well-being in middle-aged Muslim women

Adel Nikjoo, Mustafeed Zaman, Shima Salehi and Ana Beatriz Hernández-Lara

ABSTRACT

For different reasons, some women experience all-women tours in different parts of the world. Certain aspects of all-women tours, such as their mental benefits, are studied in Western and Eastern countries. However, there is a dearth of study on all-women travels in Muslim-majority countries, despite the importance of women's issues in such societies. By observing and interviewing eleven middle-aged women who participated in a 1-day, all-women, tour in Iran, we found that being away from family responsibilities, routine life, and gender-related restrictions, as well as advancement in their social and personal selves, contributed to their well-being. The findings show that the mood of women undergoing later life transitions may benefit from such commonplace tours with their peer group. We also discuss women's right to travel in the context of travel sustainability.

Introduction

In line with the general growth in tourism, the demand for same-gender travel has generated a specific market known as "girlfriend getaways," also called all-women or women-only tours. The popularity of all-women tours has motivated tourism scholars to study different aspects of this travel trend in different regions. One of the earliest research studies in this field was undertaken by Junek et al. (2006), who studied all-women city tours in Australia. In 2009, Liechty et al. (2009) explored well-being and happiness in all-women tours by older members of the American "Red Hat" society. Later, other studies in the United States shed light on the differences between all-women travel motivations at different ages (Gibson et al., 2012), its effects on well-being (Berdychevsky et al., 2013), and on couples' relationship satisfaction (Durko & Stone, 2017), as well as women's hospitality needs and preferences on all-women tours (Mirehie et al., 2018). In addition to the United States, the issue of all-women tours has been studied in Asia, including Malaysia (Khoo-Lattimore & Prayag, 2015), Hong Kong (Song, 2017), and China (Chen & Mak, 2020).

The majority of research on all-women tours is western-oriented, and there is a lack of studies investigating the motivation and experiences of Muslim and Middle Eastern women on all-women tours with their particular cultural background. The strict influence of Islamic rules on the culture of Muslim countries has resulted in many restrictions for women. In the specific case of Iran, as in our case study, following the 1979 revolution Islamists have taken control of the country and imposed strict ideological rules on many aspects of peoples' lives, especially those of women (Ehteshami, 1995). For instance, they impose compulsory veiling of women, segregate men and women at school and in other public spaces and prohibit many kinds of collective expression of fun, such as dancing (Mahdavi, 2012). Due to the extent of traditional and governmental discriminations against women, the Word Economic Forum ranked Iran 148th out of 153 countries in its 2020 gender gap report (World Economic Forum, 2020). Nevertheless, the perception of the gender gap in Iran is not very widespread, since as suggested by Eger (2020) the definitions of gender (in)equality in some Muslim-majority countries (such as Iran) are to some extent different from normative discussions of equality. The Islamic religion is strongly interwoven in many aspects of everyday life and laws in Muslim-majority countries (Moghadam et al., 2009), and therefore, it is not surprising that the authorities and part of communities see gender equality through a lens of the Islamic feminist, while in some aspects it is completely in contrast with secular feminist movements (Eger, 2020). In Iran, while a large group of people believes in the traditional and Islamic definitions of gender roles, some (particularly new generations) oppose these definitions of femininity and try to redefine the position of women on the basis of normative feminist ideas (Khalajabadi-Farahani et al., 2019).

Despite the poor condition of women's rights in Iran, there is little research that explores travel behavior and its effect on well-being among the women of this region. More specifically, while all-women tours are explored for some Western and Eastern countries, the behavior and lived experiences of Middle Eastern women who participate in this type of tours remain unstudied. By observing and interviewing middle-aged women who participated in a 1-day excursion, the present study reveals the contribution of such travels to the participants' well-being. We also discuss the need to pay attention to women's travel rights in the light of UNWTO's (2015) sustainable development goals (SDGs).

Literature review

Aging and well-being in women

Aging affects everyone's bodily and mental health. Different factors such as biology, psychology, and culture play a role in the definitions of growing old (Bullington, 2006). Although it may be a mental threat for everybody, Rutagumirwa and Bailey (2019) found that it is more challenging for women than for men. Leisure activities can facilitate successful aging through a set of benefits such as self-development, empowerment, sense of accomplishment, development of knowledge and skills, physical benefits, social engagement, sense of belonging, stress and depression reduction, positive emotions, life satisfaction, and a sense of well-being (Cheng et al., 2017; Henderson & Hickerson, 2007; Heo et al., 2010; Joseph & Southcott, 2019; Kim et al., 2017; Ku et al., 2009; Ryu & Heo, 2018). Focusing on aging women, Southcott and Joseph (2020) stated that maintaining a happy and healthy lifestyle by healthy eating, staying physically active and engaging in social activities improves life satisfaction.

Dancing is an example of such an activity through which people can combat depression, reduce their stress and anxiety, and improve their well-being (Karkou et al., 2017), since it is a multidimensional activity that offers physical, psychological, emotional, social, and spiritual benefits (Bond, 2019; Goodill, 2005; Karkou et al., 2017; Murcia et al., 2010).

Traveling is another example of active leisure that contributes to the sense of well-being and happiness. In particular, being outdoors and spending time in nature has positive both physical and mental effects (Zelenski & Nisbet, 2014). It is shown that people are happier in scenic locations (Seresinhe et al., 2019), and that pristine green or aquatic environments have restorative effects on human beings (White et al., 2010; Wyles et al., 2019).

Although many research studies focus merely on the short-term and hedonic benefits of travel, it seems that the major benefit of travelling is its contribution to long-term life satisfaction and eudaimonic well-being (Filep, 2014; McCabe & Johnson, 2013; Nawijn & Filep, 2016). Hedonic and eudaimonic are two types of well-being suggested by Seligman and Csikszentmihalyi (2000). Subjective or hedonic well-being is defined as short-term and hedonistic pleasures, while objective or eudaimonic well-being is associated with self-determination, self-realization, self-actualization, and social engagement with deeper and longer effects (Ryan & Deci, 2001). PERMA is another theory of well-being suggested by Seligman (2011), who argues that positive emotion, engagement, relationships, meaning, and accomplishment are the elements that contribute to well-being. Traveling provides opportunities for people to experience all the elements of well-being that are suggested in the PERMA theory of well-being. In the present study, we show how taking all-women tours contributes to the well-being of participants in various ways.

Motherhood, all-women tours, and well-being

Gender stereotypes are a set of social understandings about the expected roles and behaviors of men and women (Deaux & Lewis, 1984; Jackson & Henderson, 1995) and, like all stereotypes, play a significant role in the way people perceive others. Historically, men are expected to be the main economic providers, and women's main responsibilities are related to caregiving for children and elders and to fulfilling domestic duties (Choi et al., 2005; Weedon, 1997).

Although it is a growing global idea that caregiving and domestic duties should be shared between men and women to give them equal opportunities for work, leisure, and physical activities, some gender stereotypes still constrain women (McGannon et al., 2015; Miller & Brown, 2005). According to Scheiner (2014), the higher commitment of women to caregiving and household responsibilities compared with their male counterparts limits their equal economic opportunities and affects their economic independence.

Nonetheless, the expectations of women and mothers vary based on the culture in which they grew up and live (Douglas & Michaels, 2004). In many cultures, the ethic of care which is considered as a moral development is defined as women's ignorance of their own needs in order to take care of others (McGannon et al., 2015). One study on gender shows that women use relationship labels such as wife, mother, and daughter to describe themselves more frequently than men (Gonyea et al., 2008). Hence, in many cultures, an ideal mother is a woman who selflessly sacrifices her needs for her family and, consequently, a mother who prioritizes other activities such as leisure would feel an internal sense of guilt (McGannon et al., 2015; Miller & Brown, 2005).

In fact, being a woman, a mother, or a daughter, is not in itself a constraint on leisure time; rather, it is the perceived cultural expectations that women should accept more domestic and caregiving responsibilities which limit their participation in leisure activities (Petty & Trussell, 2019; Ron & Nimrod, 2018; Shaw, 1994). Samdahl (2013) suggests that the family responsibilities of women and the ethical issues of caring not only limit them in taking leisure, but also negatively affect their full enjoyment. Consequently, some women seek freedom from these responsibilities through holidays (Cockburn-Wootten et al., 2006; Junek et al., 2006; Small, 2005).

The growing demand of women for travel without family responsibilities has developed a specific travel market known as "girlfriend getaways" (Berdychevsky et al., 2013; Durko & Stone, 2017; Junek et al., 2006; Liechty et al., 2009; Mirehie et al., 2018), defined as leisure travel with a group of friends, but in the absence of their menfolk. The term "girlfriend getaway," however, does not seem to completely define the travel motivations of women participating in these types of tours. The term "girl" does not appropriately describe adult women, it is "pinky" and stereotypical, and the motivations of some women on these tours are seeking and development rather than escaping, or at least "escaping to" rather than "escaping from" (Berdychevsky et al., 2016; Durko & Stone, 2017).

The literature on all-women vacations has shed light on many aspects of this type of travelling. Studies on girlfriend getaways show that women can escape from daily routines and stress, escape from family responsibilities and the ethic of care, experience new things, understand and reconstruct themselves, bond with likeminded friends, develop friendship with others, re-connect with their "pre-mom" selves, receive free therapy from their friends to cope with negative life events, adapt to life stage transition, and feel a sense of freedom in the absence of their partner, all of which contribute to a sense of well-being and happiness (see Durko & Stone, 2017; Gibson et al., 2012; Liechty et al., 2009; Mirehie et al., 2018).

Research on all-women tours (Berdychevsky et al., 2013; Song, 2017) also suggests that the absence of men on these tours provides an opportunity for some women to express their genuine self, one which is less conscious and more comfortable when freed from the gaze of men. They can enjoy themselves instead of paying attention to their make-up, clothing, and other standards of femininity.

Shared experiences and interests, and a sense of equality and community through spending time with likeminded people, are other joyful elements of all-women tours (Berdychevsky et al., 2016; Doran, 2016; Junek et al., 2006; Mitten, 1992). It seems that these experiences shape a sense of communitas or intra-group unity, which is introduced by Turner (1974). Heimtun and Jordan (2011) show that the sense of being isolated, excluded, or conspicuous is lower in all-women tours, this being a benefit as compared to travelling in family-oriented groups. Likewise, compared with solo travelling, many women feel more security and comfort among other female travelers because the group gives them a sense of power (Junek et al., 2006; Liechty et al., 2009).

Spending time with likeminded friends is of special importance on all-women journeys. An interesting point in the research by Durko and Stone (2017) on all-women tours is that some middle-aged women used "girl" on addressing others. They explain that this use of girl reveals a desire to return to youth and a simpler time before they were labeled as a wife or mother. In another study on middle-aged women, Heimtun (2007) shows that holidays provide women with a sense of shared joy, belonging, and mattering or being important in the eyes of others.

Berdychevsky et al. (2013), in their seminal study on girlfriend getaways, explain how reconnection with female friends with this type of travel has potential eudaimonic well-being and therapeutic effects on women due to the intimacy of friendship, the voluntary nature of support, and egalitarianism, which facilitate disclosure and trust. Going even further, Durko and Stone (2017) found that a girlfriend getaway, which necessitates temporary separation from the partner, potentially strengthens a couple's relationship.

In their effort to link girlfriend getaways to well-being, Berdychevsky et al. (2013) conclude that women experience both hedonic and eudaimonic well-being. They classify four ways in which girlfriend getaways contribute to women's well-being: escapism, different gender dynamics, existential authenticity, and empowerment. Escapism refers to breaks from daily routines, social structures, and the ethic of care. Gender dynamics refers to the differences between men and women in the ways in which they enjoy their time. Existential authenticity is associated with "woman-time" and breaking away from femininity standards. Finally, empowerment is a sense of self-development which is shaped through these experiences (Berdychevsky et al., 2013).

As in the work of Berdychevsky et al. (2013), the present study aims to explore well-being among women on the all-women tour, but thousands of kilometers away in Iran, with its Muslim society and different cultural background. We explore well-being among the women of our study on the basis of hedonic and eudaimonic well-being, as explained by Seligman and Csikszentmihalyi (2000) and the PERMA well-being model of Seligman (2011).

Method

Context of the study

Iran is a large Middle Eastern country which has been governed by Islamists since the 1979 revolution. The definition of women's rights among current Iranian political leaders has its roots in

Islamic texts in which household tasks are the major responsibilities of women and they need to be protected by their men (Abedinifard, 2019). Consequently, men are prioritized in education, work, and many other sections of society which generate wealth and power (Rafatjah, 2012). This Islamic definition of women's position placed Iran among the lowest ranked countries in the World Economic Forum's gender gap report (2020). Besides the political pressures on women, the society is also highly patriarchal, mainly because men have traditionally been the financial providers for the family (Moghadam et al., 2009), and therefore, they are generally considered as the head of the family and are privileged in making decisions for themselves, their wife, and their children (Davoodi et al., 2019).

Influenced by Islamic rules, "gheyrat" (غیرت) forms an important part of Iranian men's identity (Abedinifard, 2019). Gheyrat is a behavior code among Iranian men giving them a sense of protectiveness and possessiveness toward their wife and daughters. On the other hand, women are expected to observe the behavioral codes of "effat" (عفت) or Islamic chastity and "haya" (حیا) or Islamic modesty by internalizing the monopolistic right of their men over themselves (Tizro, 2013). While, due to cultural and religious teachings, some Iranian women believe in wearing the hijab and also admire the gheyrat behavioral code, others consider these codes as limiting and working against their emancipation (Tavakoli & Mura, 2017).

Despite all the discrimination against Iranian women imposed by law, religion, and tradition, Iranian society has been changing in favor of women's positions in society (Davoodi et al., 2019). Researchers (Fadaee, 2018; Tavakoli & Mura, 2017) attribute these changes in women's rights mostly to the ubiquity of the internet and the emergence of a new middle class of educated young women who increasingly seek out a new lifestyle and leisure activities.

The tour, data collection, and analysis

A 1-day all-women excursion from the city of Isfahan was selected as a case study for the present study. Although longer all-women journeys would better reflect well-being among participants, we chose a 1-day excursion because, on the one hand, it was more accessible for the authors (the majority of longer tours include both men and women), and on the other hand, the variety of women on these tours was wider. While fewer Iranian women can obtain the permission of their husbands to leave their home for tours which require night-stays, it is easier for conservative women to participate in a 1-day excursion. In the tour we studied, around 80 women travelled by two buses to Markadeh, which is a tourist village adjacent to the Zayandeh-Rud River. While some participants regularly travelled with the tour and therefore were familiar with other loyal participants, the tour was the first all-women excursion experience for others who had come across an advertisement for the tour or were encouraged by friends to join them on the tour.

The constructive version of grounded theory (Charmaz, 2014) is applied in this study. This interpretive approach acknowledges that reality consists of multiple social constructions, and therefore, rather than generalizing their findings and shaping a universal theory, constructive grounded theory researchers try to undertake interpretive understanding of a social phenomenon (Yang et al., 2018). Based on the subjective epistemology of constructive grounded theory, researchers form part of the research, and knowledge is co-created (Rieger, 2019). To do this, one of the authors participated in the tour to observe the women's behavior, to see the world from their point of view, enter their social setting, and be perceived as an insider, as suggested by Charmaz (2014). The observer was a young woman, and she signed up for the tour as an ordinary participant. Excluding the tour guide, the driver, and herself, there were 40 women on the bus she rode on, among whom 29 were at least 50 years old, and 11 were younger. Normally, the younger women preferred to participate in mixed-gender tours or all-women student tours. The observer selected every other woman over 50 (14 women) and took their contact

Table 1. Participants' socio-demographic characteristics.

Name	Age	Children	Job	Interview	Duration
Afsaneh	59	2	Teacher	Telephone	25′
Ashraf	58	3	Retired teacher	In-person	52′
Azar	60	2	Housekeeper	Telephone	23′
Farideh	56	3	Retired teacher	In-person	64′
Jahan	60	3	Retired teacher	Telephone	26′
Manijeh	50	2	Hairdresser	Telephone	38′
Pari	58	2	Retired teacher	Telephone	88′
Shahin	63	2	Retired teacher	Telephone	27′
Shahla	63	3	Housekeeper	Telephone	33′
Shokat	60	4	Housekeeper	In-person	45′
Sima	59	3	Housekeeper	Telephone	41′

details for an interview at the end of the tour. However, she observed all the participants and took note of their behaviors. The observant contacted the selected women and interviewed 11 of them between 1 and 3 days after the tour in person, or by phone. Participants were asked to give a time at which they could talk freely. Of 14 women contacted, one did not respond to us in a timely fashion, and two said they could not find a suitable time to freely talk with us. The interviews were in Persian and progressed in a free and natural flow but, to keep focus on the aims of the study, the participants were asked questions such as: What were your major motivations for taking this all-women excursion?; Tell us about the moments that made you happy or gave you a sense of well-being?; Tell us about the most memorable experiences you had during the tour?; What would you do if you were the tour organizer?; and Compare traveling in a group of women with a family trip. The core questions were supplemented with further intrinsic probes. The interviews lasted between 23 and 88 min with an average of 42 min. Table 1 summarizes the characteristics of the participants. All interviews were recorded, transcribed verbatim, and transferred to ATLAS.ti7 software for coding procedure. Two of the authors began the coding process together through line-by-line coding and identified 68 initial codes. In the focused coding phase, the most relevant and frequent initial codes were identified and the main categories were shaped.

Findings

The findings of the present study show that Iranian women feel happy on all-women tours because, on the one hand, they can escape from family responsibilities and some women-related restrictions and, on the other hand, they can develop their social and personal selves. First, the sense of freedom from the responsibilities that they feel at home grants them a break from mundane life and makes them happy. Second, being away from some of the restrictions that women encounter in a Muslim society provides them with an exciting sense of emancipation. Third, women can establish deeper communications with their same-age peers and talk about a wider range of issues, including life-challenges, and have informative and funny womanly conversations. Fourth, being in a group of women and being away from men gives them opportunities to experience some challenges for the first time, which creates a sense of well-being and eudaimonic happiness.

Being away from responsibilities and mundane life

Becoming a mother is full of sacrifice in many cultures, and afterwards many women give up some of their hobbies, dreams, and plans. In Iran, Islamic laws and traditions impose a double pressure on women in taking on household responsibilities. It seems that being away from daily family responsibilities is one of the most important contributors to happiness among women when traveling with all-women tours. Hence, exemption from family responsibilities, concerns, and restrictions gives them a sense of relief.

For many Iranian women who feel that they devote all their life to their family responsibilities, allocating time completely to themselves seems very exciting and rejuvenating. Shahla told us about the effect of the tour on her mood: "I was relaxed because I was alone and belonged to myself for a single day, free from all home stress." Complaining about the position of women in Iran's patriarchal society, Farideh said "Unfortunately, a woman becomes very limited after getting married in our country and it ruins her temper. But these tours give her emancipation back, relieve her, and help her to come back home in a better mood." She argues that participation in all-women tours refreshes women and helps them improve their mood in their daily life after the tour. In contrast, subordinating their own preferences to those of their family members during family trips recalls negative tourism experiences in some of these women. Shokat gives an example of a family trip: "we wanted to go shopping, he [her husband] said: No, let's go to my friend's house. I said to myself that I always accepted, so I accept this time too!" Although they consciously sacrifice their preferences at the time, the reality is that disregarding themselves had negative effects on their mood. Given that our sample was of over 50 year olds, being away from home and family responsibilities gave them a sense of youth. Azar, for instance, explains: "When I am alone, I care more about myself. In particular, when I'm with my grandchildren, I don't feel good … I feel old with them, and I don't like it."

Since home responsibilities such as cooking and taking care of children are daily obligations for these women, all-women excursions that exempt them from these responsibilities have a relieving effect on them. It should be mentioned that major home responsibilities are normally extended to family excursions. In this regard, Sima explains "When you go out with the family, you have to take care of children, prepare food for one, do something for another… Traveling with friends doesn't have these hassles and limitations. In fact, the responsibilities of being a wife, a mother, and a grandmother are not on your shoulders anymore." Similarly, Manijeh complains about these responsibilities and explains how being free of them leads to a more enjoyable trip when she comments "Even when I go on an excursion with my family, I have to have my home responsibilities … it is like you take the house out! But when I am alone, my time is for myself, I enjoy more, and I laugh from my heart."

Continuous worry for the children is another matter that occupies the mind of these mothers during family travels. According to Pari "if the children were here, I would worry about them all the time and I wouldn't have been so happy and relaxed." Traveling in a group of friends gives women freedom from the mental stresses related to their family. It seems that the break from being a mother and wife, together with being alongside other women of their own age, leads these women back to the age when they were single, as implied in the following quotation from Jahan:

> We were splashing water on each other and playing. We were like a kid who, when playing, she forgets the whole world, doesn't care about the next second, whether someone needs her or she has something to do or she is late.

During lunch, one of the women said "My poor children! What are they eating now?" A second woman replied "Come on! One day you get out having fun. Stop thinking about your children! They are not 2 year olds." And a third woman added "That's right! We Iranian women infantilize our children. Look at foreigners, they stop supporting their children on their 18th birthday." The fact that this conversation continued for a while shows the importance of this issue for these women.

For some Iranian women, although traveling with the family has its advantages, it is an extension of daily life, and therefore it cannot free them from the routine things they experience at home. According to Pari, "We are at home with our husband and children; we are with the same people when we go out with them. The place changes but the people are the same!" Similarly,

Farideh explains why escaping from routines leads to a better travel experience: "The trip [on family excursions] becomes an extension of the house, but I laugh more when I travel without family. It evokes my childhood; the mind becomes free from problems and home, and cares with no grief." The women in our study experience the transition from middle age to being elderly; some are retired and have a retired husband at home. When our observer asked a woman on the tour about her motivation for coming on such tours, she said that her husband watches TV all the time at home, and she finds it boring, so she comes on these tours to be refreshed.

Being away from women-related restrictions

In Iranian society, certain restrictions are imposed on women, mainly by tradition and Islamic Law. To observe Islamic modesty, some women cannot express their complete and genuine self in the presence of men outside their immediate family. Therefore, all-women tours give them the space to express what they really are; they can wear what they want, dance with their friends, say any funny thing they wish, and laugh out loud.

According to the Quran (An-Nisa: 34) "men are guardians of women, because God has made one superior to the other, and because men spend their wealth. Therefore, the righteous women are devoutly obedient." This verse is an example of how the Quran defines the relationship between couples. Iranian culture has been affected by Islamic law for centuries, and men have learned to support and protect their women and also practice the gheyrat codes to preserve their own honor. While some Iranian women appreciate their husband's gheyrat because it implies their husband's attention to them, others dislike it and consider it to be limiting and contrary to their freedom. The following quotation represents the negative perception of gheyrat in the mind of Ashraf:

> There is greater freedom with friends than family. There is no one to tell you what do and what not to do, where to go and where not to go, don't ride the boat, don't go near the water, don't go over there because there were non-mahram men, don't go there because some people are dancing .... There [in the tour] I felt the men [there were some families in the place the tour had a stop] didn't look at us badly at all. I felt comfortable and danced with them. If my husband, my son-in-law or my son were there I wouldn't feel comfortable and I wouldn't dance so that they wouldn't think badly about me. I just enjoyed it and became happy. Indeed, we have limited ourselves because of their way of thinking.

Ashraf decides not to tell her husband about the dancing and other activities, perceived as sensitive, that she undertook during the tour "I won't tell my husband these things because he won't let me go anymore [laughing]." For some of the Iranian women who censor their strong emotions in front of male strangers in order to practice Islamic modesty or satisfy their husband's honor, all-women tours are a place in which they can easily present a more authentic version of themselves which increases their well-being. Signs of the contribution of all-women tours to happiness and the expression of a less-censored-self are noticeable in the words of Shokat "but I was my real self on this trip, not restricted, I went everywhere and laughed every second I wanted" and Farideh who said "For example, we can't laugh out loud in front of our spouses [in the presence of other men]. But we laughed easily on this trip like a kid who doesn't think she is old or that it is inappropriate. Like a light-hearted and energetic kid." Interestingly, at the times they did activities perceived as sensitive, such as dancing or laughing out loud in the presence of unknown men, some of them mocked their husbands and made fun of them and said if their husbands were there, they would limit them or glare at them. One of the women compared themselves with their daughters' generation and said "they live life. In our time, we had to say yes to our husband all the time, but the boot is on the other foot now, it is men who obey the girls today."

Apart from gheyrat behavior codes among Iranian men, many Iranian women feel more comfortable in the absence of men they do not know because of their religious beliefs, and the sense of Haya. Islam suggests women must conceal their beauty from non-mahram men in order to avoid harassment (Vanderheiden & Mayer, 2017). Haya or Islamic modesty refers to shyness or an uneasy feeling accompanied by embarrassment. A quotation from Jahan reflects this feeling:

> Well, we are definitely more comfortable with women. That is, we can remove our veil, wear whatever we want, and say any womanish words. Generally, we were freer [in the tour], and we were not under the gaze of men. We have become accustomed to not being comfortable with men because we have been separated from men during our lives [by the government and traditions].

After Iran's 1979 revolution, Islamists came to power and have worsened conditions for women. Men and women have been separated in many social spaces such as schools and transport systems, and strict laws were implemented depriving women who avoid Islamic veiling of working for the government. It should be noted that many jobs in Iran are linked to the government because power in Iran is highly centralized (Seyfi et al., 2018). When we asked Ashraf what she would do if she were the tour organizer, she responded:

> I would include water play and swimming, and prevent men from watching us by putting a fence around the river. Men are not dreadful per se but smartphones are the problem; one would film us and the [Islamic] Guidance Organization would see it [leading to unemployment].

This comment reflects the fear of being deprived of some social rights if governmental organizations notice that they did not practice Islamic dress codes in public. The tour organizers were supposed to rent a site close to the river and prevent outsiders from entering it. However, when the tour arrived in the region, some families were already present. This presence of unknown men in the area caused complaints from some of the tour participants who believed that if there were no men there, as was expected, they could remove their hijabs and feel more comfortable. One woman, however, criticized this sensitivity to the presence of men and said "good for foreigners [women] who wear everything they wish," and another woman said, "That's right, they never mind if a man looks at them." The women eventually stayed in a corner of the area far from the other families. Afsaneh's quotation "[on these tours] I can do every naughty action I want that we couldn't do in the presence of men" shows how Islamic modesty sometimes limits their happiness, and how they feel happier when they are free from this cultural constraint.

Another aspect of freedom in this all-women tour is the freedom of womanly dialogue such as jokes, gossip, nostalgia, life-stories, and making fun of others. Manijeh compares this type of travel with family trips: "The atmosphere of traveling with friends is different. We can't make these jokes on a family trip." From Ashraf "We are more comfortable on telling some jokes, laughing, and womanly words in these groups." And Farideh said: "Those womanly jokes made me laugh a lot."

Dancing in the presence of unknown men is contrary to Islamic modesty but is a joyful activity which makes people happy. Dance is so important for some of our participants that when we asked Jahan about her main motivation to attend the tour she responded:

> The main motivation? I was in an English class when one of my classmates said she planned to go Markadeh on Monday. I warned her that we had an English class on Monday. But she said that there would be so much dancing that it was worth missing the class. Then I said OK.

The Islamic Republic has imposed many restrictions on dance and music in Iran, in particular for women. Nevertheless, some women dance in the places they feel comfortable, and they found the bus aisle a suitable place for dancing. A few minutes after they boarded the bus, one of the women said, "Come on! We joined the tour to have fun, let's play joyful music so we can dance" and other women supported her. At that moment, some religious women pointed to the driver and tried to prevent dancing in the presence of a man. Then, the women began negotiating; one said "take it easy, the driver is far from us," another one said "he doesn't see us, he is looking at the road," and one joked and said "like doctors, drivers are mahram to us." Finally, some women began dancing in the middle of the bus. Dancing was not limited to the bus; women also danced at the destination. Although there were some men there many of the women danced regardless of their presence. The following quotation from Pari reflects her way of thinking and her inner dialogue before deciding to take the risk of dancing:

> It was very interesting for me that there was a female singer who sung for these groups [tours] with all of the limitations and risks. Also, those ladies who danced without a scarf were very interesting to me and I said to myself why don't we dance? So, I felt young and I forgot my sense of motherhood or being married, and I just started dancing.

The tour organizers rented pickup trucks to move women to the rafting location which was ten minutes away. On average, around ten women rode in the back of each pickup truck. The experience was exciting for many of the women because it was a rare but interesting experience, and more importantly, in the absence of men it gave the group an opportunity to sing out loud together. It is interesting to note that they normally sang children's songs, perhaps because they were the only songs that all of them knew. They laughed a lot while singing the songs and also because the bouncing of the truck caused them to lose their balance.

Develop the social self

For many of the women who we interviewed, spending time with their friends was the main reason for participation in the tour. Ashraf, for instance, told us about her positive feeling when waiting for her friends, "I had been waiting impatiently and eagerly for the bus because I knew there were people I like." Similarly, Pari explained to us why talking with friends of her age was so important to her, "I talk with my childhood friend about that time. My husband and children do not understand these words."

It seems that talking with their friends and other women of their age makes them happy, mostly because they understand each other as explained by Jahan "through conversation with people of the same age, one can open out her heart and talk about mutual difficulties … she feels that they understand her and it creates a sense of joy and calmness." Likewise, Azar explained to us how spending time with friends can make her and other women on the tour happier, "I talked to a few women on the tour, there were a lot of problems in their lives, but they came on the tour, were happy and danced, and said they came on a tour whenever they had a problem." It is also seen in a quotation from Afsaneh, when she mentioned "I was so mentally upset that I wanted to cry, so I wanted to escape that atmosphere. I thought I should go on a happy tour during which I talk and laugh with my friends so that I get out of that awful mental condition."

Besides being with their close friends, the tour without any of their family members creates a space for some women to get to know other women. As Azar said, "In fact, people on these tours meet others who can affect their lives." Farideh also mentioned "On the trip, you may meet someone who causes many new things to happen and new relationships." Also, Manijeh narrated a story of becoming familiar with a woman on the tour who looked very religious at the beginning, "there was a woman in a veil at the back of the bus, at first I thought she was too religious, but when I talked to her I realized that she was a very good and joyful woman. We even danced together." Likewise, Ashraf happily explained to us how she made new friends on the tour, "The ones we became friends with were very easy-going and joyful and I felt they were similar to us. They liked us so much too. They even invited us to their garden."

Personal development

The youth of all the women we interviewed coincided with Iran's 1979 revolution, its subsequent extreme Islamism, the 8-year Iran-Iraq war and its economic shortages, and higher levels of masculinism and tradition. As a result, many of their wishes and needs had been destroyed, and they married without experiencing a happy and free time when they were single. After becoming mothers, they tried to ease conditions for their daughters by providing them with more facilities and negotiating higher levels of emancipation with their husbands. Iranian traditional society has gradually changed, mostly because of the emergence of the Internet, and many members of the recent generations have no longer prioritized tradition. Thus, middle-aged women sometimes compare their lives with that of their freer daughters, and try to compensate for their previous restrictions.

Afsaneh, for example, said "I feel I'm single here because they [parents] didn't let us do what we wanted when we were single. That time wasn't like now. We didn't even go on a single tour because the family wouldn't allow us, so we make up for it now [laughing]." Recalling her lost wishes and presenting her real self, she also pointed out "Sometimes I go on tours to express my real self; a single naughty girl with many wishes that have not been achieved yet but she tries to use these opportunities to achieve what she wants." Arguably, this sense of satisfying a suppressed need contributes to their eudaimonic well-being. It was very interesting that the women from our study addressed other women on the tour as "Dokhtar" (which means "girl" in Persian) – a signal that these same-gender tours can mentally return them to the time they were single. The consequences of these positive emotions, even on the small scale of a 1-day excursion, can improve their general well-being.

Being in a group of same-age women encouraged some women on the tour to experience adventures they had never had before. Rafting formed part of the tour, but was optional. At first, not all participants were eager to board the boats, but they encouraged each other to the point that all of the women boarded in the end. Sima compared her level of risk-taking on this trip with family trips in the following quotation "I was less scared this time because I saw that I was the only one who remained. So I dared to go. I was really scared at first, but I calmed down when I saw everyone was comfortable." She mentioned that previously she had never accepted her children's requests to join them riding. Similarly, Pari talked to us about her reduced fear of rafting and how she enjoyed this activity this time beside her friends of the same age and gender, "my fear disappeared in the group, I mean if I was alone, I wouldn't enjoy it because of the stress." The interviews were performed 1–2 weeks after the trip, and it was interesting that when we asked them to name the most memorable experiences they had on the tour, the first response by Sima, Afsaneh, and Pari was rafting. These three women were among the last to board the boat. This suggests how the perceived risk-taking was memorable for them and may contribute to their well-being.

The tour without men was a space for learning new things for some of the women. According to Sima, "It is attractive to become familiar with new people because it improves one's public relations and also they learn new things from each other." Azar also added "For example, I learned that I should express my opinion and say what I want. I also love to learn good things from my travel companions." Azar refers to the objection of a young girl about the failure of the tour leader on missing a historical attraction, an objection that forced the leader to give all tour members a 20% discount for the next trip.

Another example appeared when we asked Shahin what she would do if she were the tour leader, and she responded "If I have pickles that I made myself, I would give them to others and teach everyone what I know, and in return, I would ask others to teach everything they know, so that we use everyone's ability. For example, anyone who knows some informative housekeeping tips teaches others."

Discussion and conclusions

The findings of this study suggest two dimensions of "escapism from" and two dimensions of "seeking for" among Iranian women in all-women tour experiences. On the one hand, being away from responsibility and mundane life as well as an escape from gender roles and other dos and don'ts, and on the other, seeking social and personal development, contribute to the happiness and well-being of these women. There is no doubt that family travels and being beside family members bring their own benefits to the well-being of middle-aged women, but the findings of this study show that occasional travel with same-aged, likeminded women can dramatically boost happiness among middle-aged women, particularly in countries like Iran in which the formal rules and social norms work against women.

We argue that fewer responsibilities and fewer concerns for others on all-women tours, as compared to family journeys, provide opportunities for women to enjoy every moment more. Forgetting their maternal roles, they can spend time with same-aged women as if they were really young girls merely labeled with wife, mother and grandmother tags. Perhaps this is one reason why many women in the study call the others "Dokhtar" meaning "girl." Interestingly, Durko and Stone (2017) noticed that women address each other as "girl" on an all-women tour in the United States, which may point to some degree of universality. Berdychevsky et al. (2013) link this behavior to the liminal space of all-women tours being considered as excluding gendered roles, and suggest that this atmosphere causes a sense of entitlement among participants. This mental return to girlhood contributes to that positive emotion which is the first dimension of well-being in the PERMA model introduced by Seligman (2011).

In addition to this, routine life for the majority of those we interviewed was too boring because they were retired and had a retired man at home. Therefore, breaking away from mundane and depressing lives has restorative effects for them. The period between 50 and 65 years old is considered a life-transition stage and it is important to deal with this period successfully (Gibson et al., 2012), especially in the case of women—who tend to care more about aging than men (Rutagumirwa & Bailey, 2019). Several studies have found that ordinary experiences, such as being out of the house, spending time with friends, and having an active life, provoke happiness, particularly when people become older (Bhattacharjee & Mogilner, 2014; Wood & Dashper, 2020). Although our case study was in no way extraordinary, surprisingly we observed that even such an ordinary 1-day excursion made an important contribution to the happiness of middle-aged women.

The greatest difference between the present study and other studies on all-women tours is related to the restrictions imposed by cultural traditions and Islamic rules on Iranian women. To abide by Islamic modesty, many Iranian women keep strict Islamic dressing codes and avoid laughing out loud and having fun in front of men outside of their immediate family. All-women tours give them a space to express a more authentic, real, happier, and funnier version of their self in the absence of men. The opportunity to freely laugh, sing, dance, and for womanly speech gives them a sense of the joy of which they are commonly deprived in Iran. Through this type of travel, some women also take a break from their husband's "dos" and "don'ts" which give them a negative sense of being someone else's possession. Although the majority of women expressed satisfaction with their husband and family, almost all of them implied that, on the tour, they felt much freer and happier in their absence. Expressions of the authentic self on all-women tours have been mentioned before in both Western and Eastern studies (Berdychevsky et al., 2013; Song, 2017) but it seems that the extent and areas of self-censorship in Muslim women are not comparable with freer Western women, since the internalized Islamic codes of modesty among pious Muslim women prevent them from representing a happy and active self. Therefore, compared to non-Muslim women, all-women tours have a greater impact on their hedonic well-being.

Similar to other studies on all-women tours (see Berdychevsky et al., 2013, 2016; Chen, & Mak, 2020; Junek et al., 2006), our findings show that spending time with friends is a common motivation among participants and contributes to their well-being because they not only have fun with their friends, but also can talk to their close friends about their problems and obtain informative advice. Many of the participants told us that spending time with likeminded women on the tour who understand them was the main factor that helped relieve them of bad moods and feel a sense of being happier and stronger. Prior to our study, Wood and Dashper (2020) suggested that spaces providing opportunities for older women to socialize have restorative effects on them and improve their sense of self-worth. It is also worth mentioning that intimacy and friendship are both described as important elements of well-being in Seligman's (2011) PERMA model.

Finally, being in a group of same-age women empowers participants to do what they previously feared. In our research, for instance, three of the participants said that they had avoided boarding the rafting boats on family trips, but when they observed that all the other women on the tour boarded the boat, they overcame their fear and decided to experience it for the first time. Furthermore, of course, this accomplishment brought great joy and memorable experiences for them. Likewise, dancing in the presence of men is against Islamic modesty for traditional or more pious Muslim women, but two of our participants who danced in the presence of men happily told us that they were proud of breaking that perceived modesty. The joy of dance and the satisfaction of breaking a perceived constraint contribute to both hedonic and eudaimonic well-being, and are a way to combat depression (Karkou et al., 2017). It is noteworthy that the nature of participation in all-women tours provoked independence and empowerment in women who felt they were highly depended on, or controlled by, their husbands.

Our study enabled participants to mentally compare traveling with and without their family, and the results suggest that women's rights to fully enjoy excursions are partially spoiled in many traditional Iranian families due to stereotypical expectations for women and to the control exerted over them. The findings also reveal that even temporary exemption from these expectations gives women a sense of freedom, happiness, empowerment, and well-being. The UNWTO (2015) suggests 17 links between tourism and the UN sustainable development goals (SDGs) that emphasize well-being and empowerment for women; for instance, SDG 3 is entitled "ensure healthy lives and promote well-being for all at all ages," and SDG 5 says "achieve gender equality and empower all women and girls."

Nevertheless, the goals are mostly host-oriented and the rights of women travelers are almost entirely neglected. This is not in line with the United Nations Development Programme (UNDP), which describes human development as granting freedom to all humans to pursue what they value. We argue that the UNWTO should strengthen the "right to enjoy travel" for women in traditional societies. Families in such communities should be educated to understand that women are not merely the service provider for the family while travelling, and that they are entitled to have adequate time for themselves during trips. We argue that the UNWTO should promote "equal emancipation" and "equal time for self" in family travels as a gender equality factor.

The present study has some practical implications; we suggest that tour operators who organize all-women tours should consider the real motivations of women instead of focusing on "girly" advertisements. On the global scale, the findings of this study and other research show that women who participate in all-women tours escape from obligatory responsibilities but, at the same time, they are interested in undertaking voluntary activities to express their skills and feel a sense of self-worth. The provision of these opportunities on all-women tours would, therefore, contribute to their sense of well-being. Second, since communicating with other women is important on this type of tour and gives women a sense of communitas, it would be beneficial for tour organizers to bring the women together and provide opportunities for them to talk and express themselves. Third, programs that set a range of challenges would provide women with a sense of accomplishment and well-being and be very worthwhile on this type of tour. Fourth, in traditional Muslim markets, tour organizers should consider activities such as group dancing and singing that are lacking in the routine lives of these women.

Although we do our best to reveal the elements of well-being on all-women tours, there are of course some limitations. First, it is probable that the women concealed some unpleasant issues about them and their family restrictions and problems in order to keep their social desirability. To deal with this problem, one of the authors travelled and communicated with them to be considered as an "insider." We also first asked them to talk about the problems that other women face in society, and then we asked them to give us personal experiences. Talking about other women helped them to feel freer and talk about the discriminations that they perceived against themselves. Second, although we chose our cases at random from participants in the most common type of all-women tour in Iran, they represent a limited socio-economic status range within a single country. We recommend that future studies should examine all-women tours in other nations to reveal the further dimensions, motivations, and consequences of this type of travel. We also believe that there is a lack of study in the field of all-men tours known as "mancation."

ORCID

Adel Nikjoo http://orcid.org/0000-0003-1468-9948

References

Abedinifard, M. (2019). Persian 'Rashti jokes': Modern Iran's palimpsests of gheyrat-based masculinity. *British Journal of Middle Eastern Studies*, *46*(4), 564–582. https://doi.org/10.1080/13530194.2018.1447440

Berdychevsky, L., Gibson, H. J., & Bell, H. L. (2013). Girlfriend getaways and women's well-being. *Journal of Leisure Research*, *45*(5), 602–623. https://doi.org/10.18666/jlr-2013-v45-i5-4365

Berdychevsky, L., Gibson, H. J., & Bell, H. L. (2016). Girlfriend getaway as a contested term: Discourse analysis. *Tourism Management*, *55*, 106–122. https://doi.org/10.1016/j.tourman.2016.02.001

Bhattacharjee, A., & Mogilner, C. (2014). Happiness from ordinary and extraordinary experiences. *Journal of Consumer Research*, *41*(1), 1–17. https://doi.org/10.1086/674724

Bond, K. (2019). (Ed.). *Dance and the quality of life*. Springer.

Bullington, J. (2006). Body and self: A phenomenological study on the ageing body and identity. *Medical Humanities*, *32*(1), 25–31. (2019). https://doi.org/10.1136/jmh.2004.000200

Charmaz, K. (2014). *Constructing grounded theory*. Sage.

Chen, X., & Mak, B. (2020). Understanding Chinese girlfriend getaways: An interdependence perspective. *Annals of Tourism Research*, *81*, 102878. https://doi.org/10.1016/j.annals.2020.102878

Cheng, E., Stebbins, R., & Packer, J. (2017). Serious leisure among older gardeners in Australia. *Leisure Studies*, *36*(4), 505–518. https://doi.org/10.1080/02614367.2016.1188137

Choi, P., Henshaw, C., Baker, S., & Tree, J. (2005). Supermum, superwife, supereverything: Performing femininity in the transition to motherhood. *Journal of Reproductive and Infant Psychology*, *23*(2), 167–180. https://doi.org/10.1080/02646830500129487

Cockburn-Wootten, C., Friend, L., & McIntosh, A. (2006). A discourse analysis of representational spaces: Writings of women independent traveller. *Turizam: međunarodni Znanstveno-Stručni Časopis*, *54*(1), 7–16.

Davoodi, Z., Fatehizade, M., Ahmadi, A., & Jazayeri, R. (2019). Culture and power: How do culture and power influence Iranian couples. *Journal of Couple & Relationship Therapy*, *18*(4), 353–313. https://doi.org/10.1080/15332691.2019.1620146

Deaux, K., & Lewis, L. L. (1984). Structure of gender stereotypes: Interrelationships among components and gender label. *Journal of Personality and Social Psychology*, *46*(5), 991–1004. https://doi.org/10.1037/0022-3514.46.5.991

Doran, A. (2016). Empowerment and women in adventure tourism: A negotiated journey. *Journal of Sport & Tourism*, *20*(1), 57–80. https://doi.org/10.1080/14775085.2016.1176594

Douglas, S. J., & Michaels, M. W. (2004). *The mommy myth: The idealization of motherhood and how it has undermined all women*. Free Press.

Durko, A. M., & Stone, M. J. (2017). Even lovers need a holiday: Women's reflections of travel without their partners. *Tourism Management Perspectives*, *21*, 18–23. https://doi.org/10.1016/j.tmp.2016.11.001

Eger, C. (2020). Equality and Gender at Work in Islam: The Case of the Berber Population of the High Atlas Mountains. *Business Ethics Quarterly*, 1–32. https://doi.org/10.1017/beq.2020.21

Ehteshami, A. (1995). *After Khomeini: The Iranian second republic*. Routledge.

Fadaee, S. (2018). Ecotours and politics of fun in Iran: From contested state-society relations to emancipatory nature-society relations. *The Sociological Review*, *66*(6), 1276–1291. https://doi.org/10.1177/0038026118774981

Filep, S. (2014). Moving beyond subjective well-being: A tourism critique. *Journal of Hospitality & Tourism Research*, *38*(2), 266–274. https://doi.org/10.1177/1096348012436609

Gibson, H. J., Berdychevsky, L., & Bell, H. L. (2012). Girlfriend getaways over the life course: Change and continuity. *Annals of Leisure Research*, *15*(1), 38–54. https://doi.org/10.1080/11745398.2012.670963

Gonyea, J. G., Paris, R., & de Saxe Zerden, L. (2008). Adult daughters and aging mothers: The role of guilt in the experience of caregiver burden. *Aging & Mental Health, 12*(5), 559–567. https://doi.org/10.1080/13607860802343027

Goodill, S. W. (2005). *An introduction to medical dance/movement therapy: Health care in motion*. Jessica Kingsley Publishers.

Heimtun, B. (2007). Depathologizing the tourist syndrome: Tourism as social capital production. *Tourist Studies, 7*(3), 271–293. https://doi.org/10.1177/1468797608092513

Heimtun, B., & Jordan, F. (2011). Wish YOU weren't here!: Interpersonal conflicts and the touristic experiences of Norwegian and British women travelling with friends. *Tourist Studies, 11*(3), 271–290. https://doi.org/10.1177/1468797611431504

Henderson, K. A., & Hickerson, B. (2007). Women and leisure: Premises and performances uncovered in an integrative review. *Journal of Leisure Research, 39*(4), 591–610. https://doi.org/10.1080/00222216.2007.11950124

Heo, J., Lee, Y., McCormick, B. P., & Pedersen, P. M. (2010). Daily experience of serious leisure, flow and subjective well-being of older adults. *Leisure Studies, 29*(2), 207–225. https://doi.org/10.1080/02614360903434092

Jackson, E. L., & Henderson, K. A. (1995). Gender-based analysis of leisure constraints. *Leisure Sciences, 17*(1), 31–51. https://doi.org/10.1080/01490409509513241

Joseph, D., & Southcott, J. (2019). Meanings of leisure for older people: An Australian study of line dancing. *Leisure Studies, 38* (1), 74–87. https://doi.org/10.1080/02614367.2018.1544655

Junek, O., Binney, W., & Winn, S. (2006). All-female travel: What do women really want? *Tourism, 54*(1), 53–62.

Karkou, V., Oliver, S., & Lycouris, S. (2017). (Eds.). *The Oxford handbook of dance and well-being*. Oxford University Press.

Kim, J., Lee, S., Chun, S., Han, A., & Heo, J. (2017). The effects of leisure-time physical activity for optimism, life satisfaction, psychological well-being, and positive affect among older adults with loneliness. *Annals of Leisure Research, 20*(4), 406–415. https://doi.org/10.1080/11745398.2016.1238308

Khoo-Lattimore, C., & Prayag, G. (2015). The girlfriend getaway market: Segmenting accommodation and service preferences. *International Journal of Hospitality Management, 45*, 99–108. https://doi.org/10.1016/j.ijhm.2014.12.003.

Khalajabadi-Farahani, F., Månsson, S. A., & Cleland, J. (2019). Engage in or refrain from? A qualitative exploration of premarital sexual relations among female college students in Tehran. *Journal of Sex Research, 56*(8), 1009–1022. https://doi.org/10.1080/00224499.2018.1546371

Ku, P.-W., Fox, K. R., & Chen, L.-J. (2009). Physical activity and depressive symptoms in Taiwanese older adults: A seven-year follow-up study. *Preventive Medicine, 48*(3), 250–255. https://doi.org/10.1016/j.ypmed.2009.01.006

Liechty, T., Ribeiro, N. F., & Yarnal, C. M. (2009). I traveled alone, but I never felt alone: An exploration of the benefits of an older women's group tour experience. *Tourism Review International, 13*(1), 17–29. https://doi.org/10.3727/154427209789130611

Mahdavi, P. (2012). Questioning the global gays (ze): Constructions of sexual identities in postrevolution Iran. *Social Identities, 18*(2), 223–237. https://doi.org/10.1080/13504630.2012.652846

McCabe, S., & Johnson, S. (2013). The happiness factor in tourism: Subjective well-being and social tourism. *Annals of Tourism Research, 41*, 42–65. https://doi.org/10.1016/j.annals.2012.12.001

McGannon, K. R., Gonsalves, C. A., Schinke, R. J., & Busanich, R. (2015). Negotiating motherhood and athletic identity: A qualitative analysis of Olympic athlete mother representations in media narratives. *Psychology of Sport and Exercise, 20*, 51–59. https://doi.org/10.1016/j.psychsport.2015.04.010

Miller, Y. D., & Brown, W. J. (2005). Determinants of active leisure for women with young children: An ethic of care prevails. *Leisure Sciences, 27*(5), 405–420. https://doi.org/10.1080/01490400500227308

Mirehie, M., Gibson, H. J., Khoo-Lattimore, C., & Prayag, G. (2018). An exploratory study of hospitality needs and preferences of US Girlfriend Getaways. *Journal of Hospitality Marketing & Management, 27*(7), 811–832. https://doi.org/10.1080/19368623.2018.1448314

Mitten, D. (1992). Empowering girls and women in the outdoors. *Journal of Physical Education, Recreation & Dance, 63*(2), 56–60. https://doi.org/10.1080/07303084.1992.10604117

Moghadam, S., Knudson-Martin, C., & Mahoney, A. R. (2009). Gendered power in cultural contexts: Part III. Couple relationships in Iran. *Family Process, 48*(1), 41–54. https://doi.org/10.1111/j.1545-5300.2009.01266.x

Murcia, C. Q., Kreutz, G., Clift, S., & Bongard, S. (2010). Shall we dance? An exploration of the perceived benefits of dancing on well-being. *Arts & Health, 2*(2), 149–163. https://doi.org/10.1080/17533010903488582

Nawijn, J., & Filep, S. (2016). Two directions for future tourist well-being research. *Annals of Tourism Research, 61*(C), 221–223. https://doi.org/10.1016/j.annals.2016.07.007

Petty, L., & Trussell, D. E. (2019). Leisure self-care, health and well-being in women's lives. *Annals of Leisure Research*, 1–12. https://doi.org/10.1080/11745398.2019.1652661

Rafatjah, M. (2012). Changing gender stereotypes in Iran. *International Journal of Women's Research, 1*(1), 61–75.

Rieger, K. L. (2019). Discriminating among grounded theory approaches. *Nursing Inquiry, 26*(1), e12261. https://doi.org/10.1111/nin.12261

Rutagumirwa, S. K., & Bailey, A. (2019). I have to listen to this old body: Femininity and the aging body. *The Gerontologist, 59*(2), 368–377. https://doi.org/10.1093/geront/gnx161

Ryan, R. M., & Deci, E. L. (2001). On happiness and human potentials: A review of research on hedonic and eudaimonic well-being. *Annual Review of Psychology, 52*, 141–166. https://doi.org/10.1146/annurev.psych.52.1.141

Ryu, J., & Heo, J. (2018). Relationships between leisure activity types and well-being in older adults. *Leisure Studies, 37*(3), 331–342. https://doi.org/10.1080/02614367.2017.1370007

Samdahl, D. M. (2013). Women, gender, and leisure constraints. In *Leisure, Women, and Gender*, edited by V. J. Freysinger, S. M. Shaw, K. A. Henderson, & M. D. Bialeschki (pp. 109–125). Venture.

Scheiner, J. (2014). Gendered key events in the life course: Effects on changes in travel mode choice over time. *Journal of Transport Geography, 37*, 47–60. https://doi.org/10.1016/j.jtrangeo.2014.04.007

Seligman, M. E. P. (2011). *Flourish: A visionary new understanding of happiness and well-being*. Free Press.

Seligman, M. E. P., & Csikszentmihalyi, M. (2000). Positive psychology: An introduction. *The American Psychologist, 55*(1), 5–14. https://doi.org/10.1007/978-94-017-9088-8_18 https://doi.org/10.1037//0003-066x.55.1.5

Seresinhe, C. I., Preis, T., MacKerron, G., & Moat, H. S. (2019). Happiness is greater in more scenic locations. *Scientific Reports, 9*(1), 4498. https://doi.org/10.1038/s41598-019-40854-6

Seyfi, S., Nikjoo, A., & Sharifi-Tehrani, M. (2018). Domestic tourism in Iran: Development, directions, and issues. In S. Seyfi & C. M. Hall (Eds.), Tourism in Iran: Challenges, development and issues (pp. 38–54). Routledge.

Shaw, S. M. (1994). Gender, leisure, and constraint: Towards a framework for the analysis of women's leisure. *Journal of Leisure Research, 26*(1), 8–22. https://doi.org/10.1080/00222216.1994.11969941

Small, J. (2005). Women's holidays: Disruption of the motherhood myth. *Tourism Review International, 9*(2), 139–154. https://doi.org/10.3727/154427205774791645

Song, H. (2017). Females and tourism activities: An insight for all-female tours in Hong Kong. *Journal of China Tourism Research, 13*(1), 83–102. https://doi.org/10.1080/19388160.2017.1327385

Southcott, J., & Joseph, D. (2020). If you can breathe, you can dance: Fine lines contemporary dance for mature bodies in Melbourne. *Journal of Women & Aging, 32*(6), 591–610. https://doi.org/10.1080/08952841.2019.1591890

Tavakoli, R., & Mura, P. (2017). Iranian women traveling: Exploring an unknown universe. In *Women and travel* (pp. 109–124). Apple Academic Press.

Tizro, Z. (2013). *Domestic violence in Iran: Women, marriage and Islam*. Routledge.

Turner, V. (1974). Liminal to liminoid, in play, flow, and ritual: An essay in comparative symbology. *Rice University Studies, 60*(3), 53–92. https://doi.org/10.1080/08952841.2019.1591890

Vanderheiden, E., & Mayer, C. H. (2017). An introduction to the value of shame – Exploring a health resource in cultural contexts. In *The Value of Shame* (pp. 1–39). Springer. https://doi.org/10.1007/978-3-319-53100-7_1

Weedon, C. (1997). *Feminist practice and poststructuralist theory* (2nd ed.). Blackwell.

White, M., Smith, A., Humphryes, K., Pahl, S., Snelling, D., & Depledge, M. (2010). Blue space: The importance of water for preference, affect, and restorativeness ratings of natural and built scenes. *Journal of Environmental Psychology, 30*(4), 482–493. https://doi.org/10.1016/j.jenvp.2010.04.004

Wood, E. H., & Dashper, K. (2020). Purposeful togetherness": Theorising gender and ageing through creative events. *Journal of Sustainable Tourism*, 1–17. https://doi.org/10.1080/09669582.2020.1803890

World Economic Forum (2020). Global Gender Gap Report 2020. Retrieved from https://www.weforum.org/reports/gender-gap-2020-report-100-years-pay-equality

World Tourism Organization (2015). *Tourism and the Sustainable Development Goals*. UNWTO. https://doi.org/10.18111/9789284417254

Wyles, K. J., White, M. P., Hattam, C., Pahl, S., King, H., & Austen, M. (2019). Are some natural environments more psychologically beneficial than others? The importance of type and quality on connectedness to nature and psychological restoration. *Environment and Behavior, 51*(2), 111–143. https://doi.org/10.1177/0013916517738312

Yang, E. C. L., Khoo-Lattimore, C., & Arcodia, C. (2018). Constructing space and self through risk taking: A case of Asian solo female travelers. *Journal of Travel Research, 57*(2), 260–272. https://doi.org/10.1177/0047287517692447

Zelenski, J. M., & Nisbet, E. K. (2014). Happiness and feeling connected: The distinct role of nature relatedness. *Environment and Behavior, 46*(1), 3–23. https://doi.org/10.1177/0013916512451901

The gendered effects of statecraft on women in tourism: economic sanctions, women's disempowerment and sustainability?

Siamak Seyfi, Colin Michael Hall and Tan Vo-Thanh

ABSTRACT

Despite sanctions being one of the most common and far-reaching forms of economic statecraft, there is a notable absence of research on the gendered effects of economic sanctions on women's empowerment in general, and more particularly in relation to tourism. This is surprising given that the burden of economic sanctions is overwhelmingly felt by women due to their vulnerable socio-economic and political status in targeted countries. Drawing upon a disciplinary base in international relations and political science and using a gendered lens via a series of interviews, this study sought to explore the gendered effects of economic sanctions on Iranian women's empowerment in the country's tourism and hospitality industry. The study's findings indicate that sanctions have negatively affected and deteriorated economic, psychological, social and political aspects of women empowerment. The results highlight the vulnerability of empowerment within the religio-patriarchal society of Iran. As such economic empowerment is recognized as a major contributor to the overall empowerment of women in Iran which is therefore severely affected by sanctions. Overall, this study fills a significant gap in tourism research by highlighting the gendered implications of a ubiquitous state tool of coercive diplomacy and foreign policy and its effects on women's empowerment.

Introduction

Economic statecraft is the use of economic means and resources to exert influence and pressure in pursuit of foreign policy objectives (Baldwin, 1985). Described as "the precision-guided munitions of economic statecraft" (Drezner, 2011, p.96) sanctions have increasingly become a ubiquitous blunt tool of coercive diplomacy and foreign policy used by states and international actors seeking to influence the behaviors of a target state or political actor (Leyton-Brown, 2017). In the aftermath of the Cold War, sanctions have widely been viewed as 'a lower-cost, lower-risk', and middle option between diplomacy and war (Masters, 2019).

Sanctions take a number of different forms and measures, ranging from comprehensive economic and trade sanctions to more targeted measures (smart sanctions) such as asset freezes, travel bans, financial or commodity restrictions, raising of customs tariffs, and arms embargo (United Nations Security Council, 2020). Nonetheless, this 'carrot-and-stick' approach in diplomacy and foreign policy (Hall, 2005) is often criticized for their devastating impacts on human rights and the well-being of civilian populations. Hatipoglu and Peksen (2018) argue that sanctions may have painful consequences for a target country's elites, and its economic and political stability, by trigging financial crisis and economic hardship. However, rather than emasculating the elites, they instead cause proportionately far greater damage to the livelihoods of the more vulnerable members of society, especially women and children, owing to their often already vulnerable socioeconomic and political status in target countries (Buck et al., 1998; Drury & Peksen, 2014; Gutmann et al., 2019). The social and political instability ensuing from sanctions may therefore lead to greater violation of women's political and social rights, further hindering their empowerment and the contributions they can make to inclusive and sustainable development (Cole, 2006; Boluk et al., 2019; Alarcón & Cole, 2019). This has previously been addressed by global institutions and initiatives, such as the third Millennium Development Goal (MDG3) and, in tourism, by the UNWTO's (2019) global report on women in tourism. Women's equality and empowerment is enshrined in goal 5 of 17 SDGs (UN Women, 2020), although it is also intrinsic and critical to the achievement of all 17 SDGs (Alarcón & Cole, 2019; Boluk et al., 2019).

Economic and financial crises have a disproportionate impact on women and the vulnerable in a society (Ruggieri, 2010; Piazzalunga & Di Tommaso, 2019) and further disempower women already affected by structural economic and cultural inequalities (Padgett & Warnecke, 2011). Sanctions are significant as they imply a deliberate attempt to cause economic hardship as part of regime change (Drezner, 2011). They are undertaken by the same governments and institutions that support the SDGs and the empowerment of women. However, the connections between policies that deliberately create economic and financial hardship and those that are aimed to respond to such difficulties and their gendered implications, is virtually missing in the analysis of sanctions.

Despite the growing number of studies on tourism and women's empowerment (Scheyvens, 1999, Cole, 2006; Tucker & Boonabaana, 2012; Moswete & Lacey, 2015; Ramos & Prideaux, 2014; Farahani & Dabbaghi, 2018; Alarcón & Cole, 2019; Figueroa-Domecq et al., 2020), significant gaps remain with respect to the political and socioeconomic conditions of countries where tourism is an economic development strategy and the often marginalized status of women in politics and policy-making. Furthermore, little research attention has been given to the gendered consequences of sanctions on tourism and its relation to women's empowerment, with the existing literature largely being either gender-blind or gender-neutral. Arguably, such issues also reflect the wider under researched role of gender in the global financial and economic policy architecture (Ruggieri, 2010). In the case of tourism studies, although the gender domain is recognized as "a critical arena for action to attain sustainable development" (Boluk et al., 2019, p.852), tourism scholars have not included the impacts of sanctions in their empirical investigations and gendered-focused analysis of sanctions has been absent in the wider scholarship. In addition to a paucity of research on gender-specific effects of sanctions and women' empowerment within a tourism context, their relationships to the SDGs has also been neglected in the literature.

Drawing upon the theoretical framework of empowerment developed by Scheyvens (1999, 2000), this study sought to explore how economic sanctions have affected the different aspects of women's empowerment in Iran in a tourism context. Iran, which is subject to one of the longest and strongest sanctions regimes in history, provides an interesting case to explore such interrelationships given the already strongly gendered nature of tourism and hospitality in the country (Farahani & Dabbaghi, 2018) and its religio-patriarchal society and masculinized political culture (Shahidian, 2002). This study is also very timely given the reimposition of sanctions by the USA in 2018 against Iran after a short spell of relaxation in the light of Iran's nuclear deal (2015-2017) and the potential implications for the county's tourism industry which witnessed a dramatic growth after the temporary lifting of sanctions (Seyfi & Hall, 2018).

The paper begins with an overview of economic statecraft and its relations to tourism and its gendered consequences. This is also contextualized by discussion of theoretical approaches to women's empowerment. The research design is then outlined and the empirical findings from qualitative research reported before conclusions are drawn.

Literature review

Economic statecraft and tourism

Sanctions have long been a popular tool of economic statecraft and have been used "in pursuit of a broad range of objectives related to international conflict prevention, conflict management, and conflict resolution" (Jentleson, 2000, p.124). In the post-Cold War era, sanctions have become a popular unilateral policy instrument (the United States being the most common user) (Farrall, 2007), and an increasingly central element of multilateral foreign security policy for international organizations (UN Security Council) or regional actors (e.g. EU). The term 'sanctions' in international law refers to "coercive measures, taken by one State or in concert by several States (the sanctioner), which are intended to convince or compel another State (the sanctionee) to desist from engaging in acts violating international law" (Ilieva et al., 2018, p.201).

The political goals behind sanctions are diverse and ambitious, while the measures themselves can take many forms. The nature of the measures imposed has shifted from comprehensive sanctions regimes (e.g. against Iraq in the 1990s) to more 'targeted' or 'smart' sanctions (e.g. against Iran, Russia, and Belarus), which are directed at specific individuals or entities (through asset freezes and travel bans) or by prohibiting particular activities (arms embargoes and export bans) (Farrall, 2007). Comprehensive sanctions are intended to restrict the trade and economic development of targeted countries, and usually negatively affect their GDP, bilateral trade, and financial services (Coleman, 2001). Neuenkirch and Neumeier (2015) suggest that multilateral sanctions (e.g. UN sanctions) and unilateral sanctions (imposed by the US) reduce, on average, the targeted country's GDP by 25% and 13% respectively.

Sanctions are a widely-researched topic in international relations, political economy and political science (Baldwin, 1985; Drezner, 2011; Leyton-Brown, 2017). Nonetheless, despite this vast and growing literature, the widespread use of economic sanctions as a coercive foreign policy tool and their immediate and multifaceted impact on the tourism industry (Seyfi & Hall, 2019b; Seyfi & Hall, 2020), there is limited research on the effect of sanctions on tourism in the target destination or firm in spite of the growing use of tourism as an instrument of geopolitical pressure (Hall, 2005; Timothy, 2017; Gillen & Mostafanezhad, 2019). Furthermore, despite the significance of politics and foreign policy in relation to tourism as well as the mobility of individuals within geopolitical systems (Hall, 2017), recent studies on the wider literature on tourism and geopolitics (Hall, 2017, Timothy, 2017; Mostafanezhad, 2018; Gillen & Mostafanezhad, 2019) have largely neglected sanctions as an important topic.

Restrictions on tourism mobilities have been used as an economically-oriented political instrument. This can be seen in a number of past and contemporary examples of mobilities restrictions. For example, these include longstanding US restrictions on its citizens traveling to Cuba (Coleman, 2001); Russian sanctions against Turkey in 2015 that restricted Russians visiting Turkey and halted tour packages (Tekin, 2015); Russian flight bans against Georgia in 2019 (Roth, 2019); travel ban placed by President Trump's administration in 2017 restricting travel to the United States for citizens of Iran, Iraq, Libya, Somalia, Sudan, Syria, and Yemen (Arafa, 2018); and more recent EU travel bans against Belarusian officials following the post-election crackdown on demonstrators in August 2020 (Barigazzi, 2020). In addition to mobility restrictions, sanctions disrupt financial investment and supply chains and impact destination image and can serve to isolate the destination from some international tourism market segments thereby affecting international visitation, investment, destination development and industry structure (Seyfi & Hall, 2019a, 2019b; 2020).

Yet, the extant studies as outlined above tend to be more state-centric and fail to examine how economic sanctions have affected specific disadvantaged groups, including women who comprise a large portion of the labor force in service-oriented industries such as tourism (Ferguson, 2011; Figueroa-Domecq et al., 2015). The gendered consequences of sanctions and how this coercive policy leads to disempowerment of women will be discussed in the next section.

Gendered instrument of statecraft

Although the sanctioning of countries is usually conceived as being a gender-neutral act, they often affect the most vulnerable populations, including women, who bear the burden of the sanctions at a higher level (Drury & Peksen, 2014). Criticism over the utility and effectiveness of sanctions consequentially resulted in a 'qualitative shift' from comprehensive embargoes towards 'smart' or 'targeted' sanctions (Cameron, 2003; De Goede, 2011). The reforms in sanctions policy supposedly enable sanctioning actors to use more focused sanctions as a means to target power elites and ruling classes while imposing less hardship on the mass public (De Goede, 2011). Nevertheless, numerous relatively comprehensive sanctions remain in place and, as noted in the introduction, the lack of consideration of gender issues in their application is remarkable given awareness of the effects of economic crisis in general, and the importance of gender empowerment as a policy response to structural economic crisis, disasters and equalities (Padgett & Warnecke, 2011).

Despite the existence of a plethora of literature on the economic impacts of sanctions, only a handful of empirical studies have focused on the impact of sanctions on marginalized and vulnerable groups such as the poor and women (Buck et al., 1998; Al-Ali, 2005; Al-Jawaheri, 2008; Drury & Peksen, 2014; Gutmann et al., 2019). When a state's overall economy is affected by sanctions, women's social, political, and economic rights are often substantially affected and the unequal effect on women may be overwhelming. In an early effort to examine the gendered consequences of sanctions, Buck et al. (1998) concluded that the costs of trade sanctions are disproportionately imposed on Iraqi women, who are the country's most marginalized political actors. Cross-national research conducted by Drury and Peksen (2014) also shows that sanctions have a gendered effect and reduce the level of respect for women's rights in the targeted countries. A more recent study by Gutmann et al. (2019) also revealed that women are affected more severely by the imposition of sanctions given that countries exposed to sanctions experience both increases in poverty and income inequality, which typically further marginalizes women.

Sanctions have also been viewed in the case of authoritarian countries such as North Korea, where women are particularly vulnerable to the effects of sanctions because of "the twin expectation that they be primary caretakers of their families and communities as well as workers fully integrated in the socialist economy" (United Nations, 2019, p. 36). A similar study of Rarick (2006) on Myanmar shows that economic sanctions exacerbated inequalities in Myanmar, with women more affected than men. Al-Jawaheri (2008) examined the gendered impact of sanctions on Iraq and found that the economic sanctions from 1990 to 2003 differentially effected female labor force participation in Iraq, their family relationship, literacy level and psychology. Taheri and Guven-Lisaniler (2018) also revealed that economic sanctions have had a significant negative impact on Iranian women's labor force participation and have inhibited valued functioning and capabilities of Iranian women. Nonetheless, to the best of our knowledge, there is no study which specifically examines the gendered consequences of sanctions in relation to tourism and female empowerment, and this study aims to fill this gap in the literature.

Women empowerment

Empowerment is a multidimensional social process to help individuals to gain control over factors that affect their lives (Scheyvens, 1999; Eger et al., 2018; Trommlerová et al., 2015). Notions of empowerment, and women's empowerment in particular, have increasingly moved to the core of understandings of sustainability and the SDGs in particular. In great part, this is because the framing of sustainability has moved from a focus on inter-generational justice, as in the Brundtland report, to a fuller concept which has an aim of sustaining human freedoms, rather than only the ability to fulfil felt needs. A concept encapsulated in Sen's (1999) notion of 'development as freedom'. As Sen (2013, p.6) observes, "Human freedoms include the fulfilment of needs, but also the liberty to define and pursue our own goals, objectives and commitments, no matter how they link with our own particular needs". This shift in focus meant, for example, that then Millennium Development Goals (MDGs) explicitly resolved to "promote gender equality and the empowerment of women as effective ways to combat poverty, hunger and disease and to stimulate development that is truly sustainable" (UN, 2000). The revision of the MDGs and preparatory work for the SDGs placed further emphasis on human rights and empowerment as being central to development (Nanda, 2016). However, there was, and still is, a substantial gap between policy and implementation. According to the United Nations Commission on the Status of Women (2014, p.12), "almost 15 years after the Millennium Development Goals were adopted, no country has achieved equality for women and girls and significant levels of inequality between women and men persist". This was particularly the case in secondary education, which is a major contributor "to the achievement of gender equality, the empowerment of women, and the human rights of women and girls and several positive social and economic outcomes" (United Nations Commission on the Status of Women, 2014, p.20). But overall progress on the MDGs was characterized by the "the lack of economic empowerment, autonomy and independence for women, including a lack of integration into the formal economy, unequal access to full and productive employment and decent work, ... and the lack of equal pay for equal work or work of equal value" (United Nations Commission on the Status of Women, 2014, p.21).

The more specific elements of women's empowerment in the SDGs (Goal 5) (UN Women, 2020), highlights the way in which the concept has become mainstreamed in development thinking, including within tourism (Nanda, 2016; Cole, 2018; Boluk et al., 2019). By generating revenue in SMEs and larger related businesses (UNWTO, 2019), it is argued that tourism can empower all individuals in a community and especially marginalized and disadvantaged groups, such as women (Ferguson, 2011; Figueroa-Domecq et al., 2015), especially by strengthening women's entrepreneurship (Kimbu & Ngoasong, 2016; Baum, 2018). As a result, women empowerment has come to be addressed by many global institutions and initiatives such as the UNWTO *Global Report on Women in Tourism* (UNWTO, 2019). Nonetheless, much rhetoric has frequently focused on tourism as a potential vehicle for women empowerment, but the highly gendered activity of tourism has often been neglected. Cole (2018) criticized the discussion on empowerment in the development literature and calls for re-conceptualisation of tourism entrepreneurship for women beyond its artificial economic, masculinist framings (Figueroa-Domecq et al., 2020). For Cole (2018, p.2), "empowerment as so frequently conceptualized deals only with productive and not reproductive labour, and fails to address the structural inequalities that lie at the base of societies built on patriarchal symbolic and normative codes". Indeed, the focus on women's entrepreneurship and access to micro-finance as empowering potentially only reinforces the existing structural gendered inequalities in global and national economic and financial systems (Gentry, 2007; Robinson et al., 2019).

Within the tourism context, the seminal conceptualisation of empowerment derives from the works of Scheyvens (1999) who examined tourism development processes via a multidimensional model that highlights economic, social, political and psychological aspects of empowerment. This framework provides a departure point for the current study given that, as Movono and

Dahles (2017) state, it provides a pathway for examination of specific marginalized groups such as women within communities dependent on tourism.

In the Scheyvens framework, social empowerment consists of improvements to community integration, collaboration and cohesion (Scheyvens, 1999; Moswete & Lacey, 2015). Scheyvens (1999) also argues that social empowerment may take different forms in response to more equitable sharing of the benefits of development. Gender-based social empowerment has been addressed in several studies (Moswete & Lacey, 2015; Ramos & Prideaux, 2014). For instance, in their study on cultural tourism initiatives in Botswana, Moswete and Lacey (2015) found that employment in tourism activities contributes to an enhanced social connection with outsiders and encouraged better lifestyle choices for women. Similarly, Peterson's (2014) research within a rural fishing village in Baja California Sur, Mexico revealed that tourism contributes to the collaboration among individuals in a community.

The political dimension of empowerment has become a significant focus of gender and empowerment studies in sustainable tourism (Scheyvens, 1999; Ferguson, 2011; Tucker & Boonabaana, 2012). One example of the political empowerment of women is the study of Moswete and Lacey (2015) in Botswana who found that after the implementation of a government-sponsored cultural tourism initiative, women had greater rates of active participation and ownership in the tourism industry as well as having their voices better represented in management and policy-making. For Cole (2006), this dimension of empowerment is at "the top end of the participation ladder where members of a community are active agents of change and they have the ability to find solutions to their problems, make decisions, implement actions" (2006, p. 631).

Economic empowerment via tourism refers to enabling the community to capture the economic benefits from tourism with the income generated being shared and distributed between a community's households thereby leading to an improved quality of life of residents (Scheyvens, 1999). Tourism therefore can potentially support economic empowerment by generating employment, income, and providing entrepreneurial opportunities. However, it is important to recognize that issues of income parity and distribution are a very significant element of empowerment processes.

Psychological empowerment relates to community members' attitudes and beliefs related to their community and organization which enables them to feel a sense of work control (Scheyvens, 1999). For Gil Arroyo et al. (2019, p.3) "a psychologically empowered individual is self-reliant and independent, whereas a psychologically disempowered individual is apathetic and submissive". The psychological dimension of empowerment focuses on the capacity of tourism development to boost confidence and self-esteem of residents and recognition of the uniqueness of natural and cultural features of their community (Scheyvens, 1999; Boley & McGehee, 2014; Ramos & Prideaux, 2014). Boosting pride and self-esteem associated with psychological empowerment is believed to be one of the most important non-economic benefits from tourism (Stronza & Gordillo, 2008), and one of the best predictors of resident support for tourism (Boley et al., 2017).

Through this lens, this study therefore aims to examine how these dimensions of tourism-related women's empowerment can be affected by the applications of economic sanctions, that are supposedly gender-neutral. This will be discussed further in the sections below.

Methods

Study context

Iran's location at the crossroads of major cultures and trade routes (e.g. the Silk Road) as well as the diverse climate and landscape has created a rich resource for tourism development (Seyfi & Hall, 2018). Iran has traditionally been a destination for cultural tourists, and in the aftermath of

Table 1. Gender gap in Iran (out of 153 countries).

Global Gender Gap Index	148
Economic participation and opportunity	147
Educational attainment	118
Health and survival	130
Political empowerment	145

Source: Global Gender Gap Report 2020, World Economic Forum (WEF), 2020.

the Islamic revolution of 1979 and the subsequent establishment of a theocracy, the country attracts many religious tourists and pilgrims from the surrounding countries. Iran has abundant tourism resources. As of August 2020, 22 historical sites and two natural sites are listed under the UNESCO World Heritage List, while 56 more sites are tentatively listed (United Nations Educational and Scientific and Cultural Organization (UNESCO), 2020). Iran also has 13 elements listed as part of the World Intangible Cultural Heritage (United Nations Educational and Scientific and Cultural Organization (UNESCO), 2020). Nevertheless, despite such abundant resources and potential, the Iranian tourism industry has suffered significantly since the upheavals of the late 1970s, with negative imagery in major tourism markets, the anti-Western stance of government, and political instability and conflicts in the Middle East (Morakabati, 2011; Seyfi & Hall, 2018). The decades-long comprehensive sanctions adversely affected Iran's economy and disrupted key trade activities including tourism. Successive waves of financial sanctions have also blocked Iran's access to the global financial system, leading to restrictions on foreign investment and inhibited Iran from purchasing new airplanes and spare parts (Seyfi & Hall, 2019a).

Although the Joint Comprehensive Plan of Action (JCPOA), a 2015 landmark agreement backed by the United Nations to resolve a long-running dispute over Iran's nuclear enrichment programme, temporarily led to an easing of sanctions (Seyfi & Hall, 2018), the unilateral withdrawal of US from the agreement and the re-imposition of sanctions in May 2018, severely hurt Iran's economy (Seyfi & Hall, 2019c). Since then, the Iranian currency has lost two-thirds of its value against the US dollar, in 2019 the inflation rate reached over 55%, unemployment rose to 16.8%, and the GDP shrank by 9.5% (SCI, 2020).

Decades of sanctions on Iran have substantial damaged the economy and adversely affected the standard of living for ordinary Iranians (Moret, 2015; Taheri & Guven-Lisaniler, 2018). Women in Iran tend to be disproportionally affected by sanctions given their already vulnerable situation in the country given its largely traditional and patriarchal nature (Fathollah-Nejad, 2014; Tahmasebi, 2018). Widespread unemployment affects entire families, but especially women, and exacerbates male dominated gender relations. The 2020 World Economic Forum Global Gender Gap report, ranks Iran 148th at global level and 16th regionally for gender equality, including equality in economic participation (World Economic Forum (WEF), 2020) (Table 1), while other indicators also show a huge gender gap in the country.

Unemployment in Iran is especially high among young people and women and 42% of Iranian women between the ages of 15 and 29 are unemployed (SCI, 2019) despite their higher education levels overall. Nonetheless, the number of women participating in Iran's job market is very low (only 17%) (SCI, 2019), due to many legal and social barriers, which restrict their livelihood and their economic contribution (Shahidian, 2002).

According to the World Travel and Tourism Council (WTTC) (WTTC, 2019) Iran's travel and tourism sector accounts for 6.5% of overall GDP, and the total contribution of travel and tourism to employment was 5.4% of total employment (1.344 million jobs) in 2019. Nonetheless, there is no reliable data on the employment rate of women in Iran's tourism and hospitality industry because of the absence of a reliable statistical system. Nevertheless, research suggests that the contribution of women to tourism-related activities increased following the lifting of economic sanctions in 2015 (Farahani & Dabbaghi, 2018). The tourism industry was among the first industries to witness an immediate growth in the light of the removal of sanctions with Iran receiving a rapid growth in international tourism. More than five million inbound tourists visited Iran in

2017, nearly three times the number in 2009, and many major airlines resumed direct flights to the country. There was also a considerable investment in the country's tourism-related infrastructure especially hotels as well as the transport (Seyfi & Hall, 2018). Yet, following US government's unilateral withdrawal from the agreement, the escalated tensions between Iran and the US and the growing regional rivalry have negatively affected the Iranian tourism industry. Many foreign companies have now left the Iranian market as they are unwilling to risk losing access to the U.S. and exclusion from the dollar-based financial system, and major airlines have halted direct flights to Iran while the country is again isolated from the global financial system.

Research design

This research used an exploratory, qualitative paradigm that originates from an interpretivist approach (Creswell & Poth, 2018) given the subjective and multifaceted nature of empowerment (Moswete & Lacey, 2015) and reflects Joo et al. (2020, p.10) observation that the "convoluted and contextual nature of empowerment may make it well suited for the qualitative approach". Given the relatively exploratory nature of the study, qualitative methods were considered appropriate as to explore the personal and subjective interpretations (Creswell & Poth, 2018) of women sharing their insights on how sanctions influenced women's empowerment. Furthermore, feminist scholarship has recommended that qualitative feminist epistemology is best positioned to give access to data on a sensitive research topic (Caprioli, 2004; Figueroa-Domecq et al., 2015).

Sampling, data collection and analysis

Given the unavailability of comprehensive established networks of women working in the tourism and hospitality industry in Iran and clear statistical data on the number of women employed in the sector, a purposeful snowball sampling technique was therefore adopted by the researchers to enable them to explore the research questions (Flick, 2009). Participants were identified by snowball sampling when those first contacted were asked to identify others who may be interested in doing an interview. This sampling technique was deemed most appropriate to attain a purposive sample suitable for this study (Flick, 2009). Although this method might entail some bias in the selection of participants, given the Iranian cultural context and the sensitivity of the topic, it is useful when the target is hard to reach (Atkinson & Flint, 2001), and where the security of respondents as well as researchers is paramount (Hall, 2011). Initial contact to request for interview was made via email and messages on LinkedIn, with an explanation of the purpose of research. Those who responded positively were contacted to explain the structure of the interview and ask their availability for a telephone, Whatsapp and Skype interview. Using emerging internet technologies as a research medium in qualitative studies has been recognised and is more conductive for some hard-to-reach and geographically dispersed groups (Hanna, 2012).

Qualitative research tends to target a relatively small and focused sample to understand the individuality of the phenomenon being explored and the uniqueness of its circumstances (Maxwell, 2008). This research followed the concept of 'data saturation', when additional interviews added little or no additional information was being generated (Creswell & Poth, 2018) and theoretical saturation was achieved after 28 interviews at which point interviewing was concluded. A pilot study was conducted with four interviewees to test the interview questions, along with interview style and approach (Kim, 2011). The sample included women working in a tourism-related role representing different sectors of tourism and hospitality industry, and women working in non-tourism jobs. A further five interviews were conducted with men working in a tourism-related job to glean some insights of masculine perspectives (Moswete & Lacey, 2015).

A semi-structured in-depth interview process was used to gather in-depth accounts of respondents' experiences (Bradford & Cullen, 2013). Questions were framed according to the focus of the research and were selected from various related studies (Scheyvens, 1999, 2000; Moswete & Lacey, 2015; Seyfi & Hall, 2020) and modified to be applicable to the Iranian context. They were open-ended in order to gain more spontaneous opinions and avoid the potential bias from restricting responses to the researcher's own fixed categories (Creswell & Poth, 2018). The interview guide inquired about female workforce characteristics, their involvement in tourism, and perceptions of women's roles and empowerment in various dimensions in pre and post sanction environments, and the effects of sanctions.

The semi-structured interviews were conducted in Persian and lasted 30 minutes, on average, covering a range of 20–60 minutes. To maintain confidentiality, the participants were assigned codes. All interviews were transcribed verbatim and were back translated into English by the lead author to ensure consistency in meaning (Creswell & Poth, 2018). To secure trustworthiness, participants were provided a copy of the transcription of their interviews and feedback was sought to increase dependability of the findings (Nowell et al., 2017). Interview transcripts were examined thematically based on the theoretical framework of the study in the psychological, political, social, economic dimensions of empowerment (Scheyvens, 1999). Qualitative thematic analysis was adopted for data analysis. Thematic analysis is the most widely used qualitative approach to analyzing interviews (Braun & Clarke, 2006) and is valuable for working within realist or constructionist social science paradigms and seeking a richer and more nuanced understanding of empirical material (Braun & Clarke, 2006; Nowell et al., 2017).

Research findings

Economic disempowerment

As noted in the discussion on sanctions as a gendered instrument of statecraft in the literature review, economic disruption caused by sanctions reduces the welfare of the target state's populace in general and often leads to greater economic discrimination against women. In the view of nearly all our interviewees, sanctions-related effects on the Iranian economy have had a direct impact on the employment and livelihoods of Iranian female workforce. One respondent commented:

> Women are more vulnerable to economic crises, often are the first to lose their jobs and have less job security as compared to men ... in our society, there is the stereotypical view that business owners are typically men who are breadwinners and are responsible for the household expenses .... sanctions have threatened the livelihoods of women and robbed women of their job security, and over the past year I have seen several instances of layoffs among my friends (interview #1)

This was echoed by another interviewee with many years' experiences in hotel industry:

> Sanctions have created a situation in which women embrace jobs without any occupational health care benefits .... this economic pain can be felt in a variety of ways such as lower pay in the hotel and accommodation sector, vulnerable employment [and] no government assistance (interview #20)

The respondents believe that the sharp reduction in tourist arrivals to Iran as result of sanctions have disrupted tourism-related services and even worsened their situation (e.g. interview #11, 12, 23). They argue that sanctions have made the inbound market very vulnerable and unstable. A tour guide explains:

> ... after the lifting of sanctions, we had a huge increase in the number of arrivals and I was always busy with the incoming tours, but after the return of sanctions, we have had less and less tours, and I just work two weeks a year ... I see this job as my main occupation and always have the stress of being fired and I can hardly foresee long-term stability in the job (interview #5)

Financial independence and women's income generation for family as the 'bread winner' were the empowerment issues most commonly identified by women in this study. The interviewees were of the belief that tourism contributed to their economic and social empowerment (e.g. interviews #4, 10, 26). This was noted by a respondent:

> Tourism arrivals in our village allowed me to rent our home to them and earn good money by providing local food and souvenirs to tourists. My daughters also helped me, and we had a good business… I felt happy and I could generate income for my family in addition to my role as the housewife and my related tasks (interview #19)

Such perspectives are significant because a woman being the main earner is not a normative gender role in household employment in Iran. The loss of employment opportunities for women therefore only serves to reinforce stereotypical gendered work and household roles in Iranian society which has economic as well as broader social implications. This was commented on by a university professor:

> Losing a job for a woman means, they not only lose their economic independence, but also their social autonomy by losing control over their lives and having fewer choices (interview #16)

Political disempowerment

Increasing female employment is central to poverty reduction, maintaining economic growth and fostering equality and independence for women. This is commented on by an interviewee who argued that tourism development had given women more voices in decision-making and tourism development:

> In our patriarchal society we always hear that men have more knowledge about political matters than do women and thus women are largely underrepresented in political arenas and policymaking… but tourism has provided a chance for women in touristic villages in Iran where women have confidence in their abilities to campaign for participation in tourism planning development. They run businesses and largely participated in the promotion of tourism in their villages by creating cooperatives for example (interview #23).

Another interviewee expressed that involvement in tourism-related activities and the income generation has changed the traditional view on women as a housewife. She commented:

> I think that the tourism industry gives a big chance for Iranian women to show their skills and it is a big opportunity for changing stereotypes about women's workforce. Just imagine how difficult it is for us to change old and traditional views on our society and we are trying to change people's minds by actively participating in tourism activities (interview #22).

Nonetheless such gains are overshadowed by sanctions. As stated by a female entrepreneur:

> … the number of tourists has declined enormously and women who were involved in tourism activities are obliged to return to their so-called 'traditional' jobs of being simply housewives (interview #17)

Social disempowerment

Some interviewees were of the belief that sanctions negatively affected their social empowerment. They believe that the growth of tourism as a result of lifting of sanctions could change the passive role of women into active role capable of managing their social life (interview #19). They also mention that this process provided them with better access to opportunities and more involvement in making decisions through social interactions (interview #15). A female tour guide stated:

> Working as a tour guide provided me with the opportunity to have interactions with foreign tourists and learning about their cultures ... now with the reduced number of tourists, I lost my job and my contacts in the network (interview #6).

This was echoed by her colleague:

> My job as a tour guide has helped me to make new foreign friends, understand other cultures, share our opinions about experiences, and extend my social networks. So, overall, I can say that my activities in tourism industry have changed my thoughts about women's roles in society. It makes me feel being an empowered woman (interview #5).

However, new difficulties have emerged for women after the reimposition of sanctions. A travel agency manager commented:

> Under normal economic situation, we have to spend three times more energy compared to men to launch our entrepreneurial activities with the male-dominated administration and more and more red tape. Imagine under economic pressure caused by sanctions, we have to spend five times more energy to prove ourselves so that they can trust our abilities (interview #12)

This was echoed by another respondent:

> During the sanctions time, the priorities changed for the government and less and less support was given for loan and banking credit, if there is any chance, priority is given to the men rather than us. The credit facilities that were given to women, especially women-headed households were restricted. This has led many women to abandon their entrepreneurial plans and stay at home (interview #10).

Our interviewees also argued that economic coercion led to higher crime rates emanating from widespread unemployment and high inflation, and therefore created an insecure business environment for women in Iran (interview #16, 18, 23). One of the respondents stated:

> I observe that sanctions directly made unsecure business environment from the societal perspective and imagine how women can continue to work under such conditions. Under the influence of sanctions, a healthy business environment has become an insecure environment, which reduces women's confidence and confidence in such an environment (interview #26).

Psychological disempowerment

In the eyes of a majority of interviewees, participation in tourism activities has produced a degree of relational satisfaction that convinces women that they have accomplished their goals. For instance, one of the respondents commented:

> in Iran, we have an idiom that says; *'Zan Rais-e Khaneh Ast'* [women are the boss of home and are not able to lead other places], but here I am working as a woman. I rent the local houses to tourists and my daughter works as a guide at the village. So, other women should not believe the proverb. I am satisfied with this job and about myself. The tourists always encouraged me, my family is proud of me ... But now, I cannot continue my business because I lose a big part of my market (interview #19)

A local hotel-restaurant owner also commented:

> When tourists were coming to our village, we could see that they are very interested in our traditions, local clothes and handicrafts. This made everyone happy in the village, old and young people were proud of their culture ... I know several women in our village decided to go back to school to continue their secondary and high school diplomas [as a result of their experiences] (interview #24).

In the eyes of many respondents tourism plays an important role in job generation for young women. However, the reduced number of tourists and the societal effects of sanctions have negatively affected such empowerment opportunities. This was noted by a university lecturer:

> I have seen many young women come to get the vocational training to enter to the job market ... having a job and earning income and interacting with others at workforce gives them build their self-confidence that they are a productive members of society .... But I have seen many of my former students are now jobless and have lost their confidence (interview #16).

The next section reflects on the implications of women's disempowerment as a result of economic sanctions on Iran in the context of the wider sanctions literature and espoused goal of the SDGs to empower women.

Discussion

Economic empowerment is central to realizing women's rights and gender equality and to the overall achievement of the SDGs (UN Women, 2020). Tourism has been viewed as a sector that can advance gender equality and women's empowerment and increased levels of economic independence for women (Kimbu & Ngoasong, 2016; UNWTO, 2019) and contribute towards SDG5 which focuses on gender equality and women's empowerment (Boluk et al., 2019). Nonetheless, as the findings of this study showed, economic sanctions has led to direct effects on employment and entrepreneurship because of fewer tourist numbers and it has also contributed to financial discrimination against women because it has only served to reinforce pre-existing structural gender discrimination in Iran's banking and financial services and other aspects of Iranian society.

Compared with other countries subjected to sanctions, the Iranian case is unique. Female empowerment via tourism is arguably more significant in Iran compared to other sectors given that tourism and hospitality employment and entrepreneurial activities serve as an 'acceptable' extension of women's emotional labor and service roles to conservative elements of a highly patriarchal society (Shahidian, 2002; Zamani-Farahani & Henderson, 2010; Farahani & Dabbaghi, 2018; Seyfi & Hall, 2019d). As the study's findings revealed, the application of sanctions removed a significant avenue of women's empowerment and only reinforces long-established conservative gender roles that frame the lack of gender equality and the capacity for women's independence in Iran's legal and political structures. Such exacerbation of female disempowerment as a result of sanctions is little discussed in the extant literature and this study has highlighted the significance of this rarely studied phenomenon in tourism.

Women in Iran face numerous cultural, religious and political barriers and are disproportionally affected by sanctions given their already vulnerable situation in a largely traditional and patriarchal society (Farahani & Dabbaghi, 2018; Tahmasebi, 2018; Taheri & Guven-Lisaniler, 2018). Women have historically been sensitive to the issue of sanctions as economic downturns and hardship lead to increased calls [by men and male religious and political leaders] for jobs to be protected for men, and for women to be forced back into "traditional" roles. This situation finds support in the work of Fathollah-Nejad (2014) on the counter-productivity of sanction who, just prior to JCPOA, suggested that "the rise in unemployment is likely to fuel regressive conservative social policies that aim at preserving the traditional social status reserved to the male population by externalizing the costs of sanctions onto the female population. These include measures that push women out of jobs, relegate them to the domestic sphere, and curtail their access to higher education" (2014, p.57).

The results of this study also highlight how women's economic empowerment is vulnerable to social and political factors. Economic empowerment is therefore not just an economic issue but needs to be placed with the broader societal context. In the case of Iran these necessitate recognition of the intensely patriarchal nature of society and the theocratic nature of governance and regulation as major factors constraining the process of women's economic empowerment. When the economy is experiencing growth, as when the JCPOA was operative and tourism was expanding, the nature of tourism service roles and the reduction in competition (with males) for tourism jobs enabled women to occupy positions and empower themselves economically. However, when the restrictions returned to the economy and tourist numbers fell so women have disproportionately lost employment. Suggesting empowerment was a temporary and vulnerable phenomenon and the empowerment cannot just be considered as an economic factor alone, without accompanying social and political change. This reflects that empowerment is a process. Without sanctions tourism provided an economic foundation for the empowerment of Iranian women. However, with their reimposition the process has stopped and gone in reverse.

As the findings of this study revealed, internal structural factors in Iran mediated disempowerment in addition to the external factor of economic sanctions. Such issues have been little previously discussed, with the empowerment literature – including in the tourism context – failing to sufficiently engage with issues of disempowerment with the factors that shape and reinforce disempowerment being rarely discussed. This would appear to be a potentially important extension to the framework of Scheyvens (1999) and research on women's empowerment in tourism.

The comments of respondents as to the societal effects of sanctions provide significant insights for gender and tourism scholarship. During economic downturns, governments tend to emphasize large infrastructure projects, generally dominated by men, to create employment, thereby neglecting the more women-oriented service industries and micro-enterprises (Drury & Peksen, 2014; Van Engeland, 2019). The Iranian situation with respect to tourism reflects the results of other studies that have found that the sectors in which women are primarily employed are also the ones most affected by sanctions (e.g., Rhodesia (Galtung, 1967), South Africa (Levy, 1999), Haiti (Gibbons & Garfield, 1999), Myanmar (Rarick, 2006), Iraq (Al-Ali, 2005; Al-Jawaheri, 2008; Buck et al., 1998), and Syria (Moret, 2015).

Other studies on the effects of sanctions have also reflected the results of this study that the target state, economic weakened by the sanctions but with domestic political support for regimes often increased and is less likely to enforce women's social and political rights (Drury & Peksen, 2014) and economic sanctions tend to reduce the target government's ability to provide welfare services (Buck et al., 1998; Gibbons & Garfield, 1999; Al-Jawaheri, 2008; Moret, 2015). As Van Engeland (2019) suggests, the economic situation has been used by conservative elements in the Iranian ruling elite to justify new laws on family or work that limit "women's contribution to the public sphere as well as their presence in society" (Van Engeland, 2019).

Other research has also found that the economic downturn triggered by the sanctions leads to greater economic grievance which incites more social disorder and instability and results in further gender-based structural limits on women's social and political rights (Buck et al., 1998; Drury & Peksen, 2014; Taheri & Guven-Lisaniler, 2018; Gutmann et al., 2019). Moreover, such impacts also have a spatial dimension as they disproportionately affect urban centres in which women has previously been able to work. Indeed, these measures not only impact women's economic empowerment but also severely affect their self-esteem and sense of wellbeing thereby contributing greatly to their psychological disempowerment.

The structural effects of political sanctions in Iran are not to be underestimated. The harsher security approach toward civil society ensuing from the sanctions have created obstacles for women in Iran, as well as minorities. Tahmasebi (2018) argues that: "women's rights advocates have often been accused by hardliners of seeking to adopt Western values, which undermine the Islamic values of the country. As such, they have worked hard to ensure their demands are rooted locally, while also reflecting universal rights principles". Fathollah-Nejad (2014, p.57) further observes, that "the sanctions can serve as the political platform on which conservative politics can go on the offensive in order to marginalize women from education and employment, consequently also limiting the space for women's rights activism". The effects of the economically coercive environment created by sanctions on women is therefore only amplified by the existing gender structures that sanctions have reinforced (Van Engeland, 2019).

One important further observation to make on the relationship between empowerment and tourism is that all notions of empowerment in tourism are inherently either explicitly or implicitly linked to its economic functions. This research identifies that, in the Iranian political, religious and cultural context, economic empowerment is the overarching contributor to empowerment to other forms of empowerment. This role as an enabling factor for other types of empowerment is arguably significant not only in terms of the sanctions placed on Iran but will potentially also apply to other societies in which women do not have equal political and human rights.

Conclusion

With women's empowerment being a core element of tourism's contribution to the SDGs, especially in a developing country context, this study has contributed to the understanding of the intersection between gender empowerment and tourism by empirically contextualizing the gendered effects of economic sanctions on empowerment. This study provides insights into the dynamics between gender empowerment and the tourism industry in Iran which are constrained by domestic (religio-patriarchal society) and external (sanctions) factors and has highlighted the key challenges Iranian women involved in tourism face as a result of sanctions.

Drawing upon Scheyvens (1999) framework that recognizes four dimensions of empowerment; social, economic, political and psychological, the findings of this study highlight the temporal aspect of empowerment and demonstrate that the tourism development process in Iran particularly in the post-sanctions era was generally posited as an empowering opportunity for women. However, it also suggests that in terms of Iran at least, the economic foundations of empowerment in tourism is critical for the other dimensions to occur within a tourism context. Unless the conservative theocratic superstructure changes the other aspects of empowerment do not effectively exist without the economic contribution of tourism. This means that the gains made during the period when sanctions were lifted are being lost. The reimposition of sanctions by the US and its negative implications for Iranian tourism, have greatly affected the empowerment of women through tourism.

This study represents the first effort to explore the gendered effects of sanctions on women's empowerment within a tourism context. It lays the groundwork for better understanding the gendered consequences of a coercive geopolitical tool of diplomacy that substantially affects tourism and indicates that such foreign policy mechanisms have implications for gender empowerment that stand in stark contrast to their intended aims. In particular it highlights that the unintended effects of sanctions on tourist flows, and therefore the tourism industry, and that the indirect mediating effects with respect to existing structural economic, cultural and political inequalities may be just as significant in (dis)empowering processes.

Limitations and directions for future research

Future research is recommended with respect to the conduct of ethnographic studies for deeper understanding of the women empowerment processes in Iran and to further explore empowerment themes. However, in the Iranian case it must be emphasized that such research has to be undertaken with great sensitivity given the broader issues surrounding women's employment and entrepreneurship. Future research should also focus on comparative studies with other countries with similar contexts to explore the factors that shape women's empowerment as well as longitudinal studies that are sensitive to understanding change. As highlighted by this research, future studies need to consider the temporality and vulnerability aspect in studying empowerment. Longitudinal studies that examine the temporal dimensions of empowerment and changing women's perceptions of empowerment in relation to tourism potentially better capture empowerment processes than single shot studies and provide more context-based insights into women's empowerment.

While these findings add new perspectives to the understanding of gender and empowerment in tourism, there are, however, limitations relating to this research. In particular, this study is constrained by the usual limitations of qualitative research and the obtained sample may not be representative of the female labor force in the Iranian tourism industry.

Acknowledgements

We would like to thank the four anonymous reviewers and the editors for their valuable comments and suggestions.

ORCID

Siamak Seyfi http://orcid.org/0000-0002-2427-7958
Colin Michael Hall http://orcid.org/0000-0002-7734-4587
Tan Vo-Thanh http://orcid.org/0000-0001-9964-3724

References

Al-Ali, N. (2005). Reconstructing gender: Iraqi women between dictatorship, war, sanctions and occupation. *Third World Quarterly, 26*(4-5), 739–758. https://doi.org/10.1080/01436590500128428

Alarcón, D. M., & Cole, S. (2019). No sustainability for tourism without gender equality. *Journal of Sustainable Tourism, 27*(7), 903–919. https://doi.org/10.1080/09669582.2019.1588283

Al-Jawaheri, Y. H. (2008). *Women in Iraq: The gender impact of international sanctions*. IB Tauris.

Arafa, M. (2018). A question to the President of the United States, Donald Trump: is it a travel ban, or a Muslim ban, or a travel Muslim ban? *Revista de Investigações Constitucionais, 5*(2), 9–33. https://doi.org/10.5380/rinc.v5i2.58990

Atkinson, R., & Flint, J. (2001). Accessing hidden and hard-to-reach populations: Snowball research strategies. *Social Research Update, 33*(1), 1–4.

Baldwin, D. A. (1985). *Economic statecraft*. Princeton University Press.

Barigazzi, J. (2020). EU takes first step toward Belarus sanctions. Politico. https://www.politico.eu/article/eu-belarus-sanctions-first-steps/.

Baum, T. (2018). Sustainable human resource management as a driver in tourism policy and planning: a serious sin of omission? *Journal of Sustainable Tourism, 26*(6), 873–889. https://doi.org/10.1080/09669582.2017.1423318

Boley, B. B., Ayscue, E., Maruyama, N., & Woosnam, K. M. (2017). Gender and empowerment: Assessing discrepancies using the resident empowerment through tourism scale. *Journal of Sustainable Tourism, 25*(1), 113–129. https://doi.org/10.1080/09669582.2016.1177065

Boley, B. B., & McGehee, N. G. (2014). Measuring empowerment: Developing and validating the resident empowerment through tourism scale (RETS). *Tourism Management, 45*, 85–94. https://doi.org/10.1016/j.tourman.2014.04.003

Boluk, K. A., Cavaliere, C. T., & Higgins-Desbiolles, F. (2019). A critical framework for interrogating the United Nations Sustainable Development Goals 2030 Agenda in tourism. *Journal of Sustainable Tourism, 27*(7), 847–864. https://doi.org/10.1080/09669582.2019.1619748

Bradford, S., & Cullen, F. (Eds.). (2013). *Research and research methods for youth practitioners*. Routledge.

Braun, V., & Clarke, V. (2006). Using thematic analysis in psychology. *Qualitative Research in Psychology, 3*(2), 77–101. [Database] https://doi.org/10.1191/1478088706qp063oa

Buck, L., Gallant, N., & Nossal, K. R. (1998). Sanctions as a gendered instrument of statecraft: The case of Iraq. *Review of International Studies, 24*(1), 69–84. https://doi.org/10.1017/S0260210598000692

Cameron, I. (2003). UN targeted sanctions, legal safeguards and the European Convention on Human Rights. *Nordic Journal of International Law*, *72*(2), 159–214. https://doi.org/10.1163/157181003322560556

Caprioli, M. (2004). Feminist IR theory and quantitative methodology: A critical analysis. *International Studies Review*, *6*(2), 253–269. https://doi.org/10.1111/j.1521-9488.2004.00398.x

Cole, S. (2006). Information and empowerment: The keys to achieving sustainable tourism. *Journal of Sustainable Tourism*, *14*(6), 629–644. https://doi.org/10.2167/jost607.0

Cole, S. (Ed.). (2018). *Gender equality and tourism: Beyond empowerment*. CABI.

Coleman, J. R. (2001). The economic impact of US sanctions with respect to Cuba. *Cuba in Transition*, *11*, 86–96.

Creswell, J. W., & Poth, C. N. (2018). *Qualitative inquiry and research design: Choosing among five approaches* (4th ed.). Sage.

De Goede, M. (2011). Blacklisting and the ban: Contesting targeted sanctions in Europe. *Security Dialogue*, *42*(6), 499–515. https://doi.org/10.1177/0967010611425368

Drezner, D. W. (2011). Sanctions sometimes smart: targeted sanctions in theory and practice. *International Studies Review*, *13*(1), 96–108. https://doi.org/10.1111/j.1468-2486.2010.01001.x

Drury, A. C., & Peksen, D. (2014). Women and economic statecraft: The negative impact international economic sanctions visit on women. *European Journal of International Relations*, *20*(2), 463–490. https://doi.org/10.1177/1354066112448200

Eger, C., Miller, G., & Scarles, C. (2018). Gender and capacity building: A multi-layered study of empowerment. *World Development*, *106*, 207–219. https://doi.org/10.1016/j.worlddev.2018.01.024

Farahani, B., & (2018). and., & Dabbaghi, H. Tourism and the empowerment of women in Iran. In S. Seyfi & C.M. Hall (Eds.). *Tourism in Iran: Challenges, development and issues* (pp. 177–192). Routledge.

Farrall, J. M. (2007). *United Nations sanctions and the rule of law*. Cambridge University Press.

Fathollah-Nejad, A. (2014). Why sanctions against Iran are counterproductive: Conflict resolution and state–society relations. *International Journal: Canada's Journal of Global Policy Analysis*, *69*(1), 48–65. https://doi.org/10.1177/0020702014521561

Ferguson, L. (2011). Promoting gender equality and empowering women? Tourism and the third Millennium Development Goal. *Current Issues in Tourism*, *14*(3), 235–249. https://doi.org/10.1080/13683500.2011.555522

Figueroa-Domecq, C., de Jong, A., & Williams, A. M. (2020). Gender, tourism & entrepreneurship: A critical review. *Annals of Tourism Research*, *84*, 102980.

Figueroa-Domecq, C., Pritchard, A., Segovia-Pérez, M., Morgan, N., & Villacé-Molinero, T. (2015). Tourism gender research: A critical accounting. *Annals of Tourism Research*, *52*, 87–103. https://doi.org/10.1016/j.annals.2015.02.001

Flick, U. (2009). *An introduction to qualitative research*. Sage.

Galtung, J. (1967). On the effects of international economic sanctions, with examples from the case of Rhodesia. *World Politics*, *19*(3), 378–416. https://doi.org/10.2307/2009785

Gentry, K. M. (2007). Belizean women and tourism work: Opportunity or impediment? *Annals of Tourism Research*, *34*(2), 477–496. https://doi.org/10.1016/j.annals.2006.11.003

Gibbons, E., & Garfield, R. (1999). The impact of economic sanctions on health and human rights in Haiti, 1991-1994. *American Journal of Public Health*, *89*(10), 1499–1504. https://doi.org/10.2105/ajph.89.10.1499

Gil Arroyo, C., Barbieri, C., Sotomayor, S., & Knollenberg, W. (2019). Cultivating women's empowerment through agritourism: Evidence from Andean communities. Sustainability, *11*(11), 3058. https://doi.org/10.3390/su11113058

Gillen, J., & Mostafanezhad, M. (2019). Geopolitical encounters of tourism: A conceptual approach. *Annals of Tourism Research*, *75*, 70–78. https://doi.org/10.1016/j.annals.2018.12.015

Gutmann, J., Neuenkirch, M., & Neumeier, F. (2019). Precision-guided or blunt? The effects of US economic sanctions on human rights. *Public Choice*, *185*(1), 161–182.

Hall, C. M. (2005). *Tourism: Rethinking the social science of mobility*. Pearson.

Hall, C. M. (2011). Researching the political in tourism: Where knowledge meets power. In C. M. Hall (Ed.), *Fieldwork in tourism* (pp. 53–68). Routledge.

Hall, C. M. (2017). Tourism and geopolitics: The political imaginary of territory, tourism and space. In D. Hall (Ed.), *Tourism and geopolitics: Issues from Central and Eastern Europe* (pp. 15–24). CABI.

Hanna, P. (2012). Using internet technologies (such as Skype) as a research medium: A research note. *Qualitative Research*, *12*(2), 239–242. https://doi.org/10.1177/1468794111426607

Hatipoglu, E., & Peksen, D. (2018). Economic sanctions and banking crises in target economies. *Defence and Peace Economics*, *29*(2), 171–189. https://doi.org/10.1080/10242694.2016.1245811

Ilieva, J., Dashtevski, A., & Kokotovic, F. (2018). Economic sanctions in international law. *UTMS Journal of Economics*, *9*(2), 201–211.

Jentleson, B. W. (2000). Economic sanctions and post-Cold War conflicts: Challenges for theory and policy. In P. Stern & D. Druckman (Eds.), *International conflict resolution after the Cold War* (pp. 123–177). National Academy Press.

Joo, D., Woosnam, K. M., Strzelecka, M., & Boley, B. B. (2020). Knowledge, empowerment, and action: testing the empowerment theory in a tourism context. *Journal of Sustainable Tourism, 28*(1), 69–85. https://doi.org/10.1080/09669582.2019.1675673

Kimbu, A. N., & Ngoasong, M. Z. (2016). Women as vectors of social entrepreneurship. *Annals of Tourism Research, 60*, 63–79. https://doi.org/10.1016/j.annals.2016.06.002

Levy, P. I. (1999). Sanctions on South Africa: what did they do? *American Economic Review, 89*(2), 415–420. https://doi.org/10.1257/aer.89.2.415

Leyton-Brown, D. (Ed.). (2017). *The utility of international economic sanctions*. Routledge.

Masters, J. (2019). *What Are Economic Sanctions?* https://www.cfr.org/backgrounder/what-are-economic-sanctions.

Maxwell, J. A. (2008). Designing a qualitative study. *The SAGE Handbook of Applied Social Research Methods, 2*, 214–253.

Morakabati, Y. (2011). Deterrents to tourism development in Iran. *International Journal of Tourism Research*, 13(2), 103–123.

Moret, E. S. (2015). Humanitarian impacts of economic sanctions on Iran and Syria. *European Security, 24*(1), 120–140. https://doi.org/10.1080/09662839.2014.893427

Mostafanezhad, M. (2018). The geopolitical turn in tourism geographies. *Tourism Geographies, 20*(2), 343–346. https://doi.org/10.1080/14616688.2018.1434820

Moswete, N., & Lacey, G. (2015). Women cannot lead": empowering women through cultural tourism in Botswana. *Journal of Sustainable Tourism, 23*(4), 600–617. https://doi.org/10.1080/09669582.2014.986488

Movono, A., & Dahles, H. (2017). Female empowerment and tourism: A focus on businesses in a Fijian village. *Asia Pacific Journal of Tourism Research, 22*(6), 681–692. https://doi.org/10.1080/10941665.2017.1308397

Nanda, V. P. (2016). The journey from the millennium development goals to the sustainable development goals. *Denver Journal of International Law and Policy, 44*(3), 389–412.

Neuenkirch, M., & Neumeier, F. (2015). The impact of UN and US economic sanctions on GDP growth. *European Journal of Political Economy, 40*, 110–125. https://doi.org/10.1016/j.ejpoleco.2015.09.001

Nowell, L. S., Norris, J. M., White, D. E., & Moules, N. J. (2017). Thematic analysis: Striving to meet the trustworthiness criteria. *International Journal of Qualitative Methods, 16*(1), 160940691773384. https://doi.org/10.1177/1609406917733847

Padgett, A., & Warnecke, T. (2011). Diamonds in the rubble: The women of Haiti: Institutions, gender equity and human development in Haiti. *Journal of Economic Issues, 45*(3), 527–558. https://doi.org/10.2753/JEI0021-3624450301

Peterson, N. D. (2014). We are daughters of the sea": Strategies, gender, and empowerment in a Mexican women's cooperative. The Journal of Latin American and Caribbean Anthropology, *19*(1), 148–167. https://doi.org/10.1111/jlca.12064

Piazzalunga, D., & Di Tommaso, M. L. (2019). The increase of the gender wage gap in Italy during the 2008-2012 economic crisis. *The Journal of Economic Inequality, 17*(2), 171–193. https://doi.org/10.1007/s10888-018-9396-8

Ramos, A. M., & Prideaux, B. (2014). Indigenous ecotourism in the Mayan rainforest of Palenque: empowerment issues in sustainable development. *Journal of Sustainable Tourism, 22*(3), 461–479. https://doi.org/10.1080/09669582.2013.828730

Rarick, C. A. (2006). Destroying a country in order to save it: The folly of economic sanctions against Myanmar. *Economic Affairs, 26*(2), 60–63. https://doi.org/10.1111/j.1468-0270.2006.00632.x

Robinson, R. N., Martins, A., Solnet, D., & Baum, T. (2019). Sustaining precarity: critically examining tourism and employment. *Journal of Sustainable Tourism, 27*(7), 1008–1025. https://doi.org/10.1080/09669582.2018.1538230

Roth, A. (2019). Putin bans Russian airlines from flying to Georgia. The Guardian. https://www.theguardian.com/world/2019/jun/22/putin-bans-russian-airlines-from-flying-to-georgia.

Ruggieri, D. (2010). Gender perspectives of the financial and economic crisis. *International Journal of Green Economics, 4*(3), 217–230. https://doi.org/10.1504/IJGE.2010.037524

Scheyvens, R. (1999). Ecotourism and the empowerment of local communities. *Tourism Management, 20*(2), 245–249. https://doi.org/10.1016/S0261-5177(98)00069-7

Scheyvens, R. (2000). Promoting women's empowerment through involvement in ecotourism: experiences from the third world. *Journal of Sustainable Tourism, 8*(3), 232–249. https://doi.org/10.1080/09669580008667360

Sen, A. (1999). *Development as freedom*. Oxford University Press.

Sen, A. (2013). The ends and means of sustainability. *Journal of Human Development and Capabilities, 14*(1), 6–20. https://doi.org/10.1080/19452829.2012.747492

Seyfi, S., & Hall, C.M. (Eds.). (2018). *Tourism in Iran: Challenges, development and issues*. Routledge.

Seyfi, S., & Hall, C. M. (2019a). Sanctions and tourism: effects, complexities and research. *Tourism Geographies, 22*(4-5), 749–767. https://doi.org/10.1080/14616688.2019.1663911

Seyfi, S., & Hall, C. M. (2019b). International sanctions, tourism destinations and resistive economy. *Journal of Policy Research in Tourism, Leisure and Events, 11*(1), 159–169. https://doi.org/10.1080/19407963.2018.1482305

Seyfi, S., & Hall, C. M. (2019c). Sanctions and tourism: conceptualisation and implications for destination marketing and management. *Journal of Destination Marketing & Management, 15*, 100381.

Seyfi, S., & Hall, C. M. (2019d). Deciphering Islamic theocracy and tourism: Conceptualization, context, and complexities. *International Journal of Tourism Research, 21*(6), 735–746. https://doi.org/10.1002/jtr.2300

Seyfi, S., & Hall, C. M. (2020). *Tourism, sanctions and boycotts*. Routledge.

Shahidian, H. (2002). *Women in Iran: gender politics in the Islamic Republic*. Greenwood.

Stronza, A., & Gordillo, J. (2008). Community views of ecotourism. *Annals of Tourism Research, 35*(2), 448–468. https://doi.org/10.1016/j.annals.2008.01.002

Taheri, E., Guven-Lisaniler, F. (2018). Gender Aspect of Economic Sanctions: Case Study of Women's Economic Rights in Iran. https://papers.ssrn.com/sol3/papers.cfm?abstract_id=3303800.

Tahmasebi, S. (2018). How US sanctions impede the women's movement in Iran. https://www.atlanticcouncil.org/blogs/iransource/how-us-sanctions-impede-the-women-s-movement-in-iran/.

Tekin, E. (2015). The impacts of political and economic uncertainties on the tourism industry in Turkey. *Mediterranean Journal of Social Sciences, 6*(2 S5), 265. https://doi.org/10.5901/mjss.2015.v6n2s5p265

Timothy, D. J. (2017). Tourism and geopolitics in the GCC region. In M. Stephenson & A. Al-Hamarneh (Eds.). *International tourism development and the Gulf Cooperation Council States*. (pp. 45–60). Routledge.

Tran, L., & Walter, P. (2014). Ecotourism, gender and development in northern Vietnam. *Annals of Tourism Research, 44*, 116–130. https://doi.org/10.1016/j.annals.2013.09.005

Trommlerová, S. K., Klasen, S., & Leßmann, O. (2015). Determinants of empowerment in a capability-based poverty approach: Evidence from The Gambia. *World Development, 66*, 1–15. https://doi.org/10.1016/j.worlddev.2014.07.008

Tucker, H., & Boonabaana, B. (2012). A critical analysis of tourism, gender and poverty reduction. *Journal of Sustainable Tourism, 20*(3), 437–455. https://doi.org/10.1080/09669582.2011.622769

UN Women (2020). *SDG 5: Achieve gender equality and empower all women and girls*. https://www.unwomen.org/en/news/in-focus/women-and-the-sdgs/sdg-5-gender-equality.

United Nations (2000). *The Millennium Declaration*. UN General Assembly, report A/res/55/2, UN.

United Nations (2019). *The human costs and gendered impact of sanctions on North Korea*. https://koreapeacenow.org/wp-content/uploads/2019/10/human-costs-and-gendered-impact-of-sanctions-on-north-korea.pdf.

United Nations Commission on the Status of Women (2014). *Challenges and achievements in the implementation of the Millennium Development Goals for women and girls*. UN Doc. E/CN.6/2014/L.7. UN.

United Nations Educational, Scientific and Cultural Organization (UNESCO) (2020). Iran (Islamic Republic of). https://whc.unesco.org/en/statesparties/ir/.

United Nations Security Council (2020). *Sanctions*. https://www.un.org/securitycouncil/sanctions/information.

UNWTO (2019). *Global report on women in tourism* (2nd ed). https://www.e-unwto.org/doi/book/10.18111/9789284420384.

Van Engeland, A. (2019). Human rights double standard: Iranian sanctions impact the most vulnerable. Jurist. https://www.jurist.org/commentary/2019/01/human-rights-double-standard-iranian-sanctions-impact-the-most-vulnerable/.

World Economic Forum (WEF) (2020). *Global gender gap report 2020*. http://www3.weforum.org/docs/WEF_GGGR_2020.pdf.

WTTC (2019). *Travel & tourism: driving women's success*. https://www.wttc.org/economic-impact/social-impact/driving-womens-success/.

Zamani-Farahani, H., & Henderson, J. C. (2010). Islamic tourism and managing tourism development in Islamic societies: the cases of Iran and Saudi Arabia. *International Journal of Tourism Research, 12*(1), 79–89.

The social, cultural, economic and political strategies extending women's territory by encroaching on patriarchal embeddedness in tourism in Nepal

Wendy Hillman and Kylie Radel

ABSTRACT

Globally, there is significant growth in women exploring entrepreneurship to disrupt poverty, notwithstanding they face more serious challenges when compared to male counterparts. For women entrepreneurs in Nepal, not only are they marginalised in their tourism business endeavours, they are also highly regulated by caste and class, and patriarchal inequality. Using a critical theoretical framework within a qualitative grounded theory approach, we revealed ways in which Nepali women subvert the institutionalised gender and power structures. The critical perspective engages with socially constructed layers of historical, political, cultural, economic and gender values that have been embedded over time. Sixteen Nepali women entrepreneurs in tourism were interviewed between 2014 and 2019. Patriarchal embeddedness in social, business and tourism environments are overcome by Nepali women through micro enterprise development. Women are engaging in female domains such as homestay, but also infiltrating male domains such as trekking and tour-guiding. Nepali women's political strategies are subtle and understated. The capacity of Nepali women to encroach on patriarchy through tourism shows that women have adapted to redress socially constructed imbalances. Tourism provided opportunities to find their agency in a world that has been largely silenced. Women's skills and roles are now infiltrating ingrained patriarchy.

Introduction

Developed and developing nations globally have seen a sharp rise in women exploring entrepreneurship as a means to disrupt poverty for themselves and their families. This is despite the fact they face more serious challenges in their activities when compared to male entrepreneurs. Research in gender studies has demonstrated that women generally still face significant obstacles to entrepreneurship (Vossenberg, 2013) in the areas of reduced human capital, limited strategic choice options, and institutionalised structural barriers including taxation and legal disparities in the governance of industry in most countries (Pinkovetskaia et al., 2019; Širec & Močnik, 2012). In addition, the socially constructed genres of 'male' and 'female' are gendered with inherent political, social, economic, and structural inequities that these signifiers entail. The resulting gendered division of labour and responsibility continues to define and restrict women's opportunities to freely engage in social and economic endeavour (Carvalho et al., 2019).

The situation becomes increasingly challenging for women in developing nations, in rural and remote locations, and for those in poverty. Women seeking entrepreneurship in developing nations such as Nepal are significantly resource constrained (Xheneti, Karki, et al., 2019; Xheneti, Madden, et al., 2019). Largely due to entrenched patriarchal systems and practices, women lack access to opportunities including education, business and entrepreneurship training, infrastructure (Panda, 2018) and access to technologies, mentoring, banking and government support mechanisms (Hillman & Radel, 2018a). They generally face challenges in balancing work and family while negotiating patriarchal social norms, and are constantly challenged by gender discrimination (Panda, 2018; Verheul et al., 2006). Nepali women generally possess a subordinate status when compared to that of their male contemporaries (Radel & Hillman, 2016, 2018). Even though gender disparities differ across ethnicity, social unit and district, overall women are ranked below males in nearly every social dimension (Government of Nepal, 2011). The male dominant inheritance and social systems, and the division of labour between females and males continues to reduce female's access to schooling, work in the community arena, skill enhancement and business management experience, irrespective of status or culture (Hillman, 2019; Hillman & Radel, 2017; Pradhan & Shrestha, 2005).

In parallel with these social, economic and cultural challenges for women in Nepal, the basis of the definition and contexts of entrepreneurship and entrepreneurial endeavour are heavily masculinised and often perceived as the male domain (Jennings & Brush, 2013; Panda, 2018). Defining characteristics of entrepreneurship describe individuals who undertake 'the organisation and management of an enterprise involving innovativeness, independence and risk, as well as opportunity for profit' (Timmons & Spinelli, 2006, p. 10). As noted by Chasserio et al. (2014), the entrepreneur is often portrayed as somehow disconnected from other social realities. Women remain suspended in the multi-dimensional social and economic situations that shape their entrepreneurial identity (Chasserio et al., 2014) and therefore struggle to establish credibility in male dominated industries as a result. Social ideologies attach insufficient and often derisory values to so-called 'independent' women (Jennings & Brush, 2013; Maden, 2015; Panda, 2018).

The strong focus on innovation and risk within the definition further highlights that successful entrepreneurs have an enviable position to be able to take advantage of new technologies, products and markets, and to accept the inherent risks associated with these advantages (Bushell, 2008). Risk, as classified by Beck (Beck, 1992, 2010; see also Calhoun, 2010; Hillman, 2019; Jarvis, 2007) is an idiosyncratic, prevailing and modifiable depiction of contemporary reality. Women are commonly considered to be low risk-takers, due to their primary responsibilities for caring duties and household roles (Bushell, 2008). With scarce or only limited industry experience or capital funds, women are consigned to a nominal risk and marginal benefit positionality (Hillman, 2019).

Entrepreneurs are described as requiring a vision for growth, a strong commitment to innovation, and an intense need for achievement (Schaper & Volery, 2007). As Amiri and Marimaei (2012, p. 152) describe, they are individuals who 'perceive a vision, commit themselves to that vision, and almost single-handedly carry the vision to its successful implementation'. The implication here is the drivers for innovation, 'vision', commitment to a singular task, and the underlying need to achieve predominates over other more mundane tasks such as reproduction, caring for families and other home-based, feminine activities (Cohen & Wolkowitz, 2018). As noted previously, such domestic tasks remain largely the domain of women in developed nations and even more so for women in developing nations such as Nepal (Acharya et al., 2010; Ahl, 2004; Hofstede, 2011). The institutionalised allocation of the positions that are subsequently available for women within any given society, therefore ensures that women remain excluded from social and economic advancement.

This article investigates the complexity of the positioning of entrepreneurship, tourism development and gender for women in Nepal. We consider the ways in which entrepreneurship is characterised through masculine domains and reproduced in multiple institutions and discourses

(Ahl, 2004) to subordinate and repress Nepali women. However, our research also shows that Nepali women are instrumental in their own empowerment. They are circumnavigating the social and cultural risks and seeking entrepreneurship opportunities through politically subtle strategies by disrupting the stringently imposed gender norms (Bennett, 2005; Subramanian, 2018). They are shaping their own significance and distinctiveness within tourism development. Our intention is to highlight Nepali women's approaches to opposing the patriarchal embeddedness within tourism in Nepal and to consider the broader complexities of gender and entrepreneurship.

Literature review

This study sought to investigate how women's entrepreneurship through tourism development effects transformation for women in Nepal. We analysed the strategies of women entrepreneurs who have sought to develop tourism enterprises to disrupt poverty for themselves and their families. It is through these strategies that women are redefining their positions within the gendered institutions in Nepal. Much has been written in relation to the politics of gender, representation and the normalisation of inequality that results as a function of the power differential created through the ideological identity. As Butler (2004, p. 2) noted, the concept of gender enables dominant groups to insist that the 'viability of our individual personhood is fundamentally dependent on these social norms' (see also Celis et al., 2013; Chasserio et al., 2014; Walker, 2016). The research aims to support the future development of evidence-based approaches to enterprise development that can effect transformation for women particularly in rural and remote Nepal; highlighting the social, cultural, economic and political inequality and gender-based repression. The strategies employed enable women to circumvent the dominant social paradigm that bounds their capacity to achieve social and economic growth within the structurally institutionalised gender restrictions of a highly patriarchal society.

In terms of tourism industry participation and entrepreneurship, for women in Nepal there are both vertical and horizontal divisions of labour as a result of continuing political and social dogma around the necessity of women's responsibilities for families (Holmes, 2007). Vertical divisions of labour by gender result in fewer women being able to access higher positions in industry and government occupations (Holmes, 2007; Rafnsdóttir & Weigt, 2019). While horizontal divisions of labour entail that work is divided by genders within occupations and worker types (Bloksgaard, 2011; Campos-Soria et al., 2011; Murphy & Cross, 2017). Together, these gendered divisions have significant implications for the ways in which wealth and power are distributed between men and women – and this remains the case in both developed and developing nations.

Entrepreneurship to disrupt patriarchal hegemony

While gendered divisions serve to continue to repress and exclude women from both taking up occupations and from succeeding within work spheres, there is little question that women in both developed and developing nations have fought to access entrepreneurial opportunities. However, their activities are obstructed by stereotyping, discrimination, patriarchy and social, cultural and economic barriers. These barriers are often based on the ideologies and social constructions of gender (Butler, 1999; Carvalho et al., 2019). However, many women have learned how to overcome the bonded realities of their own lived experiences (Greer & Greene, 2003). They have developed cooperative and flexible ways of working in their efforts to manage and negotiate the often-conflicting responsibilities of household work and external work or career activities (de la Rey, 2005; Helgesen, 1990).

Entrepreneurship is a concept that embodies the ideas of transformation, creation and vision. It is the epitome of these aspirations which should provide avenues for women to disrupt patriarchal hegemony. Reynolds et al. (1999, p. 3) describe entrepreneurship as 'new business or new venture creation, or the expansion of an existing business, by an individual, a team of individuals, or an established business'. According to Drucker (1985, p. 35), entrepreneurship requires 'a systematic innovation, ... [which] consists in the purposeful and organized search for change'. It is reliant on the 'systematic analysis of the opportunities such changes might offer for economic and social innovation'. These lofty ideals however, obscure the challenges that entrepreneurship presents as it characterises a particularly complex phenomenon. The title has been ascribed based on the entrepreneurial individuals themselves, or the business type, or indeed the environment in which the entrepreneurial endeavour occurs (Širec & Močnik, 2012; Solymossy, 1998). Further, the values of 'systematic innovation' and an organised 'search for change' (Drucker, 1985) are often far removed from the realities available to women in poverty.

The very concepts of entrepreneurship and entrepreneurial endeavour are commonly understood as being a 'man's domain' (Cesaroni et al., 2018; Das, 2000; Jennings & Brush, 2013). While gender represents a poor indicator of the potential success or otherwise of enterprises, institutionalised assumptions persist around the reduced capacity of women's ventures to perform (Gupta et al., 2019; Marlow & McAdam, 2013). Further, and perhaps more detrimentally for women seeking entrepreneurial opportunities, as Langtowitz and Minniti (2007) found, women themselves have potentially adopted the social perceptions of their unsuitability to entrepreneurship as being an accurate and appropriate reflection of reality. Indeed, they found that women across their entire study population perceived themselves to be less suited to entrepreneurship than men regardless of individual motivations. It is these assumptions and perceptions that continue to demonstrate the subtle (or not-so-subtle) gendered biases manifested within entrepreneurial fields. Many female-owned or operated enterprises are marginal and vulnerable to social and economic changes; few are considered truly innovative or have the capacity to create significant profit and wealth (Marlow & McAdam, 2013).

Women entrepreneurs are often characterised as being 'chance', 'forced' or 'created/pulled' entrepreneurs (Das, 2000; Hillman & Radel, 2018a). Chance entrepreneurs start businesses without clear goals (ie. business has grown from hobbies or was passed on by a family member). Forced or 'necessity' (Schaper & Volery, 2007) entrepreneurs are compelled to start businesses because of poor financial positions; while created or pulled entrepreneurs start businesses because they were encouraged through entrepreneurship development programmes, and/or wish 'to build their own identity, develop leadership skills, and contribute to society through their venture' (Bushell, 2008, p. 550). Alternatively, the accepted masculine traits of risk-taking, proactiveness, innovativeness (Bushell, 2008; Drucker, 1985; Solymossy, 1998) and the development of 'comprehensive entrepreneurial skills to outperform the business rivals' (Kee & Rahman, 2018, p. 700) are simply not attributable to women under these assumptions.

Indeed, the persistence of such gender-based entrepreneurial identities would be an interesting research reflexion, if in fact, the resulting consequences of gendered identities were not so grim. On the Human Development Index, Nepal currently ranks 147th of 187 countries of the United Nations member states (United Nations Development Programme [UNDP], 2019). Nepali women achieve an average of just 4.9 years of schooling and only around 35.9% of women over the age of 25 have any level of secondary education (UNDP, 2019). Nepali women are susceptible to domestic violence, sexual harassment, rape, slavery, premature matrimony, a lack of financial means, among other things (Bhushal, 2008; Hillman & Radel, 2018a). Nonetheless, women in Nepal have long been campaigning for human rights for themselves (Becker, 2015).

Even allowing for the challenges associated with overcoming barriers from socially constructed ideologies of gender (Butler, 1999, 2004; Celis et al., 2013; Chafetz, 2001; Rowlands, 1997), research indicates that there is a global growth in women entrepreneurs (Bosma et al., 2020). Research also shows that very different factors are driving this growth when comparisons are taken between developed and developing nations. As noted in much of the research into entrepreneurship in developing countries, becoming an entrepreneur is generally largely due to necessity of an individual's circumstances (Panda, 2018; Schaper & Volery, 2007). As Schoar

(2010) noted, entrepreneurs in the majority of developing countries are subsistence entrepreneurs. Yet even so, women particularly play a significant part in socio-economic development and poverty alleviation (Rosca et al., 2020). In comparison, for developed nations, entrepreneurship is largely driven by opportunity. In the absence of any practical alternatives, women in developing nations such as Nepal, therefore, seek to increase their potential household earnings with some form of entrepreneurial endeavour. They run tiny businesses that rarely grow into larger firms but provide some alternative income and employment opportunities largely for their family members (Schoar, 2010).

Attaining social and economic emancipation from the confines of gender and poverty is potentially a more difficult prospect for Nepal's women than perhaps in many other cultures. This may be attributed to the social, cultural and political structures that dictate subservience for Nepali women; as suggested by Hofstede's research (1984) for example. Hofstede (2005) believed that poor nations such as Nepal have a well-established, male dominance in the workplace, with deeply entrenched patriarchal and caste configurations. The patriarchal embeddedness of Nepal entails that it is difficult for females to achieve political, social or economic importance. The prevailing religious norms further invest authority and control overpoweringly with males. Chetterjea and Basu (1978), and Phillips et al. (2009) find that nations with such considerable social disparity have larger gaps between the rights and capacities of males and females. Hosni and Lundberg (2005) state that there continue to be strict parameters defining the social distances between males and females in Nepal. In order to better understand, document and counteract the embedded patriarchy and gender reproductions that position Nepali women as inferior, incapable and irrelevant to economic growth, our research explored women's lived experiences in Nepal, and the effects and effectiveness of Nepali women's interventions within their own lives that enhance women's rights, power and agency.

Methodology

For this research, the authors used a constructivist, grounded theory approach (Charmaz, 2006; Glaser & Strauss, 1967; Hillman & Radel, 2018b; Radel & Hillman, 2018) to design the overall research process and intersected this with a critical theory lens (Kincheloe & McLaren, 2000; Seiler, 2004). According to Charmaz and Belgrave (2012, p. 347):

> *Grounded theory* refers to a systematic method for constructing a theoretical analysis from data, with explicit analytic strategies and implicit guidelines for data collection. In addition, the term refers to the *products* of the method, the completed theoretical analysis. We emphasize the flexible strategies that constitute this method and aid the researcher to (a) study social and social psychological processes, (b) direct data collection, (c) manage data analysis and, (d) develop and test an abstract theoretical framework that explains the studied process.

We recognised that understanding the intersections between entrepreneurship in tourism and women's journeys to disrupt patriarchy was a far more complex challenge than could be interrogated by 'reducing dimensions of women's experience to a set of measurable indicators' (Cornwall, 2016, p. 343). The constructivist grounded theory method combined with a critical theory lens was used to animate the juncture where both angles of inquiry strive for understanding of culture and positionality (Daniels, 2019; Geertz, 1973); including the data collection for subsequent phases of the study. As Daniels (2019, p. 83) noted, 'adding a critical perspective to the constructivist [grounded theory] approach offers multiple perspectives to the analysis process thereby providing opportunity to increase the understanding of culture and society'. Through this interdisciplinary approach we brought together the critical social science traditions of sociology and management/marketing, to explore Nepali women's lived experiences of entrepreneurship within a patriarchal society. Our goal was to better understand the effects and effectiveness of entrepreneurship in tourism as a mechanism to disrupt poverty for Nepali women and investigate the political, social and economic strategies used to effect change on patriarchal embeddedness.

The project aimed to capture experiences from women entrepreneurs in a range of urban and rural settings. However, as critical-constructivist researchers we believe that it is imperative that research provides benefits to the communities who participate. We therefore developed a reciprocity-based methodology in the research design by offering free workshops on entrepreneurship, leadership, small business practices, and marketing for small business. Using a purposive/snowball sampling method as appropriate for difficult to reach or vulnerable communities (Valerio et al., 2016), Nepali women entrepreneurs were invited to participate in both the workshops and the research project. Participants could attend the free workshops and then they were asked if they would be willing to participate in the research interviews after the workshops were completed. Involvement in the workshops was not contingent on any further contribution to the research project and participants could withdraw from the project at any time.

Participants were recruited through local (Nepali) contacts and were selected from four locations from urban, rural and remote regions of Nepal including; Mugu district in the far north western region, Pokhara in the middle western area, Kathmandu the capital city, and Bhaktapur (a largely Newari ethic community, south east of Kathmandu). In total across the period of this stage of the study, we delivered nine workshops including: four in Mugu district at different villages; one in Pokhara; three in Kathmandu; and one in Bhakatpur. The interviews were held with women usually at or near their places of work, their homes or sites of their choosing where possible, and often interviews were held at the workshop sites. In this initial phase of the study, a total of 16 women were interviewed, using open-ended, in-depth interviewing and question prompts.

Participants were asked questions relating to their participation in the tourism industry in Nepal. Further questions regarding: how they began their enterprises; the ways they recruited other women to join them; how they responded to male interference or ridicule in the sphere of tourism; and, what strategies they employed to combat long-term and culturally embedded social, historical, cultural and economic patterns of abuse, exploitation and violence were also examined.

The critical theory lens enabled the researchers to engage with, and scrutinise, inherent socio-political conflicts and tensions embedded within the women's stories. The critical perspective enables researchers to access the lived realities of individuals and to illuminate the human dimensions of social relations that influence the ways in which we/they strive to understand our/their objective world (Bohman, 2021). Critical theory provided a standpoint from which to evaluate the narratives and give voice (Hertz, 1997) to participants' hidden positioning (Daniels, 2019; St. Louis & Calabrese Barton, 2002) within the research; engaging the researchers in examining the complex socio-cultural pluralities (Habermas, 2001) of participants' lived experiences. Following the exigencies of the critical theory standpoint, our findings seek to unpack the mechanisms for social, cultural, political, structural, economic and historic oppression being represented in the womens' lived experiences. In line with our research aim, this analysis enables future actions to be taken to further support communities to continue to disrupt patriarchal hegemony.

Interviews were transcribed by a Nepali Research Assistant to further assist researchers to engage with the nuances contained within the contexts of the data. Some women were interviewed at their business site and some of the women chose to be interviewed in groups as was culturally appropriate. Most of the women were compensated for their participation in the research by being provided with culturally appropriate food and beverages. The importance of interviewing women from across city, rural, regional and remote Nepal, who have commenced trading as 'female only' enterprises; and, who employ other women from low socio-economic circumstances, is that these Nepali women are at the frontline of negotiating and subverting the dominant structures of patriarchy, caste and class in their country. This enables them to find their own agency and voices in this oppressive climate; and, of being able to express their opinions without fear of reprisal or violence.

Following the grounded theory approach, the transcripts underwent thematic coding, gerund coding, memoing, categorising and eventual reintegration (Charmaz, 2014; Radel & Hillman, 2018b). Incorporating the critical lens, we sought to uncover women's' strategies for circumventing embedded patriarchy in the Nepal tourism industry. As shown in Table 1, a section of transcript was coded and memoed as follows. As the researchers were both English speakers with only limited Nepali language knowledge, interviews were conducted in English via a Nepali female translator where required. The excerpts from the transcripts below are direct quotes from the interviews and the transcriptions are written in situ to retain nuances of language uses.

Findings and discussion

Given the design of our study combining the constructivist grounded theory approach with the critical theory lens, our findings represent a significant challenge in terms of how we can best portray the data and contextualise that within the critical perspective. Following Daniels (2019, p. 222), critical-constructivist approach, she noted that:

> *A feature of critical constructivist grounded theory is the development of critical* [inquiry] *to aid analysis processes. … Using critical* [inquiry] *to analyse the data exposes underlying issues of power and control within the … social system. Critical inquirists examine social structures with the aim of disrupting areas of power and control. They aim to inform those impacted by oppressive practices to bring about change for the better.*

The challenge is that while the 'critical inquiry seems to provide distinct themes for analysis, social reality is not so clear-cut. Economics, politics, social and cultural spheres all overlap and intertwine. Therefore, while our findings below indicate data and discussion arranged under subheadings, there is a multiplicity of complex motivations and lived realities all interacting on individuals concurrently, making ultimate distinction of discrete evidence difficult. We therefore begin this discussion with an overview of the general characteristics of the women entrepreneurs to foreground them in the study. We then examine the women's potential agency for responding to the structural inequalities inherent in the social systems that inform and confine their capacity for entrepreneurial development. Finally, our discussion brings out women's entrepreneurial strategies for economic, political, cultural and social encroachment on patriarchal areas of power and control. Throughout this section we have included voices of the women and each narrative is identified by the participants' general characteristics in line with ethical anonymity and cultural requirements.

Participants

To date we have interviewed and analysed data from 16 Nepali women entrepreneurs. These women range in ages from 25 years up to 65 years and have varying levels of education. Most of the women have completed just a few years of schooling (if they had access to education at all). However, two of the women had completed undergraduate studies and another had attained postgraduate university studies. The participants also range in their level of business development; with some of this group being new entrepreneurs and others having been in successful business for many years.

Some of the businesses they have started include: women only trekking companies with female guides for female trekkers; a guesthouse; a small trekking hotel; a workshop for women to learn sewing, knitting and weaving skills and then sell the products onward to tourism retail markets; a not-for-profit organisation to help women lift themselves out of poverty through practical training; a community bank offering microfinance loans for members; and, a used trekking goods outlet. The women also indicated they were seeking to change social and cultural

Table 1. An example of coding, categorising and memoing of transcripts with the addition of a critical lens.

Transcript	Coding	Memos
S: In Nepal, they don't have rafting guide also. In other countries they have ladies guide. I am so exciting to see these things also. In 2008, there also I met an "Inka group" from Sweden and am just like talking with Inka-Didi. She is already a rafting guide and am so exiting to talk with Inka. After we talk rafting things also and am so happy to meet she and asked to she can you teach me like kayaking other things also. She also very happy to teach me also and this day we talk like rafting things also and there after tomorrow we decide to go to lake again and she hires kayak to go lake and she is so happy I am doing kayaking. (Here S means In Nepal, there are not any female rafting guides. In 2008, she met a lady "Inka" from Sweden who is a rafting guide. Then she asked if Inka can teach her rafting which she agreed upon and started teaching her rafting from the next day. They hired a kayak to go to the lake and from there onwards she started Kayaking and both are very happy about it). S: After that she making plans to going back Sweden and she like to teach swimming (rafting?) for Nepal and she starting to form this project. (Inka decided to go back to Sweden who wanted to teach Nepalese women swimming and rafting and started to form project)	*Developing* knowledge and passion/energy *Teaching* *"Inka-Didi"* Sister – rafting guide teacher *Transforming* Two-way transformations	Historical inequality – no female rafting guides. S – discovering alternatives to male dominated cultural traditions. A Swedish female rafting guide taught S to kayak. To subvert the social status positioning, S finds support from external agents There is an exchange of benefits here – S is taught rafting and Inka-Didi makes a change in life course to develop a Nepali Rafting training program. Two-way transformations – complex relationships that cross international boundaries producing a common perspective and (*re*)producing new ideologies
S: I am first lady from Pokhara also after I met one lady also who know kayaking and after we have only two ladies who starting. (I am the first lady from Pokhara who knows kayaking and I have also met another lady who knows kayaking so now we are two ladies who have started kayaking) Another one also coming she is not exactly ready. She is also sister from Kathmandu. She is MA. She is also support Nepalese women for other things also. After that she helping to connect to media. Inka-Didi she helping to equipment, transportation and other money also. (Another lady from Kathmandu, MA is support worker for Nepalese women has also joined them. She helps to connect to the media. The lady from INCA provides support for equipment, transportation and finances). I: Yes. Yes. S: After that other ladies also coming to join rafting things. After we have ….. we will start one to 10 ladies. We will decide to open like new NGO (Here she means: after that other ladies also came to join rafting. We were up to 10 people and then decided to open up new NGO).	*Moving forward* on the entrepreneurial journey Hiring staff *Disrupting political patriarchy.* *Expanding* into entrepreneurship *Finding voice and agency* ***Political strategy*	S – First female rafting guide in Pokhara *"Inka-Didi"* – Helping to connect the new Nepali business with media, equipment, transportation, money (resources needed for entrepreneurship and business growth). Such support mechanisms should be the domain of Nepal Government to invest in Statepreneurship. ** *Economic and social strategy* to reduce the power of patriarchy through networking outside of the dominant political constraints. Deciding to start an NGO provides Nepal women with additional avenues to disrupt embedded patriarchy.

perceptions of women in Nepal, and to help other women to *"come out of poverty"*[1]. Further, the women's motivations for entrepreneurship were all necessity driven (Pinkovetskaia et al., 2019); they were either chance or forced entrepreneurs (Das, 2000) due to their social, cultural and/or economic circumstances. Interestingly, the most common element emerging from their stories, was the desire to engage in social and cultural change in status for women. Constraints to women's agency and structural disadvantages

Social structures or institutions and our capacity to both embody and/or resist these structures simultaneously, represents our agency to respond to social and cultural hegemony. Agency may be particularly significant in an individual's private life or relating to one's social wellbeing since it is possible for an individual's welfare to be shaped by their own agency (Stampe, 2017). To put it differently, agency is about what an individual is free to do to achieve the aims or ideals that they consider essential. It is a talent to rise above difficulties, to be taken notice of in society (either alone or as part of a group), and to query and challenge circumstances of repression or denial (Hanmer & Klugman, 2016).

Hays (1994) suggests that agency can be understood in four ways and these depend on the degrees of choice considered as being available to individuals to accept or deny the social constraints. First, individuals are agents as the bearers of social structures and institutions. Second, people and these social structures are indivisible – they are constantly creating and (*re*)creating each other through socialisation processes. Third, individuals are then also agents in that they can choose to engage with social structures and transform or disrupt the institutions within which the structures are embodied. Finally, agency and the capacity to control identity is necessarily constrained by resources and access. Agency is what an individual does to reach an objective with the appropriate ethical attitude, and it can have diverse consequences for the individual using the agency attitude (Stampe, 2017). Nevertheless, how women in developing nations employ and command their own agency are particularly important questions to address – particularly in gender and entrepreneurial contexts (Chafetz, 2001).

The following participant spoke about the challenges for women in asserting their agency in a male dominated profession:

> "*… trekking is adventurous, the so-called mature man's job, you know…; and also skilled human resource, there were no other women working on that and lack of confidence, lack of knowledge, skill, training and these all [for women] … many, many challenges were there. But, once you* [were] *determined and once you do it, it's happen* [sic]" (Female, Trekking Agency Proprietor, Pokhara).

Conversely, agency symbolises the burdens of needs and considerations and meanings of cultural constructions. Agency (which can also be perceived as emancipation) is both a cause and a consequence of power, where it supports the doer's position; and, it is both a cause and a consequence of 'culture' (Chafetz, 2001). The cultural construction of power is both continually and concurrently, the cultural construction of structures of agency and value in managing dominant others (Butler, 1999; Hays, 1994). Many women have experienced male domination and patriarchal embeddedness since their entry into the tourism industry in Nepal. As noted by another participant:

> "*At the beginning, it was very, very difficult for us. They* [men] *were saying like 'why you are snatching our bread?' They were saying, 'Oh my god women guide you cannot do this all', you know, like that; also, and many times, young women when we train to the trekking area, you know that time when they come back, they almost cried and they were so confused because they were told like 'this not good job for them' you know… 'They should do something else', like that…* " (Female CEO NGO, Pokhara)

Structural disparity can also be ameliorated through basic modifications in economic and social interactions that can bring about a loss of culturally endorsed inequity. For instance, the exodus of Nepali labourers to the Middle East has empowered large numbers of citizens to bypass feudal suppression and unjust social relations in their society by functioning within a totally different financial system (Dani & de Haan, 2008). Structural disparity can also be altered by struggle and aggression, when it generates comprehensive social dislocation, which on occasion destabilises established stability and gender distinctions. Marginalised groups have need of voice, influence, and the right to use available assets (Dani & de Haan, 2008).

> "*So, it was very, very difficult at the beginning. My goodness! … and we had no idea at all, like people they suffer us* [people made us suffer]. *Even from the government, even from the local people, everybody you know! And even they took us in a court and police custody. They did not put us in the custody within, but outside - we had to go many times to explain what we are doing, why we are doing. It was very, very challenging, you know!" (Female, Travel Agency Proprietor, Pokhara)*

The experiences of structural inequities are in no way straightforward ones. They underline the need for public policy to make a distinction between deprivation and inequity or social exclusion and to integrate an appreciation of power and prejudice. They also identify the need to focus on poverty in economic, social, and cultural domains all at the same time and over an extended period. Experience in this field is consequently somewhat inadequate and will likely be a key part of policy discourse into the future (Dani & de Haan, 2008).

Subversive agency – strategies to overcome historical oppression

Historically, Nepal has witnessed significant turmoil. As Tamang (2009), Yadav (2016) and a number of other researchers (see for example Bennett, 2005; Des Chene, 1997; Grossman-Thompson, 2013; Humagai, 2012; Pant & Standing, 2011; Pherali, 2011) have shown, Nepal has suffered a 10 year long civil war, the abolishment of the monarchy, political factioning, increasing demands for inclusion from various marginalised ethnic and caste groups, along with a generalised lack of security (food, financial and physical) in many parts of the country. Within this context of political and social conflict, along with the characterisation of Nepal as a Hindu state from 1962 onwards which, according to Tamang (2009, p. 64), 'ranked the entire population in a caste hierarchy that regimented social life accordingly', have all contributed to the historical and continuing embeddedness of patriarchal legitimacy. One participant relates how she experienced marriage and learnt the skills she needed to run a small trekking hotel:

> *"Actually, we got started in 1971. It was an arranged marriage and then I was.... Here with my...* [husband]. *He was running a small restaurant. I was the only one daughter in the house. I was well trained in the cleaning and everything, washing, cooking and all".* (Female Hotel Proprietor, Kathmandu)

For the women in our study, combatting historical legacies requires forward-looking, subversive and emancipatory strategies that work to modify perceptions and expectations and also contribute directly to emancipation for women. Within the research presented here, the term 'subversive strategies' refers to the notion that those women who experience or who are oppressed via embedded hierarchies, rebel either through the polarities of meek and/or assertive approaches, and through using either approach, aim to destabilise, disrupt, subdue or weaken a prevailing system, particularly an officially established regime or collection of values. This notion is also enhanced and extended by the argument of Scheyvens (1998, p. 236) who suggested that:

> In this context it is important to remember that not all men are advantaged by existing social structures, nor are all women disadvantaged by these structures. Thus, as Kabeer (1992, p. 19) explains, men's interests will not necessarily be diametrically opposed to those of women: 'gender relations are relations of power as well as difference, of conflict as well as cooperation'.

When asked why she had become an entrepreneur, one participant who started a hostel business explained:

> *"Actually, I am also from outside of Kathmandu valley. I was staying with my relatives and some of my friends at that time also stayed at the hostel. One of my friends is also from Dharan and I am actually from Dharan. We two then made a planning and .... "Lets' start one hostel". Then I visited some* [hostels] *.... Most of the hostels around 6 or 7 I visited. Then I started from smaller one, there were 6 or 7 beds only at that time. Then after 6 months I took two floors/flat. Then* [I had] *14/15 students and then did 4–5 years. After 4–5 years I took the whole building. I am here since 14 years. We are independent and we don't want anyone's pressure... I also saw some of my friends staying in hostel and that concept raised in our mind and thought why not start a* [hostel]. *Then people like us outside of Kathmandu valley don't have to search for places where to stay. For their convenience.... only for girls..."* (Female, Women-only Hostel Owner & Operator, Kathmandu)

For this participant, the choice or necessity to become an entrepreneur was the simple answer to their feelings of isolation from their home villages. Their subversive strategy however is to provide options and opportunities for other girls and women from outside of the city area to engage with education for future emancipation. Moreover, and according to Cleves-Mosse (1993, p. 170)

…given the opportunity and support, and ways of working which respect culture and women's pace, women readily question the reasons that their lives are as they are and, far from being content, seek out ways of challenging and changing this situation (as cited in Scheyvens, 1998, p. 236).

When asked for example, how long her students stayed, the participant noted:

"Some students stay for 4–5 years also. Most of the students stay here for year or two years. Some come for just one months. But most stay here for more than a month. They don't have to worry about how to cook and where to find things to cook. We all manage for them. At lunch and dinner time, they come and eat. We provide all things like electricity, water and internet service. They finish bachelor after completing their Plus 2 [years 11 and 12 in secondary school]. *They did bachelor and went abroad. So, it obviously takes 4–5 years. So, this is their home for 4–5 years. … Yes"* (Female, Women-only Hostel Owner & Operator, Kathmandu)

To enable the autonomy of women, legislators must be willing to proceed beyond supporting women with day-to-day subsistence, and aspire to afford them abilities, expertise and conviction to establish the development direction they aspire to, and to handle the societal structures that dominate them (Scheyvens, 1998, p. 236).

In line with the research of Scheyvens (1998), it is noteworthy that the women did not endeavour to be truculent, but still efficiently and constructively altered facets of their lives and increased their own capacity (Scheyvens, 1998, p. 238). In this way, women are accepting the risks and developing entrepreneurial opportunities for the goal of enabling emancipation for more women – the opportunity to make their own choices and to have a voice in their own lived experiences.

Social strategies

Social customs and traditions at the domestic, private and social levels are key features in inequity against, and the isolation of, females and these traditions cannot be disregarded simply by installing legislation against prejudice alone (Pearse & Connell, 2016). As Butler (1999, p. 6) suggested, where gender transects with race, class, ethnicity, sexuality (or sexual orientation), and/or regionally based traditions, it becomes 'impossible to separate out "gender" [emphasis in original] from the political and cultural intersections in which it is invariably produced and maintained'. Patriarchy provides the central tenant that has deprived and repressed females within all social groupings in Nepal (and elsewhere); even though its influence is varied among caste, ethnic and religious alliances (Nazneen, 2018). Lawoti (2010, p. 21) defines patriarchy as 'a set of social relations with a material base that enables men to dominate women'. Patriarchy may be further sustained by attributes of belief systems, families and familial role requirements, constitutional structures, institutionally (*re*)produced narratives, legally framed male power over chattels (property, home, women, children, etc.), wage disparities, and female's manual labour. Patriarchy situates females in a subjugated position in family units, within communities and in society, and hinders them from contributing to political affairs, financial activities and society at large (Lawoti, 2010).

Beginning a female-owned tourism enterprise presented its specific challenges for the following participant. Understanding the anomalies in the community and social configuration of norms and mores for local females in remote locations in Nepal presented initial difficulties and obstacles.

"Just at the beginning, we liked the idea…, but we did not read the social structure. Because we were migrated from Darjeeling to here and so we can communicate in Nepali… we did not realise the cultural difference in Darjeeling and here. So, we thought, maybe this is good idea to start women trekking guide services. But there was you know… in Nepal and in Pokhara, still women were not that much you know… coming out to do something. So, that was the contribution … was the main thing, and the other thing that tourism was not that much explored for women, you know". (Female, Tourism Business Entrepreneur, Pokhara)

And, according to another female enterprise founder

> *"We started working in a tourism and guest house running* [sic]. *... But immediately, realised the women trekking guide is the need. So, we catched* [sic] *that and we worked on that you know... Like that way we did and one after another we are just trying to addressing* [sic] *the challenges you know, addressing the issues is coming up. So that's how we are developing. It is nothing like we are planning or learning".* (Female, Guest House & Trekking Guide Business Operator, Pokhara)

As Lawoti (2010) suggests, findings from affluent but conventional societies have revealed that social norms limit even educated females. For instance, females in Middle Eastern societies like Saudi Arabia are well educated but are not permitted to work in various occupations (Morrison & Jutting, 2005). Nepal does not have comparable arrangements of constraints, but it is incorrect to accept that educated females, as well as females from the Dalit[2] ('untouchable class') and other ethnic groups, do not face difficulties in politics and employment markets. Indeed, another female participant told us that, *"Here is a kind of you know social pressure, mental pressure you know like they have many different kind of* [ways to] *treat you and* [let you] *down- ... ".* When asked about how she was treated by male staff in the trekking industry, another participant spoke in more stark terms:

> *"Of course, at first they treated very badly when there were few females than males. Sometimes, they treat very good. But many times, they are jealous, they say you should be working in different sector like nursing, administration or so. I am also feeling more confident than before. Now, they are good."* (Female trekking guide and office support person, Mugu)

Lawoti (2010) has noted that patriarchy functions in two ways to replicate and support exclusion of females. Firstly, patriarchal models, for instance those that deem females to be inferior, are at odds with ineffectual governance structures and generate opposing consequences of female marginalisation. Secondly, patriarchal customs reinforce oppressive traditions, such as inequitable inheritance laws and labour rights. They support exclusionary official traditions to embed marginalisation of women in society (Lawoti, 2010). However, the women in this study, have subverted these customs and found ways to intervene and give voice to their positioning and lived experiences. As another participant said *"sometimes, male guides suggest like 'Why you are here?' I say, I am enjoying here, and you can't talk about my freedom"* (Female, Trekking Guide Entrepreneur, Mugu).

Political strategies

On attempting to disrupt poverty and social repression, the women in our study were also able to subvert patriarchy in Nepal through seemingly covert or perhaps even unconscious, subtle, politically motivated strategies. Indeed, one of the participants spoke of how it was extremely difficult to begin a female only enterprise within the tourism sector. For her and for many other women in our study, it has been a test of her abilities to perform well under pressure; presenting often frightening and isolating situations for women who dare to step outside of the historical, social, structural (patriarchal), cultural, economic and political boundaries to begin a fledgling tourism business venture. One of our participants started a trekking and guiding training enterprise and related her experiences of starting the venture:

> *"Yeah! people... started coming and they really loved our initiation. There were some women trekking with men. You know, there was no choice for women. At that time, they had some experiences. Many of them were good but some had very bad experiences from male guide, you know... they were really frightened. I said, 'OK, is there any way that we can do something in this situation?', and I thought, 'OK, once I will go to the mountain and I will see. Because I have done mountaineering training before that in 1990'."* (Female, Trekking Guide Training Enterprise Administrator & Owner, Pokhara)

As a trailblazer in women's tourism enterprises within Nepal, the participant additionally recounted how she had to come to terms with the idea that many Nepali women in remote areas of Nepal experience impoverished existences.

"... I was doing some research work and I had been on very remote areas of western Nepal. I have seen very, very, you know, poorest women, needy women over there. At that time, I was wondering why people live here because nothing is here. No ... nothing is there, and I kept asking people why these people stay here because nothing there? I was very young, and I could not understand about the sentiment of the places you know." (Female, Trekking Guide Training Enterprise Administrator & Owner, Pokhara)

The participant related that even though she did not understand why women would continue to live in remote and isolated situations, she suggests her youth and lack of experience at that time prevented her from perceiving the historical, social/structural (patriarchal), cultural and economic reasons the women remained in in villages where there was no hope for improvement in their lives. However, even though this participant belongs to a high-caste, had opportunities for higher education (having studied at university) and had opportunities to experience different parts of Nepal, her story highlights the (*re*)produced ideologies that represent substantial obstacles for women to imagine an entrepreneurial opportunity which cuts across all of these boundaries. This is more starkly illustrated in her discussion of the living conditions for Nepali women in remote locations.

When applying a critical lens to the research, it is obvious that the women the participant witnessed, living in poverty and abjection, have now been able to improve their lives and overcome their patriarchal oppression through extending their daily activities and mundane household and farming tasks to incorporate them into cottage industries and appropriate them for the tourism industry in Nepal. Nepali women are skilled at all home duty and small holding tasks. For them, this makes a perfect segue into the wider earning sphere offered by tourism activities (see Snellinger, 2010).

Cultural strategies

Taking a pragmatic stance, concepts and implications of cultural traditions offered by Rao and Walton (2004) are helpful in grasping the contexts and impacts of culture on potential development (see also Narayan, 2013). These authors view culture as essentially 'about relationality—about the dynamics of relationships between individuals within groups, between groups, and between ideas and perspectives' (Rao & Walton, 2004, p. 4). They also appreciate culture as constantly challenged and made susceptible to change. They argue that culture 'is not a set of primordial phenomena permanently embedded within national or religious groups, but rather a set of contested attributes, constantly in flux, both shaping and being shaped by social and economic aspects of human interaction' ((Rao & Walton, 2004, p. 4; Bennett, 1983, pp. 205–206). In the context of females working in the tourism industry in Nepal, and the patriarchal structures entrenched in the sector, participants spoke about how the restrictive cultural norms have shaped them and their cultural uniqueness. One of the participants recounted:

"We learned so much on this you know women empowerment and social entrepreneurs, like the cultural difference in many different [ways]*... You know how we are taking the culture, the tradition, you know very personally. But if we go back and think that is all because of our climatic condition and geographical situations and the availability of the things. They have made our culture. They practiced things and we are carrying on as a culture. And now we are somehow fighting for that, for our identity. ... That's also something, through tourism we can learn. We learn from this because we are meeting many different people. Its amazing lesson that we are learning from tourism".* (Female, Social Entrepreneur, Bhakatpur)

Another female, working in the hospitality industry in Kathmandu further explains:

"In [my culture]*, we women and man we work together. So, we are both equal. In* [Kathmandu]*, men work first, and men are superior also. They have respect in the family. Their sons will protect the family like this... But now, women are coming forward. They are fighting for their rights. In* [my culture] *as a daughter or son who can help, they can help both sides. Daughter's parents or husband or spouse's parents".* (Female, Business Operator, Kathmandu)

As previously noted, in Nepal, females are less emancipated and fall far behind their male equivalents in education, civil, financial, health, employment and many other societal areas (UNDP, 2019). Strategically, Nepali women are recognising these issues and deploying political strategies to subvert and disrupt these socio-cultural ideologies. Another participant began a weaving and clothing manufacturing enterprise specifically to raise women's status focusing on caste and status of women to disrupt the status quo. She stated:

> *"Yeah nine years ago I started. These people are really have problems… we like to* [bring them] *up* [Make them come out of poverty]. *First, we start with four ladies. Now we have around 200 people I give for training already.* [I: And now you have…?] *22 people working. Yeah… first we have to make the product… for the ladies … They don't have anything. No school, study, no education. Just we give for the education. First, we search for the family who is this kind of people and then who is this caste. My country is like so much for the caste system. After that we give for the training for them. We give for the money also. They need money to live. We sell for the products* [to bring in the money] … *We go like that."* (Female, Sewing, Weaving & Handicrafts Training Centre for Women, Proprietor, Kathmandu)

Using cultural entrepreneurship principals coupled with social and empowerment entrepreneurial goals, an encompassing desire to change cultural and social gender identities, and a great deal of hard work and tenacity, Nepali women are (*re*)writing their own cultural directions.

Economic strategies

Mountain tourism, along with other tourism activities, creates localised wages and occupations for poor communities (Lama, 1999; Lama & Sattar, 2004). Even at high altitudes, Nepali females have capacity to cater for trekkers and hire out beds for a small fee per night providing direct economic exchange. Both direct and indirect revenues are produced by the trekking industry including income from accommodation supply, and trekking house ventures; retailing of crafts; and hiring of guides, buildings, employees, and smallholders. Lama (1999) also notes a range of non-monetary benefits arising from tourism comprised of: 1) contacts with unfamiliar concepts like expertise, diverse verbal communication and viewpoints; 2) improved knowledge and enthusiasm for appropriate hygiene and refuse management; 3) prospects for foreign-subsidised tourism and schooling; 4) self-esteem in customs and self-assurance that comes from interrelating with guests; and in some instances, 5) an enriched socio-economic significance of women.

However, in terms of economic strategies to disrupt patriarchy, many of the women relate stories of their foray into the world of Non-Government Organisation (NGO) development to address social or political issues. One of the participants recalls how she and her sisters began their enterprise and concentrated on training and uplifting women in Nepal. They have since gone on to expand their business and have also created an NGO to continue to empower women.

> *"In 1998, we registered the company, a trekking company… You know we were going very well, and we were providing small trainings to the girls, so we liked to keep and continue that program and* [then] *we registered NGO,* [names NGO]. *That was solely started by…* [us], *we were looking for women guide, so we were, you know supporting each other…* [the] *trekking company is a business and make a profit and contribute to train the women. And also, they provide job on the training* [sic - on-the job training] *and the job.* [Names NGO] *is NGO that provide the training for women and also the professional other additional trainings… We are doing other programs too. So, this is how we are working together".* (Female family group of entrepreneurs, Pokhara)

And another participant stated:

> *"This hostel we are registered in government* [a registered business]. *We have to pay taxes. There is hostel association also. I am the founder and vice president of the association* [provides name of association]*."* (Female, Hostel Owner & Hostel Association Founder, Bhaktapur)

The women in our study, are not only founding enterprises to disrupt poverty and patriarchy but are subverting government structures to increase their political and legal influence from within the dominant ideologies. Other participants demonstrate economic strategies of starting their own (community-based and run) banking opportunities, and yet others are initially accessing microloans but then opting to become the patron or micro-financier for others. When asked what the most difficult part was of starting out as an emancipation entrepreneur, another participant stated:

> *"Hard things we have… main things to money.* [The hardest thing was mainly accessing money] *… my government does not help. That is one question* [sic - problem]. *And after we loan for the microloan.* [we applied/received a microloan]. *First, we start with the microloan. Still we have.* [I: So, you paid out the initial microloan and you got another one?] *Yes. We have ladies now is 28 people, we keep for the microloan also. We buy for the looms, we buy for the* [sewing] *machine, we buy for raw material."* (Female, Weaving Entrepreneur & Trainer, Bhaktapur)

Ultimately, these women are disrupting gender stereotypes through taking charge of economic means and repositioning them to work better for others. This process on a smaller scale can also be said for the women who have received their training in trekking guiding, weaving, basic hotel management etc. For over twenty years, female organisations have been training women to become independent through tourism and trekking guiding. Guides speak directly about their spending capacity as a type of self-determination; that is, the self-determination to be a contemporary participant in the new economy. Opponents of neoliberal economic strategy find 'self-determination via expenditure' a concerning result to the historical marginalisation of females from inclusion in community life (Grossman-Thompson, 2013). Yet, such philosophical allegiances should not stop recognition of the meaning of income-making and purchasing power for participants. Whereas autonomy and self-determination through improved purchasing power does not inevitably convert into political power, guides (and indeed, other women working in the tourism industry and local economy) nonetheless identify their wages as a fundamental component of acquiring increased independence, which is the basis of their individuality as contemporary women (Grossman-Thompson, 2013).

The following participant explains how she and her immediate family began in the tourism sector in Nepal:

> *"My brother used to teach me… in office. His wife used to work as cook. At that time, tourists used to climb the mountain. When they came back, they used to give us sleeping bags, jackets everything. "OK, now we are going back you can take it". We used to wash them and used to sell them. Then I invited my sister in-law. "Why don't you come here and start the trekking shop"? Whatever they gave we used to wash in the river here. That time river was very clean… So, we worked together supported by my family. Even my sister in-law. By evening to support* [us] *she had to come back, take over the cashier, check the bills, watch the bills, get the money like this".* (Female, Hotel Co-owner & Entrepreneur, Kathmandu)

Female's functions in tourism in Nepal are mainly an expansion of the domestic-manager and tourist receptionist duties. Females run trekking lodges and tea houses beside the main trekking routes, at times with their spouses or fathers, but frequently single-handedly. As cooks and principal waitstaff, they have the most interaction with trekkers and their guides and porters. They are never sedentary. Whenever they have spare time, they knit woollen hats, gloves, and slippers, weave sacks, or make souvenirs to sell to trekkers. Many females operate as porters or labourers for trekking groups, and a few have become trekking guides and even summiteers (Grossman-Thompson, 2013; Lama, 1999; Lama & Sattar, 2004). Whether employed in the sector or acting as agents of change and development through entrepreneurial activities, Nepali women are creating, (*re*)positioning and ultimately encroaching on patriarchal territories.

Implications and conclusions

By engaging the lens of a critical theoretical framework, we can uncover insights into the institutionalised gender and power structures. A critical perspective engages with the socially constructed layers of social, historical, political, cultural, economic, ethnic and gender values that have been crystallised and embedded over time (Daniels, 2019; Guba & Lincoln, 1994). Nepali women are subject to and regulated by patriarchal structures embedded both within Nepal's social and cultural contexts but also within the contexts and capacities of entrepreneurship. However, as Butler (1999, p. 4) noted, 'by virtue of being subjected to them, formed, defined, and reproduced in accordance with the requirements of those structures… the feminist subject [*sic*] turns out to be discursively constituted by the very political system that is supposed to facilitate its emancipation'. Therefore, we need to not simply describe the lived experiences of women in Nepal, but to disclose and analyse those experiences to produce alternative ways of thinking. When data analysis is enriched by adopting a critical lens, the analysis can be unpacked through multiple layers and ensure that the findings are not therefore divorced from any possibility of corrective political actions to overcome oppressions and confinement. There are many battles to be fought by women of Nepal. These include working out how to combat patriarchy in tourism through developing their own subversive and political strategies. The strategies Nepali entrepreneurial women use to contest patriarchy in tourism, and more generally in Nepal include political strategies, social strategies, subversive strategies, cultural strategies, economic strategies, and structural strategies (agency/structure debate).

Nepali women's political strategies to overcome social and cultural repression are subtle, understated and subversive. Women entrepreneurs disclose they are unable to tackle the masculine domains head on, so they negotiate around obstacles and subvert the dominant paradigms through much more sensitive practices. In this research we investigated the capacity of Nepali women to invade and disrupt patriarchal held territories in the tourism industry in Nepal. The research shows that the intersections between gender, entrepreneurship and emancipation are challenging and present significant barriers for women, but they are not insurmountable. We used a grounded theory approached informed by a critical theory lens to examine the social, cultural, economic, historical, and structural political strategies that Nepali women have adopted and adapted to access and redress the socially constructed and maintained imbalances.

There has been a lack of sensitivity towards, and action to redress, gender inequity within Nepal's cultural and social structures. Even while publicly supporting the inclusion women in economic development opportunities, typical support mechanisms have continued to constrain women from fully benefitting from tourism opportunities. On face value, this denial of gender equality to contribute to economic development creates a situation of economic irrationality on the part of governments and industry, as Nepal's female population constituted just over 54.5% of the total population in 2018 (World Bank Group, 2019). The globally recognised capacity for women to contribute to economic growth as an asset for growing labour markets, has therefore been ignored in Nepal to the detriment of the country and its peoples. The success of tourism development and growth in Nepal remains a challenging space where governments, industry and individuals must make significant efforts to support the crucial roles played by women entrepreneurs in tourism. Even allowing for increasing government interest in women's emancipation through entrepreneurship, there is yet a long way to go for Nepali women to achieve an equity akin to equality.

The study contributes to extant literature and existing knowledge through expanding the literature on women, development, patriarchy, gender inequity and inequality; and, applying women's voices and agency to lessen the domination of national structures and barriers within Nepal. The findings of the study serve to inform managerial actions and policy through highlighting the plight of women within Nepali society. While the Nepali Constitution provides for protection of women, including equal pay for equal work, over the long term, the Government of Nepal has been less than successful in realising these requirements. The position of women in Nepal persists as very deficient in terms of health, education, income, resolution-making, and access to policy programs (Secretariat & Durbar, 2015). Our research aims to inform wider Nepali society and give voice (agency) to the women for the future.

Limitations

There are some limitations to the study. One of these is that the findings are not generalisable to the female population throughout the Nepali nation. Interviewing the participants in English is also a minor limitation as, even though schooling in Nepal is conducted in English[3], and many people speak English, it is not the mother tongue and there are anomalies in the way things are expressed in English across the country. However, this has not affected the rigor or thoroughness of the data collection, transcription, and coding. Rather, it has led to a more distinctive, original and unique set of findings and conclusions regarding the pathways for women in Nepal wishing to engage within an occupation that will provide for themselves and their families. The findings and conclusions presented here will contribute to policy and legislation of government strategies to improve and expand the predicament of women seeking equality in their society.

Future research

The future of research within this study focus will need to extend and include women from all zones and districts throughout Nepal. Investigating how women extricate themselves and subvert embedded and patriarchal conditions while still maintaining, extended family, land holdings and income-based activities is a long-term project for Nepal, researchers, the government and many other nations within South Asia. Indeed, a longitudinal study would provide insights and findings to promote equality in policy and life experiences to inform government practices, legislation and understandings of equity and equality throughout the country into the future.

Notes

1. Where quotations are drawn from participants' own narratives, these quotations have been italicised and indicated with double quotations to clearly denote them as such. We have retained the language, syntax, grammar and expressions used by participants to maintain the integrity of representation of their voices and stories. To take away the expression would modify the data. Therefore, the verbatim data are their own words as far as possible.
2. In the context of Nepal, the National Dalits Commission defines 'Dalits' as "those communities who, by virtue of atrocities of caste-based discrimination and untouchability, are most backward in social, economic, educational, political and religious fields, and deprived of human dignity and social justice" (Dalit Welfare Organisation, 2018).
3. For a full discussion of English being spoken and taught in all Nepali primary schools, please see Mitchell, D. (2018) Teaching 'in our remoteness': A constructivist grounded theory of primary school teaching in a remote area of Nepal, unpublished Master of Education (Research) thesis, Central Queensland University, Rockhampton, Australia.

Acknowledgments

The authors wish to acknowledge the editors and reviewers of this article for their insightful help and patience. We also wish to acknowledge the women who participated in the study, as without them there would be no research for us to disseminate and subsequently, no contribution to knowledge.

Disclosure statement

No potential competing interest was reported by the authors.

Funding

This work was supported by the Central Queensland University Merit Grant Funding under Grant HE1308.

References

Acharya, D., Bell, J., Simkhada, P., Van Teijlingen, E., & Regmi, P. (2010). Women's autonomy in household decision-making: A demographic study in Nepal. *Reproductive Health*, *7*(1), 15. https://doi.org/10.1186/1742-4755-7-15

Ahl, H. (2004). *The scientific reproduction of gender inequality: A discourse analysis of research texts on women's entrepreneurship*. Liber, Copenhagen Business School.

Amiri, N. S., & Marimaei, M. R. (2012). Concept of entrepreneurship and entrepreneurs traits and characteristics. *Scholarly Journal of Business Administration*, *2*(7), 150–155.

Beck, U. (1992). *Risk society: Towards a new modernity*. Sage.

Beck, U. (2010). Climate for change, or how to create a green modernity? *Theory, Culture & Society*, *27*(2–3), 254–266. https://doi.org/10.1177/0263276409358729

Becker, M. (2015). Constructing SSLM: Insights from struggles over women's rights in Nepal. *Asian Studies Review*, *39*(2), 247–265. https://doi.org/10.1080/10357823.2015.1021754

Bennett, L. (2005, December 12–15). *Gender, caste and ethnic exclusion in Nepal: Following the policy from analysis to action* [Paper presentation]. Paper Presented at the Arusha Conference, New Fronrtiers of Social Policy, Kathmandu, Nepal.

Bloksgaard, L. (2011). Masculinities, femininities and work–the horizontal gender segregation in the Danish Labour market. *Nordic Journal of Working Life Studies*, *1*(2), 5–21. https://doi.org/10.19154/njwls.v1i2.2342

Bohman, J. (2021). Critical theory. In E. N. Zalta (Ed.), *The Stanford encyclopedia of philosophy*. (Spring 2021 ed.). Center for the Study of Language and Information (CSLI): Stanford University. (pp. 49–65). Routledge. https://plato.stanford.edu/entries/critical-theory/#3.

Bosma, N., Hill, S., Ionescu-Somers, A., Kelley, D., Levie, J., & Tarnawa, A. (2020). *Global Entrepreneurship Monitor (GEM) 2019/2020 global report*. London Business School. https://www.gemconsortium.org/report

Bushell, B. (2008). Women entrepreneurs in Nepal: What prevents them from leading the sector? *Gender & Development*, *16*(3), 549–564. https://doi.org/10.1080/13552070802465441

Butler, J. (1999). *Gender trouble: Feminism and the subversion of identity*. Routledge.

Butler, J. (2004). *Undoing gender*. Routledge.

Calhoun, C. (2010). Beck, Asia and second modernity. *The British Journal of Sociology*, *61*(3), 597–619. https://doi.org/10.1111/j.1468-4446.2010.01328.x

Campos-Soria, J. A., Marchante-Mera, A., & Ropero-García, M. A. (2011). Patterns of occupational segregation by gender in the hospitality industry. *International Journal of Hospitality Management*, *30*(1), 91–102. https://doi.org/10.1016/j.ijhm.2010.07.001

Carvalho, I., Costa, C., Lykke, N., & Torres, A. (2019). Beyond the glass ceiling: Gendering tourism management. *Annals of Tourism Research, 75*, 79–91. https://doi.org/10.1016/j.annals.2018.12.022

Celis, K., Kantola, J., Waylen, G., & Weldon, S. L. (2013). Introduction: gender and politics: A gendered world, a gendered discipline. In G. Waylen, K. Celis, J. Kantola, & S. L. Weldon (Eds.), *The Oxford handbook on gender and politics* (pp. 1–27). Oxford University Press.

Cesaroni, F. M., Pediconi, M. G., & Sentuti, A. (2018). *Gender stereotypes in leadership construction within family business* [Paper presentation]. Paper Presented at the International Conference on Gender Research, Reading, MA.

Chafetz, J. S. (2001). Theoretical understandings of gender: A third of a century of feminist thought in sociology. In J. H. Turner (Ed.), *Handbook of sociological theory* (pp. 613–631). Springer.

Charmaz, K. (2006). *Constructing grounded theory: A practical guide through qualitative analysis*. Sage.

Charmaz, K. (2014). *Constructing grounded theory*. Sage.

Charmaz, K., & Belgrave, L. (2012). Qualitative interviewing and grounded theory analysis. In *The SAGE handbook of interview research: The complexity of the craft* (pp. 347–365).

Chasserio, S., Pailot, P., & Poroli, C. (2014). When entrepreneurial identity meets multiple social identities: Interplays and identity work of women entrepreneurs. *International Journal of Entrepreneurial Behavior & Research, 20*(2), 128–154. https://doi.org/10.1108/IJEBR-11-2011-0157

Chatterjea, R. G., & Basu, A. (1978). The relationship between social distance and levels of conceptual integration. *The Journal of Social Psychology, 104*(2), 299–300. https://doi.org/10.1080/00224545.1978.9924074

Cho, J. Y., & Lee, E. H. (2014). Reducing confusion about grounded theory and qualitative content analysis: Similarities and differences. *The Qualitative Report, 19*(32), 1.

Cleves-Mosse, J. (1993). *Half the world half a chance: An introduction to gender and development*. Oxfam GB.

Cohen, R. L., & Wolkowitz, C. (2018). The feminization of body work. *Gender, Work & Organization, 25*(1), 42–62. https://doi.org/10.1111/gwao.12186

Corbin, J., & Strauss, A. (1990). Grounded theory research: Procedures, canons, and evaluative criteria. *Qualitative Sociology, 13*(1), 3–21. https://doi.org/10.1007/BF00988593

Cornwall, A. (2016). Women's empowerment: What works? *Journal of International Development, 28*(3), 342–359. https://doi.org/10.1002/jid.3210

Dalit Welfare Organisation. (2018). *Nepal/Dalits*. http://dwo.org.np/v1/dalit.php

Dani, A. A., & de Haan, A. (Eds.). (2008). *Inclusive states: Social policy and structural inequalities*. The World Bank Group.

Daniels, C. R. (2019). *Exploring Australian women's career transitions: A critical constructivist grounded theory study* [Doctor of philosophy]. Central Queensland University.

Das, D. J. (2000). Problems faced by women entrepreneurs. In K. Sasikumar (Ed.), *Women Entrepreneurship*. Vikas Publishing House.

de la Rey, C. (2005). Gender, women and leadership. *Agenda: Empowering Women for Gender Equity, 19*(65), 4–11. https://doi.org/10.1080/10130950.2005.9674614

Des Chene, M. (1997). Nepali women's movement: Experiences, critiques, commentaries. *Studies in Nepali History and Society, 2*(2), 291–297.

Drucker, P. F. (1985). *Innovation and entrepreneurship: Practice and principles*. Harper & Row Publishers Inc.

Geertz, C. (1973). Thick description: Toward an interpretive theory of culture. In C. Geertz (Ed.), *The interpretation of cultures: Selected essays* (pp. 3–30). Basic Books, Inc., Publishers.

Glaser, B. G. (1978). *Theoretical sensitivity*. The Sociology Press.

Glaser, B. G., & Strauss, A. (1967). *The discovery of grounded theory: Strategies for qualitative research*. Aldine Publishing Company.

Government of Nepal. (2011). *Nepal Population Report 2011*. Ministry of Health and Population - Population Division. www.mohp.gov.np/population

Greer, M. J., & Greene, P. G. (2003). Feminist theory and the study of entrepreneurship. In J. E. Butler (Ed.), *New perspectives on women entrepreneurs* (pp. 1–24). Information Age Publishing.

Gregory, R. (2010). *Design science research and the grounded theory method: Characteristics, differences, and complementary uses* [Paper presentation]. Paper Presented at the Proceedings of the 18th European Conference on Information Systems (ECIS 2010), Pretoria, South Africa.

Grossman-Thompson, B. (2013). Entering public space and claiming modern identities: Female trekking guides and wage labor in urban Nepal. *Studies in Nepali History and Society, 18*(2), 251–278.

Guba, E., & Lincoln, Y. (1994). Competing paradigms in qualitative research. In *Handbook of qualitative research* (pp. 163–194).

Gupta, V. K., Wieland, A. M., & Turban, D. B. (2019). Gender characterizations in entrepreneurship: A multi-level investigation of sex-role stereotypes about high-growth, commercial, and social entrepreneurs. *Journal of Small Business Management, 57*(1), 131–153. https://doi.org/10.1111/jsbm.12495

Habermas, J. (2001). *The postnational constellation*. MIT Press.

Hanmer, L., & Klugman, J. (2016). Exploring women's agency and empowerment in developing countries: Where do we stand?, 10.1080/13545701.2015. *Feminist Economics, 22*(1), 237–263. https://doi.org/10.1080/13545701.2015.1091087

Hays, S. (1994). Structure and agency and the sticky problem of culture. *Sociological Theory, 12*(1), 57–72. https://www.jstor.org/stable/202035 https://doi.org/10.2307/202035

Helgesen, S. (1990). *The female advantage: Women's ways of leadership*. Doubleday.

Hertz, R. (Ed.). (1997). *Reflexivity and voice*. Sage Publications, Inc.

Hillman, W. (2019). Risky business': The future for female trekking guides in Nepal. *Tourism and Hospitality Research, 19*(4), 397–407. https://doi.org/10.1177/1467358418768661

Hillman, W., & Radel, K. (2018a). Journeys of emancipation: Disrupting poverty in Nepal. In S. Cole (Ed.), *Gender equality and tourism: Beyong empowerment* (pp. 82–93). CAB International.

Hillman, W., & Radel, K. (Eds.). (2018b). *Qualitative methods in tourism research: Theory and practice*. Channel View Publications.

Hillman, W., & Radel, K. (2017). *Ameliorating poverty for village women in rural Nepal through tourism* [Paper presentation]. Paper Presented at the the 8th International Conference on Sustainable Tourism Development - Issues, Challenges & Debates, Kathmandu, Nepal.

Hofstede, G. (1984). *Culture's consequences: International differences in work-related values*. Sage.

Hofstede, G. (2005). *Cultures and organizations: Software of the mind*. McGraw-Hill Publishing.

Hofstede, G. (2011). Dimensionalizing cultures: The Hofstede Model in context. *Online Readings in Psychology and Culture, 2*(1). https://doi.org/10.9707/2307-0919.1014

Holmes, M. (2007). *What is gender? Sociological approaches*. Sage Publications Ltd.

Hosni, D., & Lundberg, C. (2005). The status of women revisited: The case of Nepal. *Equal Opportunities International, 24*(3–4), 73–85. https://doi.org/10.1108/02610150510788088

Humagai, R. P. (2012). In report on Nepal conflict, UN human rights chief voices concern over pace and extent of justice efforts. *UN News*. https://news.un.org/en/story/2012/10/422982-report-nepal-conflict-un-human-rights-chief-voices-concern-over-pace-and-extent

Jarvis, D. (2007). Risk, globalisation and the state: A critical appraisal of Ulrich Beck and the world risk society thesis. *Global Society, 21*(1), 23–46. https://doi.org/10.1080/13600820601116468

Jennings, J. E., & Brush, C. G. (2013). Research on women entrepreneurs: Challenges to (and from) the broader entrepreneurship literature? *Academy of Management Annals, 7*(1), 663–715. doi:https://doi.org/10.1080/19416520.2013.782190 https://doi.org/10.5465/19416520.2013.782190

Kabeer, N. (1992). *Triple roles, gender roles, social relations: The political sub-text of gender training*. University of Sussex, Institute of Development Studies.

Kee, D. M. H., & Rahman, N. A. (2018). Effects of entrepreneurial orientation on start-up success: A gender perspective. *Management Science Letters, 8*, 699–706. https://doi.org/10.5267/j.msl.2018.4.012

Kincheloe, J. L., & McLaren, P. L. (2000). Rethinking critical theory and qualitative research. In N. K. Denzin & Y. S. Lincoln (Eds.), *Handbook of qualitative research* (2nd ed., pp. 279–313). Sage Publications, Inc.

Lama, W. B. (1999). Valuing women as assets to community-based tourism in Nepal. *Cultural Survival Quarterly Magazine*.

Lama, W. B., & Sattar, N. (2004). Mountain tourism and the conservation of biological and cultural diversity. In M. F. Price, L. Jansky, & A. A. Iatsenia (Eds.), *Key issues for mountain areas* (Vol. 23–28 April 2002, pp. 111–148). United Nations University Press.

Langowitz, N., & Minniti, M. (2007). The entrepreneurial propensity of women. *Entrepreneurship Theory and Practice, 31*(3), 341–364. https://doi.org/10.1111/j.1540-6520.2007.00177.x

Lawoti, M. (2010). Informal institutions and exclusion in democratic Nepal. In *Ethnicity, inequality, and politics in Nepal* (pp. 18–54).

Maden, C. (2015). A gendered lens on entrepreneurship: Women entrepreneurship in Turkey. *Gender in Management: An International Journal, 30*(4), 312–331. https://doi.org/10.1108/GM-11-2013-0131

Marlow, S., & McAdam, M. (2013). Gender and entrepreneurship: Advancing debate and challenging myths; exploring the mystery of the under-performing female entrepreneur. *International Journal of Entrepreneurial Behavior & Research, 19*(1), 114–124. https://doi.org/10.1108/13552551311299288

Mitchell, D. P. (2018). *Teaching 'in our remoteness': A constructivist grounded theory of primary school teaching in a remote area of Nepal* [Doctoral dissertation]. Central Queensland University. https://www.researchgate.net/publication/338831464_Teaching_'in_our_remoteness'_A_constructivist_grounded_theory_of_primary_school_teaching_in_a_remote_area_of_Nepal

Morrison, C., & Jutting, J. P. (2005). Women's discrimination in developing countries: A new data set for better policies. *World Development, 33*(7), 1065–1081.

Murphy, C., & Cross, C. (2017). Gender, age, and labour market experiences. In *The Palgrave handbook of age diversity and work* (pp. 561–582). Palgrave Macmillan.

Narayan, U. (2013). *Dislocating cultures: Identities, traditions, and third world feminism*. Routledge.

Nazneen, S. (2018). *Gender and intersecting inequalities in local government in South Asia.* IDS Working Paper, Vol. 2018, No. 507, Institute of Development Studies and Swiss Agency for Development and Cooperation.
Panda, S. (2018). Constraints faced by women entrepreneurs in developing countries: Review and ranking. *Gender in Management: An International Journal, 33*(4), 315–331. https://doi.org/10.1108/GM-01-2017-0003
Pant, B., & Standing, K. (2011). Citizenship rights and women's roles in development in post-conflict Nepal. *Gender & Development, 19*(3), 409–421. https://doi.org/10.1080/13552074.2011.625656
Pearse, R., & Connell, R. (2016). Gender norms and the economy: Insights from social research. *Feminist Economics, 22*(1), 30–53. https://doi.org/10.1080/13545701.2015.1078485
Pherali, T. J. (2011). Education and conflict in Nepal: Possibilities for reconstruction. *Globalisation, Societies and Education, 9*(1), 135–154. https://doi.org/10.1080/14767724.2010.513590
Phillips, K., Rothbard, N., & Dumas, T. (2009). To disclose or not to disclose? Status distance and self-disclosure in diverse environments. *Academy of Management Review, 34*(4), 710–732.
Pinkovetskaia, I. S., Kryukova, L. I., Campillo, D. F. A., & Rojas-Bahamon, M. J. (2019). Female entrepreneurship: Types of economic activity. *Journal of History Culture and Art Research, 8*(2), 253–265. https://doi.org/10.7596/taksad.v8i2.2153
Pradhan, R., & Shrestha, A. (2005). *NRM Working Paper No. 4 - Ethnic and caste diversity: Implications for development.* https://www.adb.org/sites/default/files/publication/28686/wp4.pdf
Radel, K., & Hillman, W. (2016). *Disrupting poverty in rural Nepal: Transformations of women through tourism entrepreneurship* [Paper presentation]. Paper Presented at the 30th Australia and New Zealand Academy of Management (ANZAM) Conference, "under New Management - Inovating for Sustainable and Just Futures, Brisbane, Queensland University of Technology, Gardens Point. https://anzam2016.com/wp-content/uploads/ANZAM_Proceedings/ANZAM_Proceedings.pdf
Radel, K., & Hillman, W. (2018, February 5–9). *Emancipation through tourism entrepreneurship for women in Nepal – 'Empowerment Entrepreneurship* [Paper presentation]. Paper Presented at the CAUTHE Annual Conference, Newcastle, Australia, University of Newcastle, NSW.
Rafnsdóttir, G. L., & Weigt, J. (2019). Addressing the horizontal gender division of labor: A case study of support and obstacles in a heavy industry plant in Iceland. *Sex Roles, 80*(1), 91–104. https://doi.org/10.1007/s11199-018-0915-7
Rao, V., & Walton, M. (Eds.). (2004). *Culture and public action.* Stanford University Press.
Reynolds, P. D., Hay, M., & Camp, S. M. (1999). *Global entrepreneurship monitor - 1999 executive report.* Babson College. https://www.researchgate.net/publication/273704996_Global_Entrepreneurship_Monitor_1999_Executive_Report
Rosca, E., Agarwal, N., & Brem, A. (2020). Women entrepreneurs as agents of change: A comparative analysis of social entrepreneurship processes in emerging markets. *Technological Forecasting and Social Change, 157*(2020), 120067. https://doi.org/10.1016/j.techfore.2020.120067
Rowlands, J. (1997). *Questioning empowerment: Working with women in Honduras.* Oxfam Publishing.
Schaper, M., & Volery, T. (2007). *Entrepreneurship and small business* (2nd Pacific Rim Edn ed.). John Wiley & Sons Australia, Ltd.
Scheyvens, R. (1998). Subtle strategies for women's empowerment: Planning for effective grassroots development. *Third World Planning Review, 20*(3), 235. https://doi.org/10.3828/twpr.20.3.g802r410341605g2
Schoar, A. (2010). The divide between subsistence and transformational entrepreneruship. *Innovation Policy and the Economy, 10*(1), 57–81. https://doi.org/10.1086/605853
Secretariat, C. A., & Durbar, S. (2015). *Constitution of Nepal 2015.* Constituent Assembly Secretariat.
Seiler, R. M. (2004). *Human communication in the critical theory tradition.* http://www.ucalgary.ca/~rseiler/critical.htm
Širec, K., & Močnik, D. (2012, January). Gender specifics in entrepreneurs' personal characteristics. *Journal of East European Management Studies, 17*(1), 11–39. https://doi.org/10.5771/0949-6181-2012-1-11
Snellinger, A. (2010). *Transfiguration of the political: Nepali student activism and the politics of acculturation* [Unpublished PhD thesis]. Cornell University.
Solymossy, E. (1998). *Entrepreneurial dimensions: The relationship of individual, venture, and environmental factors to success* [Doctoral dissertation]. Case Western Reserve University, Weatherhead School of Management.
St. Louis, K., & Calabrese Barton, A. (2002). Tales from the science education crypt: A critical reflection of positionality, subjectivity, and reflexivity in research. *Forum Qualitative Sozialforschung/Forum: Qualitative Social Research, 3*(3), 40. http://www.qualitativeresearch.net/fqs/fqs-eng.htm
Stampe, A. Ø. (2017). *Gender and climate change - Women as agents of change in Nepal* [Master thesis]. Aalborg University. https://projekter.aau.dk/projekter/files/262066059/Gender_and_Climate_Change.pdf
Starks, H., & Trinidad, S. (2007). Choose your method: A comparison of phenomenology, discourse analysis, and grounded theory. *Qualitative Health Research, 17*(10), 1372–1380. https://doi.org/10.1177/1049732307307031
Subramanian, S. P. (2018). Feminist masculinism: Imagined response to violence against women in Pradeep Sarkar's Hindi film, *Mardaani.* In S. Jha & A. Kurian (Eds.), *New feminisms in South Asian social media, film, and literature: Disrupting the discourse.* Routledge.
Tamang, S. (2009). The politics of conflict and difference or the difference of conflict in politics: The women's movement in Nepal. *Feminist Review, 91*(1), 61–80. https://doi.org/10.1057/fr.2008.50

Timmons, J. A., & Spinelli, S. (2006). *New venture creation: Entrepreneurship for the 21st Century*. McGraw-Hill.

United Nations Development Programme [UNDP]. (2019). *Human Development Reports - Nepal*. http://hdr.undp.org/en/countries/profiles/NPL

Valerio, M. A., Rodriguez, N., Winkler, P., Lopez, J., Dennison, M., Liang, Y., & Turner, B. J. (2016). Comparing two sampling methods to engage hard-to-reach communities in research priority setting. *BMC Medical Research Methodology*, *16*(1), 1–11. https://doi.org/10.1186/s12874-016-0242-z

Verheul, I., Stel, A. V., & Thurik, R. (2006). Explaining female and male entrepreneurship at the country level. *Entrepreneurship & Regional Development*, *18*(2), 151–183. https://doi.org/10.1080/08985620500532053

Vossenberg, S. (2013). Women entrepreneurship promotion in developing countries: What explains the gender gap in entrepreneurship and how to close it? Maastricht School of Management Working Paper Series, Working Paper No. 2013/08, 1–27.

Walker, M. U. (2016). Transformative reparations? A critical look at a current trend in thinking about gender-just reparations. *International Journal of Transitional Justice*, *10*(1), 108–125. https://doi.org/10.1093/ijtj/ijv029

World Bank Group. (2019). *Population, female (% of total population) - Nepal*. https://data.worldbank.org/indicator/SP.POP.TOTL.FE.ZS?locations=NP

Xheneti, M., Karki, S. T., & Madden, A. (2019). Negotiating business and family demands within a patriarchal society – the case of women entrepreneurs in the Nepalese context. *Entrepreneurship & Regional Development*, *31*(3–4), 259–278. https://doi.org/10.1080/08985626.2018.1551792

Xheneti, M., Madden, A., & Karki, S. T. (2019). Value of formalization for women entrepreneurs in developing contexts: A review and research agenda. *International Journal of Management Reviews*, *21*(1), 3–23. https://doi.org/10.1111/ijmr.12172

Yadav, P. (2016). *Social transformation in post-conflict Nepal: A gender perspective*. Routledge.

Index

Page numbers in **bold** refer to tables and those in *italic* refer to figures.